MATERIALS, CIRCUITS & DEVICES SERIES 25

Optical MEMS for Chemical Analysis and Biomedicine

Other volumes in this series:

Volume 2	**Analogue IC Design: The current-mode approach** C. Toumazou, F.J. Lidgey and D.G. Haigh (Editors)
Volume 3	**Analogue–Digital ASICs: Circuit techniques, design tools and applications** R.S. Soin, F. Maloberti and J. France (Editors)
Volume 4	**Algorithmic and Knowledge-based CAD for VLSI** G.E. Taylor and G. Russell (Editors)
Volume 5	**Switched Currents: An analogue technique for digital technology** C. Toumazou, J.B.C. Hughes and N.C. Battersby (Editors)
Volume 6	**High-frequency Circuit Engineering** F. Nibler *et al.*
Volume 8	**Low-power High-frequency Microelectronics: A unified approach** G. Machado (Editor)
Volume 9	**VLSI Testing: Digital and mixed analogue/digital techniques** S.L. Hurst
Volume 10	**Distributed Feedback Semiconductor Lasers** J.E. Carroll, J.E.A. Whiteaway and R.G.S. Plumb
Volume 11	**Selected Topics in Advanced Solid State and Fibre Optic Sensors** S.M. Vaezi-Nejad (Editor)
Volume 12	**Strained Silicon Heterostructures: Materials and devices** C.K. Maiti, N.B. Chakrabarti and S.K. Ray
Volume 13	**RFIC and MMIC Design and Technology** I.D. Robertson and S. Lucyzyn (Editors)
Volume 14	**Design of High Frequency Integrated Analogue Filters** Y. Sun (Editor)
Volume 15	**Foundations of Digital Signal Processing: Theory, algorithms and hardware design** P. Gaydecki
Volume 16	**Wireless Communications Circuits and Systems** Y. Sun (Editor)
Volume 17	**The Switching Function: Analysis of power electronic circuits** C. Marouchos
Volume 18	**System on Chip: Next generation electronics** B. Al-Hashimi (Editor)
Volume 19	**Test and Diagnosis of Analogue, Mixed-signal and RF Integrated Circuits: The system on chip approach** Y. Sun (Editor)
Volume 20	**Low Power and Low Voltage Circuit Design with the FGMOS Transistor** E. Rodriguez-Villegas
Volume 21	**Technology Computer Aided Design for Si, SiGe and GaAs Integrated Circuits** C.K. Maiti and G.A. Armstrong
Volume 22	**Nanotechnologies** M. Wautelet *et al.*
Volume 23	**Understandable Electric Circuits** M. Wang
Volume 24	**Fundamentals of Electromagnetic Levitation: Engineering sustainability through efficiency** A.J. Sangster
Volume 29	**Nano-CMOS and Post-CMOS Electronics: Devices and modelling** Saraju P. Mohanty and Ashok Srivastava
Volume 30	**Nano-CMOS and Post-CMOS Electronics: Circuits and design** Saraju P. Mohanty and Ashok Srivastava

Optical MEMS for Chemical Analysis and Biomedicine

Edited by
Hongrui Jiang

The Institution of Engineering and Technology

Published by The Institution of Engineering and Technology, London, United Kingdom

The Institution of Engineering and Technology is registered as a Charity in England & Wales (no. 211014) and Scotland (no. SC038698).

First published 2016

The Institution of Engineering and Technology
Michael Faraday House
Six Hills Way, Stevenage
Herts, SG1 2AY, United Kingdom

www.theiet.org

British Library Cataloguing in Publication Data
A catalogue record for this product is available from the British Library

ISBN 978-1-84919-897-4 (hardback)
ISBN 978-1-84919-898-1 (PDF)

Typeset in India by MPS Limited

To my parents, Yu Wang and Xiren Jiang

— Hongrui Jiang

Contents

Preface

The research field of microelectromechanical systems (MEMS) started decades ago when conventional semiconductor fabrication techniques were modified and expanded to integrate both electronics and mechanics to realize suspended thin films, and, as a result, to develop sensors and actuators. Among a plethora of potential applications, the implementation of these components as well as their enabling technologies in optical systems, thus optical MEMS (also micro-opto-electro-mechanical systems, or MOEMS) immediately caught much attention. A well-known successful story, of course, is the invention and commercialization of the digital micromirror device (DMD) by Texas Instruments. In retrospect, it should not come as a surprise, as a miniaturized reflective surface can naturally serve as a micro-scale mirror, backbone in all kinds of optical systems, and a mechanically tunable micro-scale mirror can instantaneously provide additional benefits, in terms of more functionality and choice of configuration (and re-configuration).

Since its germination, optical MEMS remains one of the most active sub-areas under the broader umbrella of MEMS. It has been advancing in strides, from the perspectives of both product development and penetration into scientific and engineering research. The intrinsic versatility and high sensitivity/resolution afforded by optical sensing as well as the innate drive of MEMS toward miniaturization, make optical MEMS ideal to intersect with and contribute to many other fields. On the other hand, the explosive inter- and cross-disciplinary research today keeps presenting opportunities and challenges to optical MEMS researchers, and pushes the optical MEMS field forward.

This book intends to represent some most recent progress in optical MEMS research. On the applications of optical MEMS, besides the more traditional ones, this book also has a special focus on chemical analysis and biomedicine, both fields experiencing rapid growth in recent years. Of course, owing to the limited space, one book cannot cover all the exciting progress made in optical MEMS; hence, the topics were carefully selected in order to reflect the progress I personally deem urgent to be reported. Some topics covered are already established, but need an update to reflect the state of the art in the literature. Some others are new and need to be presented. The readers might note that, although the chapters cover seemingly diverse topics, the emphasis on optics and miniaturization remains strong and ties the discussed research works together.

Various aspects of optical MEMS and its applications are discussed in this book. In Chapter 1, we briefly review the history of optical MEMS devices and fabrication techniques. Moreover, biomedical and telecommunication applications of micro-optical devices are covered in this chapter. Design and fabrication methods that can particularly bring optics into MEMS are described in Chapter 2. The optical properties of both standard and exotic materials are also investigated in this chapter. Optical coherence tomography (OCT), a noninvasive imaging technique, is introduced in Chapter 3. The OCT integrated onto the tip of a probe offers a versatile diagnostic method to inspect internal organs. Micromachining plays an essential role in the modern OCT endoscopy to integrate optical and mechanical components into a small probe. In addition, a prototype of endoscopic OCT with an electrothermal scanning mirror is demonstrated for tissue imaging. Electrowetting-based tunable lenses that can adjust the location of the optical focus are described in Chapter 4. In these lenses, the curvature of the liquid interface changes with the applied voltage, and thus the focus is controlled to capture sharp images of the objects located at various distances. Miniaturization of cameras and their potential applications are discussed in Chapter 5. Such small cameras can be installed in an endoscopic probe or inserted into a laparoscopic instrument to capture images of the internal organs. Optical MEMS elements and micro/nanofabrication processes can be utilized to assemble those miniaturized cameras with high performance. The field of optics can benefit from excellent solutions found in nature. In recent years, a wide range of optical systems are inspired by diverse natural visualization systems. Various biological inspirations for optical devices are extensively discussed in Chapter 6. Nanophotonic cavities and nanoelectromechanical systems are described in Chapter 7. The recent advances in nanofabrication technology allow for scaling down electrical, optical, and mechanical devices to nanometer size. More importantly, the ability to fabricate nanostructures on silicon substrates provides an opportunity to implement optical functionalities in the monolithic electronic chips. In Chapter 8, the principle, fabrication, and applications of quantum dot (QD) nanophotonics are briefly reviewed. QDs could have a significant role toward miniaturization and integration of optical system. Unique optical properties such as photoluminescence and electroluminescence are also discussed in this chapter. The principle, physics, and applications of photothermal microfluidics are described in Chapter 9. Photothermal effect is an energy-efficient and high-speed mechanism for microfluidic actuation. Those microfluidic systems can be used for various applications such as transfection, subcellular surgery, and thermal therapy. Biomedical applications of optical manipulation are described in Chapter 10. Micro- and nano-scale objects and living cells are manipulated with the optical control. Optical manipulation can be integrated with microfluidics to achieve various functions on a micro-scale area. Potential applications include optical tweezers, cell surgery, and optically induced dielectrophoresis (ODEP). The optofluidic lenses are described in Chapter 11. The unique properties of such miniaturized lenses make them attractive for bio-imaging and electronic devices. MEMS actuation schemes are being used to change the curvature of the optofluidic lenses and achieve focus control. Fabrication methods, potential applications and challenges associated with nanostructured aluminum oxide (NAO) are discussed in Chapter 12. The possible fabrication techniques that allow for integration of NAO sensors into micro-scale optofluidics are briefly discussed. NAO structures are used for detection and sensing of various biomolecules with high sensitivity.

I want to express my sincere gratitude to many people who made this book possible. First and foremost, I thank the authors of the individual chapters for their excellent work. These authors are the top researchers in the relevant fields. Throughout the whole process from drafting to revising, these researchers demonstrated the highest scholarly quality and ethics. It has been a privilege to work with these world-class researchers, and I thank them wholeheartedly. The materials in this book are obviously drawn from the work by many researchers other than the authors of individual chapters. On behalf of all chapter authors, I thank these colleagues for their tremendous effort that has significantly advanced our field. Owing to limited space, many of them must regrettably remain unnamed; in many cases, their pertinent works must unfortunately be omitted. In fact, as mentioned before, some other important topics in optical MEMS could not be included in one single book, and should be in a future book project if possible.

Other people generously put forth their effort in the completion of this book. I wish to thank Dr. Guangyun Lin, Kari Van Grinsven, and Jayer A. Fernandes for helping to proofread the chapters, and Dr. Hewei Liu for preparing some of the drawings. I feel indebted to Christoph von Friedeburg at IET for his interest in this topic and initiation of this book project, and Paul Deards, Jennifer Grace, and Priya Devaraj for the support and assistance throughout the process. Finally, this project would not have been possible without the support from my grants from the National Science Foundation and the National Institutes of Health of the USA; to both funding agencies, I am deeply grateful.

Hongrui Jiang
Madison, WI, USA

November 2015

Chapter 1
Introduction

Mohammad J. Moghimi and Hongrui Jiang

1.1 Optical MEMS and optofluidics

Miniaturization of optical systems and components via traditional micromachining techniques has given rise to the field of optical micro-electro-mechanical systems (MEMS). Optical MEMS (sometimes called micro-opto-electro-mechanical systems, MOEMS) combine optical, electrical, and mechanical components to form a micro-scale system on a single platform. The main advantages of optical MEMS include but are not limited to low cost of manufacturing, small size, reliability, and low power consumption. Optical MEMS technology leverages batch fabrication techniques to minimize the cost of manufacturing. The micro/nanofabrication techniques such as photolithography, thin film deposition, and etching have been adopted from integrated circuits technology. Since these techniques were well developed for semiconductor industry, micro-optical systems can have higher precision than traditional macroscopic optics. In addition, micro-optical devices can use MEMS actuation schemes to control and fine tune the optical parameters of a system, such as focal length and beam direction. Miniaturized tunable lenses, mirrors, filters, and modulators have all been demonstrated and commercialized in recent years. The thriving field of optical MEMS has played a fundamental role in advancing displays, high-speed telecommunications, and biomedical devices.

Integration of microfluidics along with micro/nanoscale optics on a single chip can now be performed by exploiting micromachining and nanofabrication processes. The advent of optofluidics has also been an important step in progressing towards lab-on-a-chip devices, which have extensive applications in healthcare, environmental studies, and life science.

1.2 History

1.2.1 Processes and materials

The usage of MEMS for optical applications was initially inspired by the development of sensors and actuators on thin films in 1960s. This field started with the

Department of Electrical and Computer Engineering, University of Wisconsin–Madison, Madison, WI, USA fabrication of suspended microstructures by the selective etching of silicon substrates. One early example is the air-gap transistor with a suspended metal gate cantilever which was successfully fabricated at Westinghouse Research Labs in the mid-1960s [1]. Early micro-scale devices primarily used single crystal silicon as a mechanical structure since fabrication processes were well developed for silicon in order to manufacture microelectronic circuits on a monolithic platform [2–4]. At that time, anisotropic etching of the silicon was the dominant process used to create microstructures on silicon substrates. Boron-doped silicon was used as an etch-stop layer to precisely control the depth of etching to form micro-scale devices on the thin films. For example, at Berkeley in 1980, composite cantilever beams, composed of silicon dioxide (SiO_2) and doped silicon, were released by anisotropic etching of a silicon substrate [5].

Polycrystalline silicon is another material which is attractive for MEMS applications. Poly-silicon (poly-Si) was first used as a structural material for resonant cantilever sensors [6]. Since 1960s, poly-Si has been widely used for electrical contact in the gates of the metal-oxide-semiconductor field-effect transistors [7]. Low pressure chemical-vapor-deposition (LPCVD) process was used to deposit poly-Si [8] and surface micromachining was utilized to fabricate poly-Si cantilever beams with SiO_2 as a sacrificial layer [9]. However, non-uniform internal stress on the poly-Si layer caused the beams to curve away from the substrate. This suggested that the residual stress on the poly-Si layer needed to be controlled in order to achieve flat cantilever beams. This was a major research topic in the MEMS community for over a decade and tremendous progress was made. Several viable solutions, including heavily doping the poly-Si layer [10, 11], annealing after deposition [12], and a *multipoly* process [13], were suggested to reduce the residual stress and achieve flat cantilever beams.

Conformal deposition of poly-Si followed by thermal oxidation provided the capability of achieving a uniform coating of poly SiO_2 layer in deep trenches [14]. This was an important step towards achieving a conformal etch-resistant layer as well as electrical isolation. Conformal coating of poly-Si paved the way for fabrication of more-sophisticated structures such as bearings, linkages, and motors. In the late 1980s, micro-motors based on poly-Si were demonstrated by Professor Muller and his collaborators at Berkeley [15, 16]. Their impressive results drew large amounts of attention to the field.

1.2.2 Early devices and systems

The first market for optical MEMS devices was display technology. In the early 1970s, the fabrication of micromachined spatial light modulators (SLMs) gave rise to projection displays [17, 18]. In the 1980s, the appearance of the digital mirror device (DMD) made a significant impact on high-quality digital light processing (DLP) projections [19]. Mechanically actuated micromirrors, which were developed and commercialized by Texas Instruments, are now widely used in cinemas, TVs, and portable projectors. As shown in Figure 1.1, the DMD consists of an array of highly reflective microscopic mirrors corresponding to the pixels of the images. The bistable mirrors can be tilted $\pm 12^\circ$ to ON and OFF states (Figure 1.1(a)). When

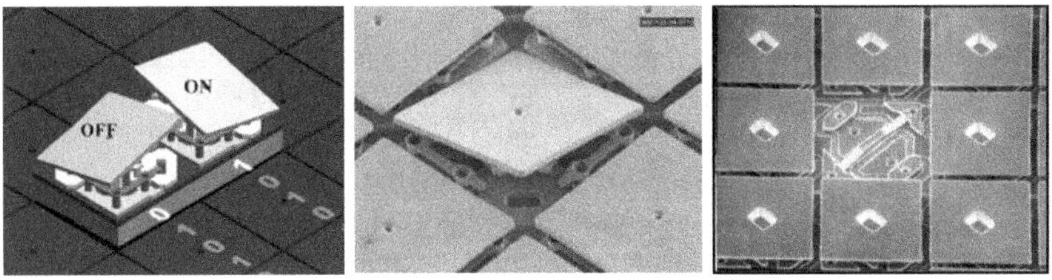

(a) (b) (c)

Figure 1.1 *Texas instrument micromirror device: (a) concept of digital micromirror arrays in operation with ON and OFF states, (b) top view of mirrors with metal coating arranged in rectangular array, the central mirror is tilted, and (c) top view of the device, central mirror was removed. Reprinted with permission from Texas Instruments*

the mirror is ON, it reflects the light to a projection lens in order to make a bright pixel on the display. For OFF state, the mirror reflects the light out of the lens area and creates a dark pixel. Because the fabrication process is compatible with complementary metal-oxide semiconductor technology, the most recent devices come with built-in static random access memories. The binary data of the displayed images are stored in the memories, which are located beneath each cell. Color images can be displayed by multiplexing multiple light sources. Advanced devices with mirror size 5.4 μm can project full-HD (1080p) videos.

The Mirasol display, the most recent display technology based on optical MEMS, was developed for portable devices by Qualcomm Incorporated and significantly minimizes the power consumption of displays. This technology eliminates the requirement for backlight illumination, which is necessary for current liquid crystal displays (LCDs). The pixels of the display consist of two plates with an adjustable gap in between. The reflected beams from the plates interfere to create various colors depending on the gap height. The interferometric display is completely visible under sunny conditions whereas visibility of current LCDs is reduced.

In the 1990s and 2000s, optical MEMS devices made a substantial impact in the field of optical telecommunications [20]. Lightwave communication is inevitable with the increasing demand for high-speed internet. The main advantage of optical communication over conventional electronic systems is that the speed of photons in fiber optics is higher than the speed of electrons in cables. Thus, electrons are replaced with photons to carry information and achieve high-speed communication. Global telecommunication systems, which connect all continents, rely solely on fiber optics and optical components. Optical MEMS technology is now a mature technology and customized optical components are available to replace their electronic analogues. For instance, optical filters, switches, spectrometers, wavelengths add/drop multiplexers, etc. have been developed for optical networks.

The usage of optical MEMS for biomedical applications was first demonstrated at Stanford in the late 1990s. A miniature confocal scanning microscope was assembled by utilizing micromachined components [21]. The microscope consisted of an off-axis grating as the objective lens and a torsional scanning mirror for real-time scanning and imaging. The overall size of the device was only 1.2 mm × 2.5 mm × 6.5 mm. Later in 2002, a three-dimensional raster scanning microlens with a low driving voltage and a large static displacement was developed at Berkeley [22]. This was an important component which allowed miniaturized microscopes to scan biological samples and provide 3D images. In 2004, two-axis gimbaled micromirrors for lateral scanning were demonstrated by Kwon *et al.* at Berkeley [23]. In that device, a large static rotation was demonstrated by the integration of vertical comb drive into gimbals.

The first miniature confocal scanning microscope with electrostatic actuation was demonstrated by Professor Luke Lee and his group in 2004 [24]. Electrostatic actuation was utilized to achieve focusing and scanning function. The overall volume of the microscope was only 2 mm^3, which was suitable for lab-on-a-chip applications. In 2004, a biaxial torsion scan mirror with deformable membrane was demonstrated for biological microscopes by Dickensheets' Group [25]. This MEMS mirror combines lateral scanning functionality along with focus control to minimize the complexity, size, and cost of the microscopes.

A fast-scanning two-photon fluorescence microscope based on a MEMS scanning mirror was developed at Stanford in 2006 [26]. This microscope was capable of capturing images with the exception resolution of 1 μm. MEMS mirrors can easily perform high-speed scanning with line acquisition rates up to 3.5 kHz. In 2007, Ra *et al.* demonstrated real-time high-resolution *in vivo* imaging with a dual axis confocal microscope [27]. This microscope comprised a two-dimensional MEMS scanner that was actuated by a vertical electrostatic comb drive. Many of these microscopes exploited near infrared (NIR) light to achieve noninvasive, *in vivo* imaging [28]. In this way, biological information can be extracted without cutting or removing tissue, which eliminated bleeding, pain and related side effects.

Miniaturization of microscopes by the use of optical MEMS components has had a profound impact on biological diagnostic systems. Micro-endoscopes for insertion into hollow tissue cavities provide minimally invasive access to internal organs [29]. Moreover, the development of miniaturized imaging modules has led to handheld microscopes for the diagnosis of diseases and early detection of skin cancer [30]. This technology has also had a significant influence on neuroscience and opto-genetics research. For example, a mini-florescence microscope has been mounted on the head of a mouse to monitor its brain activity while allow it to behave freely [31].

1.3 Progress in optical MEMS and optofluidics

In this section, we will briefly review the recent progress and applications of optical MEMS devices. Additionally, we will describe MEMS tunable optics and their potential applications and discuss biomedical and telecommunication applications of optofluidics and optical MEMS.

1.3.1 MEMS tunable optics

The optical characteristics of optical elements or systems can be precisely varied by using MEMS actuation methods to achieve tunable devices [32]. Fine-tuning of these parameters improves the functionality of optical MEMS as compared to fixed optics. Moreover, the appearance of tunable optics such as SLMs and filters has led to the development of more complex optical systems such as projectors [33] and tunable lasers [34].

Micromirrors with angular movement were some of the first actuated optical MEMS devices [35]. They consist of a reflective membrane that can continuously vibrate about the x- and y-axes for beam steering. These small devices can operate at high speeds, a requirement of video-rate scanning. Another type of micromirror is the MEMS deformable membrane mirror [36, 37], which functions by controlling the shape of the membrane—an essential feature for many applications, such as adaptive optics. One example is the deformable mirror (DM) which has been successfully developed and commercialized by Boston Micromachines Incorporated [38]. Schematics of MEMS DMs and a photograph of a completed device are shown in Figure 1.2. The diameter of these mirrors ranges from 1.5 mm to 18.6 mm and the number of actuators beneath the membrane can easily exceed 4000. The medium size devices with a diameter of 9.3 mm can operate at a maximum frame rate of 60 kHz while the smaller devices can operate even faster. This device can be used to precisely correct wavefront aberrations, and unlike piezoelectric devices, this device is free from hysteresis. Both lateral scanning mirrors and DMs are useful for many applications including optical telecommunications and biomedical imaging.

A major challenge in the realization of MEMS mirrors is the accurate positioning or shaping of the reflective membranes. Electrostatic actuation is the primary mechanism exploited in many devices to achieve angular movement of the mirrors. Deformation of the membrane is also achieved by applying a voltage to multiple electrodes distributed beneath the membrane. Electrostatic actuation is thus a simple, accurate, and low-power method; however, it requires a high voltage to operate. One example of an electrostatic scanning micromirror had a scan range as high as 20° with an angular resolution of 0.2° [39]. The area of the mirror plate was only 200 μm × 300 μm with a vibration resonant frequency of 29 kHz. The other mechanism for actuation is to use magnetic actuation to drive the mirror(s). Magnetic devices do not require a high driving voltage; however, integration of coils and permanent magnets into MEMS devices remains a substantial challenge in the field [40]. Magnetic mirrors with angular deflections of ±40° were reported in 2005 [41]. Thermal actuation has also been used in some devices to move and shape membrane mirrors. These operate at a low voltage but potential problems include limited speed of operation and high power consumption [42].

Another class of micro-optical devices is tunable fluidics optics, in which microfluidics is combined with optical systems [43]. This field has attracted significant attention for various technologies such as displays and molecular imaging. Tunable lenses, tunable photonics crystals, and fluidic lasers could also be achieved by use of optofluidics. An example of these devices is the liquid lens, in which the

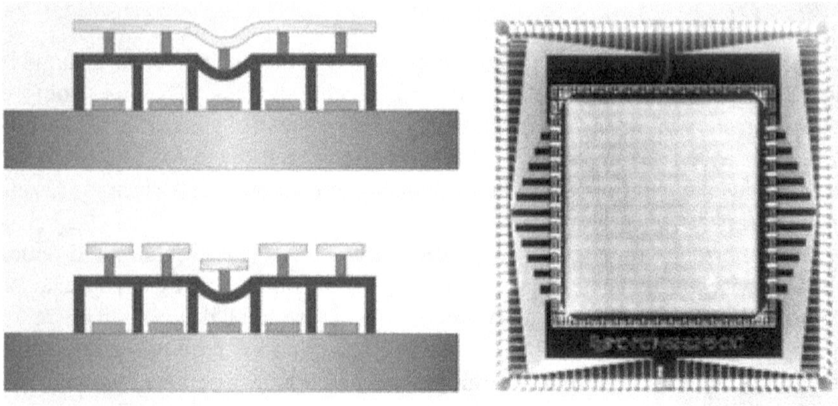

(a) (b)

Figure 1.2 The schematics of MEMS DMs. (a) Cross-section of continuous (upper) and segmented (lower) membrane. A gold-coated membrane is attached by silicon posts to an array of locally anchored compliant electrostatic actuator membranes. The membrane is shaped to cancel aberrations by applying voltage to actuator electrodes on the wafer substrate. (b) Photograph of a fabricated DM with 140 active actuators (and two inactive buffer rings) supporting a continuous DM. Reprinted with permission from Bifano [38], copyright Nature Publishing Group

curvature of the droplets can be changed to adjust the focal length. Electrowetting and pneumatic actuation have both been used to control the curvature of the droplets. The optical power of a 3-mm diameter electrowetting lens can be tuned from -100 to $+50$ diopters with a driving voltage in the order of 50–100 V [44]. Pneumatic tuning is another scheme that has been used to change the curvature of membrane-based liquid lenses. The curvature of the liquid on a single lens can be adjusted to concave, flat, and convex profiles to achieve a large focus range.

Miniaturization of variable-focus elements provides an opportunity to construct more advanced imaging systems. For instance, a compact zoom system has been demonstrated with two tunable lenses. The focus of two microlenses is changed simultaneously in order to adjust the zoom ratio of the optical system. Such a system is suitable for small-form factor cameras such as those used in cell phone cameras.

Another optical system based on liquid lenses is tunable achromatic doublet lenses, which can correct chromatic aberrations. Moreover, opaque and transparent liquids can be used to create fluidic renditions of apertures and shutters. Electrowetting and other optofluidic techniques are used to move these liquids and control the amount of light passing through an aperture. This principle is applied to either continuously move the liquids to achieve a tunable iris or switch between two defined states to create an optofluidic shutter.

1.3.2 Optical MEMS for telecommunications

Various types of optical MEMS devices are widely used in optical communication networks. Two categories of these devices are discussed in this section: broadband MEMS devices and wavelength-selective MEMS devices. Broadband MEMS devices, including data modulators, variable optical attenuators (VOAs), and MEMS optical switches, are independent of the wavelength of the incident light. Spectral equalizers, wavelength add/drop multiplexers (WADMs), wavelength-selective switches (WSSs), and wavelength selective cross-connects (WSXCs) are examples of wavelength-selective devices.

Optical modulators are important components which are necessary for encoding data on an optical signal. They modulate the parameters of the light to transmit data either in free space or within optical fibers. MEMS modulators act as optical switches that can change the intensity of the incoming light signal. For instance, the mechanical antireflection switch modulators are essentially variable Fabry–Pérot interferometers (FPIs) comprised of a silicon nitride membrane suspended over a silicon substrate. The reflection from such a structure can be tuned by changing the gap between the membrane and the substrate. This is done by applying a voltage between them to adjust the distance and modulate the intensity of the light. When the air gap is tuned to a half wavelength, the reflection and output intensity are maximized. The reflection is minimized when the distance between the plates is reduced to a quarter-wavelength.

Optical attenuators regulate optical signals to avoid power impinging on detectors. The high-speed attenuators suppress transient surges in the output of switches and other components in the network. One type of MEMS VOA consists of a sliding blade as a variable iris. The blade is driven by MEMS actuation mechanisms such as electrostatic comb-drives which open and close the iris. However, a portion of the incident beam is blocked by the blade, which reduces the optical power. Another type of VOA is made of diffractive optical elements with high contrast, high-speed operation, and low driving voltage.

MEMS optical switches can selectively route optical signals from one port to another without converting optical signals to electrical form. These switches are made of 2D arrays of micromirrors with N input ports and N output ports. MEMS mirrors are able to switch between two states: (1) direct the optical beams to the transmission path or (2) route them to the drop ports. Optical collimators are utilized in switches to collimate the beam and minimize the diffraction losses. A promising path towards achieving optical cross-connects (OXCs) is to use 3D optical switches. A port count exceeding 100×100 has been achieved by using MEMS mirrors along with collimators. The input and output ports are arranged in 2D arrays and a number of two-axis MEMS mirror routes the optical beams to the desired direction.

Dense-wavelength-division-multiplexing (DWDM) systems are an enabling technology which has been utilized to increase the capacity of optical networks. In such a system, multiple channels of data with different wavelengths are simultaneously carried in a single fiber. However, many parameters of the components in the network such as attenuation and amplification are wavelength dependent which results in fluctuating optical power of the channels in the network. MEMS-based spectral equalizers are one approach to maintain uniform optical power across different channels of the DWDM systems. The spectral channels are separated by a diffraction grating and then illuminated on arrays of MEMS equalizers. The first scheme for such an equalizer was to use a continuous etalon membrane. Another approach to achieve dynamic gain equalizers is to utilize VOAs that are spectrally uniform. These schemes for spectral equalization are required in the telecommunication network as the transmission rate is increased to 40 Gb/s.

Another versatile component in optical networks is the WADMs to manage multi-wavelength channels. WADMs can selectively insert or remove a data channel at an intermediate node of the DWDM system. The WADM is comprised of a blazed diffraction grating with an array of MEMS mirrors. The DWDM channels are separated in the grating and then illuminated on a tilt mirror. Each mirror is reserved for a single channel to either reflect back to the input port or tilt it to a new path.

Optical MEMS technology can add more flexibility to the network by providing tunable optical elements. Tunable devices simplify the optical systems by reducing the number of required elements. Wavelength management can also be improved by the use of tunable devices. Tunable filters and spectrometers are easily fabricated by MEMS fabrication technology. A tunable FPI with a multilayer dielectric mirror suspended over a rigid substrate could be fabricated to achieve adjustable filters.

Dispersion compensators provide a versatile means of avoiding unwanted variations in the amplitude of the DWDM channels. Dispersion compensation is performed in Gires–Tournois (G–T) interferometers, in which the back mirror is highly reflective. A MEMS G–T interferometer has multiple dielectric mirrors with high reflection and a single freestanding membrane as the front mirror. The air gap between two mirrors can be changed by applying a voltage between the mirrors. Such a device can behave as a linear dispersion compensator which is tunable from -100 to 100 ps/nm over 50 GHz.

Different structures have been fabricated by MEMS technology to create transform spectrometers which can measure weak signals over a wide range of wavelengths. Traditional Michelson interferometers consist of a reference mirror, a movable mirror, and a beam splitter. In MEMS interferometers, the movable mirror is actuated by an electrostatic comb drive to tune the optical path length. This tunable Michelson interferometer can provide the Fourier transform of the optical spectrum. The spectral resolution of the spectrometer is inversely proportional to the maximum range of displacement of the movable mirror. A magnetic actuation mechanism with an external magnetic field is used to achieve maximum displacement of several centimeters resulting in fractional resolution of 10^{-5} in the visible range.

Another significant advantage of micro-optics is the capability to integrate them with the electrical and mechanical components. All of the devices and linkages can be packaged and delivered as a single miniaturized chip. Tunable lasers, which consist of a semiconductor gain medium with external cavities, can be realized by using MEMS tunable filters. Accurate alignment and cavity length control are necessary for tunable lasers. Thus, MEMS actuation techniques are reliable candidates to control the length of the cavity and provide mode-hop free tuning. More complex systems can also be monolithically integrated on a single chip by combining planar lightwave circuits (PLCs) along with MEMS devices. Hybrid integration of PLCs and MEMS can yield compact optical systems with low power consumption and high-speed operation. Hybrid PLS-MEMS devices include monolithic WSSs, WSXCs, tunable dispersion compensators, and wavelength-independent switches.

1.3.3 Biology and biomedical applications

The miniaturization of biological tools provides the opportunity to realize implantable devices, biosensors, neural interface devices, etc. One of the main advantages of micro-scale analysis systems is that size of the sample under test is significantly reduced. This reduction can lower the cost of processing and enhance the user experience. This has been the primary motivation in the development of most biological MEMS (Bio-MEMS) devices, which are able to monitor and manipulate biological processes in real time. Moreover, integration of nano-structures (e.g., nano-electrodes) with mechanical and electrical components for biosensors and neural interface devices has resulted in bio-nano-electro-mechanical systems (Bio-NEMS) devices. These are valuable devices to study biological processes and metabolic disorders.

Investigation of individual molecules is of interest to biologists. Optofluidics is a promising approach to single molecule detection. One of the first attempts to achieve single molecule detection was to integrate microfluidic systems with microelectrodes. Such a system could detect fluorescently labeled DNA and single 500 nm size beads. Many optical MEMS devices have been developed for molecule detection in the field of genomic sequencing. Pacific Biosciences has developed a technology that enables the observation of DNA synthesis. Single molecule real-time sequencing is performed on a chip with zero mode waveguides (ZMWs). ZMWs are nanophotonic structures that strongly confine light for ultra-sensitive detection.

Optofluidic systems combine the advantages of advanced optics, flexible microfluidic devices, and optical MEMS/NEMS for various applications including biosensors, lab-on-a-chip instruments, and biomolecular imaging systems. Moreover, photonic circuits are integrated with fluidic systems to achieve more functional and compact tools. These devices offer unique capabilities to detect and manipulate biomolecules.

One type of optofluidic system eliminates the requirements for mechanical valves and pumps, and instead utilizes optical control mechanisms. Early work towards fabrication of all optical, large-scale, integrated microfluidic circuits was demonstrated by the Luke Lee group at Berkeley in 2006. They used light beams with sub-milliwatt power to transport biomolecules and living cells at controlled speeds along certain directions. The ability to drive and guide flow in the channels was achieved by direct conversion of optical energy to hydrodynamic energy.

A much stronger optical confinement is required for transportation and manipulation of smaller dielectric objects. Sub-wavelength liquid-core slot wave-guides are another approach to precisely transport nanoscopic material. In this method, near-field optical forces are used to confine matter inside a waveguide, while scattering/absorption forces are utilized to manipulate objects. This technique is especially useful for capturing and trapping μ-DNA. A similar concept, which was demonstrated by Tang *et al.*, is to focus light using a liquid lens in a micro-fluidic channel [45]. In that work, the dynamically reconfigurable liquid lenses for focusing were formed by laminar liquid flows.

Larger biological entities (as compared to DNA and proteins) such as *Caenorhabditis elegans* have been studied with optofluidic microscopy (OFM). OFM provides an excellent method to achieve low-cost, on-chip microscopes, which are required for biomedicine and bioscience. Conventional microscopes consist of large lenses and optical elements and are challenging to miniaturize. An alternative approach to such a bulky system is to use lens-less microscopes based on OFM. The images of the specimens are directly projected onto a CMOS sensor. A microfluidic system is utilized to position the objects in front of apertures of the sensor.

Another approach to detect biomolecules, especially at low concentrations, is to use surface-enhanced Raman spectroscopy (SERS) in a microfluidic platform. SERS is utilized to facilitate detection by amplifying biomolecular signals. Various types of biomolecules including as viruses and proteins can be detected by SERS. Cancer biomarkers can also be detected by using a SERS-based immunoassay. In this approach, a gold array-embedded gradient microfluidic chip provides various concentrations of the biomarker to eliminate tedious manual dilution process.

Modern nanofabrication techniques allow for the realization of nanoscale fluidic systems. The fluids, concentrated in the channels and pores with dimensions less than 100 nm, exhibit exotic behavior that is generally not observed in larger devices with micro-scale dimensions. Such structures benefit from an ultra-high surface to volume ratio as well as comparable scales to biomolecules. Nanofluidic systems with such excellent features offer a wide range of bio-analytical capabilities including the elongation of single DNA molecules, concentration of proteins, and separation of large and small biomolecules. They have been utilized in SERS-based systems to detect nucleic acids.

Optofluidic nanophotonic circuits are another class of highly sensitive bio-sensors. Reconfigurable photonic crystal (PhC) microcavities integrated with fluidic systems exhibit high-quality factors. However, integrating planar optical sources along with optofluidic systems is a major challenge. One approach to on-chip light source is to use chemical dyes dissolved in a liquid solution to form a gain medium. The use of liquid gain media results in a tunable dye laser. In addition, the gain media can be precisely delivered to specific locations of the device via controlled microfluidic pumping.

1.4 Brief review of book content

Various aspects of optical MEMS and their applications are discussed in this book. In this chapter, we briefly reviewed the history of optical MEMS and their fabrication techniques. Moreover, applications of optical MEMS devices and optofluidics in biomedicine and telecommunication were covered in this chapter. In the next 11 chapters, more details on concept and fabrication of particular micro-optical devices, micro/nanoscale optofluidics, and the applications of such devices are presented.

In Chapter 2, fabrication processes and manufacturing technology for optical MEMS will be reviewed. Design and fabrication methods that can specifically bring optics into MEMS and fluidic systems have been discussed. In addition, it should be noted that the optical properties of materials utilized in these devices are very important to their performance. Novel fabrication techniques are required to be developed for optical materials that can be used for the specific portion of the spectrum. Most of these materials including group III–V and II–VI semiconductors are exotic to the standard semiconductor processing. Thus, alternative materials and related manufacturing techniques are inevitable requirements for optical MEMS and optoelectronics. Fabrication of miniaturized systems with passable mechanical characteristics while maintaining optical quality is challenging. Optical properties of materials including reflectivity, transmissivity, absorption, etc. must be considered when an optical MEMS is designed and manufactured. For instance, the surface roughness of the mirrors may cause light scattering and degrade reflectivity.

In Chapter 2, the optical properties of standard materials including silicon, silicon dioxide, and silicon nitride will be investigated. These materials are compatible with established semiconductor processes such as etching and photolithography. They are widely used in optical MEMS, optoelectronics, and microfluidic systems. Furthermore, exotic materials such as germanium, zinc selenide, zinc sulfide and their optical properties have been introduced for miniaturized systems. The fabrication techniques and mechanical challenges for both standard and exotic materials will be briefly discussed.

Optical coherence tomography (OCT), which is a non-invasive biomedical imaging system, will be introduced in Chapter 3. The main advantage of OCT is its capability to provide *in vivo* images with high resolution. OCT offers a method to inspect a vital tissue without performing biopsy procedure. This is an excellent diagnostic tool for early detection of diseases. OCT endoscopy, which has the ability to reach the internal organs (gastrointestinal tract, cardiovascular artery, etc.), can be used to screen for early stages of cancer and neoplasms. This system consists of a probe to deliver light to the internal organs and collect backscattered light to create images. Miniaturization of optics and mechanical parts that can fit into a small probe is challenging. Optical MEMS devices and micro-actuators play an essential role in the modern OCT endoscopy to integrate optical and mechanical elements into the probe. A major challenge in endoscopy is to increase the field of view. One solution is to scan the samples which can be simply performed by

MEMS scanning mirrors. MEMS mirrors are small, reliable, and low cost. They can also operate at a high speed to capture high-quality images in real time. Various actuation mechanisms have been utilized to drive MEMS scanning mirrors. The mirrors can either scan a line for cross-sectional images or a surface for 3D images. In Chapter 3, principle and design of electrothermal actuated micromirrors will be explained. A prototype of an endoscopic OCT with an electrothermal scanning mirror was demonstrated for internal organ imaging. Porcine urinary bladder, a mouse ear, and a rat brain tissue have all been imaged with the OCT probes. Finally, dental, oral, and meniscus imaging was performed and the OCT images were presented.

For many imaging systems, a means must be provided to adjust the location of the optical focus and create crisp images. Such capability allows for imaging objects distributed at various locations. For laser-based scanning microscopes, that are able to capture *in vivo* images, the depth of imaging beneath the surface of the tissue is determined by the focus of the system. A traditional method to control focus is mechanical lens translation, which requires bulky motors and mechanical linkages to adjust the location of the focus. An alternative method is to adjust the focal length of a single optical element to control the focus of the system. DMs and tunable liquid lenses are two promising approaches to achieve a variable focus element. For many applications such as OCT probes and small form-factor cameras, the variable focus elements must be small and highly accurate. Optical MEMS technology offers different schemes to change the focal length of tiny lenses.

In Chapter 4, electrowetting tunable lenses will be introduced as variable-focus optics. The main advantages of these devices are short response times, low power consumption, and reliable operation. Electrowetting lenses consist of two transparent immiscible liquids placed on a hydrophobic substrate. One liquid is insulator (e.g., oil) and the other is a conductive solution (water). Actuation electrodes are placed on the outside of the substrate to create an electric field. By applying a voltage between the electrodes and the conductive liquid, the wettability of the surface as well as liquid–liquid interface is changed. This mechanism changes the curvature of the liquid lens and results in a voltage-controlled focus device. Electrowetting microlenses have a wide range of applications from displays to barcode scanners.

Miniaturization of cameras is of interest to the research community in recent years. This topic will be covered in Chapter 5. Realization of hidden cameras for security purposes was the primary motivation behind the development of tiny cameras. These cameras have found new applications in the biomedical field in the last decade. They can be installed in endoscopic probes and inserted into laparoscopic tubes for navigation inside the body. However, maintaining the optical performance of the camera, as the dimensions are significantly reduced, remains a major challenge. High resolution, low aberrations, and wide field of view must be achieved as the mini-cameras are designed and manufactured. Optical MEMS elements and micromachining fabrication techniques can be used to assemble those miniaturized cameras with acceptable optical performance. Tunable microlenses, reflective cylindrical lenses, and lens arrays on curved surfaces have been used to achieve MEMS-based cameras with excellent optical properties while the overall size of the camera is considerably reduced.

Time and again, nature has inspired the design of many engineering devices and tools. Recently, the field of optics has benefited from excellent solutions borrowed from nature. A wide range of modern optics including lighting devices, imaging systems, and displays are inspired by the plethora of vision organs. Various biological inspirations for gradient index structure, focus tunability, wide field of view imaging, antireflection coating, structural color, and illumination have been extensively discussed in Chapter 6. One example is the compound eyes in arthropods which offer outstanding features such as wide field of view, fast motion detection, polarization sensing, and high-resolution imaging. Moreover, the size of the compound eyes is small and compact. Such natural structure have inspired artificial compound eyes that have found applications in military, security, and medical fields. Micro/nanoscale structures with predefined optical properties can be realized by micro/nanofabrication processes. This outstanding capability can be used to achieve bio-inspired optical systems. Natural compound eyes comprise of an array of integrated optical units called ommatidia. Microlenses, waveguides, and sensors have been adopted from MEMS technologies to achieve artificial ommatidia. Distribution of those artificial ommatidia on either flat or curved surfaces can also be achieved by standard micromachining methods.

Nanophotonic cavities and nanoelectromechanical systems will be presented in Chapter 7. The advances in nanofabrication process allow for the scaling of electronic and mechanical devices down to the nanometer scale. This trend has also been pursued by researchers in optics to realize nanoscale optical and photonic devices. NEMS switches, PhCs, and some other nanometer scale devices have emerged in recent years. In addition, the ability to create nanostructures on silicon provides an opportunity to implement optical functionalities on the monolithic electronic chips.

A PhC is composed of periodic nanoscale structures along with intentional defects to yield specific optical functionality. Photons in PhCs interact with the nanostructures in a manner similar to the way in which electrons travel in periodic structures (semiconductor lattice). PhCs are versatile tools for light manipulation especially for wavelength selection and light confinement. These unique characteristics make them attractive for diverse applications ranging from telecommunication to chemical sensors. For example, laser cavities can be achieved with a Bragg mirror, which is essentially a 1D PhC. Another intriguing application of PhCs is to enhance light-matter interaction.

Dynamic actuation techniques, adopted from NEMS technologies, can be utilized to create tunable photonic structures. Optical properties such as transmission, reflection, and resonance frequency can be precisely changed with NEMS actuators. Comb-drive actuators can be integrated with PhC structures to generate nanoscale displacement and thus achieve a tunable nanocavity PhC. Another scheme to tune PhC structures is to exploit comb-driven dielectric probes. The optical resonance of the PhC structures can be tuned as the optical field is mechanically perturbed by the tip of the probe. Furthermore, NEMS actuators have been used to mechanically tune and control coupled PhC cavities in order to shift the resonant frequency.

Quantum dots (QDs), whose size is typically smaller than 10 nm, can play a significant role in the miniaturization and integration of optical systems. They exhibit quantum characteristics which make them attractive for many applications. QDs have been applied in transistors, solar cells, light-emitting diodes (LEDs), and diode lasers. They are also being used as a contrast agent for medical imaging. In Chapter 8, principle, fabrication, and applications of QDs nanophotonics will be reviewed. Photoluminescence and electroluminescence, in which QDs are optically or electrically excited to emit photons, are also described in Chapter 8. Many lab-on-a-chip devices rely on external light sources for illumination. Electroluminescence of QDs is a viable solution to achieve compact and on-chip light sources. Moreover, photoluminescence of QDs can be used to precisely identify cells in the multicolor fluorescence microscopy.

A medical diagnostic technique, called hyperspectral imaging, will be introduced at the end of Chapter 8. In hyperspectral microscopy, spectroscopic information is obtained from each individual spot to form a high-resolution microscopic image. This is a robust technique to identify diseases, especially isolated tumors. A wide-band light source such as white LEDs is required for the illumination of samples in hyperspectral imaging. QDs that can produce visible to NIR wavelengths are a potential alternative for compact hyperspectral microscopy. Different optical setups for hyperspectral imaging based on QDs are also presented.

The principle, physics, and applications of photothermal microfluidics will be described in Chapter 9. Light propagation is a way to transfer energy without any physical medium, eliminating the need for cables and routes in microfluidics. Energy can be compressed and then delivered to a small space through proper optical design. The photothermal effect, in which optical energy is converted to thermal energy, is a highly efficient mechanism for microfluidic actuation. Surface tension of the liquids and solids can be dynamically changed via the photothermal effect in order to drive objects in micro-scale channels. This is an energy efficient and high-speed actuation technique for fluid control. Light-heat conversion methods in microfluidics include direct water absorption, light absorbing materials, and nonlinear absorption. Photothermal microfluidics and nanofluidics are attractive tools for cell manipulation. They can be used for various applications such as transfection, subcellular surgery, and thermal therapy.

Applications of optical manipulation in biomedicine will be described in Chapter 10. Micro/nanoscale objects and living cells are manipulated with optically controlled phenomena including direct and indirect interaction. Unlike mechanical micromanipulators, optical manipulation benefits from flexibility and thus can easily be customized for specific applications. Optical manipulation can be integrated with microfluidics to achieve various levels of functionality at the microscale. Another advantage of an optofluidic system is batch fabrication which significantly reduces the cost of production of consumable lab-on-chip devices. Optical tweezers, cell surgery, and optically induced dielectrophoresis (ODEP) will be described in Chapter 10. ODEP is an indirect method which relies on a gradient in the electric field force and induced dipole on particles. In addition, optical manipulation mechanisms such as near-field particle trapping, optical cell sorters, and optical actuation of fluids have been explained in this book.

The principle of operation of the optofluidic lenses, which are used for imaging and focusing, will be described in Chapter 11. The unique properties of such miniaturized lenses make them useful for bioimaging and electronic devices. Their ability to dynamically control the focal length is an important feature for cameras and microscopes. The advent of transparent and elastic polymers such as Polydimethylsiloxane has played an essential role in the realization of optofluidic lenses. These tiny lenses (of the order of micrometer and millimeter) can be implemented on-chip and thus achieve lens functionalities in a small area. Various polymeric materials have been used to construct tunable optofluidic lenses; however, the structures of most optofluidic lenses are identical. They mostly consist of a sealed cavity with an elastic membrane. The cavity is filled with a liquid and the curvature of the membrane can be changed to tune the optical focus. This structure resembles the human eye, in which the curvature of aquula or crystalline lens is changed to achieve accommodation and form sharp images on the retina. Different schemes including pneumatic, electromagnetic, thermal, etc. have been used to drive the optofluidic lens. Other schemes and techniques have also been considered to create a tunable fluidic lens. For example, the refractive index of the liquid in the chamber is changed to achieve focusing in liquid crystal lenses.

Fabrication methods and potential applications of nanostructured aluminum oxide (NAO) are thoroughly discussed in Chapter 12. Such structures have attracted a lot of attention in recent years due to their unique optical properties. Nanoscale pores, with diameter of 10–400 nm, can be formed on aluminum sheets via the anodization technique. This method is not compatible with standard photolithography processes; however, the size and porosity of the nanostructures can be accurately changed by controlling the anodizing current, anodizing voltage, duration time, and electrolyte parameters. Porosity of NAO film can be non-destructively evaluated by measuring the intensity of the blue emission of the NAO film under UV exposure.

Recently, some fabrication techniques have been proposed to form nanopores on a thin film of aluminum deposited on different substrates. This may open up a viable path to adapt the fabrication of NAO using standard microfabrication processes. Some fabrication techniques including lift-off, microlithography, and etching process have been used to pattern NAO film. This is especially required for the integration of NAO films with micro-scale devices.

Most nanostructures including NAO films are used for detection and sensing of various materials with high sensitivity. Since NAO thin film is semitransparent, a FPI can be formed on it. Optical signals such as transmission, reflection, and interference fringes are analyzed in this thin-film interferometer for bio-sensing applications. For example, NAO film (FPI-thin film) can be micro-patterned onto an Indium tin oxide glass substrate for the detection of prostate cancer. Other optical properties such as fluorescent enhancement are used for protein and DNA sensing. NAO sensor is integrated in a microfluidic platform for both label-free bio-detection and fluorescence-based bioassay.

1.5 Conclusion

In this chapter, optical MEMS, optofluidics, and their fabrication techniques are briefly introduced. The advent of the aforementioned devices significantly changed various technologies such as displays. Devices with high precision and small size can be fabricated by the use of micro/nanofabrication processes. Currently, micro-optical devices such as MEMS-based optical switches are widely used in optical networks to achieve high-speed telecommunications. Moreover, these miniature devices are important elements in biomedical tools for diagnosis of diseases and monitoring the internal organs. Micromachined microscopes and tiny cameras could be installed in handheld devices and endoscopes, in which the available space is significantly limited. Integration of photonic nanostructures into micro-scale devices provides an opportunity to scale the biosensors and optofluidic devices down to cellular level.

References

[1] R. A. Wickstrom and J. R. Davis, "The resonant gate transistor," *IEEE Transactions on Electron Devices*, vol. 14, no. 3, pp. 117–133, 1967.

[2] K. E. Petersen, "Silicon as a mechanical material," *Proceedings of the IEEE*, vol. 70, no. 5, pp. 420–457, 1982.

[3] K. E. Petersen, "Dynamic micromechanics on silicon: Techniques and devices," *IEEE Transactions on Electron Devices*, vol. 25, no. 10, pp. 1241–1250, 1978.

[4] K. E. Petersen, "Micromechanical voltage controlled switches and circuits," in *1978 International Electron Devices Meeting,* Washington DC, 1978, vol. 24, pp. 100–103.

[5] L. A. Moudy, R. H. Bube, L. B. Fabick, R. D. Jolly, and R. S. Muller, "Miniature cantilever beams fabricated by anisotropic etching of silicon," *Electromachanical Society*, vol. 127, no. 12, December 1980, pp. 2750–2754, 1981.

[6] J. M. Jaffe and J. W. Seto, "Polycrystalline silicon pressure transducer," U.S. Patent No. 3,938,175. 10 Feb. 1976.

[7] W. R. Hunter, L. Ephrath, W. D. Grobman, *et al.*, "1 μm MOSFET VLSI technology: Part V-A single-level polysilicon technology using electron-beam lithography," *IEEE Transactions on Electron Devices*, vol. 26, no. 4, pp. 353–359, 1979.

[8] T. I. Kamins, "Resistivity of LPCVD polycrystalline silicon films," *Journal of Electrochemical Society,* vol. 126, no. 5, pp. 833–837, 1979.

[9] R. T. Howe and R. S. Muller, "Polycrystalline silicon micromechanical beams," *Journal of Electrochemical Society,* vol. 130, no. 6, p. 1420, 1983.

[10] S. P. Murarka and T. F. Retajczyk, "Effect of phosphorus doping on stress in silicon and polycrystalline silicon," *Journal of Applied Physics*, vol. 54, no. 4, pp. 957–960, 1983.

[11] M. Orpana and A. O. Korhonen, "Control of residual stress on polysilicon thin films by heavy doping in surface micromaching," in *Solid-State Sensors and Actuators, 1991. Digest of Technical Papers, TRANSDUCERS*, 91, San Francisco, CA, USA, 1991, pp. 957–960.

[12] R. T. Howe and R. S. Muller, "Stress in polycrystalline and amorphous silicon thin films," *Journal of Applied Physics*, vol. 54, no. 8, pp. 4674–4675, 1983.

[13] J. Yang, H. Kahn, A. Q. He, S. M. Phillips, and A. H. Heuer, "New technique for producing large-area as-deposited zero-stress LPCVD polysilicon films: The MultiPoly process," *Journal of Microelectromechanical Systems*, vol. 9, no. 4, pp. 485–494, 2000.

[14] I. G. Fulton, J. S. Makris, W. Falls, V. R. Nastasi, A. F. Scaduto, and C. Anne, "Trench sidewall isolation by polysilicon oxidation," U.S. Patent No. 4,666,556. 19 May 1987.

[15] L. Fan, Y. Tai, and R. S. Muller, "IC-processed electrostatic micromotors," *Sensors and Actuators*, vol. 20, no. 1–2, pp. 41–47, 1989.

[16] S. F. Bart, T. A. Lober, R. T. Howe, J. H. Lang, and M. F. Schlecht, "Design considerations for micromachined electric actuators," *Sensors and Actuators*, vol. 14, pp. 269–292, 1988.

[17] L. J. Hornbeck, "Frame addressed spatial light modulator." U.S. Patent No. 4,615,595. 7 Oct. 1986.

[18] J. Guldberg and H. C. Nathanson, "Electrostatically deflectable light valve with improved diffraction properties," U.S. Patent No. 3,886,310. 27 May 1975.

[19] L. J. Hornbeck, "Projection displays and MEMS: Timely convergence for a bright future," in *Proceedings of SPIE – The International Society for Optical Engineering*, Austin, TX, USA, 1995, vol. 2641, p. 2.

[20] M. C. Wu, O. Solgaard, and J. E. Ford, "Optical MEMS for lightwave communication," *Journal of Lightwave Technology*, vol. 24, no. 12, pp. 4433–4454, 2006.

[21] D. L. Dickensheets and G. S. Kino, "Micromachined scanning confocal optical microscope," *Optics Letters*, vol. 21, no. 10, pp. 764–766, 1996.

[22] S. Kwon, V. Milanović, and L. P. Lee, "Large-displacement vertical microlens scanner with low driving voltage," *IEEE Photonics Technology Letters*, vol. 14, no. 11, pp. 1572–1574, 2002.

[23] S. Kwon, V. Milanović, and L. P. Lee, "Vertical combdrive based 2-D gimbaled micromirrors with large static rotation by backside island isolation," *IEEE Journal of Selected Topics in Quantum Electronics*, vol. 10, no. 3, pp. 498–504, 2004.

[24] S. Kwon and L. P. Lee, "Micromachined transmissive scanning confocal microscope," *Optics Letters*, vol. 29, no. 7, pp. 706–708, 2004.

[25] Y. Shao, D. Dickensheets, and P. Himmer, "3-D MOEMS mirror for laser beam pointing and focus control," *IEEE Journal of Selected Topics in Quantum Electronics*, vol. 10, no. 3, pp. 528–535, 2004.

[26] W. Piyawattanametha, R. P. J. Barretto, T. H. Ko, B. A. Flusberg, E. D. Cocker, H. Ra, D. Lee, O. Solgaard, and M. J. Schnitzer, "Fast-scanning two-photon fluorescence imaging based on a microelectromechanical

systems two-dimensional scanning mirror," *Optics Letters*, vol. 31, no. 13, pp. 2018–2020, 2006.

[27] H. Ra, W. Piyawattanametha, M. J. Mandella, J. T. C. Liu, L. K. Wong, T. D. Wang, C. H. Contag, G. S. Kino, and O. Solgaard, "Three-dimensional in vivo imaging by a handheld dual-axes confocal microscope," *Optics Express*, vol. 16, no. 10, pp. 7224–7232, 2007.

[28] R. Weissleder, "A clearer vision for in vivo imaging," *Nature Biotechnology,* vol. 19, April 2001, pp. 316–317, 2001.

[29] B. A. Flusberg, E. D. Cocker, W. Piyawattanametha, J. C. Jung, E. L. M. Cheung, and M. J. Schnitzer, "Fiber-optic fluorescence imaging," *Nature Methods,* Washington, DC, USA, vol. 2, no. 12, pp. 941–950, 2005.

[30] H.-J. Shin, M. C. Pierce, D. Lee, H. Ra, O. Solgaard, and R. Richards-Kortum, "Fiber-optic confocal microscope using a MEMS scanner and miniature objective lens," *Optics Express,* vol. 15, no. 15, pp. 9113–9122, 2007.

[31] B. A. Flusberg, A. Nimmerjahn, E. D. Cocker, E. A. Mukamel, R. P. J. Barretto, T. H. Ko, L. D. Burns, J. C. Jung, and M. J. Schnitzer, "High-speed, miniaturized fluorescence microscopy in freely moving mice," *Nature Methods*, vol. 5, no. 11, pp. 935–938, 2008.

[32] H. H. Zappe, *Fundamentals of Micro-Optics*. Cambridge University Press, Cambridge, 2010.

[33] L. J. Hornbeck, "Current status of the digital micromirror device (DMD) fo projection television applications," in *Proceedings of IEEE International Electron Devices Meeting,* 1993, pp. 381–384.

[34] A. Q. Liu and X. M. Zhang, "A review of MEMS external-cavity tunable lasers," *Journal of Micromechanics Microengineering*, vol. 17, no. 1, pp. R1–R13, 2006.

[35] K. E. Petersen, "Silicon torsional scanning mirror," *IBM Journal of Research and Development,* vol. 24, no. 5, pp. 631–637, 1980.

[36] D. J. Dagel, W. D. Cowan, S. Member, O. B. Spahn, G. D. Grossetete, A. J. Griñe, M. J. Shaw, P. J. Resnick, and B. Jokiel, "Large-stroke MEMS deformable mirrors for adaptive optics," *Journal of Microelectromechanical Systems,* vol. 15, no. 3, pp. 572–583, 2006.

[37] M. J. Moghimi, K. N. Chattergoon, C. R. Wilson, and D. L. Dickensheets, "High speed focus control MEMS mirror with controlled air damping for vital microscopy," *Journal of Microelectromechanical Systems*, vol. 22, no. 4, pp. 938–948, 2013.

[38] T. Bifano, "Adaptive imaging: MEMS deformable mirrors," *Nature Photonics*, vol. 5, no. 1, pp. 21–23, 2011.

[39] N. C. Tien, O. Solgaard, M. Kiang, M. Daneman, K. Y. Lau, and R. S. Muller, "Surface micromachined mirrors for laser-beam positioning," *Sensors and Actuators A: Physical*, vol. 52, pp. 76–80, 1996.

[40] D. Niarchos, "Magnetic MEMS: Key issues and some applications," *Sensors Actuators A: Physical*, vol. 106, no. 1–3, pp. 255–262, 2003.

[41] T. Fujita, K. Maenaka, and Y. Takayama, "Dual-axis MEMS mirror for large deflection-angle using SU-8 soft torsion beam," *Sensors Actuators A: Physical*, Hilton Head, South Carolina, USA, vol. 121, no. 1, pp. 16–21, 2005.

[42] J. H. Comtois and V. M. Bright, "Surface micromachined polysilicon thermal actuator arrays and applications," in *Proceedings of the Solid-State Sensor and Actuator*, 1996, pp. 174–177.

[43] C. Monat, P. Domachuk, and B. J. Eggleton, "Integrated optofluidics: A new river of light," *Nature Photonics*, vol. 1, no. 2, pp. 106–114, 2007.

[44] B. H. W. Hendriks, S. Kuiper, M. A. J. van As, C. A. Renders, and T. W. Tukker, "Electrowetting-based variable-focus lens for miniature systems," *Optical Review*, vol. 12, no. 3, pp. 255–259, 2005.

[45] S. K. Y. Tang, C. A. Stan, and G. M. Whitesides, "Dynamically reconfigurable liquid-core liquid-cladding lens in a microfluidic channel," *Lab Chip*, vol. 8, no. 3, pp. 395–401, 2008.

Chapter 2

Fabrication and manufacturing technology for optical MEMS

Wing Chan, Philip Armstrong, Merlin Mah, Luke Taylor and Joseph Talghader

2.1 Introduction

To the uninitiated, the phrase "optical microelectromechanical systems" or optical MEMS must appear to refer to a field of incredible specialization. Ironically, the number of disciplines involved, optics, mechanics, and electronics, make the field most accessible to scientists of great technical breadth. This is especially true when optical MEMS is used in chemical and biological applications – the theme of this text. Underlying all of them is the technology of microfabrication.

One chapter could not possibly cover all of the techniques developed over the decades for very-large-scale integration (VLSI) and general MEMS systems. Indeed there are entire textbooks devoted specifically to both types. In this chapter then, we present the characteristics of fabrication and design that are specific to bring optics into the system. In particular, there are a number of materials and fabrication techniques that are specific to optical MEMS systems. When dealing with light, one may have to handle visible, ultraviolet, or infrared portions of the spectrum, each of which has its own special set of optimal substances. Since one often has to emit light or detect it in special wavelength regions, semiconductors other than silicon often must be incorporated, each with their own set of wet and dry chemical etching techniques and their own set of mechanical properties.

Standard mechanical characteristics that play no role in "normal" MEMS systems may prove problematic in optical MEMS. For example small size may lead to diffraction, typical surface roughness may limit optical cavity resolution, and mechanical or motion may deform mirrors to limit the number of resolvable spots. Even thermal noise may place limits on optical design. Each of these topics is covered in the pages that follow. For the reader who is interested in further exploring many of these areas, we recommend the text by Solgaard.

Department of Electrical and Computer Engineering, University of Minnesota Twin Cities, MN, USA

2.2 Optical properties of materials

Optical MEMS have a variety of uses and thus incorporate a variety of materials. Some materials, such as silicon and silicon dioxide, are standard to any semiconductor manufacturing process. Others, such as Ta_2O_5 (tantala) and Al_2O_3 (alumina), are relatively unique to optical devices. Still others have very specific roles in an optical MEMS device and are often at odds with standard processing techniques. Standard fabrication materials often fall in one of three categories: metals, semiconductors and dielectrics. Metals make good reflectors or absorbers. Dielectrics make good transmissive or lens elements as well as good reflectors when used as a part of an interference-based reflective device, such as a distributed Bragg reflector (DBR). Semiconductors are used in electronic devices, but may also be used as metals or dielectrics in certain wavelength ranges.

In addition to reflective and transmissive properties, some materials have specific properties which make them useful as optical devices. Some materials have a stronger response to electric field in certain directions making them good polarization-selective materials. Others have a strong change in optical properties due to mechanical or thermal effects in the material. Some materials exhibit gain or attenuation with pumping or when placed in a selective filter cavity. There are many more material properties which may have unique optical effects, and integrating this wide variety of materials into standard manufacturing techniques is challenging.

Among the properties of optical materials, the refractive index of the material often plays the most important role. Transmission, reflection, polarization, interference, mechanical and thermal effects are often related to optical effects via the index of refraction. The index of refraction is typically denoted by n. One way to describe refractive index is that it relates the speed of light in vacuum to the speed of light in a material, as shown in Equation (2.1). The speed of light, and therefore refractive index, is related to permittivity and permeability of the material which are measures of the electric and magnetic material properties, respectively. Since natural materials do not generally exhibit a strong magnetic response in the optical regime, the refractive index is dominated by the electric field response in the material, and is often given by the square root of the dielectric constant. The refractive index is also typically different at different electric field frequencies. Therefore, it is not always accurate to use the static-field permittivity for optical values of refractive index.

$$n = c/v = \sqrt{(\epsilon\mu)/(\epsilon_0\mu_0)} \tag{2.1}$$

$$n = n_0 \sin(\theta_i)/\sin(\theta_t) \tag{2.2}$$

An alternative formulation relates the refractive index to the angles of incidence, transmission and the refractive index of the incident medium. This results in Snell's Law, which governs the angle of transmission at a material interface, as shown in (2.2).

Finally, material loss and gain can be incorporated into the refractive index, by use of complex numbers. Imaginary index of refraction is often denoted by k. Sinusoidal light plane waves are often expressed as a complex exponentials (see Griffiths [2012] or Ulaby [2014], for examples of complex analysis of light). A complex index of refraction means that the imaginary part of the index of refraction adds an exponential decay or gain to the wave which accurately reproduces the loss in the material. Therefore, material absorption can be included in the refractive index as an imaginary number, as shown in (2.3) and (2.4).

$$\tilde{n} = n - ik \tag{2.3}$$

$$\tilde{E} = \tilde{A}e^{-i2\pi\tilde{n}z/\lambda} = \tilde{A}e^{-i2\pi nz/\lambda}e^{-2\pi kz/\lambda} \tag{2.4}$$

The imaginary refractive index is related to simple Beer–Lambert Law absorption, given by a, in the following relation:

$$\text{Absorption} = e^{-aL} = e^{-2\pi kL/\lambda} \tag{2.5}$$

$$a_{field} = 2\pi k/\lambda \tag{2.6}$$

$$a_{power} = 4\pi k/\lambda \tag{2.7}$$

In (2.6) and (2.7), the imaginary index k is related to the field absorption and the power loss, respectively. This relates to optical power being proportional to the electric field squared. The imaginary index is sometimes termed the "extinction coefficient", but context is needed since sometimes "extinction coefficient" refers to an absorption-per-mole term.

Dielectric materials often have low absorption and the imaginary index and are often negligible. Metals have high absorption, and the imaginary index is high. However, optical absorption and the imaginary index are wavelength dependent. Materials absorb optical energy if energy and momentum are conserved by the absorption process.

There are several basic processes by which materials absorb optical energy. The first is band-to-band or interband absorption by carriers (electrons or holes) in a material. Electron energy bands in materials occur naturally in materials, and lower bands are filled first. Electrons in a lower band (termed the valence band) may be excited to open states in a higher-band (termed "conduction band") if they collide with a photon and exchange the correct energy and momentum. Band-to-band absorption appears for UV light in dielectrics, such as SiO_2 (silica). While silica is transparent in the visible spectrum, the UV light is more energetic and excites electrons to available states in the higher band. Band-to-band effects are even more dominant in semiconductors, especially III–V compound semiconductors with direct bandgaps, where light efficiently excites electrons from the valence to conduction bands.

The second main absorption process is free-carrier absorption by electrons or holes. This happens primarily in metals and semiconductors where there are many free carriers to absorb optical energy. Free carriers are allowed to gain or lose energy within conservation of energy and momentum and material band structure. Light may be absorbed by these free carriers. The free carriers often require phonon momentum transfer to complete the absorption process and, in turn, dissipate energy to phonons increasing material internal energy and temperature.

The third main absorption phenomenon is phonon, lattice or vibrational absorption caused by the solid atoms, themselves, directly absorbing optical energy. Many materials exhibit lattice absorption in the IR regime where the wavelength is on the order of the vibrational resonance.

While, the imaginary index, k, may be wavelength-dependent, the absorption may be additionally wavelength-dependent depending on reflections and interference in structured materials. Structure-related dependences show up in interference coatings, filters and cavities, where wavelength is used to define material thicknesses. An antireflective coating illustrates the importance of refractive index in device design. A simple coating can minimize the reflected electric field by changing the phase of the two reflected components. If the film is $\lambda/4n_f$ thick, and $n_f \approx \sqrt{(n_s)}$, where n_f is the film refractive index and n_s is the substrate refractive index, then the reflected field is minimized. In a quantitative analysis, the reflection coefficients are given by the Fresnel equations (see Griffiths [2012] or Ulaby [2014], for example), and the wave propagates as a complex plane wave over twice the film thickness, L_f. Assuming the incident material is air with $n = 1$ and normal incidence, the total field at is given by the following.

$$\tilde{E}_{tot} = \frac{1 - n_f}{1 + n_f}\tilde{A} + \frac{n_f - n_s}{n_f + n_s}\tilde{A}e^{-i2\pi n_f 2L_f/\lambda} \tag{2.8}$$

If $n_f \approx \sqrt{(n_s)}$ and $L_f = \lambda/4n_f$, then:

$$\begin{aligned} \tilde{E}_{tot} &= \frac{1 - n_f}{1 + n_f}\tilde{A} + \frac{n_f - n_f^2}{n_f + n_f^2}\tilde{A}e^{-i2\pi n_f 2\frac{\frac{\lambda}{4n_f}}{\lambda}} \\ &= \frac{1 - n_f}{1 + n_f}\tilde{A} + \frac{n_f(1 - n_f)}{n_f(1 + n_f)}\tilde{A}e^{-i2\pi/2} = \frac{1 - n_f}{1 + n_f}\tilde{A} - \frac{1 - n_f}{1 + n_f}\tilde{A} = 0 \end{aligned} \tag{2.9}$$

This example shows the power of layered materials with different refractive indices. However, real materials may not have refractive indices exactly equal to the root of the substrate index. Additional layers may solve this problem, but the search for useful indices remains important for many device designs. Additionally, knowing the optical properties of common MEMS materials is vital for device design. More about interference devices and analysis methods can be found in a variety of places including Yeh [2005], Yariv [2007], and Macleod [2010]. For more about electromagnetic field theory and the origins of the refractive index, see Griffiths [2012], Jackson [1998], and Born [1999]. More about material refractive index and wavelength dependence can be found in Polyanskiy [2015] and Palik [1997].

2.2.1 *Thermo-optic effects*

The optical properties of materials also change with temperature. Materials may accumulate considerable internal energy during high-intensity radiation. This often

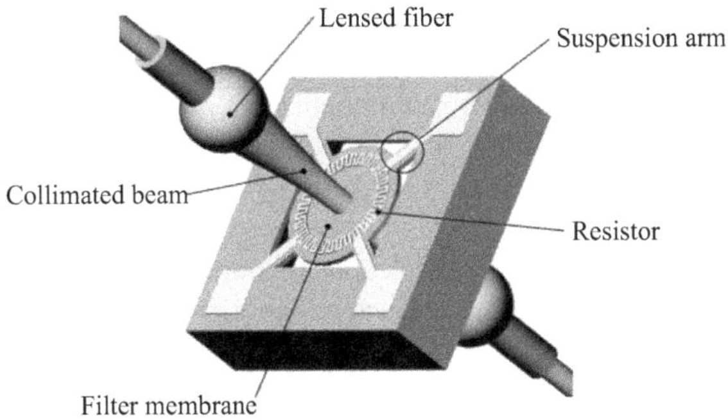

Figure 2.1 *A close up of a thermally tunable optical filter. The filter membrane consists of an optical cavity supported by a micromachined suspension system. Amorphous silicon composes the thermally active layer of the filter. The membrane is temperature-controlled by thin-film heating. Coupling in and out of the system is achieved by the use of lensed fibers. Reprinted with permission from [Hohlfeld 2004], copyright IOP Publishing*

leads to an increase in the material's temperature. This change in temperature may change fundamental material properties, including refractive index. This is typically represented by dn/dT:

$$n(\Delta T) = n_0 + dn/dT\Delta T \tag{2.10}$$

where ΔT represents the change in temperature above (or below) room temperature. This change may be small (approximately 2–10 ppm/K is typical for silica glasses [Wray 1969, Jewell 1991]), but it may be enough to have an effect on a device. Regions with increased refractive index tend to focus the light. Therefore, a positive dn/dT may lead to self-focusing and damage in moderately high power systems. On the other hand, some devices purposely change the material temperature to change the refractive index as a means of modulating a signal. Some materials, such as amorphous silicon, may have a dn/dT as high as 325 ppm/K. A device designed around this material is shown in Figure 2.1 [Hohlfeld 2004]. In this device, the amorphous silicon is placed in an optical cavity composed of dielectric mirrors. As the amorphous silicon heats up, it changes the cavity resonant frequency. The heating is controlled by a thin-film heater running around the perimeter of the filter. The resistor temperature is controlled by ohmic heating. This allows the device to select different wavelengths based on a current pulse to the resistor.

2.2.2 *Optical materials standard to MEMS fabrication*

It is important to know the optical properties of common substrate and film materials used in semiconductor and MEMS processing. This includes silicon, silicon dioxide, silicon nitride used as structural elements and for mechanical or electronic devices. Additionally, aluminum, gold, and chrome are metals used in standard processing. Standard dielectrics in addition to silicon dioxide and silicon nitride include aluminum oxide, tantalum oxide, titanium oxide, and hafnium oxide. Dielectrics and glasses used in non-integrated, standard optics include magnesium fluoride, soda–lime–silica glass, boron-doped glass, and other doped glasses, but these do not often use in MEMS devices due to their processing complications. There are many references in the literature to specific material studies, but Palik [1997] is a good starting point.

Before discussing data on material optical properties, it should be noted that data are often reported in a variety of ways with a variety of units. Some studies report the refractive index with real and imaginary parts, while others report the real part only as well as an absorption coefficient, which is usually related to the imaginary refractive index by (2.6) and (2.7). Units referring to wavelength may be given in units of energy, frequency, or wavenumber. Scales in energy, frequency, and wavenumber are all linearly related, while wavelength is inversely related to all these. An equation with conversions of the most common units and symbols is given below (Note: "frequency" is not typically found above the THz regime).

$$\lambda = 1/\tilde{\nu} = hc/E = c/f \tag{2.11}$$

where λ is the wavelength, $\tilde{\nu}$ is the wavenumber, E is the energy, f is the frequency, and h and c are Planck's constant and the speed of light in vacuum, respectively.

2.2.2.1 Silicon

The most common substrate material used in MEMS processing is silicon. Silicon is a semiconductor and is opaque to visible light and transparent in the 2- to 7-μm infrared range. It is the most common material used in electronic devices and integrated circuits. It finds many uses as a material in detectors and photovoltaic energy sources. Silicon has a high refractive index of about 3.7 in the visible. It also absorbs photon energies larger than roughly 1 eV, with an imaginary refractive index in the range of 0.3–3.0 in the visible. Figure 2.2 shows the variation of the dielectric constant of silicon at different photon energies.

Silicon is mostly opaque in the visible and shallow UV spectrum (2–4 eV in Figure 2.2), due to its high absorption and refractive index. It is typically used as a

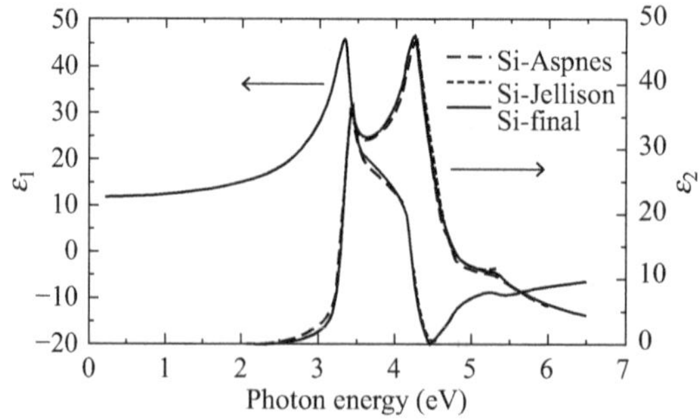

Figure 2.2 Silicon dielectric constants vs photon energy. Reprinted with permission from [Herzinger 1998], copyright AIP Publishing LLC

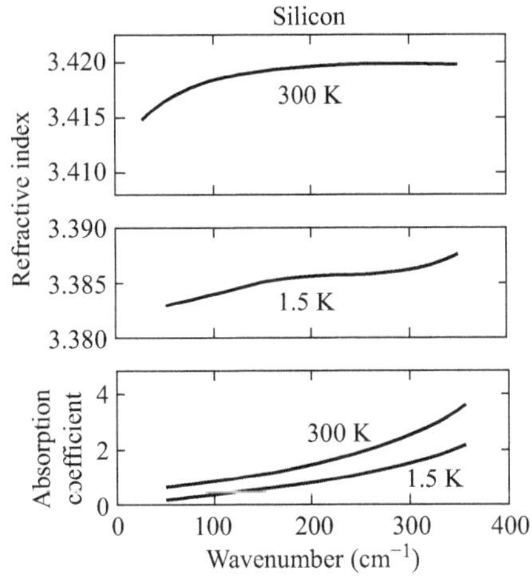

Figure 2.3 Silicon optical constants in the far-IR/THz range. Reprinted with permission from [Loewenstein 1973], copyright the Optical Society

structural element or reflector in visible devices. As previously mentioned, it is relatively transparent in the 0.1–0.6 eV range. Figure 2.3 shows the tail of vibrational absorption in the far-IR/THz range.

Although silicon was listed as a substrate, it is often deposited as a thin-film material. When thin enough, it can be used as a transmissive element as well as an electrical conductor. Characteristics may vary strongly by deposition method and thickness. A study of the optical properties of a 1-μm-thick amorphous silicon film was done in Swanepoel [1984] and found that the film refractive index at 600 nm was approximately 3.12, slightly less than that of bulk material.

2.2.2.2 Silicon dioxide

Silicon dioxide is perhaps the second most commonly used material in semiconductor and MEMS fabrication. It is often grown from a silicon substrate by an enhanced natural oxidation process. Silicon dioxide may be the most prolific optical device material due to its useful transparency in the visible and near-infrared (near-IR) regimes. The telecommunications industry often operates at wavelengths which have a minimum of loss through silicon dioxide fibers. High-quality silicon dioxide fibers can have a loss as low as 0.2 dB/km, near 1.55 μm, which is roughly a 5% signal reduction over 1 km of material! Silicon dioxide has a relatively low refractive index of approximately 1.47 in the visible. Figure 2.4 shows the refractive index of silicon dioxide at different photon energies, and Figure 2.5 shows its transmission spectrum.

Silicon dioxide has a low dn/dT, in the range of 2–9 ppm/K. It should be noted that the coefficient of mechanical thermal expansion for silicon dioxide is also on this order. Devices using silicon dioxide as a thermal modulator should take into account both index changes and mechanical changes with temperature.

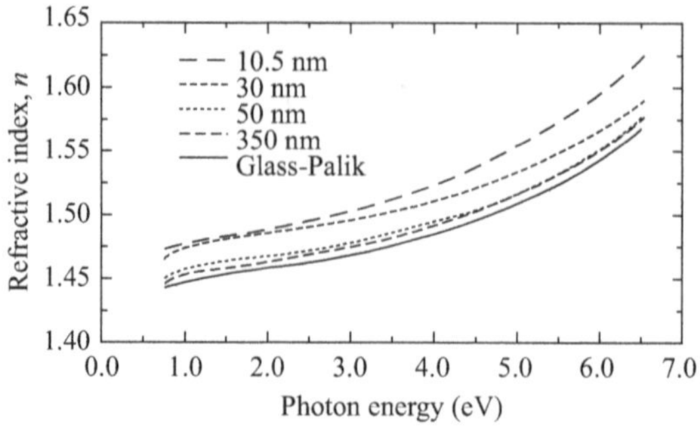

Figure 2.4 Silicon dioxide refractive index from near-IR to shallow UV of various films. Note change in the index with film thickness. Reprinted with permission from [Herzinger 1998], copyright AIP Publishing LLC

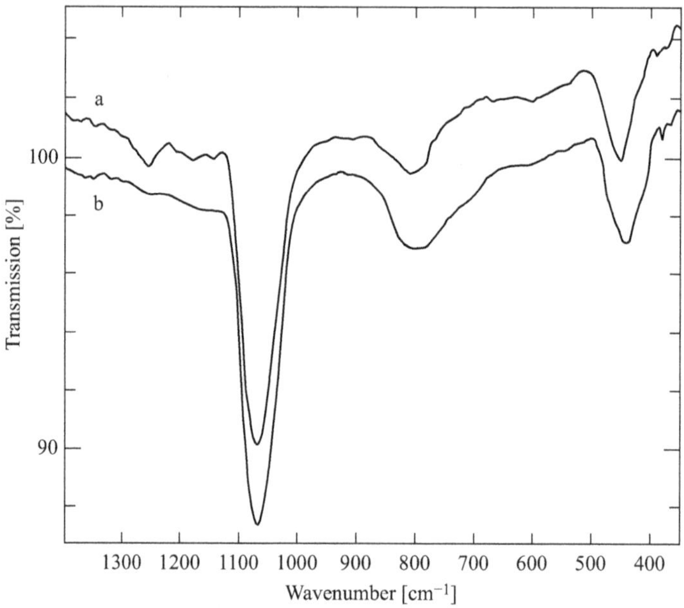

Figure 2.5 Graph of silicon dioxide transmission spectrum in the mid-IR showing phonon absorption. Films grown in (a) HCl/O_2 mixture at 850°C and (b) H_2/O_2 mixture at 950°C. Reprinted with permission from [Boyd 1982], copyright AIP Publishing LLC

2.2.2.3 Silicon nitride

Silicon nitride is another easily grown dielectric film commonly used in MEMS processing. It has a higher refractive index (around 2.02 in the visible) than silicon dioxide, and the two may be used in low-loss multilayered coatings. Silicon nitride

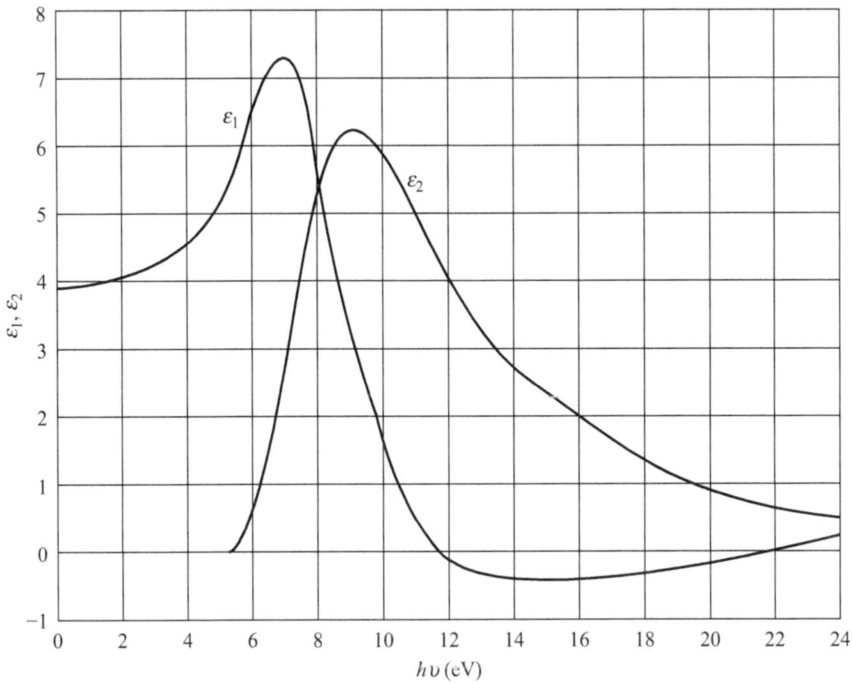

Figure 2.6 Silicon nitride real (ϵ_1) and imaginary (ϵ_2) dielectric constant spectra. Reprinted with permission from [Philipp 1973], copyright the Electrochemical Society

has an interband absorption in the UV, but closer to the visible spectrum than silicon dioxide (its bandgap is less), and therefore it may have slightly larger absorption in this regime. Figure 2.6 shows the dielectric constant of silicon nitride at different photon energies.

Silicon nitride can be deposited with a range of Si:N ratios to adjust material properties. It is often written Si_3N_4 or SiN_x, but may contain a different stoichiometry or may include hydrogen for reduced film stress.

The *dn*/*dT* of silicon nitride is similar to silicon dioxide, but tends to be higher, up to 10 ppm [Baak 1982].

2.2.3 Etching of standard MEMS materials

The use of silicon, polysilicon, silicon dioxide, and silicon nitride in a process almost guarantees that deposition and etch methods for that particular layer will be well known. Years of semiconductor and MEMS research have refined processing methods, and a listing of etches and rates for standard MEMS materials can be found in Williams [1996] and Williams [2003]. Table 2.1 shows an excerpt of the listing.

Table 2.1 List of etches and corresponding rates for standard MEMS materials. Reprinted with permission from [Williams 2003], copyright IEEE

Etchant	Etchant abbrev.	Target material
Isotropic silicon etchant "Trilogy Etch" (126 HNO_3: 60 H_2O: 5 NH_4F), −20°C	Si Iso Etch	Silicon
KOH (30% by weight), 80°C	KOH	Silicon ODE
10:1 HF (10 H_2O : 1 49% HF), −20°C	10:1 HF	Silicon dioxide
5:1 BHF (5 40% NH_4F : 1 49% HF), −20°C	5:1 BHF	Silicon dioxide
Pad Etch 4 from Ashland (13% NH_4F + 32% HAc + 49% H_2O + 6% propylene glycol + surfactant), −20°C	Pad Etch 4	SiO_2, not AI
Phosphoric Acid (85% by weight), 160°C	Phosphoric	Silicon nitride
AI Etchant Type A from Transene (80% H_3PO_4 + 5% HNO_3 + 5% HAc + 10%H_2O), 50°C	AI Etch A	Aluminum
Titanium wet etchant (20 H_2O : 1 H_2O_2 : 1 HF), −20°C	Ti Etch	Titanium
Chromium etchant CR-7 from Cyantek (9% $(NH_4)_2Ce(NO_3)_6$) + 6% $HClO_4$ + H_2O), −20°C	CR-7	Chromium
Chromium etchant CR-14 from Cyantek (22% $(NH_4)_2Ce(NO_3)_6$) +8% HAc + H_2O), −20°C	CR-14	Chromium
Molybdenum etchant (180 H_3PO_4 : 11 HAc : 11 HNO_3 : 150 H_2O), −20°C	Moly Etch	Molybdenum
Hydrogen peroxide (30wt% H_2O_2, 70wt% H_2O), 50°C	H_2O_2 50°C	Tungsten
Copper etchant type CE-200 from Transene (30% $FeCl_3$ + 3-4% HCl + H_2O), −20°C	Cu $FeCl_3$ 200	Copper
Copper etchant APS 100 from Transene (15-20% $(NH_4)_2S_2O_8$ + H_2O), 30°C	Cu APS 100	Copper
Dilute aqua regia (3 HCl : 1 HNO_3 : 2 H_2O), −30°C	Dil. Aqua regia	Noble metals
Gold etchant AU-5 from Cyantek (5% I_2 + 10% KI + 85% H_2O), −20°C	AU-5	Gold
Nichrome etchant TFN from Transene (10-20% $(NH_4)_2Ce(NO_3)_6$) + 5-6% HNO_3 + H_2O), −20°C	NiCr TFN	NiCr
1 H_2SO_4 : 1 H_3PO_4, 160°C	Phos+Sulf	Sapphire
Piranha (−50 H_2SO_4: 1 H_2O_2), 120°C	Piranha	Cleaning
Microstrip 2001 photoresist stripper, 85°C	Microstrip	Photoresist
Acetone, −20°C	Acetone	Photoresist
Methanol, −20°C	Methanol	Cleaning
Isopropanol, −20°C	IPA	Cleaning
XeF_2, 2.6 mTorr, homemade chamber	XeF_2	Silicon
HF + H_2O vapor, 1 cm over dish with 49% HF	HF vapor	Silicon dioxide
Technics plasma, O_2, 400 W @ 30 kHz, 300 mTorr	Technics O_2	Photoresist
STS ASE DRIE, mechanical chuck, high frequency, typical recipe	DRIE HF mech.	Silicon
STS ASE DRIE, electrostatic chuck, high frequency, typical recipe	DRIE HF ES	Silicon
STS ASE DRIE, mechanical chuck, stop-on-oxide (low-frequency platen), typical recipe	DRIE LF mech.	Silicon
STS ASE DRIE, electrostatic chuck, stop-on-oxide (low-frequency platen), typical recipe	DRIE LF ES	Silicon
STS 320 RIE, SF_6, 100 W @ 13.56 MHz, 20 mTorr	STS 320 SF_6	Si, SiN, metals
STS 320 RIE, SF_6 + O_2, 100 W @ 13.56 MHz, 20 mTorr	STS SF_6+O_2	Si, SiN, metals
STS 320 RIE, CF_4, 100 W @ 13.56 MHz, 60 mTorr	STS 320 CF_4	Si, SiO, SiN
STS 320 RIE, CF_4 + O_2, 100 W @ 13.56 MHz, 60 mTorr	STS CF_4+O_2	Si, SiO, SiN
Ion milling with argon ions at 500 V, ~1 mA/cm^2, normal incidence (Commonwealth data)	Ion mill	Everything

Notation:

BHF = buffered hydrofluoric acid

DRIE = deep reactive ion etch

RIE = reactive ion etch

STS ASE = Surface Technology Systems Advanced Silicon Etch

2.3 Non-standard materials incorporated into optical MEMS

2.3.1 IR materials

The purview of optical MEMS is by no means limited to the visible wavelengths. Indeed, modern optical fiber communication—historically the largest commercial application of optical MEMS technologies—is built on wavelengths in the 0.75–1.4 μm near-IR range, taking advantage of a window of low long-distance attenuation in silica fibers. Not all applications will find the "standard" microfabrication materials so readily accommodating, though: infrared sensing has historically focused on the atmospheric transmission windows in the 3–5 μm and 8–12 μm ranges, with astronomy applications pushing into the tens of micrometers; the DBRs, tunable filters, diffraction gratings, and other devices attendant to light manipulation, selection, and detection need materials which can transmit these wavelengths in order to operate.

The range of the infrared in which a semiconductor material will be strongly transmissive is determined primarily by its frequencies of lattice absorption, where light excites a vibrational mode between the atoms of a material and is thus coupled into phonons. For instance, in a linear molecule consisting of two atoms with masses m_1 and m_2, the natural frequency of oscillation will be

$$\nu = \frac{1}{2\pi}\sqrt{\frac{k}{m_r}} \tag{2.12}$$

where k is the "spring constant" of the bond and m_r is the reduced mass given by $\frac{1}{m_r} = \frac{1}{m_1} + \frac{1}{m_2}$. In more complex molecules, other vibrations will be present, as well as harmonics thereof; the summed phonon interactions can cause the infrared absorption edge to appear much earlier than the first harmonic of the fundamental lattice absorption [Musikant 1990]. Other mechanisms such as free-carrier absorption may also help determine the degree of transmittance.

Unfortunately, this cause of absorption also implies that a material which is transparent deep into the infrared also stands a good chance of being mechanically weak and a poor heat conductor. Even with careful selection, the materials chosen for their optical properties are not always the ones we would choose for ease of fabrication and may display some less-than-desirable properties when being integrated into MEMS processes and devices. Materials also often behave differently as thin films than they do in bulk. Crystal and grain structure, density and porosity, precise stoichiometry, and the presence of contaminants can significantly affect the refractive index, extinction coefficient, transmission, and other parameters. This abundance of variables can sometimes be used to good advantage, but it also means that great care must be used in film deposition and processing. Table 2.2 shows a list of optical and mechanical properties of some common materials used in optical MEMS.

2.3.1.1 Ge

Germanium is perhaps the best-known IR material; it is used widely for IR lenses and windows thanks to a transmittance that does not begin to attenuate until wavelengths past 10 μm and remains usefully high to double that. Other attractive

Table 2.2 Selected optical and mechanical properties of some common materials used in optical MEMS. The refractive index and reflectance data correspond to a wavelength of 1 μm (approximately). [Adachi 1999, Wakaki 2007, Harman 1994, Heavens 1957, Wortman 1965, Petersen 1979, Grillo 2002, Khenata 2006, Madelung 2012, Oliver 1992, Jacobs 1986, Fontanella 1974, Jiang 2003, Ghosh 1999 and Bennett 1963]

Material	Chemical formula	Refractive index (n)	Dielectric constant (ϵ)	Reflectance R%	Young's modulus (GPa)
Barium fluoride	BaF_2	1.47	2.17	–	53.1
Magnesium fluoride	MgF_2	1.39	4.8–5.5	–	93
Quartz	SiO_2	1.54	4.5–4.6	–	95
Sapphire	Al_2O_3	1.75	8.9–11.1	–	403
Zinc sulphide	ZnS	2.29	5.26	15.4	70
Zinc selenide	ZnSe	2.48	6.13	18	72
Silicon	Si	3.56	12.69	31.5	169
Germanium	Ge	4.87	23.5	43.8	138
Gallium arsenide	GaAs	3.49	12.19	30.8	123
Cadmium telluride	CdTe	2.70	7.1	22.8	20–54

characteristics for conventional optics include a refractive index that hovers just over 4.0 between 3 and 12 μm, relatively decent mechanical robustness, and low dispersion in the range of 8–12 μm long-wave IR [Musikant 1990].

Evaporated germanium films display good adhesion and surface roughness, but only for thicknesses in the hundreds of nanometers; as thickness increases past 1 μm, the high intrinsic stress causes pinholes, cracks, and eventually widespread adhesion problems to manifest. Assistance by substrate heating, ion bombardment, or sputtering is preferred to improve adhesion [Gawarikar 2013, Hirsch 1978]. Fortunately, the amorphous Ge films deposited by thermal and e-beam evaporation are optically very similar to those deposited by sputtering [Theye 1980], and the large refractive index of Ge can help reduce the required film thickness in many designs. Figure 2.7 compares the transmittance of a 12-mm-thick piece of silicon with a 3-mm-thick piece of germanium.

Obtaining single- or polycrystalline Ge films is possible by preparing an underlying template to promote epitaxial growth and sputtering [Wang 2013] or by chemical vapor deposition (CVD)/molecular beam epitaxy (MBE) techniques. A post-deposition anneal may be necessary to restore tensile stress [Franke 1999]. Oxidized germanium is water-soluble, so wet etching germanium is a simple matter of using an aqueous strong oxidizer such as H_2O_2; bases and non-oxidizing acids do not attack it. Dry etching can be done with the fluorine-based plasmas used on silicon. Franke [1999] gives a convenient summary of Ge and $Si_{1-x}Ge_x$ processing techniques.

2.3.1.2 ZnSe

Zinc selenide will admit the entirety of the near-visible to 12 μm infrared range before its transparency begins to roll off, and it remains usefully transmissive out to ~21.5 μm [Musikant 1990, Kaye 1995]. Unlike its fellow high-temperature

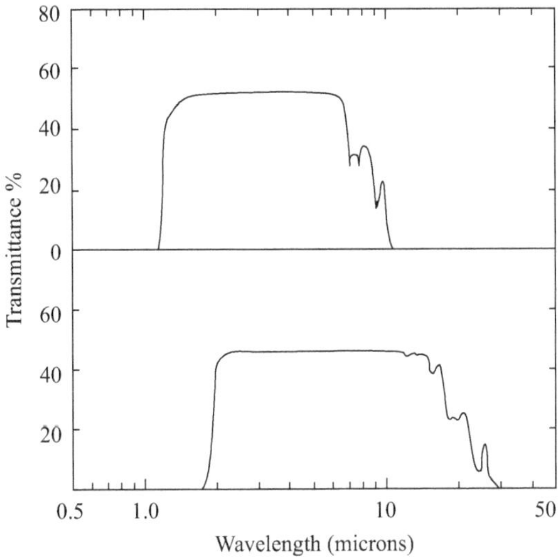

Figure 2.7 The transmittance vs wavelength of (top) a 12-mm-thick piece of silicon, and (bottom) 3 mm of germanium. Reprinted with permission from [Musikant 1990], copyright Taylor and Francis Group LLC

wide-bandgap semiconductors GaN and SiC, ZnSe's 2.71 eV bandgap [CRC 2015] belies a relatively weak ionic bond and poor thermal conductivity. It happens to have nearly the same lattice spacing and structure as GaAs, making it a popular substrate material.

ZnSe can be wet etched by combining an oxidizing agent, e.g. HCl or H_2O_2, with NaOH or NOCl (the product of nitric and chloric acid interaction), although this produces roughened or textured surfaces. On crystalline ZnSe, Tamura *et al.* found that a $KMnO_4/H_2SO_4$ solution could be used to yield a qualitatively mirror-like finish, an important quality when device dimensions are measured in fractions of a wavelength. All three of the etchant combinations were anisotropic between the (100), (111)A, and (111)B orientations, although the proportions vary significantly [Tamura 1994]. A CH_4-/H_2-based plasma etch may also be used [Pearton 1993].

2.3.1.3 ZnS

Zinc sulfide is actually the form in which naturally occurring zinc is most likely to be found, albeit with impurities. Although its reach into the infrared falls short of that of zinc selenide, it is still transmissive out to 12 μm [Musikant 1990]. Its refractive index of around 2.3 means that it, along with ZnSe (2.57), also finds use as a high-index material for visible wavelengths [Ohring 2001]. Figure 2.8 shows the transmittance of zinc sulfide and zinc selenide.

Besides the usual (metal-organic) CVD and MBE growth methods used for epitaxial semiconductors, ZnS can be deposited via evaporation and sputtering; thermal evaporation reliably produces stable, reproducible films and is as a result the most common method [Wu 2008]. Thermally evaporated ZnS tends to exhibit a

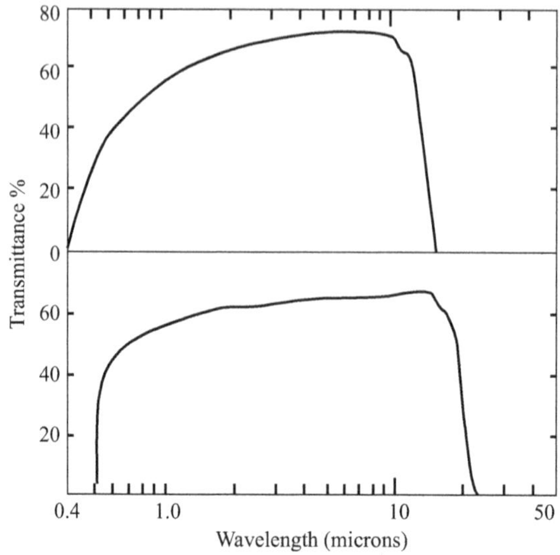

Figure 2.8 The transmittance in the infrared of (top) 3 mm of bulk zinc sulfide and (bottom) 3 mm of zinc selenide. Reprinted with permission from [Musikant 1990], copyright Taylor and Francis Group LLC

compressive stress which can be effectively controlled by substrate heating [Gawarikar 2013] or electron bombardment, and if tuned properly will yield a mostly cubic zincblende crystalline film which is durable enough for a laboratory optic coating (i.e. withstanding water exposure and immersion, mild detergents, polishing with cotton, and other forms of light mechanical abuse) [Macleod 2010, Cox 1958]. Sputtering may use either ZnSe targets or reactive Zn sputtering in H_2Se environment [Nouhi 1986, Franke 1999].

2.3.1.4 Sol–gel/spin-on-glass

Spin-on-glass (SOG) is an aptly named category of substances composed of silicates or organic siloxane dissolved in a volatile solvent. The liquid is spun on to a wafer, where it fills in any exposed voids present and makes a more-or-less level top surface. When the coated wafer is subsequently heated, the solvent is driven off and the silicates left behind form Si–O–Si bonds to produce a layer of SiO_2. While this process sounds like an almost magical shortcut to CVD- or thermal-quality silica, there are some drawbacks: leftover organics from the solvents can cause cracks, cavities, and pores in the SOG as it cures or be trapped in voids and contaminates subsequent layers deposited into them (when the void is to be used as an electrical via, the spike in resistivity is known as "via poisoning") or cause unwelcome absorption peaks; the layer's shrinkage as the solvent and water depart can cause high tensile stress and consequently cracking; and annealing can still require temperatures high enough to be incompatible with some materials such as aluminum [Plummer 2000 p. 723]. (Ion implantation can be used as an alternative to annealing, but causes its own complications.) Still, the sol–gel, liquid-dispensed nature of this substance is unique enough to offer interesting applications such as (relatively) low-temperature wafer bonding [Lin 2002] and mold-castable microchannels [Liu 1999]. Well-cured SOG is similar in transmission range to thermal SiO_2, although the much larger porosity is believed to contribute to higher absorption in areas of the 8–12 μm wavelength regime [Chou 1994]; the presence of leftover solvents will add a scattering of absorption peaks at 3.3 μm and longer wavelengths [Harkness 1998, Nakano 1995].

2.3.1.5 NaCl/KCl

Sodium chloride and potassium chloride are great optical IR materials for high-power applications: they are transparent to wavelengths past 12 μm and have high laser damage thresholds [Wood 1993]. Bulk crystals or deposition source materials are non-toxic (perhaps unsurprising for a major food additive) and inexpensive to obtain. Unfortunately, sodium and potassium chlorides are also mechanically soft, aggressively absorb moisture from ambient atmosphere, and readily dissolve in water. The resulting requirements of water-free processing and total encapsulation for almost any terrestrial application tend to limit their use.

2.3.2 UV materials

On the opposite side of the visible wavelengths lies the ultraviolet regime. Whereas a material's infrared absorption edge is set by its earliest lattice absorption frequency, the minimum wavelength for strong transmission of UV is determined by the energy needed to excite electronic band-to-band transitions.

Applications for UV include using the higher photon energies to induce luminescence or trigger chemical reactions; for example, amino acids, proteins, and DNA/RNA all absorb strongly in the 200–300 nm range, making this an important neighborhood for biological studies [Rondi 2009]. (And of course, MEMS fabrication itself makes use of UV to selectively trigger photoresist polymerization.)

2.3.2.1 Fluorides

Fluorides exhibit the most impressive UV transmission range of perhaps any family of thin-film material: standouts MgF_2 and LiF reaching down to 110 nm, in the vacuum ultraviolet (so named because it is strongly absorbed by the oxygen in the atmosphere) [Kaye 1995, Ohring 2001]. The low refractive index of MgF_2—around 1.38 in the visible—earns its widespread usage as an inexpensive single-layer antireflection coating [Glocker 2002]. Unfortunately, high tensile stress often prevents its use in structures with a large number of layers [Macleod 2010].

Fluorides tend to disassociate when sputtered, leaving evaporation as the physical deposition method of choice. Heated substrates or ion-assisted deposition help to yield durable high-quality films [Glocker 2002]. The incorporation of oxygen can increase refractive index in finished films. It can also cause trouble during deposition: magnesium oxide "tends to explode" during heating, which propels bits of the MgF_2 source material out of the evaporation hearth. Wet etching of fluorides uses nitric acid [Edstrom 2010].

2.3.2.2 Oxides

Semiconductor- and metal-oxide compounds constitute many of nature's most common substances. They tend to be robust and amorphous thin films [Ohring 2001]. The most famous of these is perhaps SiO_2, discussed earlier in Section 2.2.2.2; others include the highly chemically resistant HfO_2, durable Al_2O_3, and high-index TiO_2 [Macleod 2010].

Oxides can be deposited by e-beam and thermal evaporation, sputtering, and CVD techniques; silica, thanks to its relationship to the "fundamental" MEMS substrate material, often has the added option of thermal oxidation, which produces diffusion-limited thicknesses of high-quality films [Hsu 2008, Lee 1986]. Evaporation of oxides tends to be semi-reactive—i.e. starting with suboxides such as SiO or Ti_2O_3, and evaporating in an oxygen environment—to ensure that the film stoichiometry does not wander. Suboxides also tend to have a lower melting point than dioxides, so source sublimation can proceed at a lower temperature than if the target material was used [Macleod 2010].

Oxides (and some nitrides) can be wet etched by hydrofluoric acid, generally buffered by NH_4F to increase controllability; this etch can be masked by a conveniently wide variety of materials, including photoresist [Williams 1996]. Plasma etching is done with fluorine, chlorine, or bromine. The choice of reactive gas is informed in part by the volatility of the non-oxygen etch product, which determines how much ion bombardment or plasma heating is required to clear it from the surface and which can in turn be used to tune the etch anisotropy [Plummer 2000, Coburn 1982].

2.3.3 III–V semiconductors

The major difference in optical properties between III–V semiconductors and elemental semiconductor materials like Si and Ge is that the former has a direct bandgap while the latter has an indirect one. The maximum of the valence band occurs at the same crystal momentum value as the minimum of the conduction band, which means carriers can recombine and emit photons without significant changes in their momentums. III–V materials are therefore good candidates in applications where emission of high-density photons is needed.

Commonly used III–V semiconductors include binary alloys such as InP, InAs, GaAs, GaP, ternary alloys such as $In_{1-x}Ga_xAs$, and quaternary alloys such as $In_{1-x}Ga_xAs_yP_{1-y}$. Composition parameters x and y can be adjusted during crystal growth, which is mostly done using MBE in research settings and MOCVD on industrial scales.

2.3.3.1 Optical properties

For optical MEMS applications, we are concerned about how our material choices would affect the performance of the devices. Design choices can be made during the crystal growing phase on the alloy composition and doping levels, which could adjust the optical properties such as bandgap energies [Bugajski 1985, Jain 1990], absorption coefficient [Ballman 1983], and refractive indices [Burkhard 1982].

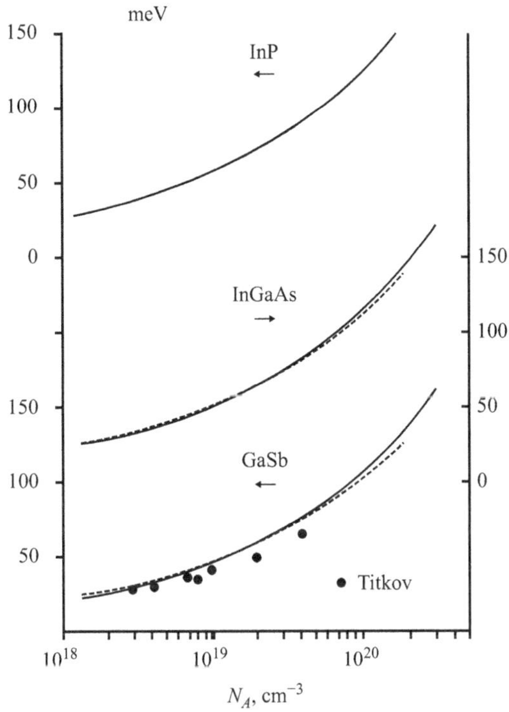

Figure 2.9 Bandgap narrowing of p-type InP, InGaAs, and GaSb at different doping levels. Solid lines are theoretical calculations and dotted lines are fittings. The solid dots are luminescence measurement of GaSb samples by Titkov et al. Modified with permission from [Jain 1990], copyright AIP Publishing LLC

Figure 2.9 shows the bandgap energy variations of InP, InGaAs, and GaSb at different doping levels.

For example, the band structure of the materials can be significantly changed by the alloy composition. GaAs is a material with a direct bandgap at the Γ point in the momentum space (or the *k*-space). Al can be added in the composition to form AlGaAs, but as the percentage of Al increases to replace Ga, the band structure of the alloy changes. The valence band edge at the Γ point rises and the one at the χ point lowers, and they cross when the Al reaches about 45%. The bandgap of the alloy therefore becomes indirect.

The refractive index of the materials can also be changed via doping. This effect can be explained by a combination of bandfilling, bandgap shrinkage, and free-carrier absorption [Bennett 1990]. For bandfilling, the lowest energy states in the conduction band are filled with free carriers, and electrons from the valence band would need greater energy to jump into the available but higher energy states. For bandgap shrinkage, the high free carrier concentration at the bottom of the conduction band would result as a screening of electrons which lowers the energy of the

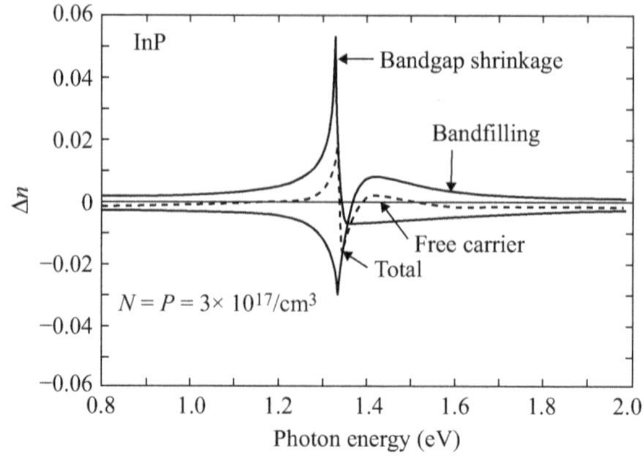

Figure 2.10 Predicted changes in refractive index of InP with electron and hole concentration at $3 \times 10^{17}/cm^3$. *Reprinted with permission from [Bennett 1990], copyright IEEE*

conduction band edge. For free-carrier absorption, free carriers absorb a photon and move to a higher energy state within the same band. Here the change in refractive index is modeled to be directly proportional to the concentration of electrons and holes in the Drude model [Henry 1981] and can be described by the equation

$$\Delta n = -\left(\frac{e^2\lambda^2}{8\pi^2 c^2 \epsilon_0 n}\right)\left(\frac{N}{m_e} + \frac{P}{m_h}\right) \tag{2.13}$$

where N and P are the concentrations of free electrons and holes, respectively; m_e and m_h are the effective masses of electrons and holes, respectively; λ is the photon wavelength; c is the speed of light in vacuum; n is the refractive index; e is the electron charge; and ϵ_0 is the permittivity of free space. Figure 2.10 shows the predicted change in refractive index of InP at different photon energies.

Besides the material properties, the way the materials are put together could also change the optical properties of the devices. For optoelectronics applications, III–V materials are often layered as a heterostructure where we can take advantage of their bandgaps to create quantum wells, wires, and dots. The behavior of these quantum structures is predominantly defined by the thicknesses of the materials after considering the material choice itself. One can essentially see this as a resonance cavity for electrons, similar to the way Fabry–Perot cavity works for photons. Photoluminescence and cryogenic measurement are often used to characterize the energy states of these structures [Moroni 1987, Temkin 1987, Makowski 2010]. Figure 2.11 shows the photoluminescence of InGaAs/InP quantum wells of different thicknesses measured at cryogenic temperature.

2.3.3.2 Processing

The processing methods for III–V semiconductors are well explored. III–V compounds such as InP are often more fragile than silicon, so care should be exercised when handling them. Below we provide an overview of etching techniques for the materials.

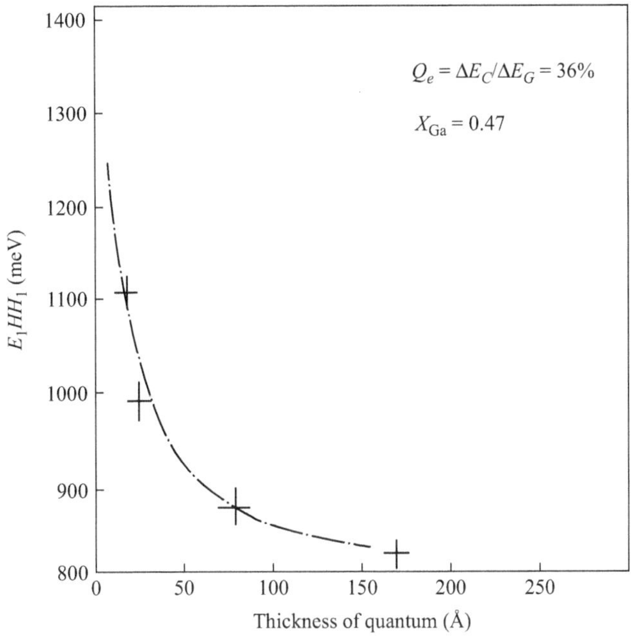

Figure 2.11 Photoluminescence of InGaAs quantum wells of different thicknesses with InP barriers. The curve is from calculations and the crosses are measurements done at 4 K. Reprinted with permission from [Moroni 1987], copyright AIP Publishing LLC

Wet etching is the most prevalent etching method. Etchants with good selectivity are readily available. For example, InGaAs can be etched by sulfuric acid and InP by hydrochloric acid, but not the other way around. This makes it easy to etch specific layers of InGaAs/InP heterostructures, a popular combination of materials for their ease of epitaxial growth due to closely matched lattice spacing. Anisotropicity plays an important role for many etchants, especially in geometrically sensitive processes such as undercutting and etch releases. In cases such as an etch release of a cantilever, attention needs to be paid to the alignment of structures with respect to the substrate orientation to avoid incomplete releases [Makowski 2008, Chan 2013]. Figure 2.12 shows an undesirable etch release of a cantilever structure due to misalignment. Localized etching is also made possible by means of photochemical etching [Ehrlich 1980, Talapin 2002]. Further information can be found in references such as Pearsall [2000] and Baca [2005].

Dry etching with reactive ion etching provides a more anisotropic etch for the materials. The most commonly used gases are based on chlorine, bromine, or iodine since they form volatile products with III–V materials, as oppose to non-volatile products with fluorine-based gases. Some chlorine gases contain carbon and might form polymer residue on the substrate [Abernathy 1996]. Fluorine-based gases such as SF_6 or CCl_2F_2 (Freon-12) may be added sometimes to form a thin non-volatile compound as an etch-stop on InAlAs or AlGaAs [Pearton 1990].

Figure 2.12 *Structure misalignment leading to undesirable etch release. A cantilever was aligned along the <100> direction of the InP sacrificial layer, which is the fast-etching facet in HCl-based etchants. The etch rate was faster along the length than along the width of the cantilever, resulting in a pointed edge of InP underneath the cantilever that was difficult to remove. Reprinted with permission from [Makowski 2008]*

Methane and hydrogen may also be used in place of the halogen-based gases for a more controlled etching. Clawson [2001] has compiled a list of references on various chemical etchants for III–V semiconductors.

2.3.4 Birefringent materials

Birefringence is the optical effect of a material for exhibiting different refractive indices for light propagating in different directions. A simple case of such effect would involve an uniaxial material, whose refractive index structure can be described by an ellipsoid. In this ellipsoid, there is an "extraordinary axis" or "*c*-axis" where the light travels along it would see one refractive index (n_e). Light traveling along the plane spanned by the other two orthogonal axes, namely the "ordinary axes", would see another refractive index (n_o). Any light traveling in any other direction would experience a composition of both refractive indices depending on its polarization state. As a result, part of the light may travel faster than another part of the light, thereby altering its polarization state. This effect is termed "retardance," and it can be described by a general equation

$$\delta = \frac{2\pi d(n_e - n_o)}{\lambda} \tag{2.14}$$

where δ is the phase retardance, n_e is the refractive index along the extraordinary axis, n_o is the refractive index along the ordinary axis, λ is the wavelength, and d is the distance that the light traveled [Yariv 2007]. Common uniaxial materials

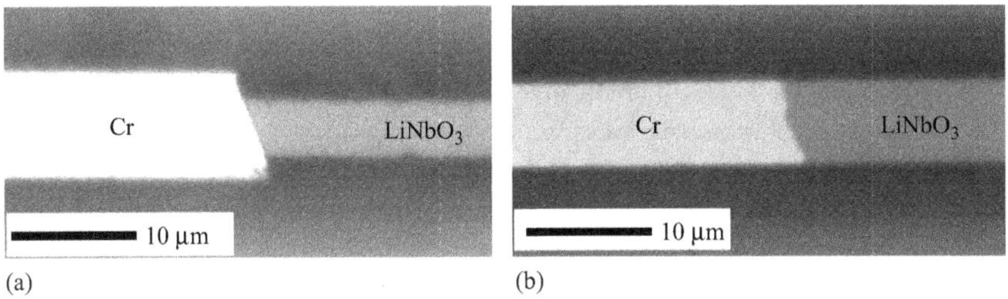

Figure 2.13 Wet etching of $LiNbO_3$ ridge waveguides with a HF–HNO_3 etchant using a Cr mask (a) without annealing and (b) with annealing after mask definition. The Cr mask was partially removed. Undercutting is significantly reduced by annealing before etching. Reprinted with permission from [Hu 2007], copyright IEEE

include quartz, calcite, and $LiNbO_3$. These materials have found uses in polarization-related optical elements such as waveplates and electro-optic modulators.

Microfabrication techniques for these materials become important when such materials are integrated into MEMS. $LiNbO_3$ is of particular interest as a candidate for active optical components because of its high electro-optic coefficient and low optical loss [Kawabe 1978]. Unfortunately, it is not an easy material to microfabricate. Etch rates are slow, and surfaces are rough due to its high resistance to standard chemical etchants.

Efforts have been made toward better etching techniques for $LiNbO_3$. Reactive ion-beam etching [Zhang 1984] and focused ion beam milling [Ulliac 2012] are both used as dry etching techniques for the material. Laurell [1992] proposed a wet etching method for the material based on proton exchange (PE). However, lateral movement of the protons results in round edges [Ting 2006], and this is a significant drawback, especially in ridge waveguide applications. There has been improvement to the technique by deploying different proton sources and using electric-field-assisted PE etching [Ting 2006, Wang 2004]. Other innovations include using an annealed Cr mask before wet etching to reduce undercutting [Hu 2007], as shown in Figure 2.13. For waveguide patterning, Ti is often diffused into $LiNbO_3$ as films [Krauss 1988] or ribbons [Ulliac 2012] to increase the extraordinary refractive index of the material [Burns 1979].

2.3.5 Reflective materials

2.3.5.1 Optical properties

Reflective materials, primarily metals, are commonly employed in optical MEMS as mirrors. The high reflectivity of metals in the visible spectrum comes largely from the behavior of a free electron. Metals have a sea of free electrons surrounding the fixed positive ions. By applying the Drude–Lorentz model, we can calculate for the plasma frequency ω_p:

$$\omega_p = \sqrt{\frac{Ne^2}{\epsilon_0 m_0}} \tag{2.15}$$

where N is the free electron density, e is the electronic charge, ϵ_0 is the electric permittivity of free space, and m_0 is the electron mass.

For frequencies below the plasma frequency, the theoretical reflectivity is 100%. For full derivation, see Jackson [1998].

Due to the high density of free electrons, the plasma frequencies of metals usually lie in the ultraviolet range. Visible lights, particularly red and yellow, which are of lower frequencies, reflect strongly as a result. However, if a free electron plasma fully described metal behavior, one would expect all metals to look the same except for the position of ω_p. We of course know that this is not true. Silver and aluminum look "whitish" while gold and copper look "yellowish". The individual characteristic hues are due to their corresponding band structures. Interband transitions occur at different frequencies for different metals. For example, copper has an interband transition edge around 560 nm which causes wavelengths below that to be absorbed, giving it the yellowish color. Silver, on the other hand, has its interband transition edge in the ultraviolet range. The whole visible spectrum is equally reflected, and it therefore appears whitish [Fox 2001].

2.3.5.2 Processing of metals

Metals are usually deposited by physical deposition methods, such as evaporation and sputtering. In evaporation, a crucible containing the metal is heated, and atoms sublimate and travel ballistically to the substrate. In sputtering, a metal target is hit with high energy ions, and metal atoms are ejected onto the substrate. For both cases, the process is performed at low pressures. If an alloy is desired, multiple crucibles/targets can be used. The quality of the films is affected by the vacuum, the deposition rates, and the geometry of the apparatus. For more information on technologies of evaporation and sputtering, please refer to Campbell [2008] and Ohring [2001].

When patterned, many metals require harsh etchants that are incompatible with other materials. For this reason, photolithography is often performed using a lift-off process as shown in Figure 2.14. Lift-off resists are spun on and patterned on the substrate using specialized developers. "Lips" would be formed at edges of the developed resist to interrupt the film and ensure a clean lift-off. Metals are then deposited, and the unwanted deposition is simply lifted off with a chemical stripper. At times when etching of metals is needed, there are a number of wet etchants that will do. Williams has compiled a list of etch rates for metals in different etchants.

2.3.5.3 Thin-film materials vs bulk materials

Metals are very reflective, but can exhibit interesting optical properties when film thickness is extremely thin. This could lead to partial absorption of incoming light, and some applications such as optical cavities where partial light absorption is desirable can take advantage of this phenomenon [Gawarikar 2013]. The optical constants of metals in literature are often obtained from bulk materials, which does not reflect the effect of thickness in thin metal films. It is therefore most reliable to measure the absorption of a sample directly. Fourier transform infrared measurement and ellipsometry are commonly used methods to determine absorbance of a film. Figure 2.15 shows the absorbance of gold nanoparticle films in the UV/visible spectra.

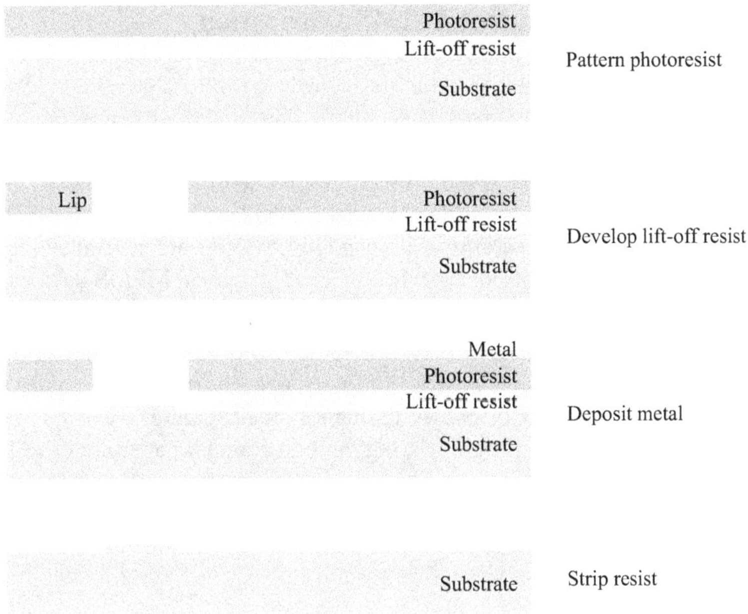

Figure 2.14 The lift-off process for depositing a metal layer

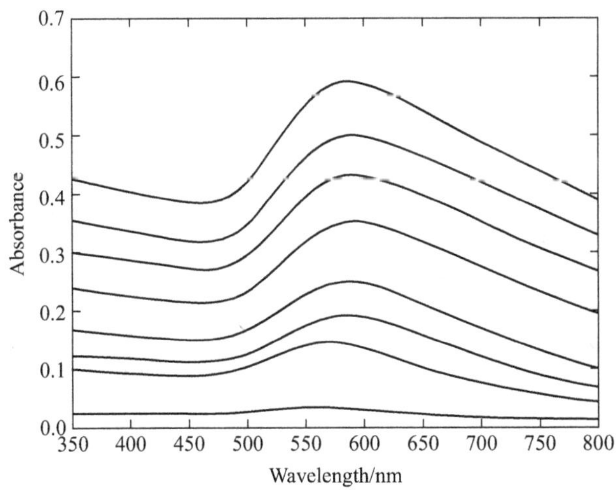

Figure 2.15 Absorbance of multilayer 6-nm gold nanoparticle films in the UV/visible spectra. Eight layers of gold nanoparticles linked with 1,9-nonanedithiol, a linking agent, were deposited on glass, and the absorbance was measured at each deposition. Deposition of further layers led to the formation of a brass-colored mirror. Reprinted with permission from [Brust 1998], copyright American Chemical Society

2.4 Challenges in optical MEMS fabrication

There are many challenges specific to optical MEMS fabrication. Some are mechanical in nature, due to small device dimensions and the fabrication processes for optical MEMS. Others are thermal, due to heat generated during fabrication or device operation. In addition, some difficulties are due to the wave nature of light. The interaction of optical waves with features that are microns in size is different than with larger optics where these effects are negligible. Below are several examples of challenges that should be considered when fabricating optical MEMS devices.

2.4.1 Diffraction

In micro-scale optical systems, diffraction plays a larger role than in macro-scale optics. As light reflects off or passes through a small feature, wave interference occurs. An example of this is light passing through a circular aperture or reflecting off of a small surface of width w. The effect of diffraction will be the same for transmission through a small feature or reflection off of a small feature. The light of wavelength λ diverges by an angle θ over a distance z as shown in (2.16).

$$\sin\theta = \frac{d}{z} = 1.22\frac{\lambda}{w} \tag{2.16}$$

The diffraction pattern in this case is known as an airy pattern with the central lobe having a diameter of d, such as the one shown in Figure 2.16. The diffraction of

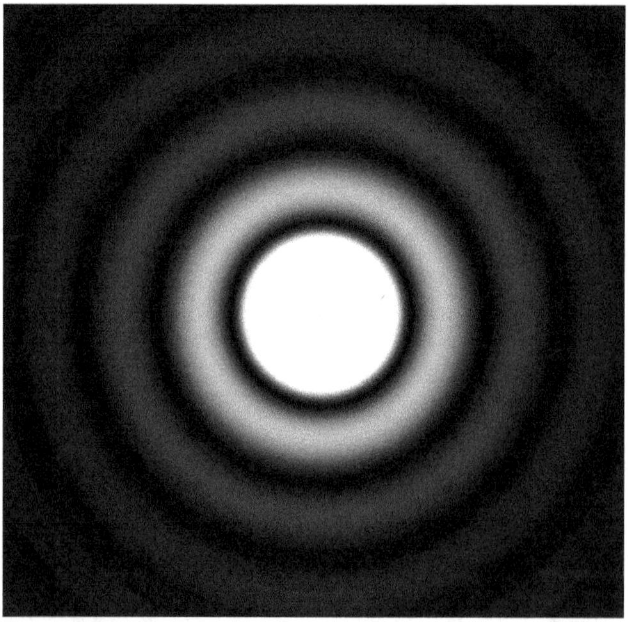

Figure 2.16 Diffraction pattern from a circular aperture

light can reduce the resolution of an optical system, and the effect becomes greater with smaller aperture sizes. This limits device miniaturization. Examples of these limitations can be seen with actuated micromirrors [Solgaard 1995, Kiang 1996, Werth 2011] and scanning microlenses [Dickensheets 1996, Kwon 2002]. In each of these cases, it is best to fabricate optics with sizes larger than 100 μm for light with a wavelength on the order of 1 μm. The diffraction angle is also strongly affected by the working distance z between the aperture and the plane of observation; for shorter working distances the feature sizes can be smaller because the light will not diverge as much.

Diffractive effects are not always detrimental in optical MEMS and can actually be useful in many cases. An example of a device that uses diffraction is a deformable grating [Godil 2002, Hagelin 2000, Solgaard 1992]. This device uses the divergence of light caused by diffraction to modulate light. When the device is actuated, the light will diverge less than when the device is not actuated, leading to a change in contrast. In the fabrication of these devices, the dimensions will be on the order of the wavelength of light.

2.4.2 Dynamic mechanical effects

Some optical MEMS devices, such as the actuated micromirror as shown in Figure 2.17, are used in high-scan-speed applications such as barcode scanners and raster scanning displays [Conant 2000, Woltar 1998, Kiang 1996, Khechana 2005, Sourani 2010, Wolter 2005, Niesten 2010, Ohira 2009, Milanović 2007, Hofmann 2012, Park 2006, Tsai 2009]. These micromirrors may need to be actuated in the kilohertz regime or faster. As the micromirror rotates back and forth, the rapid movement can begin to deform the mirror surface [Ji 2006, Conant 2000, Conant 2000 2], leading to added divergence of reflected light. Like diffraction, this divergence of light causes limitations in the resolution of the micromirror. The main factors that lead to this deformation are the material properties and dimensions of the micromirror. The density ρ of the material being used as either the support for the micromirror, or the mirror substrate can lead to deformation. For the same dimensions, higher density means greater mass. So if the mirror is actuated rapidly, the greater mass at the edge will cause more inertia. As the mirror is actuated, this inertia will slow down the movement at the edge of the mirror leading to deformation.

As an example of design challenges, the choice of mirror width and thickness can have a strong impact on the mirror deformation described in the preceding paragraphs. The mirror width, or more precisely, the distance from the center of the mirror to the edge, will be termed L. With increased L there will be more mass, which means more inertia. Additionally, as L increases, there will be more torque from the edge of the mirror. Therefore, mirror width will have the greatest effect on the amount of deformation because of both the added mass and torque. The thickness is another trade-off in the design of the micromirror. When the thickness increases then there is added mass, but the mirror also will be more rigid and less likely to deform.

Material stiffness also plays a role in mirror deformation. The stiffness of the mirror and its supports are determined by the elastic modulus E of the material

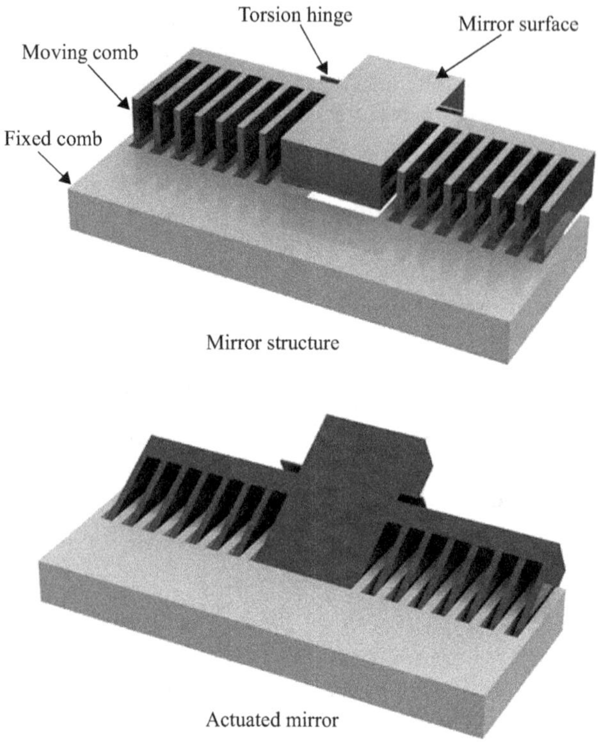

Figure 2.17 Typical actuated micromirror design using a comb drive for actuation

used. In the case of the rapidly actuated micromirror, if the elastic modulus is low, the mirror will bend more readily due to inertia from its edges. Choosing a material with the right material properties is important to help reduce deformation during operation. An equation for dynamic deformation of a rectangular micromirror which combines the effect of the material properties and device dimensions is shown in (2.17) [Conant 2000], where δ is the non-planar deformation, ν is Poisson's ratio, and $(2\pi f)^2\theta$ is the angular acceleration at scanning frequency f.

$$\delta = 0.183\frac{\rho(1-\nu^2)(2\pi f)^2\theta}{Et^2}L^5 \tag{2.17}$$

Note that the value L, for the half-width of the rectangular mirror, is to the power of 5, so this will have the greatest impact of all variables. The micro-fabrication techniques available will often limit the choice of materials. The three material properties of interest here are Poisson's ratio, elastic modulus, and density of the material. A common structural material is silicon because of its relatively good mechanical properties and availability. The mirror surface will often have metal deposited on it to make it more reflective. An example of this type of structure can be seen in Ji [2006]. The reflective coating deposited onto the rectangular area is usually much thinner than the base, so most of the mechanical properties are dominated by the base material. Also important to note: with higher mass and scanning speed, more torque is required to operate the micromirrors. More torque means more power is needed in order to actuate. This means that material properties affect not only deformation but also the power consumption of devices. Therefore, device dimensions and choice of materials are both important when considering dynamic mechanical effects.

2.4.3 Multilayer stress and strain effects

When discussing any MEMS device, stress and strain should always be considered. Here, stress refers to the internal forces within a region of material, while strain refers to deformation. The inherent stress in a film will affect its mechanical properties along with its shape. The simplest example is when a thin film is deposited onto the substrate. There are two causes of stress. The first is thermal expansion coefficient mismatch, which leads to a difference in strain between film and substrate and results in stress along the interface. The second is a difference in lattice parameters, the spacing between a material's constituent atoms.

$$\begin{aligned} &\text{1D in plane stress}-\text{strain} \qquad \sigma = \epsilon E \\ &\text{2D in plane biaxial stress}-\text{strain} \qquad \sigma = \epsilon \frac{E}{1-\nu} \end{aligned} \tag{2.18}$$

The greatest stress in optical MEMS devices is usually due to the thermal expansion coefficient difference between substrate and film. Lattice parameter mismatch is taken into account when growing a crystalline film on a crystalline substrate such as in epitaxial thin-film growth, a much less common technique in MEMS. The relationship between stress and strain for the 1D and 2D cases is shown in (2.18) [Senturia 2001] where σ and ϵ stand for stress and strain, respectively. The variable ν is Poisson's ratio, which is the ratio between axial and transverse strain. Stress is applied to an object, which causes strain. Stress and strain are related to each other through the elastic modulus (also known as Young's modulus) for the 1D case. In the 2D case Poisson's ratio is also required because as an object is deformed along one direction due to stress, it will also be deformed along the direction perpendicular to the applied stress. Examples of the 1D and 2D cases are a cantilever and membrane, respectively. For cantilevers that are much longer than they are wide, the biaxial stress is negligible. However, in the case of a membrane, the biaxial stress will have a significant effect.

In the simple case mentioned before, where a thin film is deposited onto a thicker substrate, a strain is induced. This is because during a thin-film deposition process the substrate and film are usually raised to a high temperature. As the thin film is deposited, it goes down stress free. After deposition the film and substrate will cool down, and the difference in the thermal coefficient of expansion between film and substrate will cause the two materials to expand or contract by different

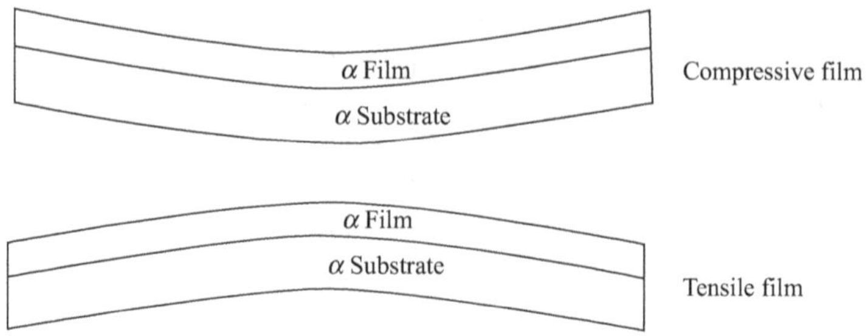

Figure 2.18 Difference between the effect of compressive and tensile films on a substrate

amounts, thereby leaving a residual strain. The strain can be described by (2.19) [Senturia 2001],

$$\epsilon = (\alpha_f - \alpha_s)\Delta T \tag{2.19}$$

with α_f and α_s standing for the coefficient of thermal expansion for the film and substrate, respectively. ΔT is the difference in temperature during the thin-film deposition and room temperature. When the coefficient of thermal expansion is smaller for the film than it is for the substrate, the film will not contract as much as the substrate, resulting in a tensile film stress. In the opposite case (when the coefficient is bigger for the film than the substrate), the film will contract more and experience compressive stress. A basic representation is shown in Figure 2.18. In a plate-shaped substrate such as a wafer, the correlation between thin-film stress and curvature is given in (2.20), known as Stoney's formula [Feng 2007]:

$$\sigma_f = \frac{E_s t_s^2 R}{6t_f(1 - \nu_s)} \tag{2.20}$$

with E being the elastic modulus, t the thickness, R the radius of curvature, σ the stress, and ν being Poisson's ratio. The subscripts s and f stand for substrate and film, respectively. The material parameters such as Poisson's ratio and elastic modulus are limited by the choice of materials, which is often determined by microfabrication limitations. The most common substrate material by far is silicon due to its widespread use in VLSI fabrication. The thickness of the film is important as the thicker it is, the greater the resulting curvature.

It is easy to see that a mismatch between two coefficients of thermal expansion can lead to the deformation of optical MEMS structures. This can be detrimental to optical systems that require very flat surfaces such as a micromirror or resonant cavity. Micromirror curvature is a common problem in optical MEMS fabrication [Sasaki 2008, Sangameswaran 2008, Gass 2003, Ilic 2004, Cowan 1997, Qiao 2004, Jin 2003]. A micromirror device usually consists of a substrate layer, which is part of the actuation mechanism, and a reflective coating. In many cases, there is at least some mismatch in the thermal coefficient of expansion. This leads to a deformation in the micromirror surface. A commonly used criterion is that a micromirror must be flat to within ~$\lambda/10$ in order to preserve the phase of reflected

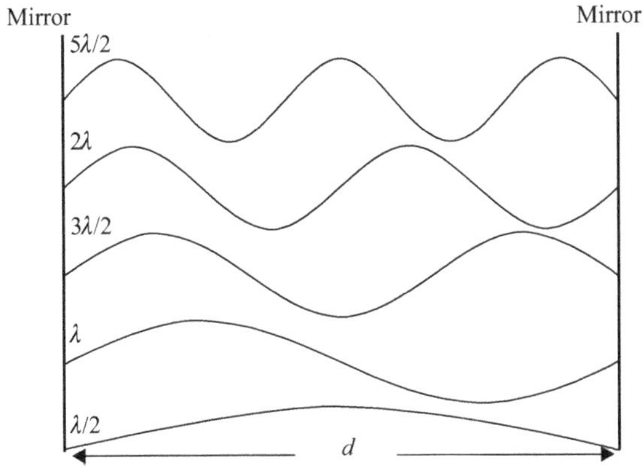

Figure 2.19 Basic structure of a Fabry-Perot cavity

light [Sasaki 2008]. For a system working at the smallest visible wavelengths around 400 nm, this means that the micromirror needs to have a total deformation no greater than 40 nm. Equation (2.20) suggests that the radius of curvature may be increased, i.e. the micromirror made flatter, by simply increasing the thickness of the substrate while keeping the film thickness constant; however, recall from Section 2.4.2 that adding a thicker substrate will lead to more mass, more inertia, and a reduced operating frequency.

Optical resonators, such as the Fabry–Perot cavities used in optical filters and spectrum analyzers [Morley 2014, Stupar 2009, Gupta 2011, Keating 2006, Ramam 2005, Rissanen 2015, Antila 2012, Takahashi 2012, Tuohiniemi 2014], are also dependent on the flatness of their reflective surfaces. The Fabry–Perot cavity has two reflective surfaces facing each other with a gap between them. The resonant condition for the cavity is that the fields go to zero at the mirrors. This means only integral multiples of half wavelengths will resonate, as shown in (2.21) and Figure 2.19. The most important value here is *d*, the distance between the mirrors. Since *d* determines the resonant conditions of the cavity, if it changes then the wavelengths that resonate in the cavity will change. The mirror will often be a transparent substrate material, like silicon, with a reflective metal coating. If there is a thermal expansion mismatch between these two materials, the mirrors can begin to deform. The deformation of a mirror will mean a change in *d* along the mirror in the cavity. This will lead to a continuum of modes in the cavity and a break down of intended operation.

$$d = m\frac{\lambda}{2} \qquad m = 1, 2, 3 \ldots \tag{2.21}$$

A more general way of looking at the resonance of the cavity is to consider light incident at some angle θ' on a mirror from outside, as shown in Figure 2.20. Equation (2.22) [Yariv 2007] gives the resonant wavelength as

$$\lambda = \frac{2nd\cos\theta}{m} \qquad m = 1, 2, 3 \ldots \tag{2.22}$$

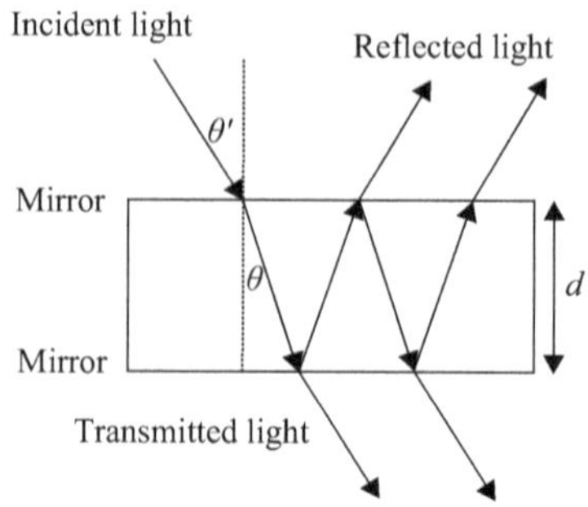

Figure 2.20 Optical path of light for off angle incident light

where n is the index of refraction of the material between mirrors, θ is the angle of internal reflections, and m stands for the mode (just an integer, in this case). At normal incidence where $\theta' = \theta = 0$, (2.22) becomes (2.21). Now if the mirror surfaces in Figure 2.20 begin to deform, not only will d change but also the angle θ for internal reflection will change, adding to the difference in resonant wavelengths.

The wavelengths of each resonance point are related to an optical frequency by the speed of light in vacuum and the mirror-enclosed material's index of refraction, the relative speed of light propagation in a certain medium. It is usually mathematically simpler to express resonant wavelengths as frequencies instead. The difference in optical frequency between the cavity's resonant modes is described by (2.23) [Yariv 2007],

$$\Delta f = \frac{c}{2nd \cos\theta} \qquad c = \frac{\lambda}{n} f \tag{2.23}$$

where c is the speed of light in vacuum.

Some important features of the mirrors that have not been mentioned yet are the reflectance, transmission, and absorption. For the sake of simplicity, only reflectance and transmission will be considered here. In a real Fabry–Perot cavity, the actual resonant mode does not occur at a single wavelength, but rather in a small range of wavelengths centered at the wavelength calculated using (2.22). The width of the range is determined by the reflectivity of the mirrors of the cavity. Equation (2.24) [Yariv 2007] is the spectral full-width at half-maximum (FWHM), the width of the resonant peak at half the maximum intensity value. Figure 2.21 shows how the width of the peaks changes with reflectance. When the reflectance of the two mirrors is equal to 0.85 then the two peaks are isolated from each other, but as the reflectance goes down the two peaks will begin to overlap and become unresolvable. In applications such as a spectrum analyzer [Hill 2007, Lammel 2002], it is important to have no overlap

$$\Delta f_{1/2} = \frac{c}{2nd \cos\theta} \frac{1 - R}{\pi\sqrt{R}} \tag{2.24}$$

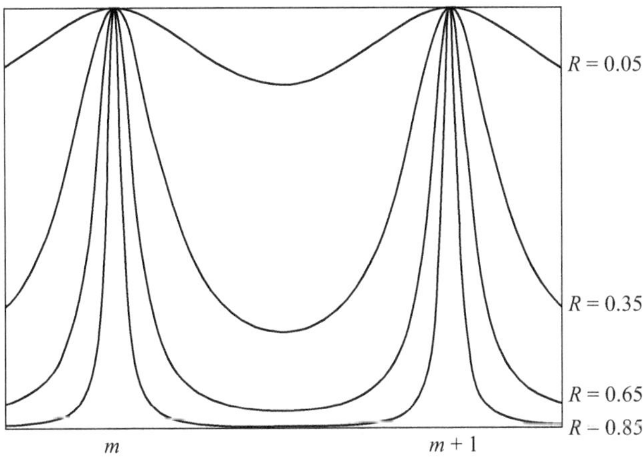

Figure 2.21 Resonance curves for a Fabry-Perot at different values of reflectance

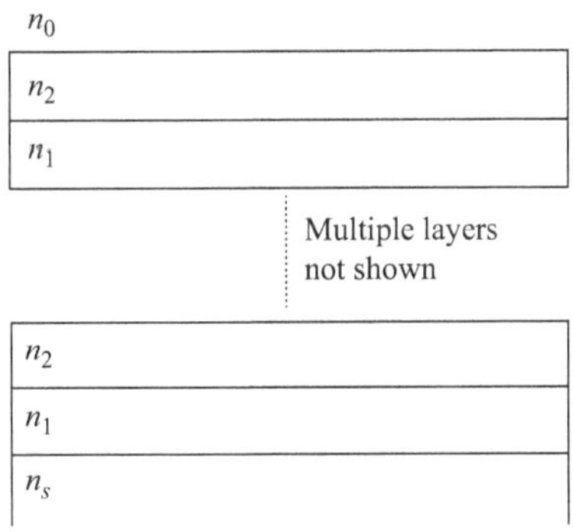

Figure 2.22 Basic structure of a DBR

between peaks. For instance, the application may call for detecting fluorescence from multiple sources, requiring the resolution of several emission peaks; the Fabry–Perot cavity should be designed by choosing the appropriate values in (2.23) and (2.24) to give as much separation between the peaks as possible. If the reflectance is too low, such as 0.05 as seen in Figure 2.21, then the two signals could overlap leading to a false reading. For instance, if we are looking for a resonance that corresponds to a mode *m*, there could be some overlap from $m + 1$. And if there was actually no signal at *m* but there was a noticeable measurement this could lead to a false positive reading.

So far the only mirrors mentioned were made of metal. There is also a DBR. These are multilayer structures which have alternating films of differing index of refraction, such as the one shown in Figure 2.22. They work by constructive and destructive interferences of light in each layer. The full description of how they work is beyond the scope of this chapter.

$$R = \left[\frac{n_0(n_2)^{2N} - n_s(n_1)^{2N}}{n_0(n_2)^{2N} + n_s(n_1)^{2N}}\right]^2 \tag{2.25}$$

$$\frac{\Delta f}{f_o} = \frac{4}{\pi} arcsin\left(\frac{n_2 - n_1}{n_2 + n_1}\right) \tag{2.26}$$

Equations (2.25) and (2.26) [Born 1999] describe how reflective a DBR is at certain wavelengths where n_0 is the index of air or other medium from which light is entering the DBR. The DBR consists of alternating layers of n_1 and n_2 which are low and high indices of refraction, respectively. Each layer of low and high indices is $\lambda/4n$ thick, with λ being the wavelength to be reflected by the DBR, n being the index of refraction of the material, and n_s the index of refraction of the substrate. The low- and high-index materials are always in pairs, and the number of pairs is represented by N. The value of Δf is the range of optical frequencies in which (2.25) is valid; outside that range the reflectance drops and begins to oscillate with optical frequency. f_o is the optical frequency which is related to λ used to determine the thickness of each layer. DBRs only reflect light over a range of wavelengths, but metals are broadband reflectors and reflect wavelengths over a much larger range.

An advantage to a DBR is that it can have a higher reflectance than metals by adding more layers, but the thermal coefficient of expansion difference between each layer of a DBR can compound to cause so much stress that the film can delaminate. There are some possible solutions to the stress-induced deformation of a mirror. One of the solutions mentioned before for the metal-on-substrate mirrors was to make the substrate thicker to reduce the curvature, but this leads to lower operating frequencies and higher power requirements. Another solution is to deposit an additional layer of a stress compensating film on the top or bottom of the mirror [Cao 2001, Tarraf 2004]. For example, if the mirror has a tensile stressed metal layer deposited on it, then it will tend to make it convex. If a compressively stressed dielectric layer is deposited on the metal, it will counteract the stress of the metal. The difficulty with this method of reducing the curvature of a mirror is that the film being used to flatten the mirror needs to be of a precise thickness. Stress reduction is important for DBRs, because when trying to get a high reflectance mirror it may require as many as 120 layers [Rabaste 2002]. By depositing alternating layers of compressive and tensile stress films, the total amount of stress can be reduced.

2.4.4 Surface roughness

Surface roughness is also an issue for mirrors and Fabry–Perot resonators. An unexpectedly rough surface can cause light scattering, phase change, and in the case of optical resonators, change in distance between mirrors. If the light becomes scattered, then this will lead to a decrease in the reflectance of a mirror [Ohlidal 1974, Huang 2006, Su 2001, Pruessner 2007]. A decrease in reflectance will lead to a smaller signal, which is undesirable for just about any optical system. For the Fabry–Perot cavity, we have seen in (2.24), a decrease in reflectance will cause the width of the resonant peaks to increase, eventually running the risk of overlap. The change in reflectance with respect to surface roughness is shown in (2.27) [Huang 2006]. R_0 is the reflectance without roughness, σ is the r.m.s. value of the roughness, n_0 is the index of refraction of the surrounding medium, and λ is the wavelength of light. This equation is derived from Kirchhoff's scalar theory of diffraction and assumes a normally incident beam on a clean surface.

$$R = R_0 \exp\left(\frac{-16\pi^2}{\lambda^2} {n_0}^2 \sigma^2\right) \tag{2.27}$$

Surface roughness can also lead to a phase change in the light. This is because since the surface is uneven the light will reflect off of the peaks and troughs of the rough surface leading to a change in the optical path length. This will cause a change in phase that is different with each reflection. For applications such as optical switches, this effect is not important, but for optical MEMS interferometers, which depend heavily on the phase of light, it can reduce the resolution by basically introducing noise into the system. In the case of resonant cavities the roughness will change the gap distance of the resonators. Unlike stress-induced curvature the gap will not change continuously but will be related to the r.m.s. value of roughness. Even with the difference there is a single equation that is useful in examining how the resonant peaks change with gap variation. Equation (2.28) shows how the resonant peak in optical frequency will change given a change in gap spacing [Yariv 2007]. The change in gap is dl, resonant peak spacing Δf, change in resonant peak df, wavelength λ, and index of refraction n which the light is passing through.

$$\frac{df}{\Delta f} = -\frac{dl}{(\lambda/2n)} \tag{2.28}$$

The value of dl can be the r.m.s. value of roughness or the change in gap spacing due to stress-induced deformation. In either case, the effect of the gap variation will be a broadening of the resonant peaks and a change in the separation between resonant peaks. Similar to the case of reflectance, broadening peaks may begin to overlap. Reducing surface roughness is difficult; some proposed solutions are nanolaminates [Gabriel 2011], vapor etch [Chandrasekaran 2006], and wet etch [Mackay 2011]. The nanolaminates work by depositing a multilayer stack of different materials which stop crystal grain growth, a major cause of surface roughness. The vapor etch is an isotropic etch for silicon which will etch the peaks in the surface more quickly than the bulk, leading to less roughness. The wet etch methods are hydroxides that preferentially etch along crystal planes of silicon, leading to atomically smooth surfaces.

2.4.5 Thermomechanical challenges

Most optical MEMS devices are released from the substrate, which reduces the mechanisms of heat transfer. For released devices, these mechanisms are: conduction through the air to the substrate, conduction through the supports of the device to the substrate, convection to the surrounding environment, and radiation. The dominant mechanism will depend on the support size, distance from the substrate, ambient environment, and temperature of the device. For instance, if the device is only a few microns from the surface, then heat conduction through the air to the substrate will dominate; if the supports are large, they can dominate heat transfer; and at high operating temperatures, radiative losses will dominate. But in optical MEMS, radiative losses are usually the smallest mechanism of heat transfer. If the device is operated in a vacuum, convection is eliminated, and heat transfer through the supports will dominate. Recall from the section on stress induced by mismatch in coefficient of thermal expansion that there is a value of ΔT in (2.19). This means that with an increasing temperature difference from room temperature (300 K), the strain between films becomes larger [Knoernschild 2010, Yeh 2011, Talghader 2004, Liu 2002, Talghader 2001]. Many optical MEMS devices are highly thermally isolated and will readily heat up. In the previous section, a solution to unwanted deformation due to stress was to deposit a stress canceling film, but that assumes room temperature operation; here, instead of just choosing a material or deposition technique that introduces a compensating stress, it will also be necessary to consider the effect the thermal coefficient of expansion will have at different temperatures. In order to offset the thermal expansion of the substrate, a film with a greater coefficient of thermal expansion can be deposited on the device [Liu 2002, Talghader 2001], so that as the substrate begins to expand, leading to deformation, the added layer will expand even more to offset the expansion of the substrate. This has the downside of possibly introducing more stress by depositing a new layer, so the intrinsic stress and coefficient of thermal expansion need to be taken into account in the choice of film. Another method for reducing thermal deformation is to increase the thermal conduction of the device. Decreasing the gap between the device and substrate allows for more heat transfer by air conduction to the substrate. The support legs could also be made larger or have a high thermally conductive material deposited on them to increase conduction to the substrate, but as before, it is important to note that adding another layer could possibly add more stress to the system. Thermal isolation may also be a desired or required quality for the device to function, as in IR-sensing microbolometers, and thus its intentional reduction might cause unacceptable sacrifices in performance.

Another issue with thermally isolated devices is thermal noise. Equation (2.29) [Senturia 2001] is the equation for thermal noise; k_B is Boltzmann's constant, T is the temperature, f_0 the resonant frequency, m is mass, and Q is the quality factor. Thermal noise is a fundamental noise and is usually much smaller than the noise introduced by the circuitry used to operate the device.

$$\text{Thermal noise} = \sqrt{\frac{8\pi k_B T f_0}{mQ}} \tag{2.29}$$

References

[Griffiths 2012] D. J. Griffiths, *Introduction to Electrodynamics*, 4th edition. Boston: Addison-Wesley, 2012.

[Ulaby 2014] F. T. Ulaby and U. Ravaioli, *Fundamentals of Applied Electromagnetics*, 7th edition. Prentice Hall, 2014.

[Yeh 2005] P. Yeh, *Optical Waves in Layered Media*, 2nd edition. Hoboken, NJ: Wiley-Interscience, 2005.

[Yariv 2007] A. Yariv and P. Yeh, *Photonics: Optical Electronics in Modern Communications*, vol. 6. New York: Oxford University Press, 2007.

[Macleod 2010] H. A. Macleod, *Thin-Film Optical Filters*, 4th edition. Boca Raton, FL: CRC Press, 2010.

[Jackson 1998] J. D. Jackson, *Classical Electrodynamics*, 3rd edition. New York: Wiley, 1998.

[Born 1999] M. Born, E. Wolf, A. B. Bhatia, P. C. Clemmow, D. Gabor, A. R. Stokes, A. M. Taylor, P. A. Wayman, and W. L. Wilcock, *Principles of Optics: Electromagnetic Theory of Propagation, Interference and Diffraction of Light*, 7th edition. Cambridge, NY: Cambridge University Press, 1999.

[Polyanskiy 2015] M. N. Polyanskiy, Refractive index database. Available at http://refractiveindex.info (accessed Feb. 29, 2015).

[Palik 1997] E. D. Palik, *Handbook of Optical Constants of Solids*, 1st edition. San Diego, CA: Academic Press, 1997.

[Wray 1969] J. H. Wray and J. T. Neu, "Refractive index of several glasses as a function of wavelength and temperature," *Journal of Optical Society of America*, vol. 59, pp. 774–776, 1969.

[Jewell 1991] J. M. Jewell, "Thermooptic coefficients of some standard reference material glasses," *Journal of the American Ceramic Society*, vol. 74, pp. 1689–1691, 1991.

[Hohlfeld 2004] D. Hohlfeld and H. Zappe, "An all-dielectric tunable optical filter based on the thermo-optic effect," *Journal of Optics A: Pure and Applied Optics*, vol. 6, no. 6, p. 504, Jun. 2004.

[Herzinger 1998] C. M. Herzinger, B. Johs, W. A. McGahan, J. A. Woollam, and W. Paulson, "Ellipsometric determination of optical constants for silicon and thermally grown silicon dioxide via a multi-sample, multi-wavelength, multi-angle investigation," *Journal of Applied Physics*, vol. 83, no. 6, pp. 3323–3336, Mar. 1998.

[Loewenstein 1973] E. V. Loewenstein, D. R. Smith, and R. L. Morgan, "Optical constants of far infrared materials 2: crystalline solids," *Applied Optics*, vol. 12, no. 2, p. 398, Feb. 1973.

[Swanepoel 1984] R. Swanepoel, "Determination of surface roughness and optical constants of inhomogeneous amorphous silicon films," *Journal of Physics E: Scientific Instruments*, vol. 17, no. 10, p. 896, Oct. 1984.

[Boyd 1982] I. W. Boyd and J. I. B. Wilson, "A study of thin silicon dioxide films using infrared absorption techniques," *Journal of Applied Physics*, vol. 53, no. 6, pp. 4166–4172, Jun. 1982.

[Philipp 1973] H. R. Philipp, "Optical properties of silicon nitride," *Journal of the Electrochemical Society*, vol. 120, no. 2, pp. 295–300, Feb. 1973.

[Baak 1982] T. Bååk, "Silicon oxynitride; a material for GRIN optics," *Applied Optics*, vol. 21, no. 6, p. 1069, Mar. 1982.

[Williams 1996] K. R. Williams and R. S. Muller, "Etch rates for micromachining processing," *Journal of Microelectromechanical Systems*, vol. 5, no. 4, pp. 256–269, Dec. 1996.

[Williams 2003] K. R. Williams, K. Gupta, and M. Wasilik, "Etch rates for micromachining processing – Part II," *Journal of Microelectromechanical Systems*, vol. 12, no. 6, pp. 761–778, Dec. 2003.

[Chou 1994] J.-S. Chou and S.-C. Lee, "Effect of porosity on infrared-absorption spectra of silicon dioxide," *Journal of Applied Physics*, vol. 77, no. 4, pp. 1805–1807, Feb. 1995.

[Cox 1958] J. T. Cox and G. Hass, "Antireflection coatings for germanium and silicon in the infrared," *Journal of the Optical Society of America*, vol. 48, no. 10, pp. 677–678, Oct. 1958.

[Franke 1999] A. E. Franke, D. Bilic, D. T. Chang, P. T. Jones, T.-J. King, R. T. Howe, and G. C. Johnson, "Post-CMOS integration of germanium microstructures," in *Twelfth IEEE International Conference on Micro Electro Mechanical Systems, 1999. MEMS '99,* 1999, pp. 630–637.

[Haynes 2015] W. M. Haynes, Ed., *CRC Handbook of Chemistry and Physics*, CRC Press 96th edition, 2015.

[Gawarikar 2013] A. S. Gawarikar, "Spectrally Selective High Detectivity Uncooled Detectors for the Long Wave Infrared." PhD Dissertation, University of Minnesota, Twin Cities, 2013.

[Harkness 1998] B. R. Harkness, K. Takeuchi, and M. Tachikawa, "Demonstration of a directly photopatternable spin-on-glass based on hydrogen silsesquioxane and photobase generators," *Macromolecules*, vol. 31, no. 15, pp. 4798–4805, 1998.

[Kaye 1995] *Tables of Physical & Chemical Constants*, 16th edition 1995. 2.1.4 Hygrometry. Kaye & Laby Online. Version 1.0 (2005). www.kayelaby.npl.co.uk

[Lin 2002] H. C. Lin, K. L. Chang, G. W. Pickrell, K. C. Hsieh, and K. Y. Cheng, "Low temperature wafer bonding by spin on glass," *Journal of Vacuum Science & Technology B*, vol. 20, no. 2, pp. 752–754, Mar. 2002.

[Liu 1999] R. H. Liu, M. J. Vasile, and D. J. Beebe, "The fabrication of nonplanar spin-on glass microstructures," *Journal of Microelectromechanical Systems*, vol. 8, no. 2, pp. 146–151, Jun. 1999.

[Musikant 1990] S. Musikant, Ed., *Optical Materials: A Series of Advances*, 1st edition, vol. 1. New York, NY: Marcel Dekker, 1990.

[Nakano 1995] T. Nakano, K. Tokunaga, and T. Ohta, "Effects of Si—C bond content on film properties of organic spin-on glass," *Journal of The Electrochemical Society*, vol. 142, no. 4, pp. 1303–1308, 1995.

[Nouhi 1986] A. Nouhi and R. J. Stirn, "Highly conducting ZnSe films by reactive magnetron sputtering," *Journal of Vacuum Science & Technology A*, vol. 4, no. 3, pp. 403–407, May 1986.

[Pearton 1993] S. J. Pearton and F. Ren, "Plasma etching of ZnS, ZnSe, CdS, and CdTe in electron cyclotron resonance $CH_4/H_2/Ar$ and H_2/Ar discharges," *Journal of Vacuum Science & Technology B*, vol. 11, no. 1, pp. 15–19, Jan. 1993.

[Plummer 2000] J. D. Plummer, M. D. Deal, and P. B. Griffin, *Silicon VLSI Technology*, 1st edition. Upper Saddle River, NJ: Prentice Hall, 2000.

[Tamura 1994] H. Tamura, Y. Okuno, and H. Kato, "Chemical etching of ZnSe crystals." *Journal of electronic materials*, 23, no. 8, pp. 835–838, Aug. 1994.

[Theye 1980] M.-L. Theye, A. Gheorghiu, T. Rappeneau, and A. Lewis, "Transport properties of evaporated versus sputtered amorphous germanium films," *Journal de Physique*, vol. 41, no. 10, pp. 1173–1181, 1980.

[Wang 2013] R. Wang, S. N. Sambandam, G. Majkic, E. Galstyan, and V. Selvamanickam, "High mobility single-crystalline-like germanium thin films on flexible, inexpensive substrates," *Thin Solid Films*, vol. 527, pp. 9–15, Jan. 2013.

[Wood 1993] R. M. Wood, *Optical Materials*, 1st edition. London: The Institute of Materials, 1993.

[Wu 2008] X. Wu, F. Lai, L. Lin, J. Lv, B. Zhuang, Q. Yan, and Z. Huang, "Optical inhomogeneity of ZnS films deposited by thermal evaporation," *Applied Surface Science*, vol. 254, no. 20, pp. 6455–6460, Aug. 2008.

[Coburn 1982] J. W. Coburn, "Plasma-assisted etching," *Plasma Chemistry and Plasma Processing*, vol. 2, no. 1, pp. 1–41, Mar. 1982.

[Edstrom 2010] C. Edström, "Wet Etching of Optical Thin Films," Master's, Jönköping University, 2010.

[Glocker 2002] D. A. Glocker and S. I. Shah, Eds., *Handbook of Thin Film Process Technology*, 1st edition, vol. 2, 2 vols. Bristol: Institute of Physics Publishing, 2002.

[Hsu 2008] T.-R. Hsu, *MEMS and Microsystems: Design, Manufacture, and Nanoscale Engineering*, 2nd edition. Hoboken, NJ: John Wiley & Sons, 2008.

[Lee 1986] J. Lee, I.-C. Chen, and C. Hu, "Comparison between CVD and thermal oxide dielectric integrity," *IEEE Electron Device Letters*, vol. 7, no. 9, pp. 506–509, Sep. 1986.

[Rondi 2009] A. Rondi, J. Extermann, L. Bonacina, S. M. Weber, and J.-P. Wolf, "Characterization of a MEMS-based pulse-shaping device in the deep ultra-violet," Applied Physics B, vol. 96, no. 4, pp. 757–761, May 2009.

[Bugajski 1985] M. Bugajski and W. Lewandowski, "Concentration-dependent absorption and photoluminescence of n-type InP," *Journal of Applied Physics*, vol. 57, no. 2, pp. 521–530, 1985.

[Jain 1990] Jain, S. C., J. M. McGregor, and D. J. Roulston, "Band-gap narrowing in novel III–V semiconductors," *Journal of Applied Physics*, vol. 68, no. 7, pp. 3747–3749, 1990.

[Ballman 1983] A. A. Ballman, *et al.*, "Double doped low etch pit density InP with reduced optical absorption," *Journal of Crystal Growth*, vol. 62, no. 1, pp. 198–202, 1983.

[Burkhard 1982] H. Burkhard, H. W. Dinges, and E. Kuphal, "Optical properties of $In_{1-x}Ga_xP_{1-y}As_y$, InP, GaAs, and GaP determined by ellipsometry." *Journal of Applied Physics*, vol. 53, no. 1, pp. 655–662, 1982.

[Henry 1981] C. H. Henry, R. A. Logan, and K. A. Bertness, "Spectral dependence of the change in refractive index due to carrier injection in GaAs lasers," *Journal of Applied Physics*, vol. 52, no. 7, pp. 4457–4461, 1981.

[Bennett 1990] B. R. Bennett, R. A. Soref, and J. del Alamo, "Carrier-induced change in refractive index of InP, GaAs and InGaAsP," *IEEE Journal of Quantum Electronics*, vol. 26, no. 1, pp. 113–122, 1990.

[Temkin 1987] H. Temkin, *et al.*, "Low-temperature photoluminescence from InGaAs/InP quantum wires and boxes," *Applied Physics Letters*, vol. 50, no. 7, pp. 413–415, 1987.

[Moroni 1987] D. Moroni, *et al.*, "Photoluminescence investigation of InGaAs–InP quantum wells," *Journal of Applied Physics*, vol. 62, no. 5, pp. 2003–2008, 1987.

[Makowski 2010] J. D. Makowski, *et al.*, "Mechanical construction of semiconductor bandgaps," *IEEE Journal of Quantum Electronics*, vol. 46, no. 9, pp. 1261–1267, 2010.

[Makowski 2008] J. D. Makowski, *Band-Gap Tuning through Mechanical Semiconductor Heterostructures*. PhD Dissertation, University of Minnesota, Twin Cities, ProQuest, 2008.

[Chan 2013] W. S. Chan, M. J. Saarinen, and J. J. Talghader, "Fabrication and operation of an electrostatic actuator for controlling nanometer-scale gaps in collapsed cantilever heterostructures," *Applied Physics Letters*, vol. 102, no. 24, 243508, 2013.

[Ehrlich 1980] D. J. Ehrlich, R. M. Osgood Jr, and T. F. Deutsch, "Laser-induced microscopic etching of GaAs and InP," *Applied Physics Letters*, vol. 36, no. 8, pp. 698–700, 1980.

[Talapin 2002] D. V. Talapin, *et al.*, "Etching of colloidal InP nanocrystals with fluorides: photochemical nature of the process resulting in high photoluminescence efficiency," *The Journal of Physical Chemistry B*, vol. 106, no. 49, pp. 12659–12663, 2002.

[Pearsall 2000] T. P. Pearsall, Ed., *Properties, Processing and Applications of Indium Phosphide*, No. 21. The Institute of Electrical Engineers, United Kingdom, 2000.

[Baca 2005] A. G. Baca, and C. Ashby. *Fabrication of GaAs Devices*, No. 6. The Institute of Electrical Engineers, United Kingdom, 2005.

[Abernathy 1996] C. R. Abernathy and F. Ren, *Topics in Growth and Device Processing of III–V Semiconductors*, vol. 1. World Scientific Publishing Co. Pte. Ltd. Singapore, 1996.

[Pearton 1990] S. J. Pearton, *et al.*, "Cl_2 and $SiCl_4$ reactive ion etching of In-based III–V semiconductors," *Journal of The Electrochemical Society*, vol. 137, no. 10, pp. 3188–3202, 1990.

[Clawson 2001] A. R. Clawson, "Guide to references on III–V semiconductor chemical etching," *Materials Science and Engineering: R: Reports*, vol. 31, no. 1, pp. 1–438, 2001.

[Kawabe 1978] M. Kawabe, M. Kubota, K. Masuda, and S. Namba, "Microfabrication in $LiNbO_3$ by ion-bombardment-enhanced etching," *Journal of Vacuum Science & Technology*, vol. 15, 1096–1098, 1978.

[Krauss 1988] T. Krauss, A. Speth, M. M. Oprysko, B. Fan, and K. Grebe, "Laser microfabrication of optical circuits in $LiNbO_3$," *Applied Physics Letters*, vol. 53, pp. 947–949, 1988.

[Zhang 1984] B. Zhang, S. Forouhar, S. Huang, W. S. C. Chang, "C2F6 reactive ion-beam etching of $LiNbO_3$ and Nb_2O_5 and their application to optical waveguides," *Journal of Lightwave Technology*, vol. 2, no. 4, pp. 528–530, Aug. 1984.

[Ulliac 2012] G. Ulliac, *et al.*, "Fabrication of Bragg gratings on $LiNbO_3$ optical waveguides," *Microelectronic Engineering*, vol. 97, pp. 185–188, 2012.

[Laurell 1992] F. Laurell, *et al.* "Wet etching of proton-exchanged lithium niobate: a novel processing technique," *Journal of Lightwave Technology*, vol. 10, no. 11, pp. 1606–1609, 1992.

[Ting 2006] T.-L. Ting, L.-Y. Chen, and W.-S. Wang, "A novel wet-etching method using joint proton source in $LiNbO_3$," *Photonics Technology Letters, IEEE*, vol. 18, no. 4, pp. 568–570, 2006.

[Wang 2004] T.-J. Wang, *et al.* "A novel wet-etching method using electric-field-assisted proton exchange in $LiNbO_3$," *Journal of Lightwave Technology*, vol. 22, no. 7, pp. 1764–1771, 2004.

[Hu 2007] H. Hu, R. Ricken, W. Sohler, R. B. Wehrspohn, "Lithium niobate ridge waveguides fabricated by wet etching," *Photonics Technology Letters, IEEE*, vol. 19, no. 6, pp. 417–419, March15, 2007.

[Burns 1979] W. K. Burns, *et al.*, "Ti diffusion in Ti: $LiNbO_3$ planar and channel optical waveguides," *Journal of Applied Physics*, vol. 50, no. 10, pp. 6175–6182, 1979.

[Fox 2001] A. M. Fox, *Optical Properties of Solids*, vol. 2010. New York: Oxford University Press, 2001.

[Campbell 2008] S. A. Campbell, *Fabrication Engineering at the Micro and Nanoscale*, vol. 4. New York: Oxford University Press, 2008.

[Ohring 2001] M. Ohring, *Materials Science of Thin Films*. Academic Press, San Diego, California, USA, 2001.

[Brust 1998] M. Brust, *et al.*, "Self-assembled gold nanoparticle thin films with nonmetallic optical and electronic properties," *Langmuir*, vol. 14, no. 19, pp. 5425–5429, 1998.

[Solgaard 1995] O. Solgaard, M. Daneman, N. C. Tien, A. Friedberger, R. S. Muller, and K. Y. Lau, "Optoelectronic packaging using silicon surface micromachine alignment mirrors," *Photonics Technology Letters, IEEE*, vol. 7, no. 1, pp. 41–43, Jan. 1995.

[Werth 2011] N. Werth, "Diffraction errors in micromirror-array based wavefront generation," *Optics Communications*, vol. 284, no. 9, pp. 2317–2322, 2011.

[Dickensheets 1996] D. L. Dickensheets and G. S. Kino, "Micromachined scanning confocal optical microscope," *Optics Letters*, vol. 21, no. 10, pp. 764–766, 1996.

[Kwon 2002] S. Kwon, V. Milanovic, and L. P. Lee, "Vertical microlens scanner for 3D imaging," Tech. Dig. 2002 Solid-State Sens. Actuator Workshop, Hilton Head Island, SC, USA, pp. 227–230.

[Godil 2002] A. Godil, "Diffractive MEMS technology offers a new platform for optical network," *Laser Focus World*, vol. 38, no. 5, p.181, May 2002.

[Hagelin 2000] P. M. Hagelin, U. Krishnamoorthy, J. P. Heritage, and O. Solgaard, "Scalable optical cross-connect switch using micromachined mirrors," *Photonics Technology Letters, IEEE*, vol. 12, no. 7, pp. 882–885, Jul. 2000.

[Solgaard 1992] O. Solgaard, "Deformable grating optical modulator," *Optics Letters* vol. 17, no. 9, pp. 688–690, 1992.

[Woltar 1998] A. Woltar, H. Schenk, E. Gaumont, and H. Lakner, "The MEMS microscanning mirror for barcode reading: from development to production," *Proceedings of the SPIE*, vol. 3513, pp. 182–190, 1998.

[Kiang 1996] M.-H. Kiang, O. Solgaard, R. S. Muller, and K. Y. Lau, "Micromachined polysilicon microscanners for barcode readers," *Photonics Technology Letters, IEEE*, vol. 8, no. 12, pp. 1707–1709, Dec. 1996.

[Khechana 2005] F. Khechana, et al., "Thermally actuated micro scanner for barcode reader applications," *Optical MEMS and Their Applications Conference*, 2005. *IEEE/LEOS International Conference*. IEEE, 2005.

[Sourani 2010] S. Sourani, "Resonance MEMS mirrors design considerations." *MOEMS-MEMS*. International Society for Optics and Photonics, San Francisco, California, USA, 2010.

[Wolter 2005] A. Wolter, *et al.* "Applications and requirements for MEMS scanner mirrors." *MOEMS-MEMS Micro & Nanofabrication*. International Society for Optics and Photonics, San Jose, California, USA, 2005.

[Niesten 2010] M. Niesten, *et al.*, "Scanning laser beam displays based on a 2-D MEMS."*SPIE, Optics, Photonics, and Digital Technologies for Multimedia Applications*, vol. 7723, 77230U-1, 2010.

[Ohira 2009] Y. Ohira, *et al.*, "Hybrid MEMS optical scanner for volumetric 3-D displays." *Journal of the Society for Information Display*, vol. 17, no. 5, pp. 419–422, 2009.

[Milanović 2007] V. Milanović, "Improved control of the vertical axis scan for MEMS projection displays," *Mirror*, vol. 306, no. 100, p. 1000, 2007.

[Hofmann 2012] U. Hofmann, J. Janes, and H.-J. Quenzer, "High-Q MEMS resonators for laser beam scanning displays," *Micromachines* vol. 3, no. 2, pp. 509–528, 2012.

[Park 2006] Y.-H. Park, *et al.* "Perspective of MEMS based raster scanning display and its requirements for success," *MOEMS-MEMS 2006 Micro and Nanofabrication*. International Society for Optics and Photonics, San Jose, California, USA, 2006.

[Tsai 2009] J.-C. Tsai, "The evolution of MEMS displays," *IEEE Transactions on Industrial Electronics*, vol. 56, no. 4, pp. 1057–1065, 2009.

[Ji 2006] C.-H. Ji, *et al.* "An electrostatic scanning micromirror with diaphragm mirror plate and diamond-shaped reinforcement frame," *Journal of Micromechanics and Microengineering*, vol. 16, no. 5, p. 1033, 2006.

[Conant 2000] R. A. Conant, *et al.*, "A flat high-frequency scanning micromirror," *Proceedings of Solid-State Sensor and Actuator Workshop*, 2000.

[Conant 2000 2] R. A. Conant, *et al.* "Dynamic deformation of scanning mirrors." 2000 *IEEE/LEOS International Conference on Optical MEMS.* IEEE, 2000.

[Senturia 2001] S. D. Senturia, *Microsystem Design*, vol. 3. Boston: Kluwer Academic Publishers, 2001.

[Feng 2007] X. Feng, Y. Huang, and A. J. Rosakis, "On the Stoney formula for a thin film/substrate system with nonuniform substrate thickness," *Journal of Applied Mechanics*, vol. 74, no. 6, pp. 1276–1281, 2007.

[Sasaki 2008] M. Sasaki, *et al.*, "An optically flat micromirror using a stretched membrane with crystallization-induced stress," *Journal of Optics A: Pure and Applied Optics*, vol. 10, no. 4, 044004, 2008.

[Sangameswaran 2008] S. Sangameswaran, *et al.*, "ESD reliability issues in microelectromechanical systems (MEMS): a case study on micromirrors," *Electrical Overstress/Electrostatic Discharge Symposium, 2008. EOS/ESD 2008. 30th.* IEEE, 2008.

[Gass 2003] F. R. Gass, *et al.*, "Stress and curvature in MEMS mirrors." *Micromachining and Microfabrication*. International Society for Optics and Photonics, San Jose, California, USA, 2003.

[Ilic 2004] A. Ilic, and K. W. Goossen, "Phase transition vs. thickness in stress-induced curvature on Cr/Au MEMS mirror layers," *Micromachining and Microfabrication.* International Society for Optics and Photonics, San Jose, California, USA, 2004.

[Cowan 1997] W. D. Cowan, *et al.* "Modeling of stress-induced curvature in surface-micromachined devices," *Micromachining and Microfabrication*. International Society for Optics and Photonics, Austin, Texas, USA, 1997.

[Qiao 2004] D.-Y. Qiao, W.-Z. Yuan, and W.-J. Li, "Residual stress induced deformation of bilayer plate with patterned additional layer," *Intelligent Mechatronics and Automation, 2004. Proceedings. 2004 International Conference*. IEEE, 2004.

[Jin 2003] S. Jin, *et al.*, "Control of microelectromechanical systems membrane curvature by silicon ion implantation," *Applied Physics Letters*, vol. 83, no. 12, pp. 2321–2323, 2003.

[Morley 2014] M. L. Morley, *et al.* "Design and fabrication of Fabry–Perot filters for infrared hyperspectral imagers." *Journal of Micro/Nanolithography, MEMS, and MOEMS*, vol. 13, no. 1, 011116, 2014.

[Stupar 2009] P. A. Stupar, *et al.*, "MEMS tunable Fabry–Perot filters with thick, two sided optical coatings," *Solid-State Sensors, Actuators and Microsystems Conference, 2009. TRANSDUCERS 2009. International*. IEEE, 2009.

[Gupta 2011] N. Gupta, S. Tan, and D. R. Zander, "MEMS-based tunable Fabry–Perot filters," *SPIE Defense, Security, and Sensing*. International Society for Optics and Photonics, Orlando, Florida, USA, 2011.

[Keating 2006] A. J. Keating, *et al.* "Optical characterization of Fabry–Perot MEMS filters integrated on tunable short-wave IR detectors," *Photonics Technology Letters, IEEE*, vol. 18, no. 9, pp. 1079–1081, 2006.

[Stupar 2009] P. A. Stupar, *et al.* "MEMS tunable Fabry–Perot filters with thick, two sided optical coatings," *Solid-State Sensors, Actuators and Microsystems Conference, 2009. TRANSDUCERS 2009. International*. IEEE, 2009.

[Ramam 2005] A. Ramam, *et al.* "MEMS tunable Fabry–Perot optical filter." *Optical MEMS and Their Applications Conference, 2005. IEEE/LEOS International Conference*. IEEE, 2005.

[Rissanen 2015] A. Rissanen, *et al.*, "Large-aperture MOEMS Fabry–Perot interferometer for miniaturized spectral imagers," *SPIE OPTO*. International Society for Optics and Photonics, San Francisco, California, USA, 2015.

[Antila 2012] J. Antila, *et al.* "Spectral imaging device based on a tuneable MEMS Fabry-Perot interferometer," *SPIE Defense, Security, and Sensing*. International Society for Optics and Photonics, Baltimore, Maryland, USA, 2012.

[Takahashi 2012] K. Takahashi, *et al.*, "Highly-sensitive label-free protein sensor using MEMS Fabry–Perot interferometer," *Optical MEMS and Nanophotonics (OMN), 2012 International Conference*. IEEE, 2012.

[Tuohiniemi 2014] M. Tuohiniemi, A. Näsilä, and A. Akujärvi, "MEMS Fabry–Perot interferometer with Si-air mirrors for mid-and thermal infrared." *SPIE MOEMS-MEMS*. International Society for Optics and Photonics, San Francisco, California, USA, 2014.

[Hill 2007] G. C. Hill, *et al.* "SU-8 MEMS Fabry–Perot pressure sensor," *Sensors and Actuators A: Physical*, vol. 138, no. 1, pp. 52–62, 2007.

[Lammel 2002] G. Lammel, *et al.* "Tunable optical filter of porous silicon as key component for a MEMS spectrometer," *Journal of Microelectromechanical Systems*, vol. 11, no. 6, pp. 815–828, 2002.

[Cao 2001] K. Cao, W. Liu, and J. J. Talghader, "Curvature compensation in micromirrors with high-reflectivity optical coatings," *Journal of Microelectromechanical Systems*, vol. 10, no. 3, pp. 409–417, 2001.

[Tarraf 2004] A. Tarraf, *et al.* "Stress investigation of PECVD dielectric layers for advanced optical MEMS," *Journal of Micromechanics and Microengineering*, vol. 14, no. 3, p. 317, 2004.

[Rabaste 2002] S. Rabaste, *et al.*, "Sol–gel fabrication of thick multilayers applied to Bragg reflectors and microcavities," *Thin Solid Films*, vol. 416, no. 1, pp. 242–247, 2002.

[Ohlidal 1974] I. Ohlidal, F. Lukeš, and K. Navratil, "Rough silicon surfaces studied by optical methods," *Surface Science*, vol. 45, no. 1, pp. 91–116, 1974.

[Huang 2006] B. Huang, *et al.*, "Investigation of the optical properties of sputtered ZnO films by reflectance spectroscopy," *Optical Engineering*, vol. 45, no. 12, 123801, 2006.

[Su 2001] G.-D. J. Su, H. Toshiyoshi, and M. C. Wu, "Surface-micromachined 2-D optical scanners with high-performance single-crystalline silicon micromirrors," *Photonics Technology Letters, IEEE*, vol. 13, no. 6, pp. 606–608, 2001.

[Pruessner 2007] M. W. Pruessner, T. H. Stievater, and W. S. Rabinovich, "Process development, design, and characterization of high-finesse micro-machined optical Fabry–Perot microcavities," *MOEMS-MEMS 2007 Micro and Nanofabrication*. International Society for Optics and Photonics, San Jose, California, USA, 2007.

[Gabriel 2011] N. T. Gabriel, and J. J. Talghader, "Thermal conductivity and refractive index of hafnia-alumina nanolaminates," *Journal of Applied Physics*, vol. 110, no. 4, 043526, 2011.

[Chandrasekaran 2006] A. Chandrasekaran, *et al.*, "Hybrid bulk micro-machining process suitable for roughness reduction in optical MEMS devices," *International Journal of Manufacturing Technology and Management*, vol. 9, no. 1, pp. 144–159, 2006.

[Mackay 2011] R. E. Mackay, N. Lionis, and H. R. Le, "3D surface topography and reflectivity of anisotropic etched silicon micromirrors for BioMEMS," *Microsystem Technologies*, vol. 17, no. 12, pp. 1763–1770, 2011.

[Knoernschild 2010] C. Knoernschild, *et al.* "Investigation of optical power tolerance for MEMS mirrors," *Journal of Microelectromechanical Systems*, vol. 19, no. 3, pp. 640–646, 2010.

[Yeh 2011] M. K. Yeh and C. L. Lu, "The Effect of thermal prestress on the deformation of micromirror chip embedded with through-silicon vias," *Key Engineering Materials*, vol. 462, pp. 563–568, 2011.

[Talghader 2004] J. J. Talghader, "Thermal and mechanical phenomena in micromechanical optics," *Journal of Physics D: Applied Physics*, vol. 37, no. 10, p. R109, 2004.

[Liu 2002] W. Liu and J. J. Talghader, "Thermally invariant dielectric coatings for micromirrors," *Applied Optics*, vol. 41, no. 16, pp. 3285–3293, 2002.

[Talghader 2001] J. J. Talghader, "Thermal management in optical MEMS," *Micromachining and Microfabrication*. International Society for Optics and Photonics, San Francisco, California, USA, 2001.

[Adachi 1999] S. Adachi, *Optical Constants of Crystalline and Amorphous Semiconductors: Numerical Data and Graphical Information*. Springer Science & Business Media, 1999.

[Wakaki 2007] M. Wakaki, T. Shibuya, and K. Kudo, *Physical Properties and Data of Optical Materials*. CRC Press, 2007.

[Harman 1994] A. K. Harman, S. Ninomiya, and S. Adachi, "Optical constants of sapphire (α-Al2O3) single crystals," *Journal of Applied Physics*, vol. 76, no. 12, pp. 8032–8036, 1994.

[Heavens 1957] O. S. Heavens and S. D. Smith, "Dielectric thin films," *Journal of the Optical Society of America*, vol. 47,no. 6, pp. 469–472, 1957.

[Wortman 1965] J. J Wortman and R. A. Evans, "Young's modulus, shear modulus, and Poisson's ratio in silicon and germanium," *Journal of Applied Physics*, vol. 36, no. 1, pp. 153–156, 1965.

[Petersen 1979] K. E. Petersen and C. R. Guarnieri, "Young's modulus measurements of thin films using micromechanics," *Journal of Applied Physics*, vol. 50, no. 11, pp. 6761–6766, 1979.

[Grillo 2002] S. E Grillo, *et al.*, "Nanoindentation of Si, GaP, GaAs and ZnSe single crystals," *Journal of Physics D: Applied Physics*, vol. 36, no. 1, p. L5, 2002.

[Khenata, 2006] R. Khenata, *et al.*, "Elastic, electronic and optical properties of ZnS, ZnSe and ZnTe under pressure," *Computational Materials Science*, vol. 38, no. 1, pp. 29–38, 2006.

[Madelung 2012] O. Madelung, *Semiconductors: Data Handbook.* Springer Science & Business Media, 2012.

[Oliver, 1992] W. C. Oliver and G. M. Pharr, "An improved technique for determining hardness and elastic modulus using load and displacement sensing indentation experiments," *Journal of Materials Research*, vol. 7, no. 6, pp. 1564–1583, 1992.

[Jacobs 1986] S. D. Jacobs, *et al.*, "Adhesion enhancements and internal stress in MgF2 films deposited with ion beam assistance," *Thin Solid Films*, vol. 144, no. 1, pp. 69–76, 1986.

[Fontanella 1974] J. Fontanella, C. Andeen, and D. Schuele. "Low-frequency dielectric constants of α-quartz, sapphire, MgF2, and MgO," *Journal of Applied Physics*, vol. 45, no. 7, pp. 2852–2854, 1974.

[Jiang 2003] H. Jiang, *et al.*, "First-principles study of structural, electronic and optical properties of BaF2 in its cubic, orthorhombic and hexagonal phases," *Journal of Physics: Condensed Matter*, vol. 15, no. 4, p. 709, 2003.

[Ghosh 1999] G. Ghosh, "Dispersion-equation coefficients for the refractive index and birefringence of calcite and quartz crystals," *Optics Communications*, vol. 163, no. 1, pp. 95–102, 1999.

[Bennett 1963] H. E. Bennett, M. Silver, and E. J. Ashley, "Infrared reflectance of aluminum evaporated in ultra-high vacuum," *Journal of the Optical Society of America*, vol. 53, no. 9, pp. 1089–1095, 1963.

Chapter 3

Electrothermally actuated MEMS scanning micromirrors and their applications in endoscopic optical coherence tomography imaging

Can Duan and Huikai Xie

3.1 Introduction

Cancer is the second most common cause of death and accounts for nearly one of every four deaths in the United States [1]. The American Cancer Society estimates about 585,720 deaths of cancer and 1,665,540 new diagnosed cancer cases in the U.S. in 2014 [1]. Understanding and eventually curing cancer has become a global research focus for several decades, and the overall cancer death rate has declined steadily since early 1990s. The current 5-year survival rate is improved to 68% from 49% in the 1970s, mainly attributed to the improvement on the prevention, early detection, and treatment techniques. Especially, the early cancer detection plays an important role in dramatically increasing the cancer survival rate for most cancers including breast, colon, rectum, cervix, prostate, testis, oral cavity, and skin cancers [2]. For instance, the survival rate of breast cancer, which is the second most common cancer in women in U.S., has been estimated up to 99% if diagnosed in the early localized breast cancer stage. While the survival rate falls to 24% when the cancer has spread to lymph nodes around the collarbone or organs at the distant stage [1, 3]. Moreover, colorectal cancer, which is the third leading cancer in both men and woman, has a decreasing incidence rate about 3.0% per year in women and 2.5% per year in men from 2006 to 2010 [1, 4]. The declining mortality rates reflect the improvements in early diagnosis and timely removal of precancerous polyps attributed to colorectal cancer screening. Therefore, methodologies for effective and accurate cancer diagnosis at early stage are critical in decreasing the mortality caused by cancers.

Traditionally cancer diagnosis relies on medical imaging followed by biopsy, which is an invasive procedure posing various problems and risks such as bleeding, trauma, infection, and random sampling [5]. Conventional bio-imaging

Department of Electrical and Computer Engineering, University of Florida, Gainesville, FL, USA

Table 3.1 Overview of traditional and optical biomedical imaging systems

Technique	Resolution	Depth	Real-time	Label-free
CT	0.1–1 mm	No limit	No	Yes
MRI	100–500 μm	No limit	No	Yes
Ultrasound	100 μm	10 mm	Yes	Yes
Photoacoustic	15 μm–1 mm	3 mm	Yes	Yes
Confocal microscopy	0.4–1 μm	200–500 μm	Yes	Yes
Fluorescence microscopy	1–5 μm	300–800 μm	Yes	No
NLOM	1 μm	1 mm	Yes	Limited
OCT	1–10 μm	1–3 mm	Yes	Yes

techniques, including X-ray computed tomography (CT), magnetic resonance imaging (MRI), ultrasound imaging, and positron emission tomography, have been well-developed and extensively used for clinical cancer diagnosis and intervention. However, these biomedical imaging techniques suffer from low resolution which makes them inadequate to detect precancerous lesions at early stage [6]. Fortunately, several new high-resolution optical imaging techniques have emerged in the last two decades, including optical coherence tomography (OCT), laser scanning confocal microscopy, nonlinear optical microscopy (NLOM), and photoacoustic microscopy. A brief comparison of these emerging optical imaging techniques to the conventional biomedical imaging systems in terms of imaging resolution and penetration depth is listed in Table 3.1 [7, 8]. The tradeoff between resolution and depth of different imaging modalities is clearly shown in this table. Optical imaging techniques have high resolution but small imaging depth, while MRI and CT are exactly opposite. As indicated in the table, OCT is capable of providing high imaging resolution in the range of 1–10 μm and penetration depth of 1–3 mm, which is an interesting compromise of both resolution and depth. Study has found that over 85% of cancers are originated in the thin epithelium layers (~2 mm thick) of internal organs [9]. Thus, OCT can be used for early cancer detection *in vivo*. The challenge then becomes how to apply OCT imaging inside human body. This is exactly the focus of this chapter.

In the following section, OCT principle and the challenges of using OCT for endoscopic imaging will be introduced. Then a class of electrothermally actuated MEMS scanning mirrors suitable for endoscopic OCT imaging will be discussed in detail in Section 3.3. Section 3.4 focuses on electrothermal MEMS micromirrors-based OCT endoscopic probes and their applications.

3.2 Optical coherence tomography and endoscopic imaging

3.2.1 Optical coherence tomography

OCT is an emerging optical biomedical imaging modality which is capable for real-time *in vivo* noninvasive cross-sectional imaging. It was first demonstrated by Huang *et al.* in 1991 [10]. OCT has high resolution of micrometers or even sub-micrometers within imaging depth of 1–3 mm underneath biological tissue surface.

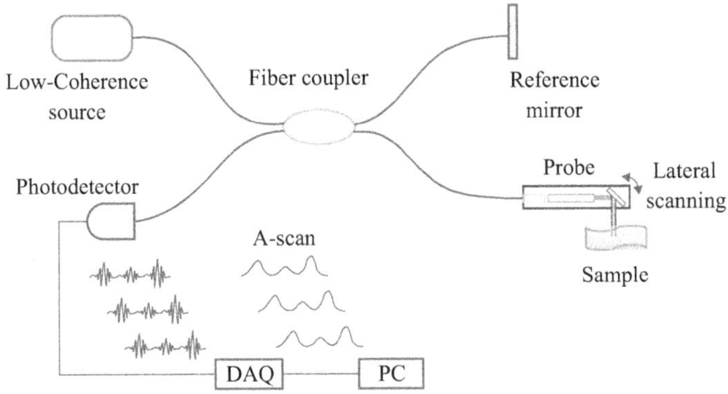

Figure 3.1 Schematics of fiber-optic time domain OCT system utilizing a broadband light source and a scanning reference arm

OCT utilizes the scattering light from a specific tissue site, similar to ultrasound imaging except for using light instead of sound. As shown in Figure 3.1, the optical scattering property of the sample at a specific site is obtained by the interference signal formed between the scattering light from the sample and the reflected light from a reference arm. Typical axial resolution of current OCT technology ranges from 1 μm to 15 μm, which is almost 10–100 times higher than that of standard ultrasound imaging [11]. As compared in Table 3.1, OCT technology bridges the standard ultrasound imaging and the optical microscopy and stands out as the most suitable intravital optical imaging approach for early cancer detection. In addition, low numerical aperture lenses can be employed by OCT imaging, which provides relatively long working distances and simplifies the optical design for endoscopic imaging. Also, optical fibers can also be incorporated into OCT systems to facilitate the miniaturization of the catheters/endoscopes for imaging in internal organs. With the help of two-axis optical microscanners, OCT can provide 3D images of precancerous lesions, which may not be identifiable by traditional imaging modalities [5]. These factors have combined to adopt OCT for an abundance of diagnostic applications, especially for preclinical intravital tumor screening, diagnosis, and postoperative treatment monitoring. OCT has a wide range of diverse clinical application areas such as ophthalmology, cardiology, GI tract, dentistry, and so on.

Time domain OCT (TDOCT) is the first generation of OCT systems. Figure 3.1 illustrates a schematic of a typical TDOCT system. The light emitting from a low-coherence light source is directed into a Michelson interferometer with the incident optical power divided evenly into the sample and reference arms by a 2×2 fiber-optic coupler. A tunable optical delay line is included in the reference arm to generate a range of optical delay, while a scanning mechanism is utilized in the sample arm to rotate the light beam to scan the sample. The light beams in the reference arm and the sample arm both partially return to the fiber coupler and then interfere on the surface of a photodetector. The photodetector picks up the optical interfere signal and converts it into an electrical signal. At each spot, the optical delay line is fully scanned so that a series of interference signals are captured and processed into a depth-resolved amplitude signal scattering back from the sample, which is called A-scan, representing the depth-resolved reflectivity profile of the sample. Multiple A-scans along one direction can be obtained by sweeping the light beam in a lateral direction (B-scan) on the sample surface and a two-dimensional cross-sectional image can be obtained. Furthermore, a collection of B-scans, acquired by the two-axis scanning mechanism in the sample arm, can be used to construct the 3D rendering of the sample. The major advantage of TDOCT is that the focus position of the light beam into the sample can be synchronously adjusted with the optical delay in the reference arm, yielding high transverse resolution over the total depth range. On the other hand, TDOCT systems typically have slow imaging speed due to the limited scan speed of the optical delay in the reference arm.

In the past few years, there is a general shift from TDOCT to frequency domain OCT (FDOCT) [12]. Compared to TDOCT, FDOCT systems are mechanically more stable due to the stationary reference mirror and a dramatic increase of imaging speed; the sensitivity is also increased by a factor of over 100. FDOCT is based on the fact that the intensity of each wavelength of the light in the Michelson interferometer is a superposition of all the interferences of that wavelength from different depths of the sample. Therefore, the inverse Fourier transformation of the spectrum yields the backscattering amplitude as a function of depth. Two distinct methods have been developed to implement FDOCT, i.e., spectral-domain OCT (SDOCT) and swept-source OCT (SSOCT), as shown in Figure 3.2(a,b). In both systems, the reference arm is fixed without scanning. The fundamental difference of the two FDOCT systems is the method to record the spectral interferogram. An SDOCT system utilizes a broadband continuous light source, same as that in a TDOCT system, and a spectrometer to record the spectrum of the interference signal between the reference arm and all depths reflections inside the sample. Typically the spectrometer is composed of a charge-coupled device (CCD) or a photodiode array. On the other hand, SSOCT records the spectral interferogram by using a single photodetector to detect the interference signals of successive wavelengths generated by a swept light source with narrow instantaneous linewidth and rapid scan speed in a wide wavelength range. SSOCT has become widely used in the last few years due to its higher signal-to-noise ratio (SNR), higher imaging speed, and deeper penetration depth [13, 14].

Numerous progresses have been achieved in OCT system design and associated key components as well during the last decade. These technological breakthroughs include but are not limited to: ultrahigh-speed pulsed lasers [15]; high-power, high-speed, high-bandwidth swept source lasers [16]; fast spectrometers [17]; functional OCT imaging – Doppler OCT [18] and polarization sensitive OCT [19]; image processing [20]; and endoscopic OCT [21].

The applications of OCT in ophthalmology have long been established for providing both structural and quantitative information of retina and anterior segment [22]. For instance, Carl Zeiss launched the first clinical OCT system for diagnosis of retinal diseases as early as 1997 [23]. Since then, OCT technology has been extensively investigated for the clinical applications in dermatology,

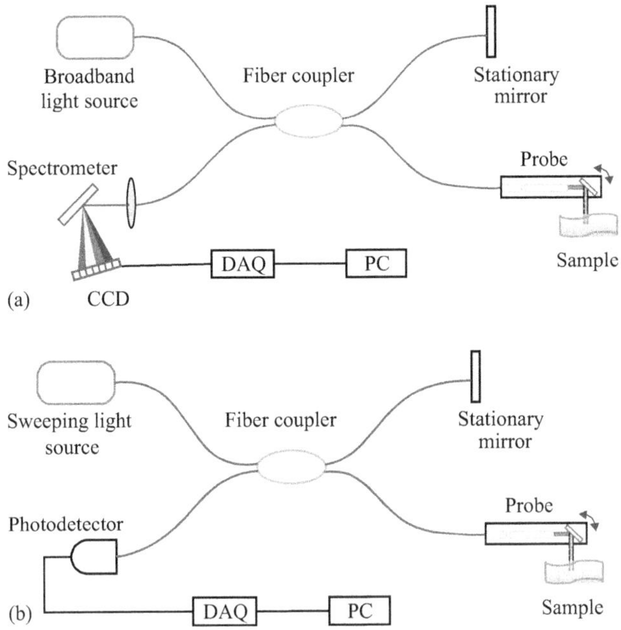

Figure 3.2 Schematics of fiber-optic FDOCT systems. (a) SDOCT system utilizing a broadband light source and a spectrometer. (b) SSOCT system utilizing a sweeping light source

cardiology, and gastroenterology [11]. Recently, OCT has been investigated for endoscopic diagnosis and image guided microsurgery [24].

3.2.2 OCT endoscopic imaging

OCT endoscopy may play a vital role in the screening of early stage cancer or other neoplasms in internal organs such as gastrointestinal tract, cardiovascular artery, bladder, and bronchia [25]. An OCT catheter or probe is utilized to deliver light into and collect the back-scattering optical signal from the internal organs or other biological specimens. A typical OCT catheter contains a probe mount used to fix an optical fiber and other optical components, an outer tubing, and external motors or micro-optical scanners. Optical fibers and optical components such as microprisms and GRIN lenses are used to direct and focus the light beam. Tubing provides protection of internal optical components from the surrounding medium. Scanning mechanisms are crucial for endoscopic OCT probes, which may make a big difference in terms of the imaging range, speed, and the size of the probe.

Endoscopic OCT probes can be divided into two types based on the mechanical scan driving modes as proximal scanning and distal scanning. Proximal scanning employs either a motor at the proximal end of the probe to rotate a long, slender fiber-prism assembly [26, 27] or an optical scanner at the proximal end plus relay optics [28]. For instance, the FDA-approved Lightlab's (now St. Jude Medical)

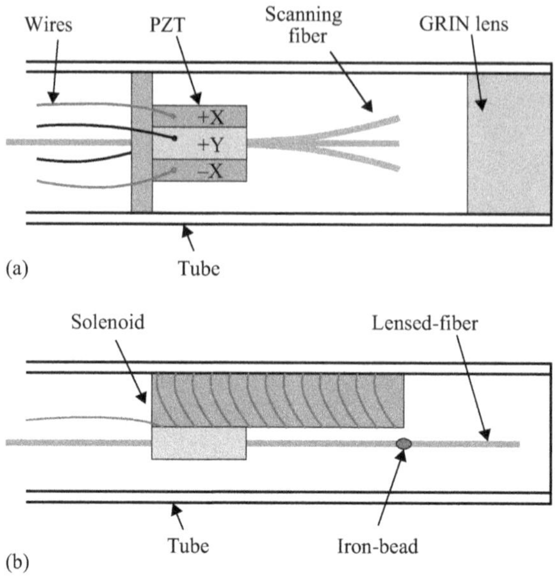

Figure 3.3 Distal scanning endoscopic OCT probe using (a) PZT tube and (b) magnetic force for scanning

cardiovascular OCT probe employs a thin (0.36 mm in diameter) and long (1.35 m) guide wire rotating in a 0.9 mm-diameter sheath driven by an external motor [29]. A similar scanning technology has been adopted by Ninepoint for OCT imaging in GI tract [30]. Rotating a long thin wire is very challenging and may be costly. This chapter will focus on distal scanning which has potential for low cost. Distal scanning utilizes a microactuator to tether the fiber tip, a micromotor to rotate a prism, or a micromirror to steer the light beam at the distal end of the probe [31, 32].

Oscillating fiber tips can be done by various microactuators, such as lead zirconate titanate (PZT) tubes [33–36], electrostatic actuators [37], electromagnetic cantilevers [38, 39], and electrothermal actuators [40], as shown in Figure 3.3. For instance, Munce *et al.* constructed an electrostatically actuated 2.2 mm probe by placing an etched optical fiber strung within platinum alloy coil and then plug into the lumen of a dissipative polymer [37]. However, constant high voltage of 1–3 kV was required to generate oscillation of 33°. Boppart *et al.* implemented the first piezoelectric cantilever actuated forward-viewing probe [33]. Li *et al.* developed a series of forward-viewing probes using a miniature tubular PZT actuator to rapidly scan a fiber cantilever and a GRIN lens [34–36]. The drive voltage has been reduced to 35 V to obtain a 0.9 mm fiber deflection using a 17-mm-long fiber. To overcome the high drive voltage required by electrostatic and piezoelectric cantilevers, Min *et al.* reported a compact cantilever actuated OCT probe by attaching a ferromagnetic iron bead on the optical fiber and fixing a solenoid coil inside the probe head [38, 39]. The drive voltage is low but their outer diameters of 4.5 mm were relatively large. All of the scanning methods described above are based on

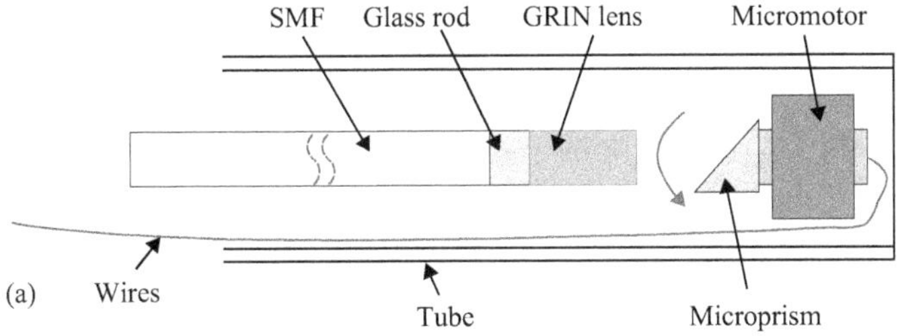

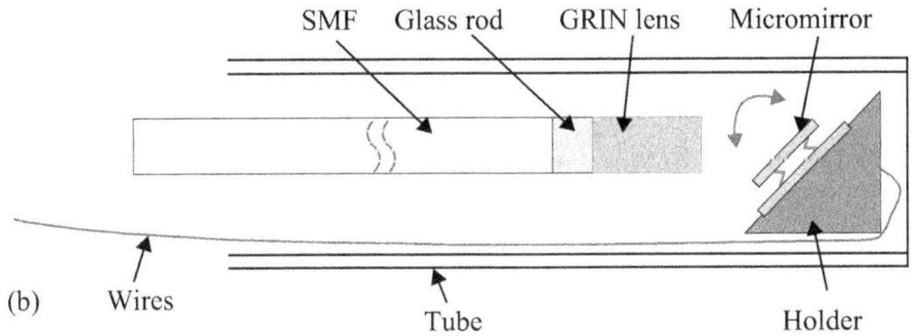

Figure 3.4 Distal scanning endoscopic OCT probe using (a) microprism and (b) micromirror for scanning

resonant scanning. Recently, Zhang *et al.* reported an electrothermally actuated fiber scanner achieving an optical scanning range of 1 mm at nonresonant operation [40].

Tethering fiber tips may create nonuniform optical coupling issues and often require resonant scanning. Fortunately, micromotors and micromirrors can be used to manipulate light beam in free space, as shown in Figure 3.4. Micromotors are usually used for full circumferential scanning, while MEMS mirrors are most suitable for side-view or forward-view areal scanning. Three-dimensional OCT images can be obtained through either translationally moving a micromotor-driven prism back and forth or utilizing a two-axis scanning MEMS mirror.

Micromotor-based OCT probes have been explored since 1996 [41]. Tran *et al.* constructed a 2.4 mm outer diameter circumferentially-scanning probe employing a 1.9 mm MEMS micromotor in 2004 [42]. Herz *et al.* reported a similar rotational scanning micromotor-based endoscopic OCT probe with outer diameter of 5 mm at the same year [43]. A more compact MEMS micromotor-based circumferential scanning probe with an outer diameter of 2.7 mm was demonstrated by Su *et al.* in 2006, enabling 30 frames/s real-time OCT imaging frame rate [44]. The smallest distal micromotor scanning side-viewing probe, which was reported in 2013, has an outer diameter of 1.1 mm, using a customized two-phase synchronous micromotor [45]. Chang *et al.* demonstrated a 2.4 mm motor wires shadow-free side-viewing probe by using a squiggle micromotor consisting of four PZT plates bonded to a threaded screw shaft at the distal end of the OCT probe [46]. To the best of our knowledge, the smallest size of reported shadow-free circumferential scanning probes is 1.5 mm using a hollow ultrasonic motor demonstrated by Chen *et al.* in 2013 [47].

Compared to micromotors, MEMS micromirrors provide better flexibility and versatility for OCT probe designs. Micromirror-based OCT was first demonstrated in 2001 [48]. After that, numerous micromirror-based OCT probes were reported [49–57]. For instance, Sun *et al.* demonstrated a series of electro-thermal MEMS micromirrors-based OCT side-viewing probes including one-axis scanning MEMS mirrors and two-axis scanning MEMS mirrors for 3D in situ imaging ability [49–52]. Similar designs employing two-axis MEMS mirrors of other actuation mechanisms such as electrostatic actuators [53, 54], electro-magnetic actuators [55], and piezoelectric actuators [56] have also been developed and extensively studied.

3.2.3 Challenges in endoscopic OCT

Rotating OCT needles at the proximal end has the advantages of small size and full circumference coverage, but it requires rotating the whole tethered probe using external proximal motors. Thus the rotational torque is transferred from the proximal end to the distal needle, which leads to unstable vibration and uneven rotational speed [57]. For applications requiring long fibers, the nonlinear effects will be even more severe. Other deficiencies also remain such as low scan speed and polarization distortions introduced by the twisting of optical fibers.

Distal scanning eliminates the fiber twisting and nonlinear problems of proximal rotation probes. By scanning at the distal end, this design avoids rotating the whole probe or using fiber-optic junction to couple light from stationary fibers to rotational fibers. It also enables rapid scanning speed without the torque transfer from the proximal end. However, one remaining challenge for distal scanning probes is the probe miniaturization, which is limited mainly by the size of the optical scanner. MEMS micromirrors provide a feasible and ideal solution due to their small size and the capability of arbitrary pointing within the scan range. In addition, among three different architectures for distal scanning probes, scanning MEMS micromirrors are the only scanning mechanism that can achieve 2D *en face* scanning without the need of pulling the whole fiber back and forth. Thus, the following sections will be focused on discussing endoscopic OCT probes using MEMS mirrors as the distal scanning mechanism.

3.3 MEMS scanning micromirrors

MEMS technology is a widespread technology making small-sized devices and systems ranging from micrometers to millimeters. MEMS accelerometers, gyroscopes, microphones, and pressure sensors have a vast application in cell phones, automotive, inkjet print heads, and video game controllers [58]. Optical MEMS mirrors have been widely employed in displays and optical switches and have been demonstrated with great potential for optical biomedical imaging, especially for endoscopic imaging applications. The main advantages of MEMS scanning mirrors include rapid scanning speed, small chip size, low cost, and easy integration. The MEMS technology provides OCT imaging with miniaturization of bulky two-axis optical scanners [5]. Overall, two-axis scanning MEMS micromirrors are the most suitable scan engines for 3D endoscopic OCT imaging [59].

In general, there are four types of actuation mechanisms for MEMS scanning mirrors, including electrostatic, electromagnetic, piezoelectric, and electrothermal actuators. Among them, electrostatic actuators are most attractive thanks to their fast scan speed and low power consumption. However, it requires high drive voltage in the order of 100 V to achieve 20° optical scan [60]. The high drive voltage required causes potential risk for *in vivo* applications. Electromagnetic micromirrors are capable to generate large optical scan angle of 40° at low drive voltage of 3 V [55]. However, high power consumption, packaging complexity caused by the required permanent magnet, and electromagnetic interference constrain their applications in endoscopic imaging. Piezoelectric MEMS mirrors demonstrate advantages of fast scan and low power consumption [56], but the weaknesses such as hysteresis and charge leakage problem need to be overcome for practical endoscopic imaging. Another common factor that hinders electrostatic, electromagnetic, and piezoelectric micromirrors for the miniaturization of endoscopic probes is their relatively small fill factor, usually less than 10%. In contrast, electrothermal MEMS mirrors have been investigated to further extend the optical scanning range with low drive voltage smaller than 10 V [49]. In addition, electrothermal MEMS mirrors have good linear scan and high fill factor (~25%). These features make electrothermal MEMS mirrors very suitable for miniature OCT probes.

In this section, an overview of electrothermally actuated single-crystal silicon (SCS) MEMS micromirrors is discussed in detail. Design, fabrication, and characterization of various electrothermal MEMS micromirrors with 1D and 2D scanning ability are presented.

3.3.1 Electrothermal bimorph actuation principle

Electrothermal actuation mechanism utilizes the bending motion of bimorph beams caused by temperature change. Specifically, the initial displacement of the bimorph actuator is determined by intrinsic stress, while the bending motion is generated by the thermally induced extrinsic stress through Joule heating. As shown in Figure 3.5 [61], a bimorph beam is a cantilever consisting of two layers of thin-film materials with different coefficients of thermal expansion (CTEs). A heating layer is embedded in the bimorph, which serves as a resister to generated Joule heating when an electrical current passes through it. The bimorph cantilever will bend toward the layer which has smaller CTE when the temperature is increased. The tip tilt angle is proportional to the temperature change which is determined by the power consumed by the resistor. Such a cantilevered bimorph beam acts as the basic building block for forming electrothermal bimorph actuators with more sophisticated structure designs.

The stresses inside a bimorph cantilever are resulted from both intrinsic stress $\Delta\varepsilon_{in}$ and extrinsic stress $\triangle\varepsilon_{ex}$, where the intrinsic stress consists of the intrinsic stress mismatches $\triangle\varepsilon_{in\_T}$ due to the difference of the deposition temperatures of the two bimorph layers and all other factors introduced intrinsic stress $\triangle\varepsilon_{in\_U}$. These

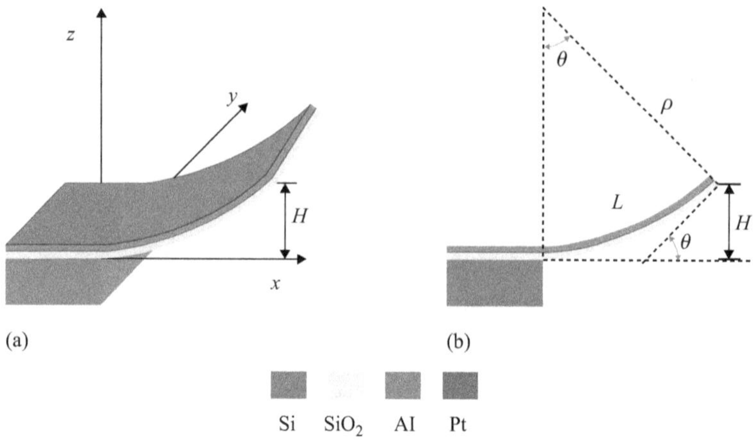

Figure 3.5 Illustration of the electrothermal bimorph beam. (a) 3D model and (b) cross-sectional schematic of a curved bimorph cantilever

unknown factors include the grain coalescence, vacancy annihilation, impurity incorporation, and structure damage due to energetic deposition process [61]. On the other hand, the extrinsic stress $\triangle\varepsilon_{ex\_J}$ is purely caused by the Joule heating. Therefore, the total stress encompassing both the intrinsic stress and the extrinsic stress is represented by:

$$\triangle\varepsilon = \triangle\varepsilon_{in} + \triangle\varepsilon_{ex} = \triangle\varepsilon_{in\_T} + \triangle\varepsilon_{in\_U} + \triangle\varepsilon_{ex\_J} \tag{3.1}$$

The bending motion when the drive voltage is added to the bimorph actuator is determined by the extrinsic thermally induced strain $\triangle\varepsilon_{ex\_J}$. Assuming the temperature change introduced by the Joule heating along the cantilever is uniform, the extrinsic strain difference caused by the difference between the CTEs of the two layers is given by:

$$\triangle\varepsilon_{ex\_J} = (\alpha_1 - \alpha_2)\triangle T_J \tag{3.2}$$

where, α_1 and α_2 represent the CTEs of the two materials on the top and bottom layers. And $\triangle T_J$ represents the temperature change of the bimorph introduced by the Joule heating.

Then the angular actuation due to extrinsic strain is shown as:

$$\theta_{ex\_J} = \frac{l_b}{r_{ex\_J}} = \frac{\beta_b}{t_1 + t_2}\triangle\varepsilon_{ex\_J} l_b = \frac{\beta_b l_b}{t_1 + t_2}(\alpha_1 - \alpha_2)\triangle T_J \tag{3.3}$$

where, l_b and r_{ex_J} represent the length and the radius of curvature of the bimorph respectively. Furthermore, β_b represents the curvature coefficient and t_1 and t_2 are the thicknesses of the two layers.

However, the temperature distribution along the bimorph is nonuniform in real case. Then the radius of curvature varies along the bimorph and the tangential tilting angle on the bimorph tip is an accumulation of changing radius of curvature, given by:

$$\theta_{ex\_J} = \int_0^{l_b} \frac{dx}{r_{ex\_J}(x)} = \frac{\beta_b l_b}{t_1 + t_2}(\alpha_1 - \alpha_2)\overline{\triangle T_J} \tag{3.4}$$

where, $\overline{\triangle T_J} = \frac{1}{l_b}\int_0^{l_b} \triangle T_J(x)dx$ is the average temperature change along the bimorph. Equation (3.4) shows that the thermal actuation angle is linearly proportional to the average temperature change on the bimorph.

3.3.2 Material selection

The actuation responsivity is defined as the ratio of the thermal induced deflection angle to the temperature change, shown as:

$$S = \frac{\theta_{ex\_J}}{\overline{\triangle T}} = \frac{\beta_b l_b}{t_1 + t_2}\Delta\alpha \tag{3.5}$$

where, the curvature coefficient β_b is dependent on the thickness ratio of two thin layers. In addition to the curvature coefficient, the responsivity is linearly dependent on the bimorph beam length and the CTE difference between the two layers.

The curvature coefficient can be simplified as the following form by defining thickness ratio $\zeta = \frac{t_1}{t_2}$ and biaxial elastic modulus ratio $\chi = \frac{E_1'}{E_2'}$:

$$\beta_b = \frac{6(1+\zeta)^2}{\frac{1}{\chi\zeta} + \chi\zeta^3 + 2(2\zeta^2 + 3\zeta + 2)} \tag{3.6}$$

By setting the partial derivative of β_b to zero, the maximum value of curvature coefficient is calculated to be 1.5. Then the biaxial modulus ratio should satisfy the following equation:

$$\frac{t_1}{t_2} = \left(\frac{E_1'}{E_2'}\right)^{-\frac{1}{2}} \tag{3.7}$$

Therefore, the geometry of a bimorph will be determined by the properties of the selected materials so that the biaxial elastic moduli of the two selected materials should be close to prevent a dramatic thickness difference. In addition, the ideal materials should have a high CTE difference in order to optimize the actuation responsivity. Table 3.2 shows a summary of commonly used MEMS materials with their thermal and mechanical properties [62, 63].

Polymer materials such as PDMS have much larger CTE than semiconductor or metal materials. However, the Young's moduli of polymer materials are usually very low, which will result in very large thickness of the bimorph cantilever. In addition, the mechanical properties of polymer materials are highly dependent on temperature, which hinders their applications as the electrothermal bimorph actuator materials. Al and SiO_2 have relatively large CTE difference and similar

Table 3.2 Thermal and mechanical properties of popular MEMS materials

Materials	CTE (10^{-6}/K)	Thermal conductivity ($W/m^{-1}K^{-1}$)	Young's modulus (GPa)	Poisson ratio
Si	3.0	150.0	179	0.27
Poly-Si	1.6	–	160	0.22
SiO_2	0.4	1.4	70	0.17
Si_3N_4	3.3	30.0	310	0.24
SiC	3.5	86.5	457	0.14
Al	23.6	237.0	70	0.35
Au	14.5	318.0	78	0.44
Cu	16.9	401.0	120	0.34
Pt	8.9	71.6	168	0.38
Pb	28.7	35	160	0.42
Cr	5.0	93.9	279	0.21
Ti	8.6	21.9	116	0.32
Ni	12.8	90.9	200	0.31

Young's modulus. Furthermore, both Al and SiO_2 are the most commonly used MEMS materials and their fabrication processes are well-developed and cost-effective. Al can also serve as the mirror coating material due to its high reflectance in a wide wavelength range. Therefore, Al and SiO_2 are a good pair of very suitable bimorph materials. The optimal thickness ratio of two Al and SiO_2 layers for maximum responsivity is 0.88.

The temperature change along the bimorph cantilever is introduced by Joule heating of an embedded resistor. Poly-silicon has been used as the resistor in early versions of electrothermal CMOS-MEMS mirrors [62, 64]. However, the embedded Poly-Si heater suffers from the hysteresis problems and self-annealing effects that limit the stable scanning range [65, 66]. In addition to Poly-Si, Pt is a better alternative as the heater material due to its high stability at high temperature [67]. Pt also has larger temperature coefficient of resistivity than Poly-Si, which provides a means of measuring temperature in real time. A thin SiO_2 layer is typically deposited between the Pt layer and Al layer for insulation. Cr serves as the adhesion material between the metal layers and SiO_2. The final structure of an Al/SiO_2 bimorph cantilever is shown in Figure 3.6. Recently, Cu and W were employed as the bimorph materials for better robustness and faster thermal response [68]. In this case, W is the heater material.

3.3.3 Electrothermal MEMS mirror designs

Various planar electrothermal MEMS micromirrors have been successfully developed, demonstrating great potential for endoscopic imaging applications [65–77]. At first, a CMOS-compatible MEMS process was used for fabricating electrothermal MEMS mirrors [65]. In that case, Al/SiO_2 was used as bimorphs and Poly-Si was used as the heater. SCS coated with Al was employed as the mirror plate to maintain the flatness of the mirror as well as to achieve high reflectivity in broadband. Later on, more sophisticated SOI-based MEMS mirrors with the lateral shift free (LSF)-large vertical

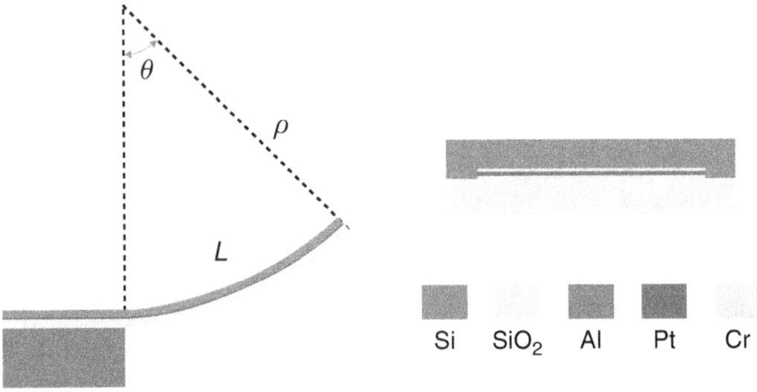

Figure 3.6 Schematic view of an Al/SiO$_2$ cantilever beam with embedded Pt heater

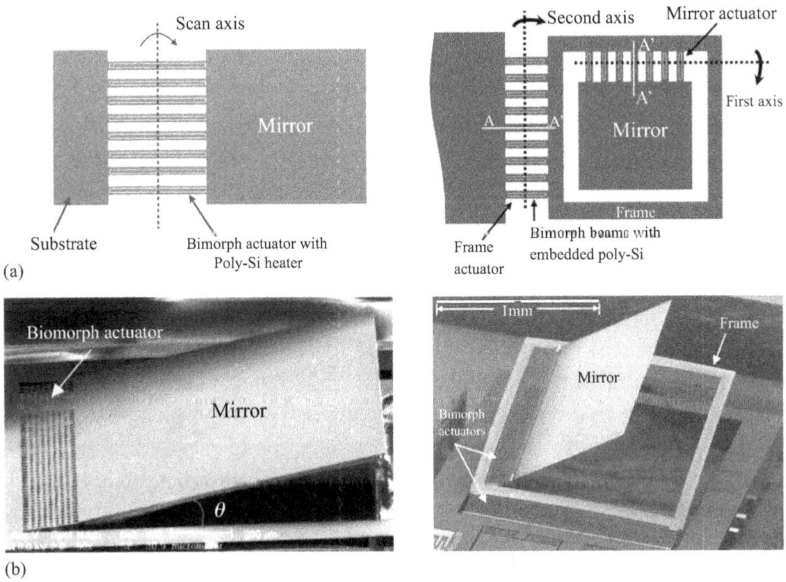

Figure 3.7 Schematics and SEMs of the one-axis and two-axis electrothermal MEMS designs. (a) Top view schematic and (b) SEM of one-axis scanning mirror [48]. ©[2001] OSA. Reprinted with permission. (c) Top view and (d) schematic of two-axis scanning mirror [66]. ©[2004] IEEE. Reprinted with permission

displacement (LVD) actuator design and the folded-dual-S shaped actuator design bring forth the benefits of large scan range, high fill factor, and large aperture [73–77]. Also, noble metals such as Pt, which are more stable than Poly-Si, were used as heaters in this SOI process. Detailed designs and their characteristics are presented in the following paragraphs.

The schematic and SEM images of the 1D electrothermal micromirror design are shown in Figure 3.7 [48, 65]. The 1D micromirror consists of a bulk SCS mirror plate coated with Al and a mesh of Al/SiO_2 bimorph beams the actuators. The Al/SiO_2 cantilevers are connected in parallel and Al is used as the top layer, providing an upward initial curling generated by the residual intrinsic stress. The bimorph actuator mesh unfolds downward when a drive voltage is applied to the embedded Poly-Si heater; consequently, the mirror plate rotates around the bimorph. Therefore, a light beam can be scanned laterally in one dimension. The resonant frequency of this 1D mirror was 165 Hz and the maximum mechanical rotation angle was 17° at the drive voltage of 33 Vdc.

Based on this 1D design, a 2D MEMS mirror was developed by simply adding another array of bimorph beams with orthogonal orientation [66]. The mirror frame was connected to the outer set of bimorphs, shown as the frame actuators in Figure 3.7(c). And the mirror plate was supported by the inner set of bimorphs, shown as the mirror actuators. Two sets of drive voltages were added respectively to the inner and outer actuators in order to rotate the mirror plate in two perpendicular axes. The resonant frequencies of this 2D MEMS mirror were 445 Hz for the inner actuators and 259 Hz for the outer actuators. The maximum mechanical rotation angle of the mirror plate was 40° at the drive voltage of 15 Vdc and the frame rotated up to 25° with 17 Vdc [66].

The MEMS mirrors shown in Figure 3.7 suffer from two drawbacks, large initial tilt angle and nonstationary center of rotation. The large initial tilt angle complicates the packaging process and the optical design. And a stationary rotation axis is highly desired for optical scanners used for imaging purpose to achieve high image quality and fidelity.

In order to solve the first drawback, large-vertical-displacement (LVD) bimorph actuators-based MEMS mirrors have been demonstrated [69], shown in Figure 3.8. Two sets of bimorph actuators are placed in the opposite orientations to compensate the initial tilting. Both sets have Al as the top layer so the cantilevers all bend upward due to the residual intrinsic stress. The bimorph actuators have the same length so that the tilting angles of the frame and mirror plate compensate each other completely, leading to a mirror plate parallel to the substrate. The 1D LVD mirror achieved 26.5° maximum rotation angle when 3 Vdc was applied to the mirror actuator, and a 200 μm piston displacement with less than 0.03° tilting was also obtained at 6 Vdc [69]. The measured resonant frequencies of the mirror actuator and the frame actuator were 2.62 kHz and 1.18 kHz, respectively [69]. This type of MEMS mirrors without the large initial tilting angle is easier to be assembled and the optical design is simplified as well.

Furthermore, a 2D LVD MEMS mirror composed of two cascaded, orthogonally oriented 1D LVD designs was developed [70]. By compensating the initial tilting through two sets of actuators with same length on each axis, the mirror plate of the 2D LVD MEMS mirror was also parallel to the substrate with negligible initial tilting angle, shown in Figure 3.8(d). The 2D LVD MEMS mirror achieved $\pm 30°$ maximum mechanical rotation angle in both axes at 12 Vdc. The resonant frequencies for the inner and outer actuators in the two axes were 870, 312, 452, and 170 Hz [70].

Although the LVD MEMS mirrors overcome the drawback of large initial tilting angle of the mirror plate and enable bidirectional scanning as well as vertical piston actuation, the issues of the nonstationary rotation axes and large lateral shift

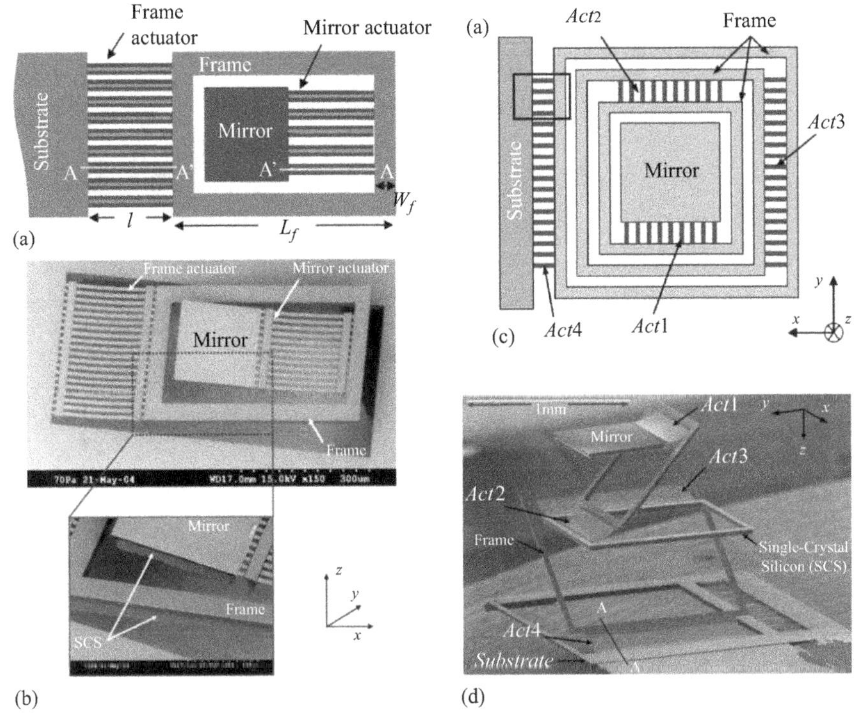

Figure 3.8 Schematics and SEMs of the LVD electrothermal MEMS design. (a) Top view schematic and (b) SEM of bidirectional one-axis scanning mirror [69]. ©[2005] Elsevier. Reprinted with permission. (c) Top view and (d) schematic of bidirectional two-axis scanning mirror [70]. ©[2006] Elsevier. Reprinted with permission

in the piston mode still remain unfixed. Furthermore, the gimbaled structure design results in small fill factor of the device.

Thus, a lateral-shift-free (LSF)-LVD MEMS mirror was proposed and demonstrated by Wu *et al.* [73]. The schematic and SEMs of the LSF-LVD micromirror are shown in Figure 3.9. The mirror plate is supported by four identical actuators with one three-segment-bimorph actuator on each side. Each actuator consists of three Al/SiO_2 bimorph segments and two rigid silicon-backed beams. Each actuator is connected to an Al coated bonding pad on the frame of the substrate. All of the four actuators share a common ground but can be actuated individually by adding drive voltages respectively. The folded three-segment design eliminates the gimbal and thus achieves high fill factor (mirror area/chip area) up to 25%. The lengths of the five segments of each actuator are designed to compensate the lateral shift and tilting during vertical displacement. Then the rotation center is fixed along the center of the mirror plate, which dramatically simplifies the optical alignment and eliminates the artifacts in imaging.

When the same voltages are applied on the four actuators simultaneously, the mirror plate will scan up and down as the piston mode. When the differential

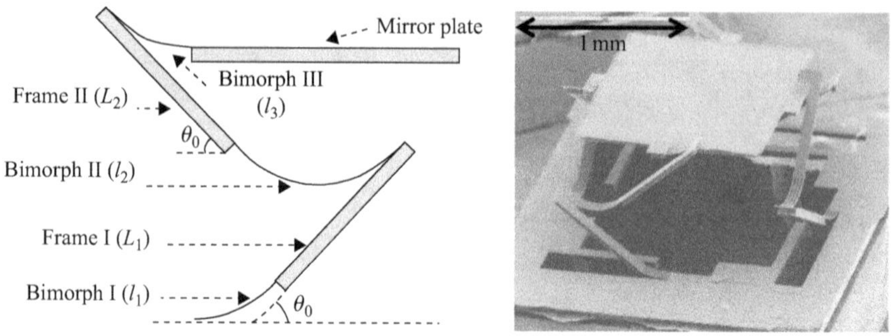

(a) (b)

Figure 3.9 Schematics and SEMs of the LSF-LVD electrothermal MEMS design. (a) Side view of LSF-LVD actuator. (b) SEM of fabricated device [49]. ©[2010] OSA. Reprinted with permission

voltages with the same amplitude and a phase difference of 180° are applied to the two opposing actuators, the mirror plate will scan in the tip-tilt motion along this axis. When the two sets of actuators on both axes are driven simultaneously, the mirror pate will perform 2D scanning.

As shown in Figure 3.9, the initial elevation of the flat mirror pate parallel to the substrate was 646 μm, which left sufficient space for the mirror plate to perform tip-tilt scanning or piston scanning. The measured characteristics of the LSF-LVD MEMS mirror showed a large linear piston displacement range of 620 μm and a linear optical scan range from ±5° to ±31°, obtained between 1.5 V and 4.5 V [73]. A raster scan pattern with the scan range of 30° and 36° in the two axes was achieved by differentially driving two sets of actuators with the ramp waveforms of 0.5~3.8 V at 320 Hz and 8 Hz for two orthogonal axes respectively [73].

The LSF-LVD design overcomes the lateral shift and nonstationary rotation axis to a large extent. However, still small rotation axis shift exists and slight skewed scanning remains due to the structural asymmetry. The resonant frequency and the 3 dB thermal cut-off frequency are relatively low due to the large thermal capacitance of the rigid silicon beams.

Thus, a folded-dual-S-shaped bimorph (FDSB) MEMS mirror was proposed and demonstrated by Jia *et al.* to further increase the fill factor and overcome the rotation axis shift by employing symmetric FDSB actuators [50, 74, 78]. The structure of a FDSB actuator is shown in Figure 3.10(a). A FDSB actuator consists of two folded double-S-shaped bimorph cantilevers, and each S-shaped cantilever employs two inverted Al/SiO_2 bimorph cantilevers in series with an overlap section. For each S-shaped cantilever, the length ratio of L_1 and L_3 has been optimized to achieve a horizontal tangential tip. Since the two S-shaped cantilevers have the same curling angle and geometrical symmetric parameters, the lateral shift and curling angle have been compensated to realize pure vertical displacement. Furthermore, a hidden actuator design has been proposed using the FDSB actuator to further increase the fill factor by hiding the actuators underneath the mirror plate

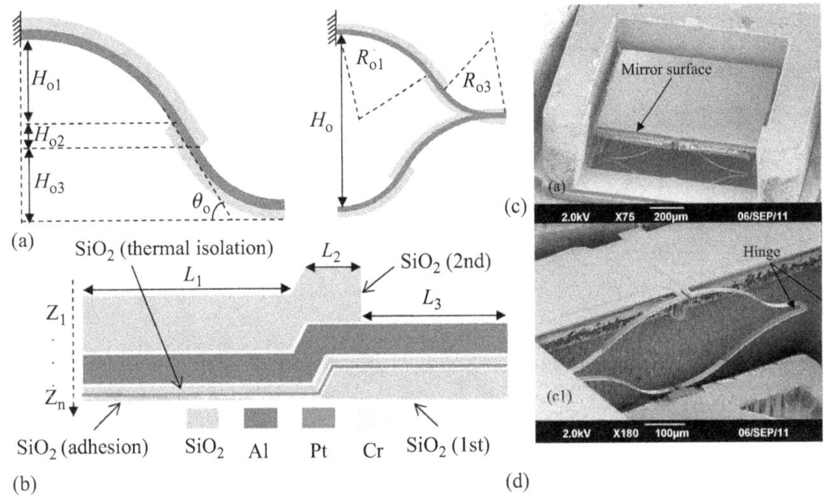

Figure 3.10 Schematics and SEMs of the FDSB electrothermal MEMS design [50]. (a) Conceptual illustration of a single and double S-shaped bimorph actuator. (b) Cross-sectional illustration of all layers of the actuator. (c) SEM of the two-axis MEMS mirror. (d) A close-up SEM showing one FDSB actuator. ©[2012] IEEE. Reprinted with permission

[50, 78]. At the same time, the mirror plate is designed to have a large initial downward displacement so that the mirror plate is protected by the surrounding walls in the substrate. Figure 3.10(c) shows an SEM picture of a FDSB MEMS micromirror with the aperture size of 1 mm × 1 mm and the footprint of 1.55 mm × 1.7 mm. The fill factor of this device is 38%. The mirror plate has a large initial downward displacement of 340 μm. The maximum piston displacement of the mirror plate was 317 μm by driving the four actuators simultaneously at 4.8 V. Maximum optical scan angle of ±23° was achieved for tip-tilt motion. The resonant frequency of the tip-tilt motion was around 406 Hz with a 1 dB (~80%) thermal cutoff of 200 Hz.

3.4 MEMS-based endoscopic OCT imaging

Since the very first demonstration of employing MEMS micromirror for lateral scan in a forward-viewing OCT endoscopic probe by Pan *et al.* in 2001 [48], a variety of electrothermal MEMS mirror-based endoscopic probes/catheters have been reported for volumetric OCT imaging [49–52]. General side-viewing and forward-viewing probe designs using MEMS micromirror are shown in Figure 3.11. MEMS mirrors are usually placed after optical components such as a GRIN lens at the distal end [79]. One-axis rotation of the MEMS mirror can direct the incident light for lateral scanning on samples, while two-axis rotation of the MEMS mirror can generate 2D scanning patterns such as a raster scan or a

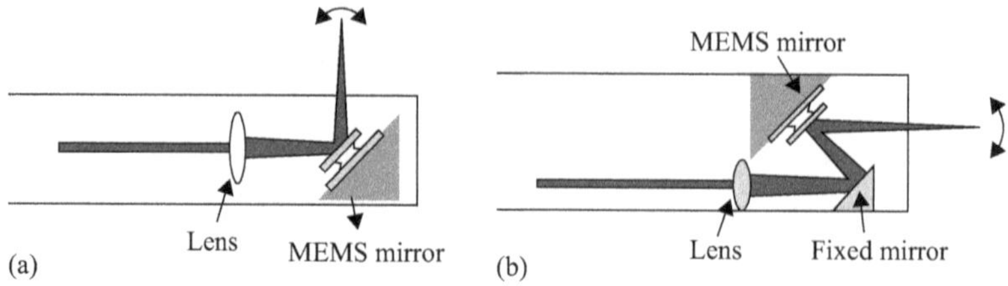

Figure 3.11 Schematic of generic MEMS-based endoscope designs. (a) Side-viewing probe. (b) Forward-viewing probe

Lissajous scan pattern. Endoscopic OCT probes based on the electrothermal MEMS mirrors described above and their imaging applications are reviewed in the following sections.

3.4.1 Internal organ imaging

Pan *et al.* demonstrated the first prototype of an electrothermal MEMS micromirror-based OCT endoscope in 2001 [48]. As the schematic in Figure 3.12(a) shows, this front-view endoscope contains a single mode fiber (SMF), a GRIN lens, and a MEMS mirror at the distal end to focus and scan the light beam. The 1D MEMS mirror shown in Figure 3.7(a,b) is employed in this system. The MEMS mirror is fixed at a tilted ferrule inside a stainless-steel mount, and the whole assembly is then contained in a transparent outer tubing for protection. The MEMS micromirror generates 1D transverse scanning, and the measured optical scan angle is $\pm 15°$, enabling a maximum scan range of 2.9 mm on the sample when a GRIN lens with a focal length of 10 mm is used. The outer diameter of this probe is 5 mm, which is small enough to fit in the internal clearance of a 22 Fr conventional endoscope. OCT imaging of a porcine urinary bladder was performed *in vivo* with this forward-view probe in the working channel of a standard endoscope. The OCT image shown in Figure 3.12(d) delineated the morphological layers of the bladder wall, including the urothelium (U), submucosa (SM), and muscularis (MS) layer. This result demonstrates the potential of the MEMS-based endoscopes for early detection of bladder cancer since most transitional carcinomas originate in the urothelium layer of the bladder wall [48].

After that, Xie *et al.* performed a comprehensive study of rat bladder cancer using both a MEMS-based OCT probe and a bench-top OCT system in 2003 [80]. Detailed morphologic changes, including identification of normal and hyperplastic structure layers, backscattering increases in the area of early neoplastic urothelium and heavy vascularization, are clearly shown in the OCT images. However, the endoscope has slightly degraded image fidelity compared to the bench-top results, such as decreased SNR, slightly degraded lateral resolution, and limited lateral scan range (Figure 3.13(a)). Later on, Wang *et al.* reported an *in vivo* animal study of bladder imaging diagnosis based on a porcine acute inflammation model in 2007 [81]. A MEMS-based SDOCT system was used to visualize bladder morphology and inflammatory lesions. The system was capable of providing high-fidelity

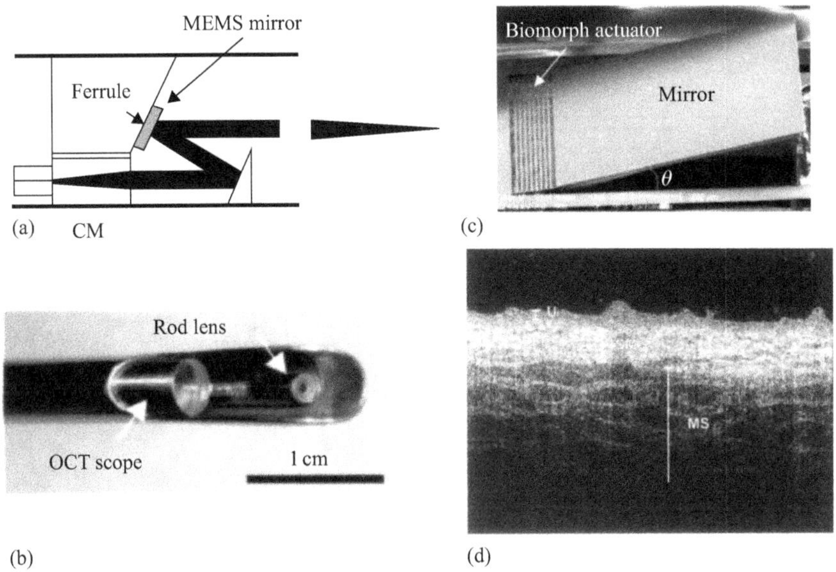

Figure 3.12 *Endoscopic OCT probe based on 1D scanning electrothermal MEMS mirror [48]. (a) Schematic of the probe. (b) Picture of an assembled probe. (c) SEM of the 1D MEMS mirror. (d) OCT images of the bladder wall. ©[2001] OSA. Reprinted with permission*

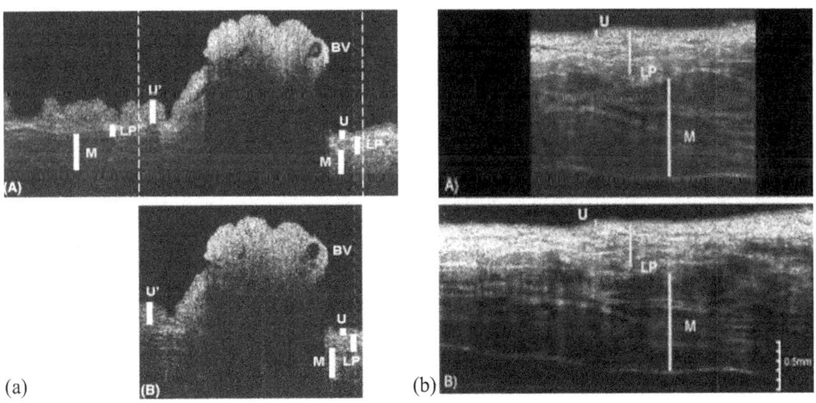

Figure 3.13 *Comparison of images of rat bladder cancer with larger papillary hyperplasia obtained by (a) bench-top OCT [80] and (b) endoscopic OCT [81]. ©[2003] OSA. Reprinted with permission. ©[2007] SPIE. Reprinted with permission*

morphological details of the bladder comparable to those obtained from a bench-top OCT system, with axial/lateral resolution of 10/12 μm and high sensitivity of 108 dB. The OCT images match well with histological images and morphological changes can be clearly identified by OCT. These experiments suggest that the

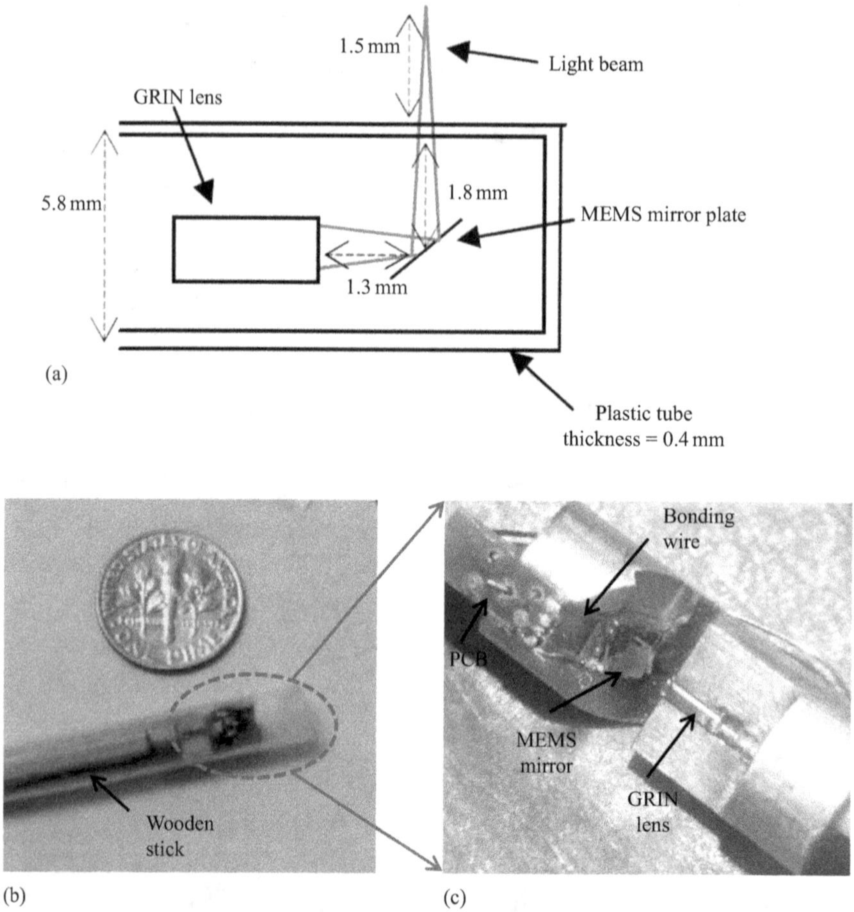

Figure 3.14 Two-axis LSF-LVD MEMS mirror-based endoscopic OCT probe [49]. (a) Schematic of the probe. (b) Fully assembled probe. (c) Close-up of MEMS mirror inside the probe mount. ©[2010] OSA. Reprinted with permission

MEMS OCT endoscopic imaging has great potential for early detection of transitional bladder cell carcinomas.

3.4.2 In vivo animal imaging

OCT endoscopes with 2D lateral scanning capability received considerable attention due to the requirement of 3D *in vivo* endoscopic imaging and the increase of imaging speed of OCT systems. Sun *et al.* reported a 2D scanning endoscope with an outer diameter of 5.8 mm in 2011 [79]. Figure 3.14 shows the general schematic of the side-view OCT endoscope, including a SMF, a GRIN lens, and a 2D MEMS mirror fixed on a 45° slope inside a probe mount. The electrical connection was provided by either wire bonding or silver epoxy bonding. The MEMS mirror shown

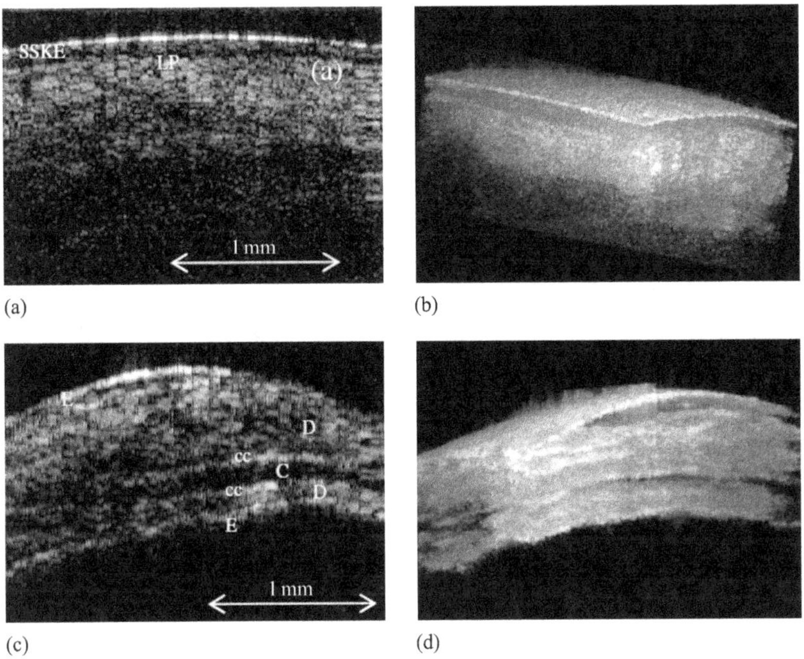

Figure 3.15 *(a) 2D and (b) 3D OCT images of a mouse tongue. (c) 2D and (d) 3D OCT images of a mouse ear. ©[2010] OSA. Reprinted with permission*

in Figure 3.9 was employed, whose optical scan range reached to ±31°. *In vivo* imaging experiments were performed on a female athymic nude mouse. Figure 3.15 shows both 2D and 3D images of the mouse ear and tongue. As shown in Figure 3.15(c,d), the subsurface morphological layers of the mouse ear were delineated clearly, representing the epidermis (E), dermis (D), cartilage (C), and conjunctive capsule (cc) structure layers corresponding to the dark and bright bands. Similarly, the OCT images of the mouse tongue show relatively strong signal bands near the upper layer, indicating the stratified squamous keratinized epithelium (SSKE) layer and lamina propria (LP) layer.

A two-axis electrothermal MEMS mirror with through-silicon vias (TSVs) was designed and assembled into a new MEMS-based OCT endoscope by Liu *et al.* [51]. Figure 3.16(a) shows an SEM of a fabricated device, with four TSVs located at the four corners symmetrically to allow electrical connections to pass through from the back side to be connected to the pads on the front side. The TSV design for interconnection enables ultracompact probe mount design and leads to a dramatic decrease of the probe size to 2.6 mm in diameter. *Ex vivo* sliced rat brain tissues were used to test the imaging performance.

Recently, a two-axis endoscopic probe with outer diameter of 2.8 mm using a unique high-fill-factor electrothermal MEMS mirror enabled by hidden actuators was reported by Samuelson *et al.* [50]. The device has a footprint of

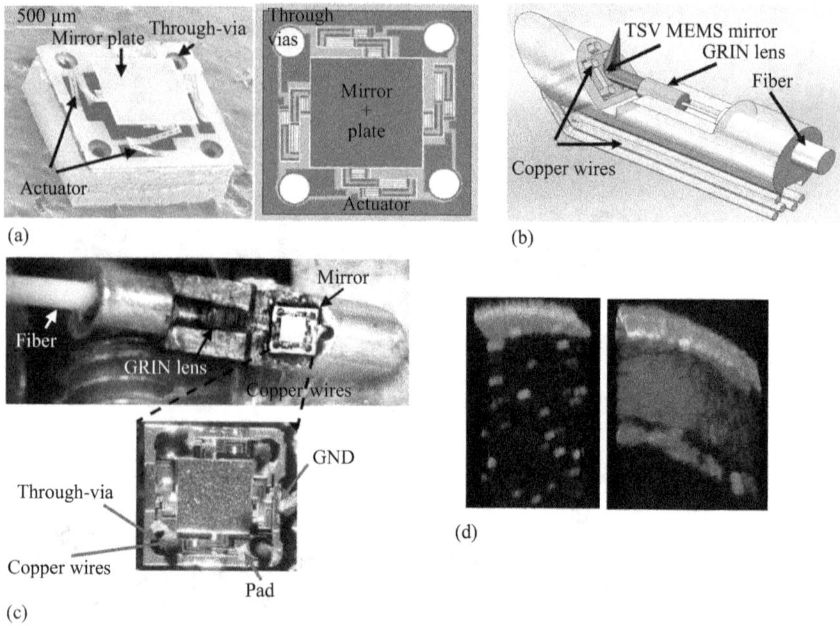

Figure 3.16 Endoscopic OCT probe using 2D electrothermal MEMS mirror with TSVs [51]. (a) SEM and schematic of the 2D MEMS mirror with four TSVs on four corners. (b) 3D model of the probe. (c) Fully packaged probe showing MEMS mirror assembled on the probe mount. (d) 3D cross-sectional OCT images of microspheres embedded in PDMS and rat brain tissue. ©[2011] SPIE. Reprinted with permission

1.55 mm ×1.7 mm and a large mirror size of 1 mm × 1 mm, shown in Figure 3.10. The MEMS mirror contains four symmetrically placed FDSB actuators, achieving large optical scan angle up to 46° at only 4.8 V. A side-viewing OCT imaging endoscope was assembled (Figure 3.17) and the ability for early cancer detection was demonstrated by diagnosing and monitoring the growing tumor mass on a nude mouse model. A clearly visible layered structure of the normal tissue and a firm delineation between the normal tissue and the tumor area were identified in the obtained 2D and 3D OCT images, shown in Figure 3.18.

3.4.3 Oral and teeth imaging

A similar probe design with a two-axis MEMS mirror developed by Jia *et al.* was employed to demonstrate the potential of endoscopic OCT for dental imaging in 2009 [82]. Figure 3.19(a,b) shows a prototype probe in imaging operation with mock-real human mouth (Figure 3.19(c)). A SSOCT system with an axial resolution of 10 µm was employed to perform dental OCT imaging experiments. The lateral resolution was measured to be 35 µm over a large 6 mm × 6 mm field of view, allowing the OCT to provide more evident microstructure details than

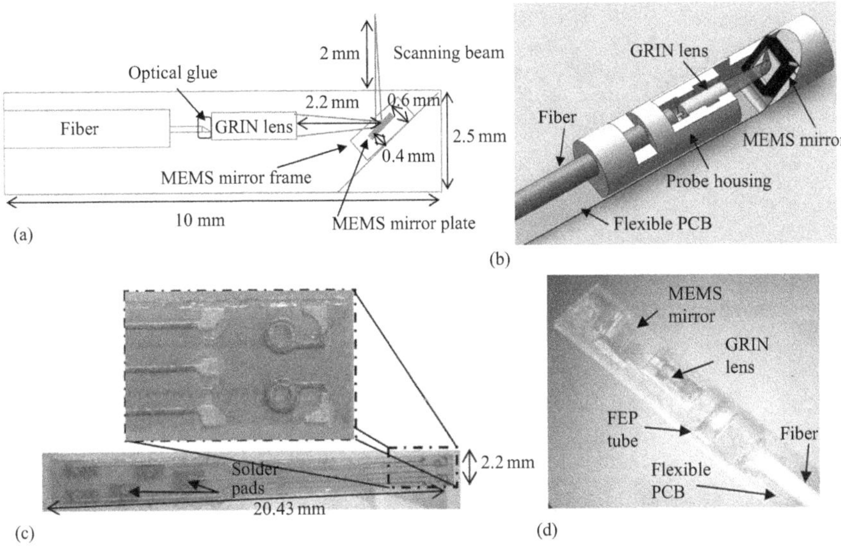

Figure 3.17 *Endoscopic OCT probe using 2D electrothermal MEMS mirror with FDSB actuators [50]. (a) Schematic and (b) 3D model of the probe. (c) Flexible PCB for electrical connection to the MEMS actuators. (d) Assembled prototype of the probe. ©[2012] IEEE. Reprinted with permission*

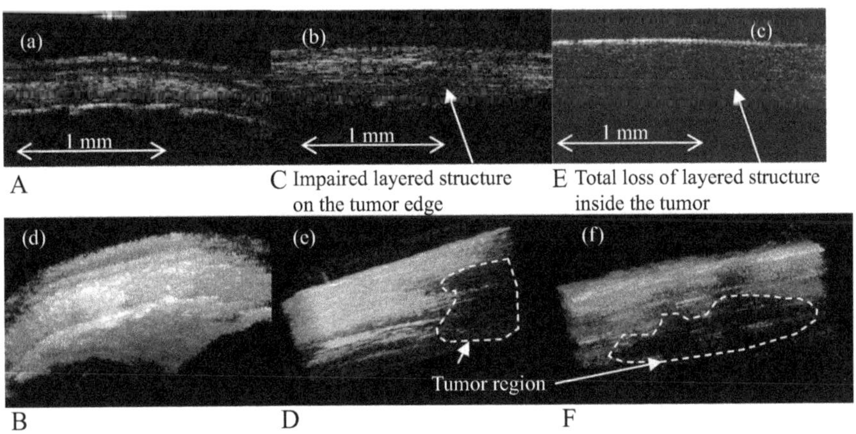

Figure 3.18 *2D and 3D OCT images of normal tissue and tumor region [50]. (a) 2D and (b) 3D images of normal tissue structure region. (c) 2D and (d) 3D images of half-tumor half-normal tissue region. (e) 2D and (f) 3D images of tumor region. ©[2012] IEEE. Reprinted with permission*

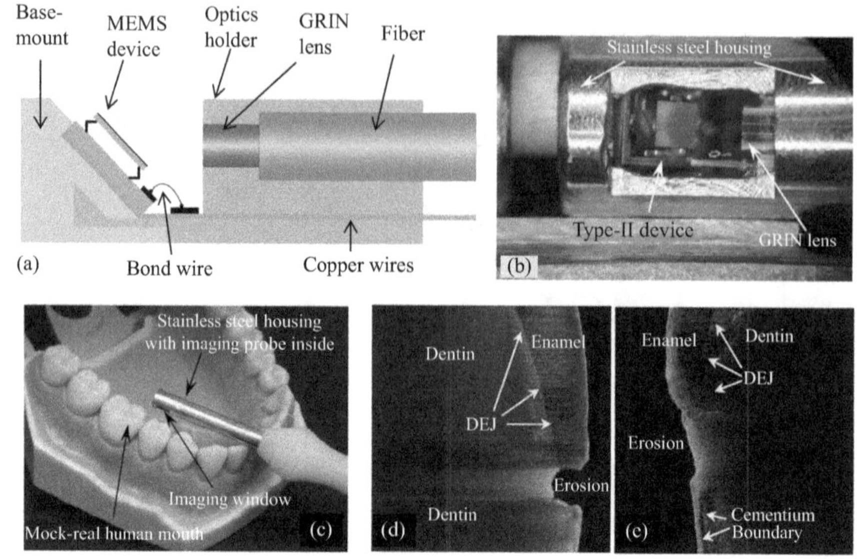

Figure 3.19 Endoscopic OCT probe based on 2D MEMS mirror [82]. (a) Schematic of the probe. (b) An assembled 2D MEMS mirror with LSF actuators on MEMS base-mount. (c) Optical image of the prototype MEMS-based dental OCT imaging probe in operation. Dental OCT images showing erosion of (d) DEJ and (e) adult incisor with erosion. ©[2009] IEEE. Reprinted with permission

conventional X-ray images. As shown in Figure 3.19(d), erosion to the enamel layer is clearly visible and the dentin enamel junction (DEJ) is already exposed. Figure 3.19(e) also clearly shows DEJ with Ebner striations of an adult incisor. The tooth surface has been eroded, as shown in the middle of the OCT image.

In 2013, Wang *et al.* developed a 3.5 mm endoscopic probe for an SSOCT system with imaging speed of 50 frames/s and imaging depth of 1.5 mm inside the oral mucosa [52]. The prototype is shown in Figure 3.20. A fiber-GRIN lens single module was employed to eliminate the effort for manual alignment. The probe consists of two parts, a stainless steel protection tube and a separate plug holder. This design effectively enhanced the robustness and reliability of the packaging and dramatically simplified the assembly process by avoiding dealing with the MEMS mirror inside a narrow tubular probe mount. Both suspicious oral leukoplakia tissue and normal mucosa tissue were imaged on a human patient with suspicious oral cancer. The OCT images of normal buccal mucosa is shown in Figure 3.20(f) delineating three distinct layers of epithelium (EP), lamina propria (LP), and submucosa (SM) from the top to the bottom layers. The boundaries, especially the basement membrane (BM) between EP and LP layers, are clearly shown for the normal mucosa tissue. In contrast, the OCT image of the suspicious region contains no identifiable boundaries between different layers inside the tissue, as shown in Figure 3.20(h). The histopathological image shown in Figure 3.20(g) agrees well

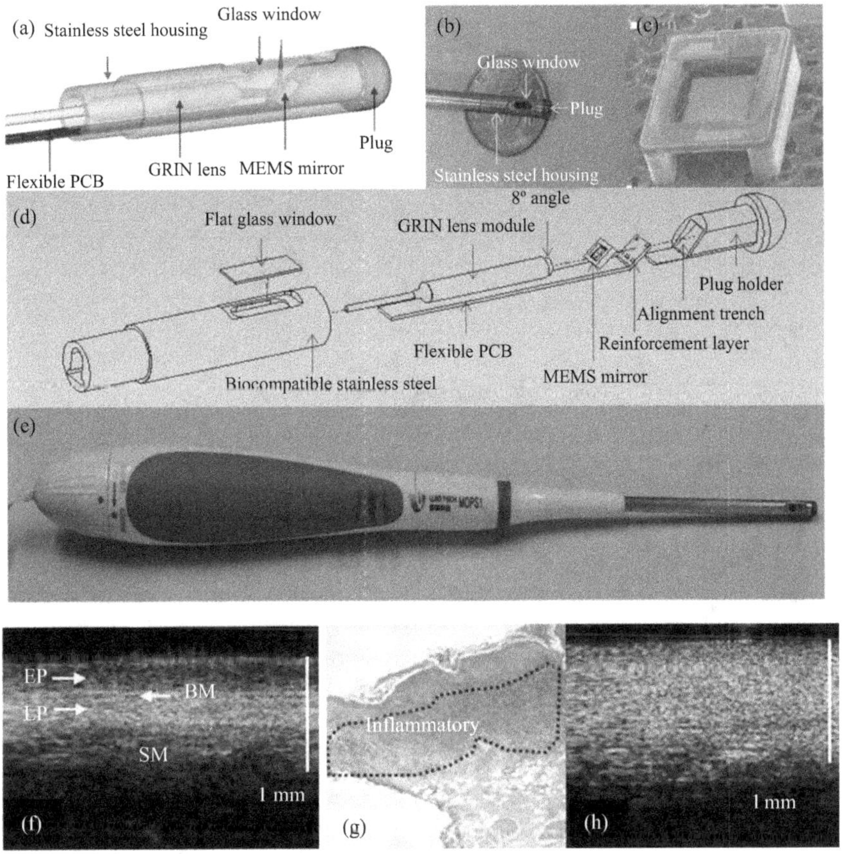

Figure 3.20 2D scanning side-viewing OCT probe [52]. (a) 3D model of the probe. (b) A packaged probe. (c) SEM of the S-shape bimorph MEMS mirror. (d) Exploded view of the probe assembly. (e) Fully packaged probe with handler. (e) OCT image of normal buccal mucosa. (g) Histopathological image and (h) OCT image of oral leukoplakia mucosa showing inflammatory cells infiltrated into nearby tissues. ©[2013] SPIE. Reprinted with permission

with the OCT findings, both showing no clear BM boundary. The histopathological image also shows inflammatory cells infiltrated into nearby layers.

3.4.4 Meniscus and brain tissue imaging

Although not yet done *in vivo*, electrothermal MEMS-based OCT endoscope prototypes have been employed in imaging cadaver canine meniscus tissue and mouse brain tissue to study their potential and feasibility for brain imaging [83] and non-invasive joint diagnosis [84]. Sun *et al.* measured the refractive index (RI) of acute brain tissue slices obtained from both white-matter and gray-matter regions [83].

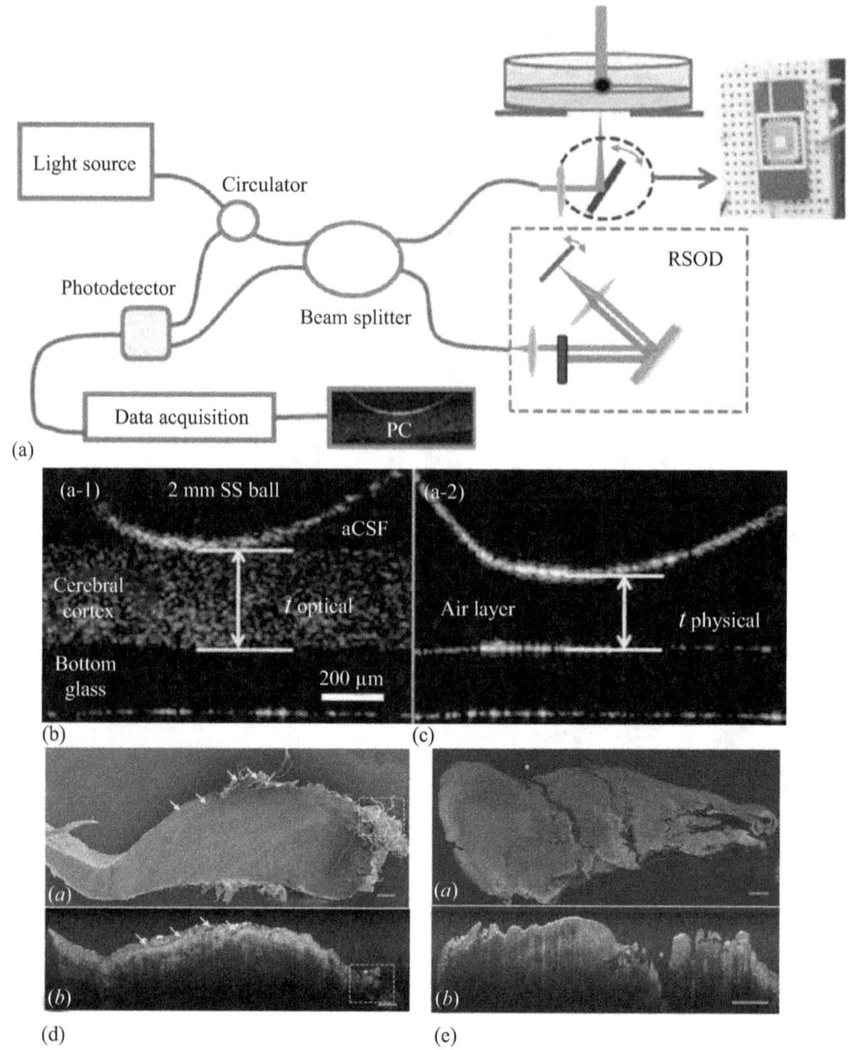

Figure 3.21 Bench-top OCT system using MEMS mirror for lateral scanning. (a) Schematic of the OCT system. (b) and (c) Measurement of optical and physical thickness of rat brain tissue slices [83]. ©[2012] OSA. Reprinted with permission. (d) SEM and OCT images of a degenerative meniscus. (e) SEM and OCT images of a meniscus with a buckle handle tear [84]. ©[2010] IOP Publishing. Reproduced with permission. All rights reserved

A free-space MEMS-based TDOCT system was used, shown in Figure 21(a). Optical and physical thicknesses of rat brain tissues from various anatomical regions were measured from the OCT images to calculate the refractive indices, shown in Figure 21(b,c). Ling *et al.* presented the ability of OCT for the detection and diagnosis of meniscal pathology by imaging collected torn medial menisci from dogs with anterior cruciate ligament insufficiency [84]. A similar free-space MEMS mirror-based TDOCT system was used. Gross and microstructural abnormalities identified from OCT images were verified by comparing with SEM images. In the OCT images of the meniscus with a degenerative tear shown in Figure 3.21(d), several small horizontal cleavages near the surface and irregularities with wrinkles are noticed, as corresponding to the SEM image. An obvious buckle handle tear and laminar separation of the collagen fibrils are shown in both OCT and SEM images of another meniscus sample (Figure 3.21(e)). In addition, OCT also provides reduction and loss of birefringence of meniscal tears in the torn menisci.

3.5 Summary

This chapter is focused on introducing a unique class of electrothermal bimorph MEMS mirrors and their applications in endoscopic OCT imaging. The novel bimorph actuator designs provide the electrothermal MEMS mirrors with unrivaled combination of large linear scan range, small drive voltage, and high fill factor, making them especially suitable for endoscopic *in vivo* imaging of internal organs. Using the electrothermal MEMS mirror-based OCT probes, clinical imaging experiments have been performed to detect and stage early bladder cancer and oral cancer. Those MEMS OCT probes have also been successfully applied to brain tissue imaging, tooth imaging, and meniscus imaging. All these imaging experiments have demonstrated the potential and feasibility of the electrothermal MEMS-based OCT endoscopic imaging. With further improvement of the robustness of the MEMS mirrors, the clinical use of this technology is coming in the foreseeable future.

References

[1] "Cancer facts and figures 2014," Atlanta: American Cancer Society; 2014

[2] Smith R.A., Cokkinides V., von Eschenbach A.C., *et al.* "American Cancer Society guidelines for the early detection of cancer." *CA: A Cancer Journal for Clinicians.* 2002;52(1):8–22

[3] DeSantis C., Ma J., Bryan L., Jemal A. "Breast cancer statistics, 2013." *CA: A Cancer Journal for Clinicians.* 2014;64(1):52–62

[4] Siegel R., DeSantis C., Jemal A. "Colorectal cancer statistics, 2014." *CA: A Cancer Journal for Clinicians.* 2014;64(2):104–17

[5] Jung W., McCormick D.T., Ahn Y.-C., *et al.* "In vivo three-dimensional spectral domain endoscopic optical coherence tomography using a micro-electromechanical system mirror." *Optics Letters.* 2007;32(22):3239–41

[6] Tearney G.J., Brezinski M.E., Bouma B.E., *et al.* "In vivo endoscopic optical biopsy with optical coherence tomography." *Science.* 1997;276(5321):2037–9

[7] Vakoc B.J., Fukumura D., Jain R.K., Bouma B.E. "Cancer imaging by optical coherence tomography: preclinical progress and clinical potential." *Nature Reviews Cancer.* 2012;12(5):363–8

[8] Weissleder R., Pittet M.J. "Imaging in the era of molecular oncology." *Nature.* 2008;452(7187):580–9
[9] Backman V., Wallace M. B., Perelman L.T., *et al.* "Detection of preinvasive cancer cells." *Nature.* 2000;406(6791):35–6
[10] Huang D., Swanson E.A., Lin C.P., *et al.* "Optical coherence tomography." *Science.* 1991;254(5035):1178–81
[11] Drexler W., Fujimoto J.G. *Optical coherence tomography: technology and applications*. Springer; 2008
[12] Yaqoob Z., Wu J., Yang C. "Spectral domain optical coherence tomography: a better OCT imaging strategy." *Biotechniques.* 2005;39: S6–13
[13] Mao Y., Chang S., Murdock E., Flueraru C. "Simultaneous dual-wavelength-band common-path swept-source optical coherence tomography with single polygon mirror scanner." *Optics Letters.* 2011;36(11):1990–2
[14] Sun J. "MEMS based optical coherence tomography imaging." Ph.D. dissertation; University of Florida; 2012
[15] Nishizawa N. "Ultrashort pulse fiber lasers and their applications." *Japanese Journal of Applied Physics.* 2014;53(9):090101
[16] Gora M., Karnowski K., Szkulmowski M., *et al.* "Ultra high-speed swept source OCT imaging of the anterior segment of human eye at 200 kHz with adjustable imaging range." *Optics Express.* 2009;17(17):14880–94
[17] Yun S., Tearney G., Bouma, B., Park B., de Boer J. "High-speed spectral-domain optical coherence tomography at 1.3 μm wavelength." *Optics Express.* 2003;11(26):3598–604
[18] Vuong B., Lee A., Luk T.W.H., *et al.* "High speed, wide velocity dynamic range Doppler optical coherence tomography (Part IV): split spectrum processing in rotary catheter probes." *Optics Express.* 2014;22(7):7399–415
[19] Bonesi M., Sattmann H., Torzicky T., *et al.* "High-speed polarization sensitive optical coherence tomography scan engine based on Fourier domain mode locked laser." *Biomedical Optics Express.* 2012;3(11): 2987–3000
[20] Chan K.K.H, Tang S. "High-speed spectral domain optical coherence tomography using non-uniform fast Fourier transform." *Biomedical Optics Express.* 2010;1(5):1309–19
[21] Kang J.-U., Han J.H., Liu X, Kang Z. "Common-path optical coherence tomography for biomedical imaging and sensing." *Journal of the Optical Society of Korea.* 2010;14(1):1–13
[22] Potsaid B., Baumann B., Huang D., *et al.* "Ultrahigh speed 1050 nm swept source/Fourier domain OCT retinal and anterior segment imaging at 100,000 to 400,000 axial scans per second." *Optics Express.* 2010;18(19):20029–48
[23] Fujimoto J.G. "Optical coherence tomography for ultrahigh resolution in vivo imaging." *Nature Biotechnology.* 2003;21(11):1361–7
[24] Sun C.-W., Lee S.-Y., Lin K.-F. "Review: Optical scanning probe for optical coherence tomography." *Journal of Medical and Biological Engineering.* 2014;34(1):95–100

[25] Yaqoob Z., Wu J., McDowell E.J., Heng X., Yang C. "Methods and application areas of endoscopic optical coherence tomography." *Journal of Biomedical Optics.* 2006;11(6):063001
[26] Li X., Chudoba C., Ko T., Pitris C., Fujimoto J.G. "Imaging needle for optical coherence tomography." *Optics Letters.* 2000;25(20):1520–2
[27] Quirk B.C., McLaughlin R.A., Curatolo A., Kirk R.W., Noble P.B., Sampson D.D. "In situ imaging of lung alveoli with an optical coherence tomography needle probe." *Journal of Biomedical Optics.* 2011;16(3):036009
[28] Fu H.L., Leng Y., Cobb M.J., Hsu K., Hwang J.H., Li X. "Flexible miniature compound lens design for high-resolution optical coherence tomography balloon imaging catheter." *Journal of Biomedical Optics.* 2008;13(6):060502
[29] Dragonfly Duo OCT Imaging Catheter, St. Jude Medical. Available from http://professional.sjm.com/products/vas/intravascular-diagnostics-imaging/oct/dragonfly-duo-oct-imaging-catheter#tech-specs
[30] Kang W., Wang H., Wang Z., *et al.* "Motion artifacts associated with in vivo endoscopic OCT images of the esophagus." *Optics Express.* 2011;19(21):20722–35
[31] Atif M., Ullah H., Hamza M.Y., Ikram M. "Catheters for optical coherence tomography." *Laser Physics Letters.* 2011;8(9):629
[32] Sun C.-W., Lee S.-Y., Lin K.-F. "Review: Optical scanning probe for optical coherence tomography." *Journal of Medical and Biological Engineering.* 2014;34(1):95–100
[33] Boppart S.A., Bouma B.F., Pitris C., *et al.* "Forward-imaging instruments for optical coherence tomography." *Optics Letters.* 1997;22(21):1618–20
[34] Liu X., Cobb M.J., Chen Y., Kimmey M.B., Li X. "Rapid-scanning forward-imaging miniature endoscope for real-time optical coherence tomography." *Optics Letters.* 2004;29(15):1763–5
[35] Wu Y., Leng Y., Xi J., Li X. "Scanning all-fiber-optic endomicroscopy system for 3D nonlinear optical imaging of biological tissues." *Optics Express.* 2009;17(10):7907–15
[36] Zhang N., Tsai T.-H., Ahsen O.O., *et al.* "Compact piezoelectric transducer fiber scanning probe for optical coherence tomography." *Optics Letters.* 2014;39(2):186–8
[37] Munce N.R., Mariampillai A., Standish B.A., *et al.* "Electrostatic forward-viewing scanning probe for Doppler optical coherence tomography using a dissipative polymer catheter." *Optics Letters.* 2008;33(7):657–9
[38] Min E.J., Na J., Ryu S.Y., Lee B.H. "Single-body lensed-fiber scanning probe actuated by magnetic force for optical imaging." *Optics Letters.* 2009;34(12):1897–9
[39] Min E.J., Shin J.G., Kim Y., Lee B.H. "Two-dimensional scanning probe driven by a solenoid-based single actuator for optical coherence tomography." *Optics Letters.* 2011;36(11):1963–5
[40] Zhang X., Duan C., Liu L., Li X., Xie H. "A non-resonant fiber scanner based on an electrothermally-actuated MEMS stage." *Sensors and Actuators A: Physical.* 2015;233:239–45

[41] Tearney G.J., Brezinski M.E., Fujimoto J.G., *et al.* "Scanning single-mode fiber optic catheter–endoscope for optical coherence tomography." *Optics Letters.* 1996;21(7):543–5

[42] Tran P.H., Mukai D.S., Brenner M., Chen Z. "In vivo endoscopic optical coherence tomography by use of a rotational microelectromechanical system probe." *Optics Letters.* 2004;29(11):1236–8

[43] Herz P.R., Chen, Y., Aguirre, A. D., *et al.* "Micromotor endoscope catheter for in vivo, ultrahigh-resolution optical coherence tomography." *Optics Letters.* 2004;29(19):2261–3

[44] Su J., Zhang J., Yu L., Chen Z. "In vivo three-dimensional microelectromechanical endoscopic swept source optical coherence tomography." *Optics Express.* 2007;15(16):10390–6

[45] Wang T., Wieser W., Springeling G., *et al.* "Intravascular optical coherence tomography imaging at 3200 frames per second." *Optics Letters.* 2013;38 (10):1715–7

[46] Chang S., Murdock E., Mao Y., Flueraru C., Disano J. "Stationary-fiber rotary probe with unobstructed 360 view for optical coherence tomography." *Optics Letters.* 2011;36(22):4392–4

[47] Chen T., Zhang N., Huo T., *et al.* "Tiny endoscopic optical coherence tomography probe driven by a miniaturized hollow ultrasonic motor." *Journal of Biomedical Optics.* 2013;18(8):086011

[48] Pan Y., Xie H., Fedder G.K. "Endoscopic optical coherence tomography based on a microelectromechanical mirror." *Optics Letters.* 2001;26 (24):1966–8

[49] Sun J., Guo S., Wu L., *et al.* "3D in vivo optical coherence tomography based on a low-voltage, large-scan-range 2D MEMS mirror." *Optics Express.* 2010;18(12):12065–75

[50] Samuelson S.R., Wu L., Sun J., Choe S., Sorg B.S., Xie H. "A 2.8-mm imaging probe based on a high-fill-factor MEMS mirror and wire-bonding-free packaging for endoscopic optical coherence tomography." *Journal of Microelectromechanical Systems.* 2012;21(6):1291–302

[51] Liu L., Wu L., Sun J., Lin E., Xie H. "Miniature endoscopic optical coherence tomography probe employing a two-axis microelectromechanical scanning mirror with through-silicon vias." *Journal of Biomedical Optics.* 2011;16(2):026006

[52] Wang D., Fu L., Wang X., *et al.* "Endoscopic swept-source optical coherence tomography based on a two-axis microelectromechanical system mirror." *Journal of Biomedical Optics.* 2013;18(8):086005

[53] Jung W., McCormick D.T., Zhang J., Wang L., Tien N.C., Chen Z. "Three-dimensional endoscopic optical coherence tomography by use of a two-axis microelectromechanical scanning mirror." *Applied Physics Letters.* 2006;88 (16):163901

[54] Aguirre A.D., Hertz P.R., Chen Y., *et al.* "Two-axis MEMS scanning catheter for ultrahigh resolution three-dimensional and en face imaging." *Optics Express.* 2007;15(5):2445–53

[55] Kim K.H., Hyle Park B., Maguluri G.N., *et al.* "Two-axis magnetically-driven MEMS scanning catheter for endoscopic high-speed optical coherence tomography." *Optics Express.* 2007;15(26):18130-40
[56] Gilchrist K.H., McNabb R.P., Izatt J.A., Grego S. "Piezoelectric scanning mirrors for endoscopic optical coherence tomography." *Journal of Micromechanics and Microengineering.* 2009;19(9):095012
[57] Su J., Zhang J., Yu L., Chen Z. "In vivo three-dimensional microelectromechanical endoscopic swept source optical coherence tomography." *Optics Express.* 2007;15(16):10390–6
[58] Senturia S.D. *Microsystem design.* Vol. 3. Boston: Kluwer academic publishers; 2001
[59] Atif M., Ullah H., Hamza M.Y., Ikram M. "Catheters for optical coherence tomography." *Laser Physics Letters.* 2011;8(9):629
[60] Jung W., McCormick D.T., Zhang J., Wang L., Tien N.C., Chen Z. "Three-dimensional endoscopic optical coherence tomography by use of a two-axis microelectromechanical scanning mirror." *Applied Physics Letters.* 2006;88 (16):163901
[61] Jia K. "The development and applications of high fill-factor, small footprint MEMS micromirrors and micromirror arrays." Ph.D. dissertation; University of Florida; 2009
[62] Sharpe W.N., Yuan B., Vaidyanathan R., Edwards R.L. "Measurements of Young's modulus, Poisson's ratio, and tensile strength of polysilicon." *Proceedings of the IEEE, Tenth Annual International Workshop on Micro Electro Mechanical Systems 1997. MEMS'97. IEEE*; 1997. pp. 424–9
[63] Wu L. "Low-voltage, large-range MEMS optical scanners and their applications." Ph.D. dissertation; University of Florida; 2009
[64] Geisberger A.A., Sarkar N., Ellis M., Skidmore G.D. "Electrothermal properties and modeling of polysilicon microthermal actuators." *Journal of Microelectromechanical Systems.* 2003;12(4):513–23
[65] Xie H., Pan Y., Fedder G.K. "A SCS CMOS micromirror for optical coherence tomographic imaging." *The Fifteenth IEEE International Conference on Micro Electro Mechanical Systems, 2002. IEEE*; 2002. pp. 495–8
[66] Jain A., Kopa A., Pan Y., Fedder G.K., Xie H. "A two-axis electrothermal micromirror for endoscopic optical coherence tomography." *IEEE Journal of Selected Topics in Quantum Electronics.* 2004;10(3):636–42
[67] Qu W., Wlodarski W., Austin M. "Microfabrication and reliability study of sapphire based Ti/Pt-electrodes for thin-film gas sensor applications." *Microelectronics Journal.* 2000;31(7):561–7
[68] Xiaoyang Z., Li B., Li X., Xie H. "A robust, fast electrothermal micromirror with symmetric bimorph actuators made of copper/tungsten." *Tech. Digest, 15th IEEE International Conference on Solid-State Sensors, Actuators and Microsystems (Transducers' 15)*; Anchorage, AK, USA, June 2015
[69] Jain A., Qu H., Todd S., Xie H. "A thermal bimorph micromirror with large bi-directional and vertical actuation." *Sensors and Actuators A: Physical.* 2005;122(1):9–15

[70] Jain A., Xie H. "A single-crystal silicon micromirror for large bi-directional 2D scanning applications." *Sensors and Actuators A: Physical*. 2006;130: 454–460

[71] Wu L., Xie H. "124 rotation angle electrothermal micromirror with integrated platinum heater." *IEEE Journal of Selected Topics in Quantum Electronics*. 2007;13(2):316–21

[72] Fu L., Jain A., Cranfield C., Xie H., Gu M. "Three-dimensional nonlinear optical endoscopy." *Journal of Biomedical Optics*. 2007;12(4):040501

[73] Wu L., Xie H. "A large vertical displacement electrothermal bimorph microactuator with very small lateral shift." *Sensors and Actuators A: Physical*. 2008;145: 371–9

[74] Jia K., Xie H. "Multi-axis electrothermal scanning micromirror with low driving voltage." *In Conference on Lasers and Electro-Optics, p. CThJJ1*. Optical Society of America; 2007

[75] Jia K., Pal S., Xie H. "An electrothermal tip–tilt–piston micromirror based on folded dual S-shaped bimorphs." *Journal of Microelectromechanical Systems*. 2009;18(5):1004–15

[76] Jia K., Samuelson S.R., Xie H. "High-fill-factor micromirror array with hidden bimorph actuators and tip–tilt-piston capability." *Journal of Microelectromechanical Systems*. 2011;20(3):573–82

[77] Samuelson S.R., Xie H. "A large piston displacement MEMS mirror with electrothermal ladder actuator arrays for ultra-low tilt applications." *Journal of Microelectromechanical Systems*. 2014;23:39–49

[78] Wu L., Samuelson, S.R., Sun J., *et al.* "A 2.8-MM imaging probe based on a high-fill-factor MEMS mirror and wire-bonding-free packaging for endoscopic optical coherence tomography." *IEEE 24th International Conference on Micro Electro Mechanical Systems (MEMS). IEEE*; 2011. pp. 33–6

[79] Sun J., Xie H. "MEMS-based endoscopic optical coherence tomography." *International Journal of Optics. 2011*;2011:12 pages

[80] Xie T., Xie H., Fedder G.K., Pan Y. "Endoscopic optical coherence tomography with a modified microelectromechanical systems mirror for detection of bladder cancers." *Applied Optics*. 2003;42(313):6422–6

[81] Wang Z., Lee C.S.D., Waltzer W.C., *et al.* "In vivo bladder imaging with microelectromechanical-systems-based endoscopic spectral domain optical coherence tomography." *Journal of Biomedical Optics*. 2007;12(3):034009

[82] Jia K., Pal S., Wu L., Hamilton D., Xie H. "Dental optical coherence tomography employing miniaturized MEMS-based imaging probe." *In 2009 IEEE/LEOS International Conference on Optical MEMS and Nanophotonics*. 2009

[83] Sun J., Lee S.J., Wu L., Sarntinoranont M., Xie H. "Refractive index measurement of acute rat brain tissue slices using optical coherence tomography." *Optics Express*. 2012;20(2):1084–95

[84] Ling C.H.-Y., Pozzi A., Thieman K.M., *et al.* "The potential of optical coherence tomography for diagnosing meniscal pathology." *Measurement Science and Technology*. 2010;21(4):045801

Chapter 4
Electrowetting-based microoptics

Chenhui Li and Hongrui Jiang

4.1 Brief history of electrowetting

In device miniaturization and microtechnology in general, the control of surface and surface energies poses one of the most important challenges. Capillary forces dominate for liquid droplets of submillimeter dimension. Therefore, controlling interfacial energies has become a widely used method for manipulating liquid droplets. Electrowetting controls interfacial energies through a phenomenon called electrocapillary, and it was first described in detail by Gabriel Lippmann in 1875. He found that the capillary depression of mercury in contact with electrolyte solution could be varied by applying a voltage between the mercury and the electrolyte. He subsequently formulated a theory of the electrocapillary effect and also developed several applications, including a very sensitive electrometer and a motor. This concept was further developed in the early 1990s when Berge introduced the idea of using a thin insulating layer to separate the conductive liquid from the metallic electrode in order to eliminate the electrolysis problem. This recent development is now known as electrowetting on a dielectric (EWOD) [1].

4.2 Surface tension

Surface tension is a property of the surface of a liquid, and it leads to many behaviors of liquids. Surface tension has a Système International (SI) unit of force per unit length, N/m, and the commonly used centimeter-gram-second (cgs) unit is dyne per centimeter, dyn/cm. Water at 20°C has a surface tension of 72.8 dyn/cm. It is also sometimes useful to consider surface tension in terms of work (or energy) per unit area.

Liquid molecules experience cohesive forces from their neighbor molecules. The spherical shape of a liquid droplet is due to intermolecular forces between the molecules of the droplet. As shown in Figure 4.1, a molecule inside the droplet is equally attracted in all directions by the molecules surrounding it, and therefore

Department of Electrical and Computer Engineering, University of Wisconsin – Madison, Madison, WI, USA

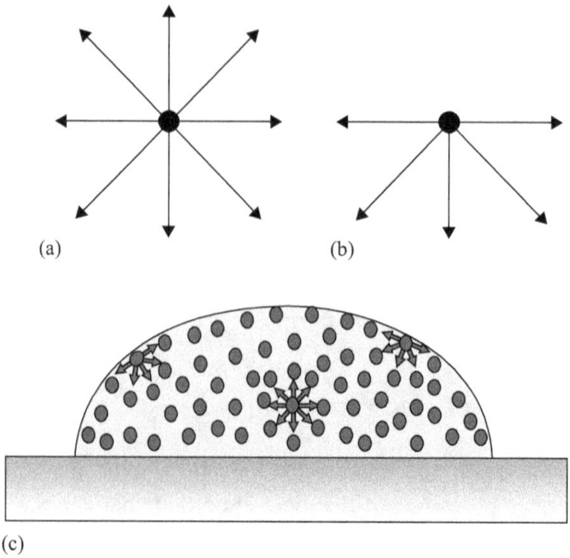

Figure 4.1 Liquid molecules exhibiting cohesive forces from neighbor molecules. (a) A molecule is attracted equally in every direction, resulting in a net force of 0. (b) A molecule is subject to net forces pulling them inward toward the bulk of the liquid. (c) Molecules inside a liquid droplet are attracted equally while molecules at the surface of the droplet are attracted inward

the total force exerted on it is zero. However, molecules located at the surface are attracted only by its inner neighbors, and the asymmetric attraction results in a force in the direction of its neighbor molecules, assuming the forces exerted from the molecules in the air onto the liquid molecules can be neglected as they are much smaller. The surface tension is created by this effective attraction of a molecule on the surface.

Surface tension creates some internal pressure and forces liquid surfaces to contract to a minimal area, reducing system energy. Therefore, surface tension determines the shapes of liquid droplets. For example, droplets of water tend to form spherical shapes because of the cohesive forces exerted on the surface layer. A molecule in contact with a neighbor is in a lower energy state than if it were isolated. The boundary molecules have less neighbors than the interior ones and therefore have higher energy. For a liquid to minimize its energy state, the number of boundary molecules must be minimized, and as a result, the surface area is minimized. A liquid surface will form a shape that is as smooth as possible, because any curvature in the surface profile leads to both a greater area and a higher energy. Therefore, the liquid surface tends to resist any curvature, just like an object lifted upward tends to go down to reduce its gravitational potential energy.

For a liquid surface, when no normal force is applied to a tensioned surface, the surface remains flat. However, if the pressure on one side of the surface differs

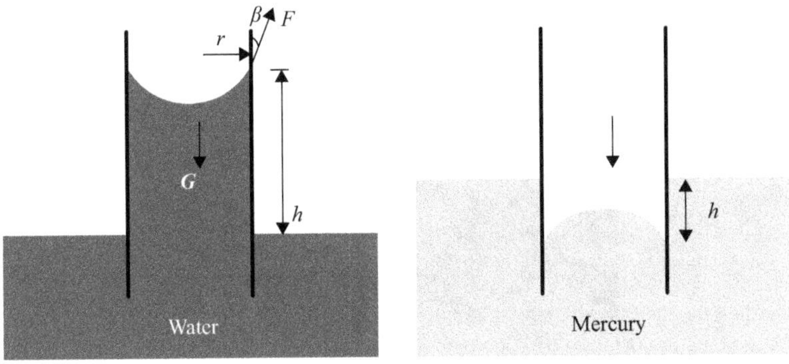

(a) (b)

Figure 4.2 *Capillary action. (a) A glass tube is inserted into water. The water–air meniscus bends downward and h > 0; (b) a glass tube is inserted into mercury. The mercury–air meniscus bends upward and h < 0*

from that on the other side, the pressure difference times the surface area provides a normal force. At equilibrium, the surface tension balances the normal force, resulting in a curved surface. The balanced forces and the surface shape satisfy the Young–Laplace equation:

$$\Delta p = \gamma\left(\frac{1}{R_x} + \frac{1}{R_y}\right) \tag{4.1}$$

where Δp is the pressure difference, γ is the surface tension, and R_x and R_y are radii of curvature in the axes that are parallel to the surface.

Surface tension at the interface of two materials depends on both of their properties. Attractive forces between different molecules, such as a liquid molecule and a solid molecule, are called adhesive forces. For example, the adhesive forces between water molecules and glass surfaces are stronger than the cohesive forces between the water molecules. Therefore, water tends to be pulled along the glass surfaces, leading to a phenomenon known as capillary action. Figure 4.2(a) shows a glass tube inserted into water. The adhesive forces between water and glass molecules pull water upward until they are balanced by gravity. Because of surface tension, the water-to-air interface forms a curved meniscus to minimize its surface energy. The height of the water column is:

$$h = \frac{2F\cos\beta}{\rho g r} \tag{4.2}$$

where F is the net force pulling the water meniscus, β is the angle between the glass surface and the tangent of the water meniscus, ρ is the density of water, r is the radius of the glass tube, and g is the standard gravity.

If the adhesive forces between the liquid and solid molecules are less than the cohesive forces between the liquid molecules, the solid surface will tend to push the liquid away. If mercury replaces water, as shown in Figure 4.2(b), the

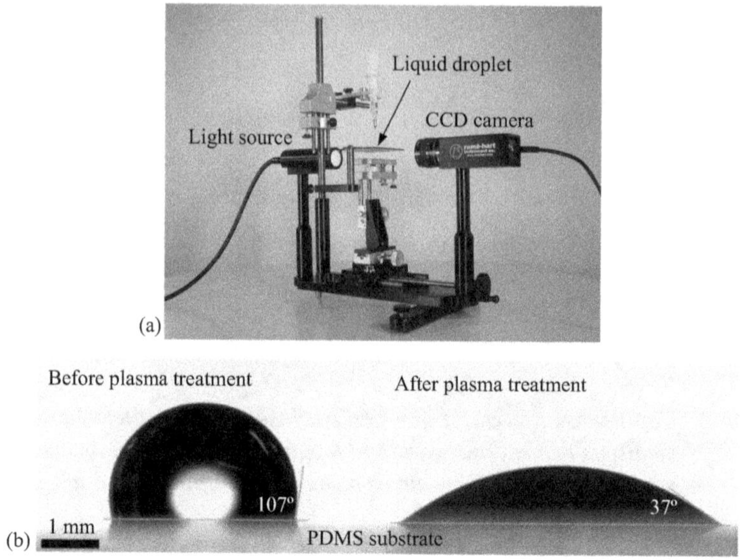

Figure 4.3 *(a) Setup of a goniometer. (b) Images recorded by a goniometer when a water droplet is placed on a PDMS surface before and after oxygen plasma treatment. Reprinted with permission from Zeng et al. (2010) [3], copyright IOP Publishing*

mercury-to-air interface curves upward and the mercury in the glass tube is lower than that outside ($h < 0$).

4.3 Contact angle

The contact angle is the angle at which a liquid–vapor interface meets a solid surface. The equilibrium contact angle is determined by the molecular interactions across the liquid–vapor, solid–vapor, and solid–liquid interfaces. The shape of a droplet on a solid surface is determined by the Young–Laplace equation, and the contact angle quantifies boundary conditions. Contact angle can be measured using a contact angle goniometer [2]. Figure 4.3(a) illustrates the setup of ramé-hart Model 200 Standard Contact Angle Goniometer. When this goniometer is in use, samples like liquid droplets are illuminated by a light beam, and the shadow of the droplet is recorded by a CCD camera and then used to calculate its contact angle. Figure 4.3(b) shows the goniometer-recorded images of a small water droplet resting on a flat polydimethylsiloxane (PDMS) surface before and after surface modification. The surface of PDMS film is naturally hydrophobic and becomes much more hydrophilic after an oxygen plasma treatment. Therefore, its pre-treatment contact angle (107°) is much larger than the post-treatment one (37°).

The contact angle is determined by Young's equation:

$$\cos\theta_Y = \frac{\gamma_{sv} - \gamma_{sl}}{\gamma_{lv}} \tag{4.3}$$

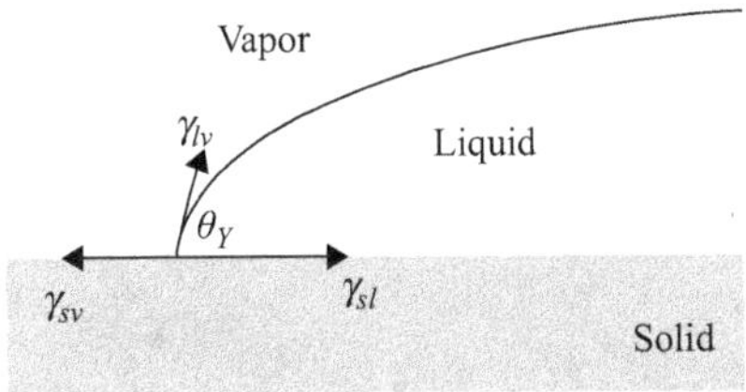

Figure 4.4 Boundary of liquid droplet on solid surface. At equilibrium, the net force at the boundary must be 0, leading to Young equation

where γ_{sl} is the solid–liquid interfacial energy (i.e. the surface tension), γ_{sv} is the solid–vapor interfacial energy, and γ_{lv} is the liquid–vapor interfacial energy. Figure 4.4 illustrates the boundary and interfacial energies of a droplet on a flat surface.

Contact angle quantifies the degree of wetting. The commonly referred two different regimes of wetting, "partial wetting" and "total wetting", are distinguished by the spreading parameter which is defined as $S = \gamma_{sv} - (\gamma_{lv} + \gamma_{sl})$. If $S > 0$, it is total wetting, and the liquid completely spreads to reach the minimum surface energy; otherwise, it is partial wetting, and a liquid droplet in general forms a spherical cap with contact angle θ_Y on a solid surface at equilibrium. In Figure 4.4, θ_Y is an acute angle and it is sometimes referred to as "mostly wetting". If $\theta_Y > 90°$, it is "mostly non-wetting". For example, solvents such as ethanol and toluene will spread on clean glass surfaces. A solid surface that tends to have high affinity energy for water is often called hydrophilic and water spreads on hydrophilic surfaces. On the other hand, if a water droplet forms a spherical cap on a solid surface with a large contact angle, the surface is called hydrophobic, because it tends to repel water. In practice, the terms "hydrophobic" and "hydrophilic" are often used relatively rather than distinguished by an absolute quantity.

4.4 Focal length of a liquid lens

A liquid–liquid interface forms a lens when the two liquids have different refractive indices and for a spherical interface, the focal length is approximately [3]:

$$f = \frac{R}{n_1 - n_2} \tag{4.4}$$

where R is the radius of curvature of the interface, n_1 and n_2 are the refractive indices of the two liquids. Figure 4.5 illustrates a convex lens formed by a liquid-liquid interface when $n_1 > n_2$.

4.5 Principles of electrowetting

In electrowetting, one generally deals with droplets of partially wetting liquids on planar solid substrates, as shown in Figure 4.6. In most applications of interest, the

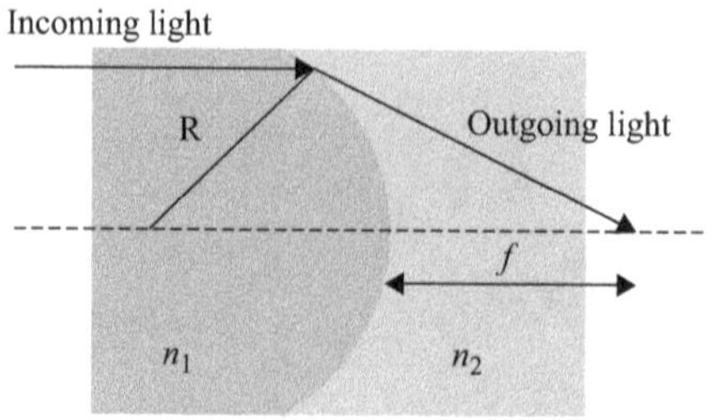

Figure 4.5 A convex lens formed by a liquid–liquid interface with $n_1 > n_2$

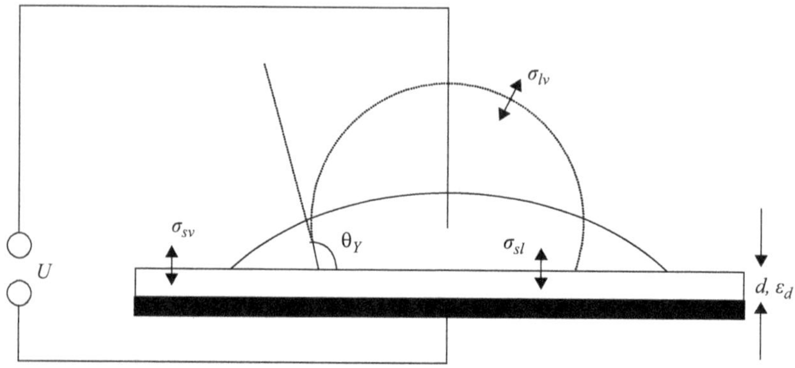

Figure 4.6 Generic electrowetting setup. Partially wetting liquid droplet at zero voltage (dashed line) and at high voltage (solid line). Modified with permission from Mugele et al. (2005) [1], copyright IOP Publishing

droplets are aqueous salt solutions with a submillimeter size (~1 mm or smaller), and therefore, gravity is often neglected because of surface tension's dominance. The ambient medium is either air or an immiscible liquid, like oil.

The free energy F of a droplet is a function of the droplet's shape, and it is the sum of the area A_i of the interfaces between the three phases involved: the solid substrate (s), the liquid droplet (l), and the vapor (v), weighted by their respective interfacial energies σ_i (σ_{sv} for solid-vapor, σ_{sl} for solid–liquid, and σ_{lv} for liquid–vapor):

$$F = F_{if} = \sum_i A_i \sigma_i - \lambda V \tag{4.5}$$

where λ is a Lagrangian variable that satisfies the constant volume (V) constraint and equals Δp, the pressure drop across the liquid–vapor interface. Due to variational minimization, any liquid shape has to satisfy two conditions at equilibrium. The first is the Laplace equation:

$$\Delta p = \sigma_{lv}\left(\frac{1}{r_1} + \frac{1}{r_2}\right) = \sigma_{lv} k \tag{4.6}$$

where r_1 and r_2 are the two principal radii of curvature of the surface, and k is the constant mean curvature. The pressure drop Δp is a constant, independent of the position on the interface. For homogeneous substrates, it means that droplets will adopt the shape of a spherical cap at equilibrium. The second condition is given by Young's equations in (4.3), which relates Young's equilibrium contact angle θ_Y to interfacial energies. Both equations are derived through approximations for mesoscopic scales.

For an electrolyte droplet placed directly on an electrode surface, as shown in Figure 4.6, the contact angle changes with the voltage applied between the electrolyte droplet and the electrode surface:

$$\cos\theta = \cos\theta_Y + \frac{\varepsilon_0\varepsilon_l}{2d_H\sigma_{lv}}(V - V_0)^2 \tag{4.7}$$

where θ_Y and θ are the contact angles before and after applying a voltage V, ε_0 is vacuum permittivity, ε_l is the dielectric constant of the liquid, d_H is the distance from the counter-ions in the electrode to the surface (typically a few nanometers), and V_0 is the initial voltages. Thus, if V_0 is 0, the contact angle decreases when V is applied. Equation (4.7) is only applicable when the applied voltage is below the onset of electrolytic process, typically up to a few hundred millivolts. In modern electrowetting applications, this problem is solved by adding a thin dielectric film to insulate the droplet from the electrode, also known as EWOD. Under the EWOD configuration, the contact angle equation becomes:

$$\cos\theta = \cos\theta_Y + \frac{\varepsilon_0\varepsilon_1}{2d\sigma_{lv}}(V - V_0)^2 \tag{4.8}$$

where d is the thickness of the thin dielectric film, and ε_1 is its dielectric constant. Equation (4.8) holds as long as the applied voltage is below the dielectric breakdown voltage and a system-dependent threshold voltage beyond which the contact angle becomes almost saturated.

At high applied voltages, the contact angle changes slowly when the voltage keeps increasing, and it saturates before reaching total wetting. The mechanism of contact angle saturation is different for alternating current (AC) and direct current (DC) voltages. For AC, the saturation occurs concomitantly with air ionization in the vicinity of the sharp droplet edge [4]. The air ionization generates charges crossing through the ionized air, from the edge to the insulator surface surrounding the droplet. These charges reduce then stop the electrowetting effect. An hydrophilic ring marked on the insulator at the outside of the droplets indicates damages of the hydrophobic surface due to the charges. For DC, charges trapped in the insulator are invoked, which is likely to cause the hydrophobic surface a damage at the exact location of the droplet [5].

4.6 Tunable liquid microlens utilizing electrowetting

In one of the earliest reported liquid lenses utilizing electrowetting, a droplet of hexadecanethiol (HDT, ~50 pL) was used as a planar convex lens [6]. It was positioned on a transparent gold electrode and immersed in aqueous sodium

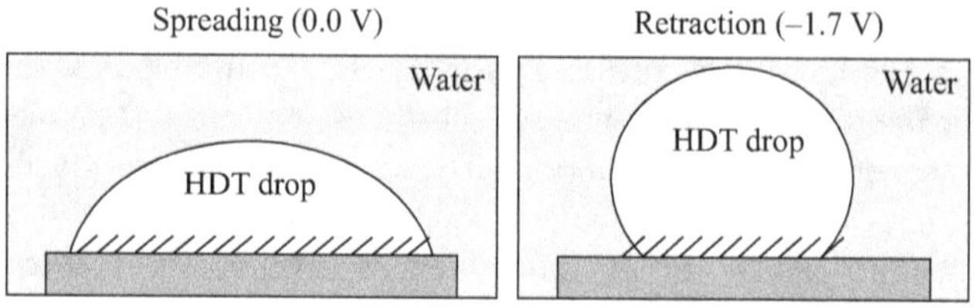

Figure 4.7 Behavior of drop of HDT on gold surface at 0 and at −5.7 V in sodium perchlorate electrolyte solution. At 0 V, the HDT reacted with the gold surface and formed a monolayer wet by the HDT liquid. At −5.7 V, the monolayer reductively desorbed and the HDT drop retracted. Modified with permission from Gorman et al. (1995) [6], copyright American Chemical Society

perchlorate solution, as illustrated in Figure 4.7. The drop spread on the gold surface at 0 V (in reference to a silver electrode) and formed a hydrophobic monolayer under the drop. When the applied voltage changes to −1.7 V, the electrodesorption of the hydrophobic monolayer caused a retraction of the droplet. Compared to the droplet at 0 V, the droplet at −1.7 V formed a lens with a much shorter focal length.

Rapid progress in performance of electrowetting-based liquid lens has been achieved in the last two decades due to improvements in materials and processing. Electrowetting lenses have the advantages of short response time, low electrical power consumption, and robustness under voltage cycling, and they have been utilized for an increasing number of applications, such as adaptive lenses, mobile phone cameras, optical switch, and display technology.

Berge *et al.* reported an electrowetting lens using two liquids [7]. Figure 4.8 illustrates two practical realizations of this lens. The lens cell is filled with two transparent immiscible liquids – one was insulating and nonpolar oil, and the other was conductive water solution. They had different refractive indices but the same density. Therefore, the liquid-to-liquid interface was not deformed by the liquids' gravity, and it remained spherical under any cell orientation. The oil droplet was placed on a thin insulating substrate, and the surface of the substrate was hydrophobic. Thus, the oil droplet naturally spread on it. A transparent counter-electrode was deposited on the outer side of the substrate. When voltage was applied between the counter-electrode and the water solution, the wettability of the surface increased, and the contact area between water and the substrate expanded, reshaping the liquid–liquid interface. Therefore, the focal length changed with the voltage applied. Figure 4.9 shows the variation of the reciprocal of the focal length as a function of the voltage applied. The power of the lens is relatively stable at low voltages and starts to rise when the voltage exceeds 90 V. For a typical electrowetting lens, the actual electrical power dissipated in the system is very low, because it is only consumed when the focal length changes. In this example, the power dissipation was a few milliwatts at 250 V.

For an electrowetting lens, when its contact angle is reversibly altered, at a given voltage the equilibrium value of the contact angle is not attained in most cases.

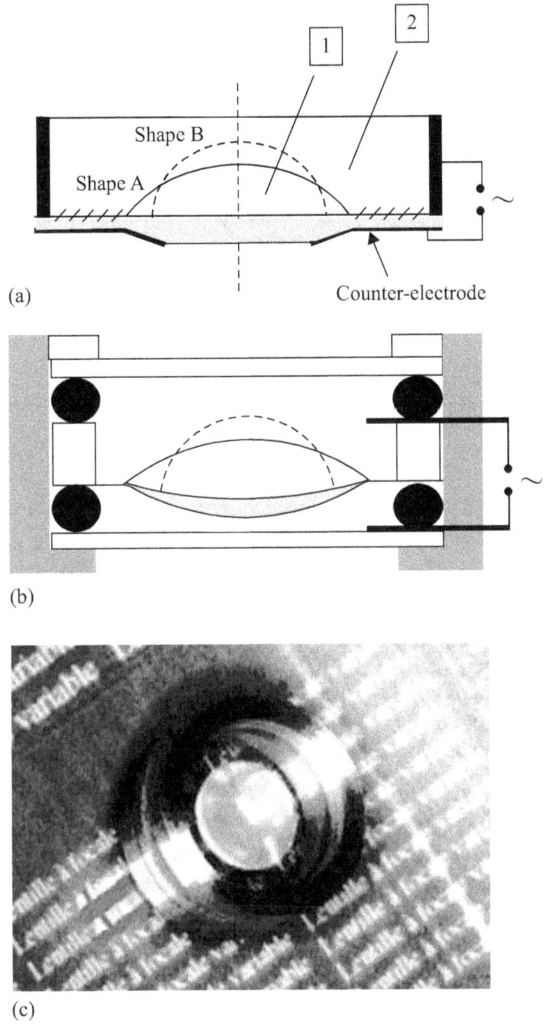

Figure 4.8 *(a) Electrowetting-driven liquid lens. The cell is filled with water. A drop of an insulating nonpolar liquid is placed on the bottom wall made of transparent insulating material. The central disk on the bottom surface is hydrophobic and the outer zone is hydrophilic. Therefore, the drop is trapped at the center. The optical axis is shown as a short long-dashed line. (b) Variable-focus liquid lens. The cell is sealed by o-rings. The central part diameter is 5 mm. (c) Photo of one lens sample. Diameter of the stainless steel outer case is around 12 mm, and the oil drop appears to be white at the center of the case. Reprinted with permission from Berge et al. (2000) [7], copyright European Physical Journal*

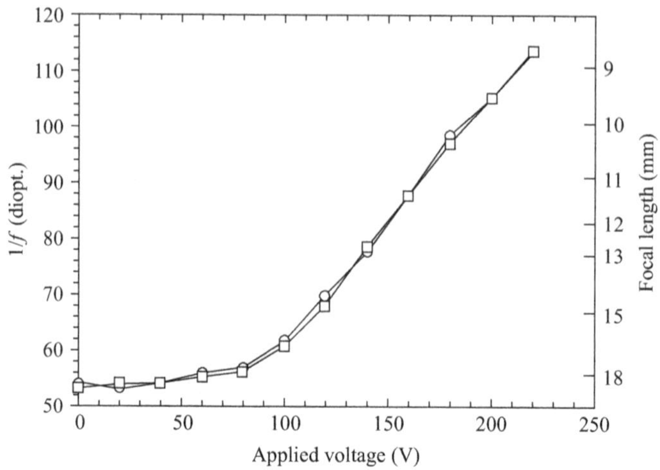

Figure 4.9 Focal length of a variable-focus lens which is 6 mm in diameter. The two curves correspond to increasing and decreasing applied voltages. Reprinted with permission from Berge et al. (2000) [7], copyright European Physical Journal

Instead, the contact angle reached is within a small angle range, and this phenomenon is called contact angle hysteresis [5]. One way to mitigate the hysteresis is to impregnate the substrate surface with an oil as the lubricant [6, 8]. Another way is to separate the droplet from the substrate surface by creating conditions under which a lubricant would penetrate spontaneously under the droplet. Figure 4.10 illustrates one of such lens designs reported by Krupenkin *et al.* [9]. The lubricant and the liquid droplet are two immiscible liquids. A thin layer of the lubricant prevents the droplet from directly interacting with any contaminant on the substrate, and it also increases the homogeneity of the substrate. Therefore, the contact angle hysteresis is almost eliminated. A hole made in the dielectric film allows electrical connection to the droplet without distorting its shape, and the center of the droplet is in direct contact with the ground electrode beneath the dielectric film.

The liquid droplet in Figure 4.10(b) is immersed into a secondary immiscible liquid which serves multiple purposes – (1) it prevents the microlens droplet from evaporating; (2) its chemical composition is adjusted to match its density with that of the lens droplet, and thus the gravity does not affect the lens operation; (3) it also serves as the lubrication liquid. The lubricant would form a thin layer spontaneously if forming two interfaces, primary liquid–lubricating liquid and lubricating liquid–solid, has lower total free energy of the system than that of one solid–primary liquid interface.

The focal length of this microlens is approximately determined by:

$$f^3 = \frac{3\Omega}{\pi(1-\cos\theta)(2-\cos^2\theta-\cos\theta)(n_L-n_V)^3} \tag{4.9}$$

where Ω is droplet volume, θ is the contact angle, n_L and n_V are the refractive indices of the primary lens liquid and the secondary surrounding liquid, respectively. Figure 4.11 shows the theoretical and measured focal lengths of this tunable lens

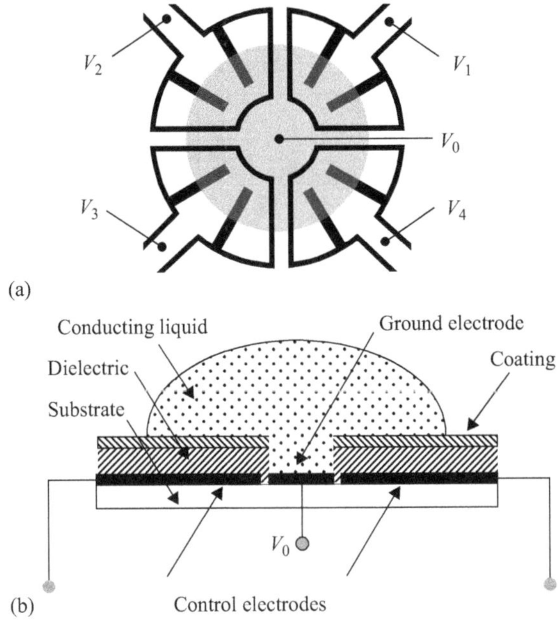

Figure 4.10 Tunable liquid microlens. (a) Electrode design. Black lines indicate etched areas on ITO. Applied voltages are indicated as V_0 through V_4. Lightly shaded area represents approximate droplet position. (b) Device cross-section. Voltage applied to ground electrode is indicated as V_0. Reprinted with permission from Krupenkin et al. (2003) [9], copyright American Institute of Physics

when the voltage is up to ~100 V. For the lens without lubricant, the experimental focal length saturates at higher voltages, unlike the theoretical values which continue to grow with the voltage. At around 80 V, the diameter of the microlens contact spot started to approach the outer diameter of the control electrode, thus preventing further spread of the droplet. Utilizing a similar setup of split electrodes, Yang *et al.* [10] achieved liquid microlens alignment through electrowetting.

Mobile phones with integrated cameras are an important application of electrowetting-based variable-focus microlens. However, this application faces several challenges, including evaporation of lens droplet, sustaining mechanical shock and a wide range of operating temperatures. Kuiper and Hendriks demonstrated an electrowetting liquid lens consisting of two immiscible liquids, and it is intended for miniature systems [11]. In contrast to previous designs, this lens wets the sidewall, allowing the formation of a plano-concave lens. Figure 4.12(a,b) illustrates the cross-sections of the variable lens without and with voltage, respectively. Figure 4.12(c,d) are pictures of the liquid interface in the lens without and with voltage, corresponding to Figure 4.12(a,b), respectively. This electrowetting liquid lens is used to design and construct a camera module for use in a mobile phone. Figure 4.13 illustrates the cross-section of the assembled camera module and a picture of it. The liquid lens is integrated with a commercial VGA CMOS sensor

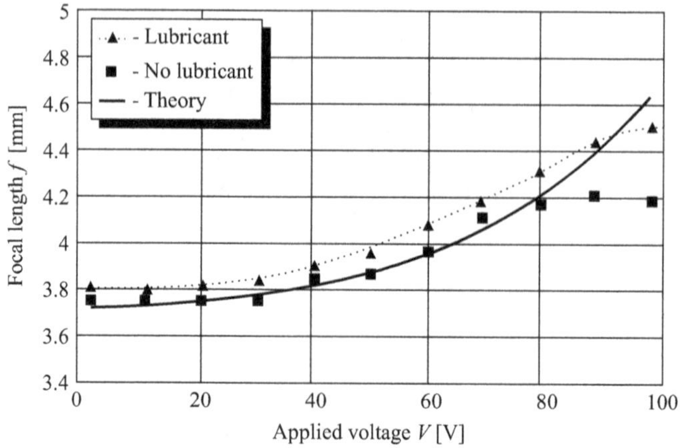

Figure 4.11 *Microlens focal length versus applied voltage. Filled squares indicate data obtained without lubricating liquid. Filled triangles indicate data obtained with lubricating liquid. Dashed line is a guide for the eyes. Solid curve corresponds to theoretical prediction. Reprinted with permission from Krupenkin et al. (2003) [9], copyright American Institute of Physics*

(Philips OM6802, 640 × 480 pixels). The focal length can be adjusted between 2.85 and 3.55 mm. The lens has an f number of 2.5, a field view of 60° and an entrance pupil of 1.43 mm. The liquid lens cell is set between two plastic injection molded lenses. The achromatized lens stack has a high optical quality, and the liquid lens is able to focus faster than the refresh rate of the CMOS sensor.

Despite the advantages of variable-focus electrowetting liquid lenses, a liquid lens system must be carefully designed to achieve its performance, and a cheap, reliable, and reproducible process is needed to make this technology mass-producible. First of all, liquid lens positioning is very important. The liquid lens has to be permanently centered in order to prevent the liquid from moving out of the optical axis under electrowetting actuation or mechanical shock. Three common ways to center a liquid lens are – (1) placing the liquid droplet in a recessed structure like holes or grooves; (2) variation of the thickness of the dielectric layer. The thicker the dielectric layer, the lower the electric field and the larger the contact angle. If the outside of the lens working area has thicker dielectric layer, the droplet is forced to stay in the center area when the voltage is applied. However, this mechanism does not work without a voltage applied; (3) variation of the wettability of the substrate. For a water-based liquid droplet, a hydrophilic central area and a hydrophobic outer area can force the droplet to stay in its initial position.

The materials of the liquid lens are also critical to the imaging quality. Mechanical wave or shock causes surface waves at the liquid-to-liquid interface and deforms its spherical shape, decreasing the imaging quality. Using two density-matched liquids and sealing the system hermetically can address the shock and vibration sensitivity.

When the droplet is made of a conductive water-based liquid with higher refractive index than the surrounding nonconductive liquid, the contact angle can

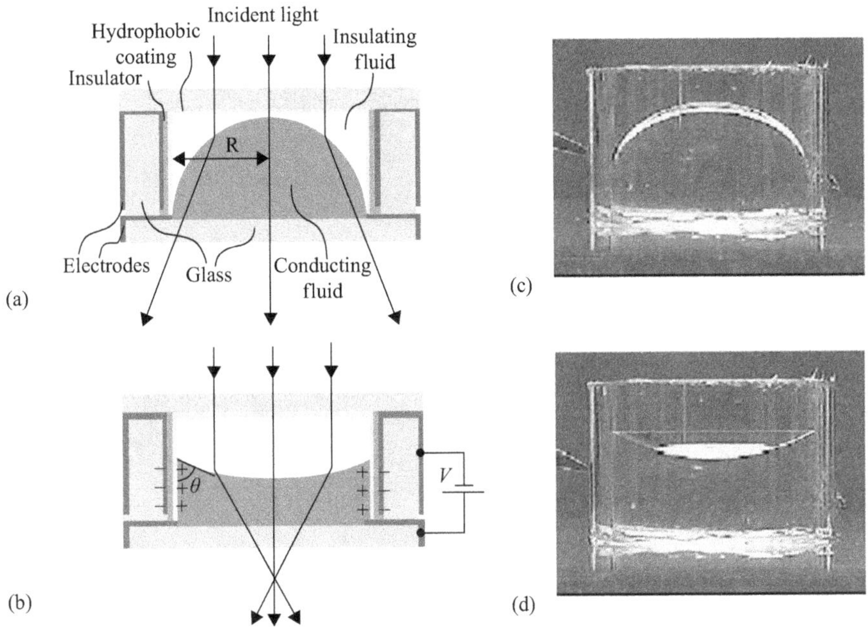

Figure 4.12 *(a) Cross section of electrowetting-based variable lens in cylindrical glass housing. Transparent electrodes were formed of 50-nm ITO. Insulator is a 3-mm parylene-N layer; hydrophobic top coating is 10-nm AF1600 dip-coated fluoropolymer (Dupont Chemicals, Willington, DE). Top and bottom glass plates were glued onto glass cylinder with epoxy. (b) When voltage is applied, charges accumulate in wall electrode, and opposite charges are collected near the solid–liquid interface in conducting liquid. Resulting electrostatic force effectively lowered the solid–liquid interfacial tension and contact angle θ. (c,d) Video frames of a 6-mm-diameter liquid lens at 0 and 120 V, respectively. Reprinted with permission from Kuiper et al. (2004) [11], copyright American Institute of Physics*

only decrease when a voltage is applied. Therefore, the focal length is higher than the initial value when the droplet is actuated. The minimal focal length of an electrowetting liquid lens system is limited by its initial contact angle. This problem can be addressed by using two liquids with high difference in refractive indices. A high refractive index difference results in a low initial back focal length, which can then be increased by applying a voltage, therefore extending the system focal range.

Using silicon as a base material for electrowetting lens systems can benefit system properties such as actuation voltage, dielectric breakdown voltage, and surface roughness and also facilitate manufacturing process of the system. Krogmann *et al.* reported an electrowetting liquid microlens system with an aperture between 300 and 800 μm [12]. The fabrication of this microlens system is based on a highly parallel and reproducible process using standard silicon MEMS processes. Figure 4.14 shows the schematic of the liquid microlens. The center of the lens

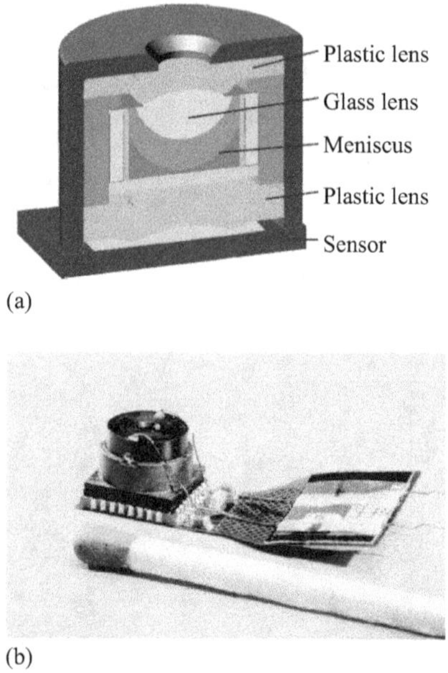

Figure 4.13 *(a) Optical design of a camera module containing liquid lens. (b) Picture of an assembled camera module using an electrowetting liquid lens. Reprinted with permission from Kuiper et al. (2004) [11], copyright American Institute of Physics*

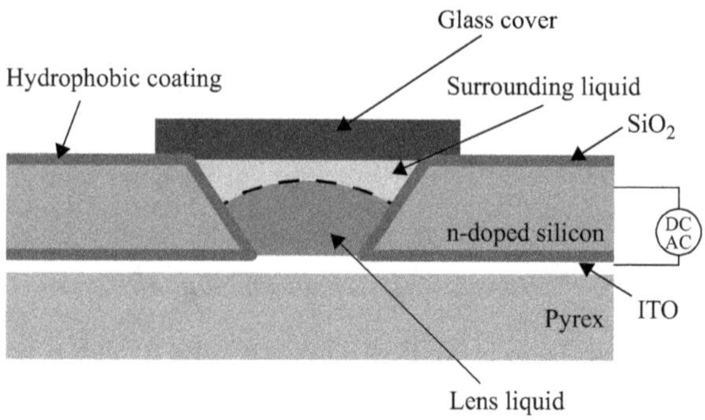

Figure 4.14 *Cross-section view of an electrowetting liquid lens system fabricated by MEMS process. A KOH-etched groove inside a silicon wafer contains the lens liquid and the surrounding liquid. The system is sealed by a glass cover. Reprinted with permission from Krogmann et al. (2006) [12], copyright IOP Publishing*

structure is a quadratic V-groove-defined hole formed inside a standard silicon wafer. All surfaces are coated with silicon dioxide and hydrophobic coating as an insulating layer. The bottom of the lens structure is bonded to an indium tin oxide (ITO)-coated Pyrex wafer for bringing the droplet and an electrode into contact. The middle of the V-groove is filled with lens liquid, which is then embedded in the surrounding liquid. As the final step, the system is sealed by a glass cover. The aperture of the lens, which is determined by the size of the bottom of the V-groove, varies between 300 and 800 μm. The thickness of the system is 1.525 mm, including 525-μm thick silicon wafers and 500-μm thick Pyrex wafers and the top glass. Figure 4.15 illustrates the fabrication process of the system,

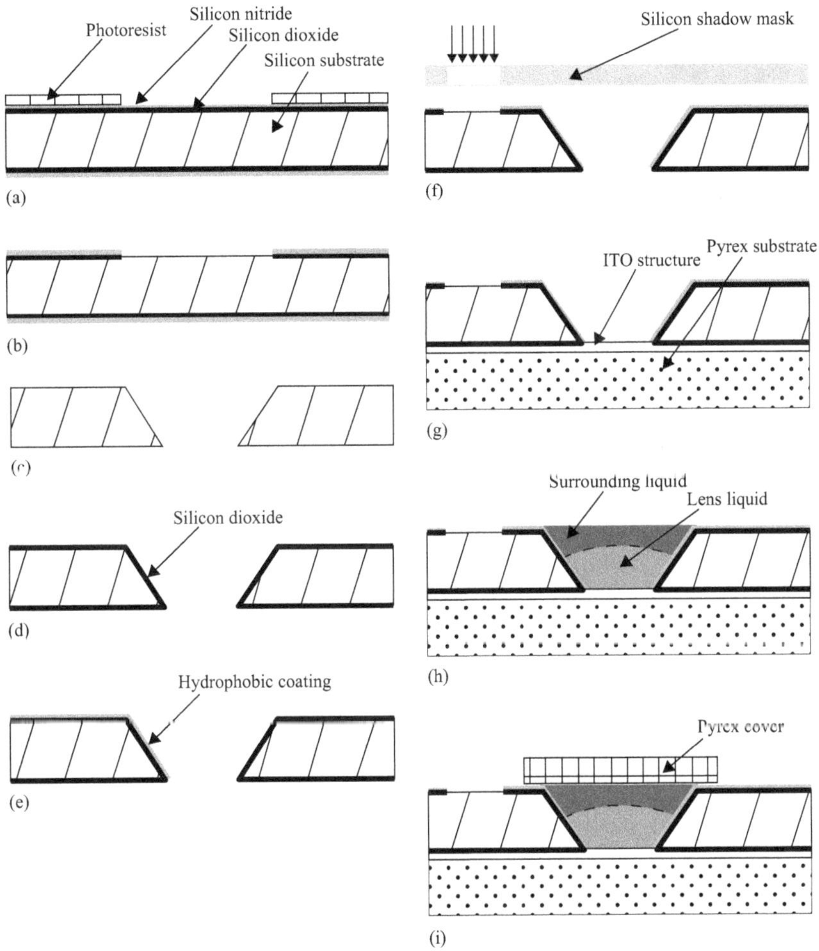

Figure 4.15 *(a) Deposition of masking layers (silicon dioxide and silicon nitride); (b) structuring of the silicon nitride and silicon dioxide layer; (c) KOH etching and removal of the masking layer; (d) deposition of a thermally grown silicon dioxide layer; (e) deposition of a hydrophobic layer by using octafluorocyclobutane; (f) opening of the silicon dioxide layer by use of a shadow mask; (g) bonding an ITO-structured Pyrex substrate to the bottom of the substrate; (h) filling the device with surrounding and lens liquid; (i) closing the system with a top glass cover. Reprinted with permission from Krogmann et al. (2006) [12], copyright IOP Publishing*

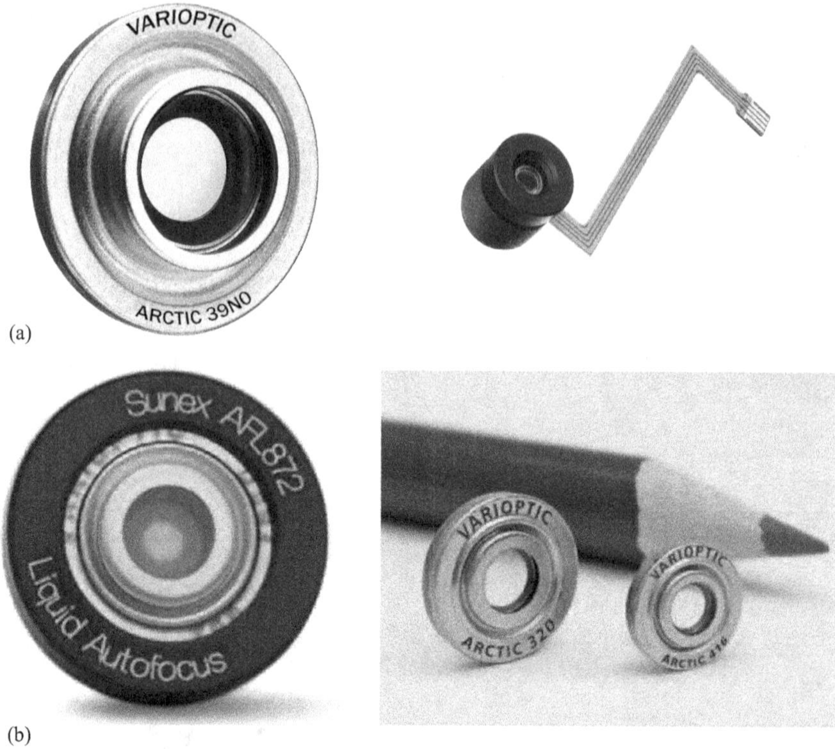

Figure 4.16 *(a) A commercial mobile phone camera module, Varioptic Arctic 39N0. This camera uses an electrowetting-based liquid lens, and it has 8 mm in diameter and 2 mm of thickness. (b) Varioptic Arctic (b) AFL872 autofocus module. It uses an Arctic 320 electrowetting lens and a Sunex DSL 872 fixed optical lens (Courtesy of Varioptic)*

which is based on MEMS processes for achieving high reproducibility and production accuracy.

Figure 4.16(a) shows Arctic 39N0 by Varioptic Co., a commercial electrowetting-based liquid lens designed for imaging and laser applications. This electrowetting-based liquid lens allows a very simple structure of miniature cameras with no moving parts. It is a key advantage for the durability of cameras in mobile applications which are often subject to mechanical shock. Another advantage of liquid lenses is their high optical quality although the solid supporting parts are processed by conventional machining techniques. For a liquid lens, its surface has a self-healing property. Due to surface tension, a liquid surface remains smooth on a nanometer scale (10^{-9} m) even if the supporting solid surface is polished on a micrometer scale (10^{-6} m). Another desired property of these lenses is that when they are perturbed (e.g. mechanical agitation), the perturbation does not propagate easily from the lens outside the perimeter toward the central region. As a result, liquid lenses have a highly precise optical surface, equivalent to that obtained in glass lenses after elaborate and delicate polishing work. Arctic 39N0 has a clear aperture of 3.9 mm at 0° field of view and its low and high optical power is −5 and 15 diopters, respectively. The reaction time is <20 ms, making it suitable for the applications of specific features recognition, such as scanning barcodes or biometric characteristics (e.g. iris or finger prints).

The autofocus ability of electrowetting liquid lens provides new economic solutions for monitoring system. Figure 4.16(b) shows a picture of AFL872, an autofocus module consisting of an Arctic 320 liquid lens and of a Sunex DSL 872 fixed optical lens. AFL872 is a complete autofocus module with M8 × 0.35 thread connection. The trend of system miniaturization also creates new opportunities for liquid lenses in the area of imaging. For instance, electrowetting lenses are suitable for installation inside endoscope [13] and confocal microscope [14].

4.7 Electrowetting-based microlens on a flexible curvilinear surface

The electrowetting-based lenses are normally placed on flat and mostly rigid substrates. Li *et al.* extended the technology to flexible polymer surfaces [15]. Figure 4.17(a) shows an electrowetting-based tunable liquid lens made of flexible materials. The lens chamber is bonded to a thin polymer substrate. Both the chamber and the substrate are made of PDMS, a flexible transparent polymer. The whole structure could then be wrapped conformally onto a curvilinear surface. Figure 4.17(b) shows a flexible electrowetting lens wrapped onto a commercial contact lens. In contrast to flat rigid substrates, soft polymers like PDMS are more likely to be damaged in fabrication processes. To prevent cross-linked PDMS from deformation, a low temperature (30°C) fabrication process is developed to reduce the stress on the PDMS substrate. In addition, the dielectric and the hydrophobic

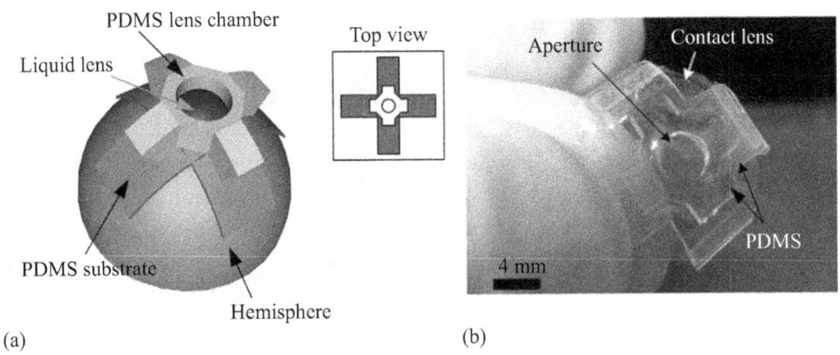

Figure 4.17 (a) Electrowetting-based liquid lens fabricated on curved surface. (b) Integration of a flexible electrowetting lens and a commercial contact lens. Reprinted with permission from Li et al. (2012) [15], copyright American Institute of Physics

coating materials are critical. Instead of silicon nitride (SiN), silicon oxide (SiO_2) film is deposited as the dielectric layer. When SiN was coated on PDMS, the stress on the PDMS surface induced in the cooling step led to the cracking of the SiN film. Teflon, a usual hydrophobic coating, requires a 165°C surface treatment during which PDMS is vitrified. Therefore, trichlorosilane is used as the hydrophobic coating since it requires lower processing temperature.

4.8 Arrayed electrowetting prism and switchable microlens

Beam-steering technology has been a limiting factor for laser radar, agile imaging, 3D displays, and numerous other applications, because current approaches cannot yet provide large-area, thin, transmissive, and wide-angle beam deflection. Although not switchable, prismatic plates like those long used in lighthouse Fresnel lenses provide most of the desired features for a wide-angle beam-steering element.

Smith *et al.* have fabricated arrayed electrowetting prism that has greater than 20° total beam deflection [16]. Electrowetting microprisms operate by modifying the physical geometry between two immiscible liquids. In an electrowetting microprism, low index water ($n \sim 1.3$) and high index oil ($n > 1.6$) are confined by at least two electrowetting sidewalls, as shown in Figure 4.18(a). These sidewalls are connected to two distinct voltages (V_L, V_R) that independently control the water contact angle at each sidewall (θ_L, θ_R). As long as the voltages are selected such that $\theta_L + \theta_R = 180°$, the meniscus is held flat and a variable prism is created.

Using a simplified version of the fabrication technique used to make microprism, Smith *et al.* has demonstrated switchable microlens arrays [17], as shown in Figure 4.18(b). They are controlled by a single electrode and can be inverted from convex to concave. The lens array has myriad applications, such as stereoscopic displays, adaptive optics, laser array collimators, and portable wave-front sensors. Compared to other microlens array technology, electrowetting microlens array is relatively simple to fabricate, and they can invert focal length and achieve nearly hemispherical lens curvature.

4.9 Electrowetting-controlled liquid mirror

An electrowetting liquid lens is often made of two immiscible liquids, and their densities are matched to resist shock, vibration, and distortion by gravity. The difference between the refractive index of the two liquids is often limited, and to control chromatic aberrations, the curvature of the liquid-to-liquid interface cannot be too large. To enhance the optical performance of electrowetting liquid lenses, the liquid–liquid interface can be covered with optically active solid particles. Bucaro *et al.* reported an electrowetting-controlled liquid mirror in which a liquid lens surface is covered with silicon hexagons [18]. Figure 4.19(a) shows an scanning electron microscope (SEM) picture of a silicon hexagon which is 8 μm in diameter and 1 μm in thickness. One side of the silicon hexagon is coated with gold, making it significantly more hydrophobic than the other bare surface side.

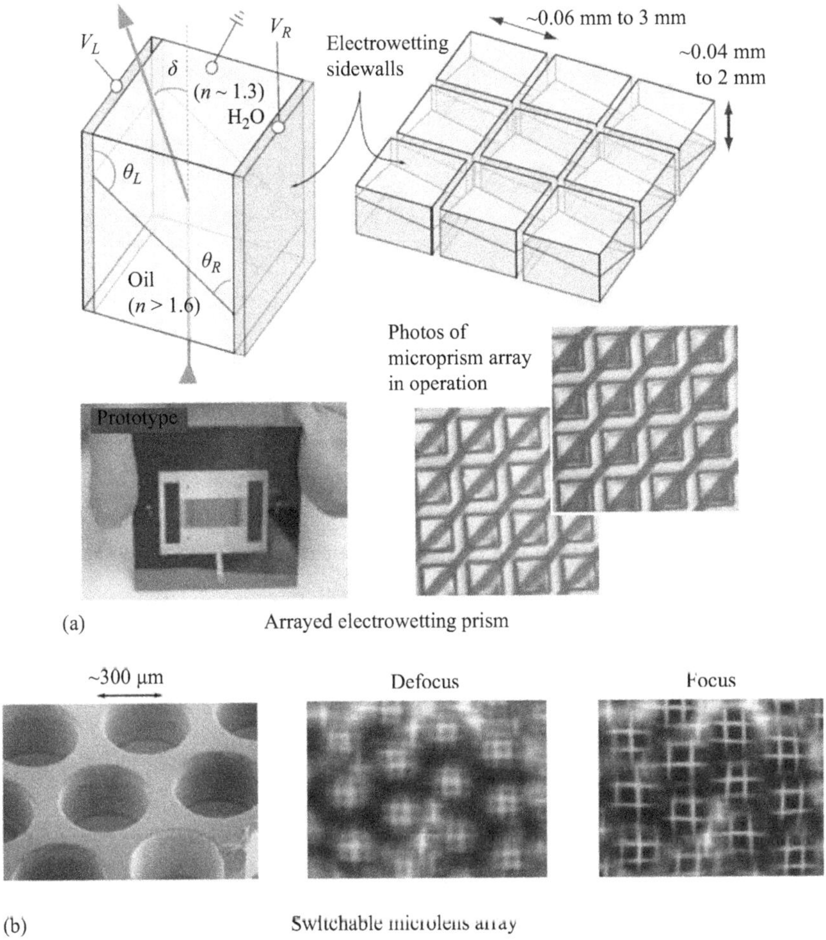

Figure 4.18 *(a) Arrayed electrowetting prism. Low index water (n ~ 1.3) and high index oil (n > 1.6) are confined by at least two electrowetting sidewalls. Reprinted from Reference 16 with permission.*
(b) Switchable electrowetting microlens array. Pictures of the lens array structure and images formed by the lens array when the object is in defocus or in focus. Reprinted with permission from Smith et al. (2009) [17], copyright the Optical Society

When a series of silicon hexagons are released onto an oil–water interface, they align parallel to the interface, and the sides coated with gold are in contact with the oil, forming a concave gold mirror, as shown in Figure 4.19(b). Figure 4.20(a) illustrates the cross-section of the whole device. The top of the oil droplet is in contact with a transparent electrode. The water–oil interface is controlled by applying voltages between the top electrode and the water, like a two-liquid electrowetting lens. Therefore, the shape of the water–oil interface and the focal length

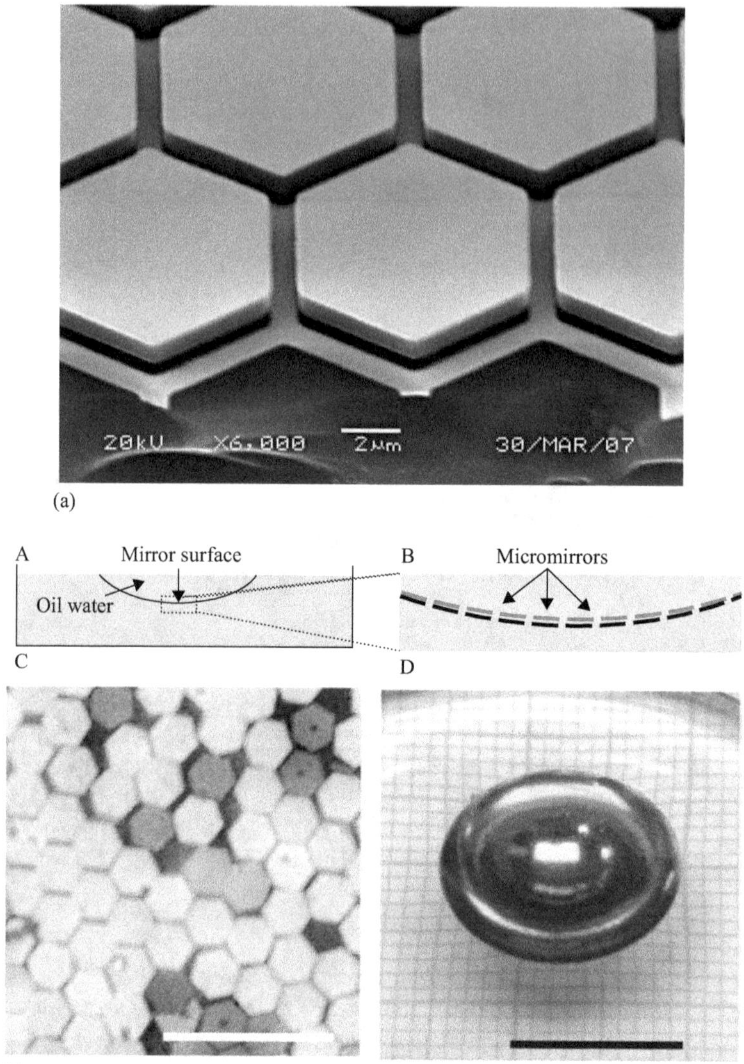

Figure 4.19 *(a) SEM picture of a silicon hexagon which has 8 µm in diameter and 1 µm in thickness. (b) Schematic and picture of the electrowetting-controlled liquid mirror. Gold-coated silicon hexagons form a mirror at the liquid interface which is controlled by electrowetting mechanism. Reprinted with permission from Bucaro et al. (2008) [18], copyright American Chemical Society*

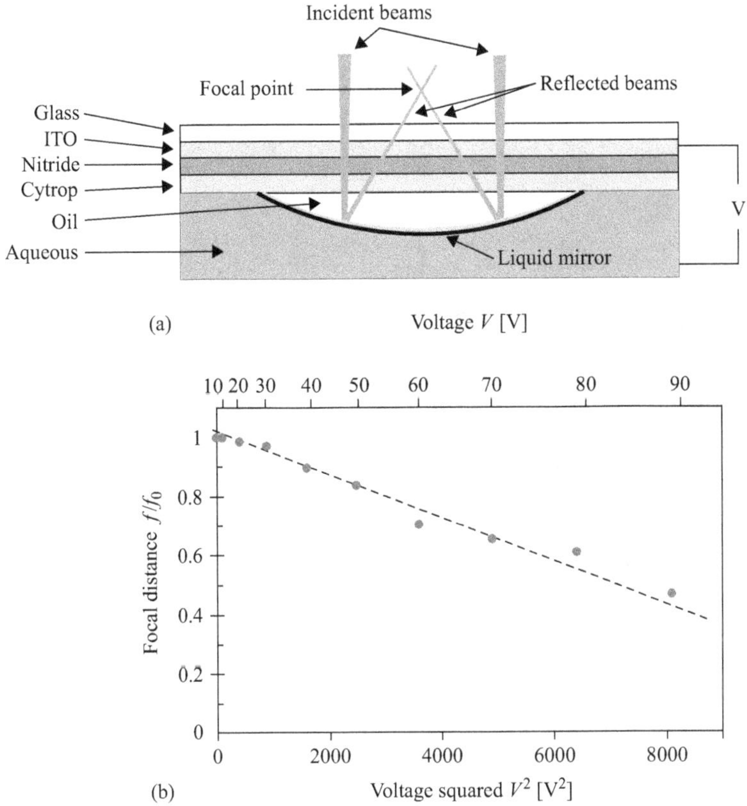

Figure 4.20 *(a) Cross-section of the electrowetting-controlled mirror device. (b) Focal length of the mirror versus the voltage applied. Reprinted with permission from Bucaro et al. (2008) [18], copyright American Chemical Society*

of the mirror change with the voltage applied. Figure 4.20(b) shows the measured focal length at different voltages applied. f_0 is approximately 8.3 mm. The focal length of the mirror is between 8.3 and 4.1 mm when the voltage applied changes between 0 and 90 V.

4.10 Electrowetting-driven optical switch and aperture

Optical switches and tunable optical apertures are traditionally controlled by the use of mechanical parts. In potential applications such as microscopy, display, and lab-on-chip, small-scale optical switches are needed. However, operating small-scale mechanical parts must deal with challenges like inaccurate control, increased friction, and long-term wear. Murade *et al.* demonstrated an electrowetting-based optical switch with tunable aperture [19]. Figure 4.21(a,b) illustrates the schematic

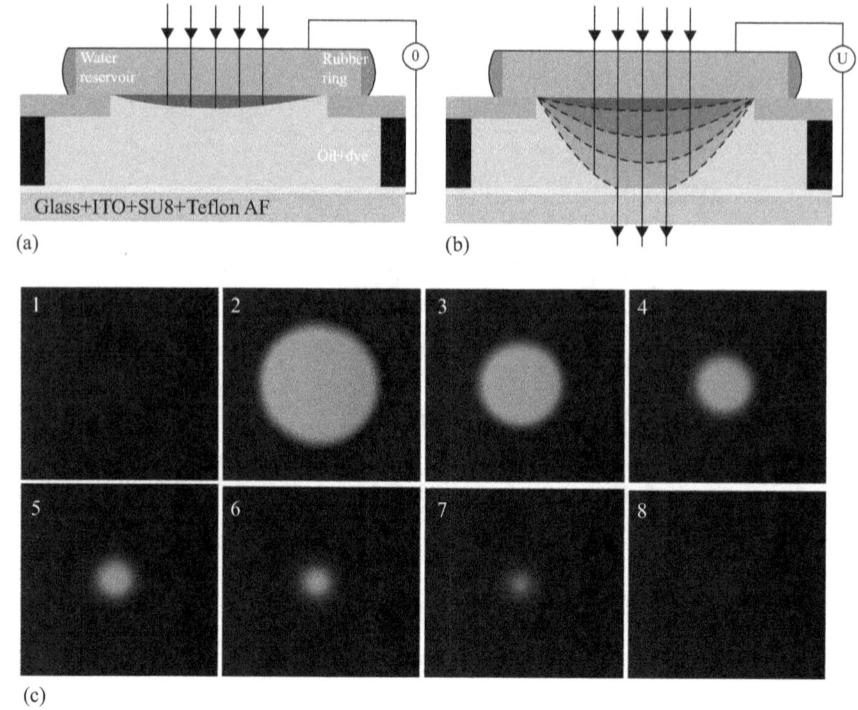

Figure 4.21 *(a) Electrowetting optical switch at off-state. (b) Electrowetting optical switch at on-state. (c) Images recorded by a CCD camera when the optical switch is tuned between off-state and on-state. Images 1 and 8 show the off-state; Image 2 shows the on-state. All the other images show the intermediate states. Reprinted with permission from Murade et al. (2011) [19], copyright the Optical Society*

of the cross-section of their electrowetting switch, which consists of two glass substrates and a nontransparent oil in between. The top and bottom substrates are separated by a glass spacer (60 μm), and they are both coated with a thin Teflon AF film (~20 nm) to make the surface more hydrophobic. The top glass substrate has a hole of 1.2 mm diameter, and on top of it there is a water reservoir confined by a rubber ring. The bottom glass substrate is coated with a transparent ITO layer to make it conductive. The oil is mixed with dye so that light cannot pass through the device. When a voltage is applied between the water phase and the ITO layer, an electric field is generated within the oil and the electric forces push the water–oil interface downward. As the voltage increases, the interface bends further downward until it snaps down to the bottom substrate when the voltage goes above a threshold value. This threshold is determined by the balance of Maxwell stress, surface tension, and viscous stresses. Figure 4.21(c) shows images recorded by a CCD camera when the optical switch is in "off"-state, "on"-state, and

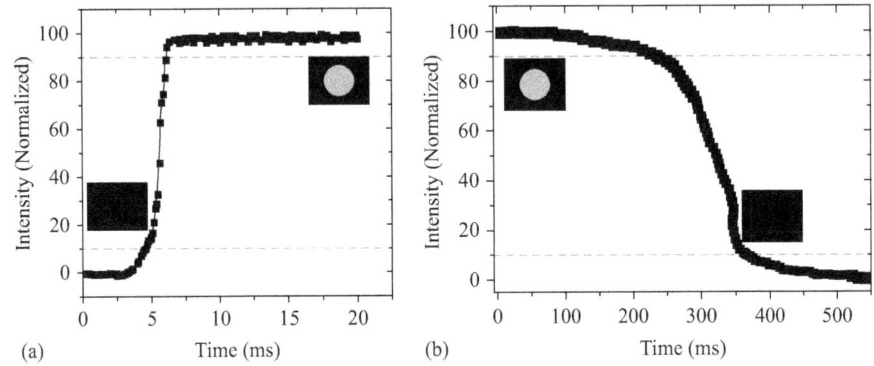

Figure 4.22 Response time of the electrowetting optical switch. (a) off-state to on-state. (b) on-state to off-state. Reprinted with permission from Murade et al. (2011) [19], copyright the Optical Society

intermediate states. This optical switch can attain 100% intensity attenuation, and the diameter of its aperture can be tuned between 0.2 and 1.2 mm. Response time is another key parameter defining the performance of an optical switch. This electrowetting optical switch has a response time of ~2 and ~120 ms for switching on and off, respectively, as shown in Figure 4.22.

4.11 Electrowetting display

In electrowetting displays, a simple optical switch is obtained by contracting a colored oil film electrically. This electrowetting switch has many attractive properties, such as color brightness, video speed, and low power consumption – all of which make it a viable technology as a display. Figure 4.23 illustrates the mechanism of a reflective electrowetting display [20]. In each pixel, the optical function module consists of a transparent electrode, a hydrophobic insulator, a colored oil layer, and water. Without any voltage, the colored oil forms a film between the water and the hydrophobic insulator at equilibrium, because the system is at the lowest energy state. As a result, the pixel is at a colored off-state. In a typical display, pixel size is at most 200 μm, and surface tension is much larger than the gravitational force. Therefore, gravitational force can be ignored, and the oil film is continuous and stable in any orientation.

When a voltage is applied between the water and the electrode, an electric field is formed between them and an electrostatic term is added to the energy balance. As a result, the original state is no longer the lowest energy state, and the system energy can be lowered by moving the water into contact with the insulator and reshaping the oil layer. The oil is moved to the side, and the distance it travels is determined by the equilibrium state at which surface tension balances electrostatic force. If the pixel is small enough, human eyes can only perceive the average optical intensity through the pixel. Therefore, the pixel can be tuned between the

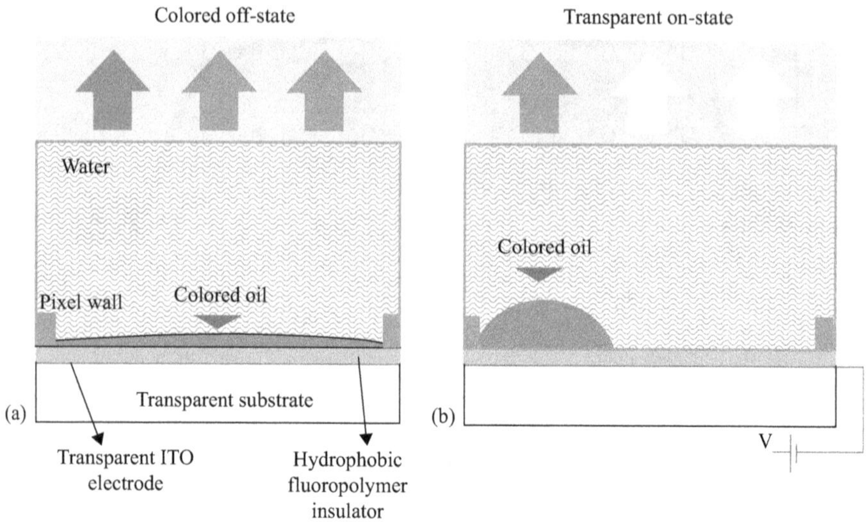

Figure 4.23 Mechanism of electrowetting display. (a) Without voltage, the substrate surface is hydrophobic, and colored oil forms a film on it. The pixel is at colored off-state. (b) With a voltage applied between water and the ITO electrode, the substrate becomes relatively hydrophilic, and the oil is contracted to the side wall. The pixel is at transparent on-state. Reprinted with permission from Hayes et al. (2003) [20], copyright Nature Publishing Group

colored off-state and a transparent state continuously by changing the voltage applied. The white reflector is integrated into each pixel by coating a white polymer foil (substrate) with a thin patterned ITO electrode and a fluoropolymer insulator.

Figure 4.24 shows the electro-optic behavior of an electrowetting display pixel. The reflectivity and contrast are a function of the DC voltage applied. The oil film starts to be displaced when the applied voltage goes below −5 V. At around −20 V, the oil has contracted dramatically, and the reflectivity is near 35%. When the voltage further increases, the reflectivity does not change much given the oil is already at its maximum contraction state. The maximum contrast of the electrowetting display is around 15.

The thickness of the oil film and the thickness of the fluoropolymer are two critical parameters for the performance of electrowetting display. The thicker the oil film, the stronger electric field needed to reshape the oil film. Therefore, the oil film should be thin enough so that the device can be operated at low voltages. On the other hand, it cannot be too thin. Otherwise, the off-state of the device is not satisfactory. The oil film is typically around 10 μm. At the on-state, the strength of the electric field is determined by the thickness of the insulator, which should be as thin as possible but thick enough to avoid dielectric breakdown.

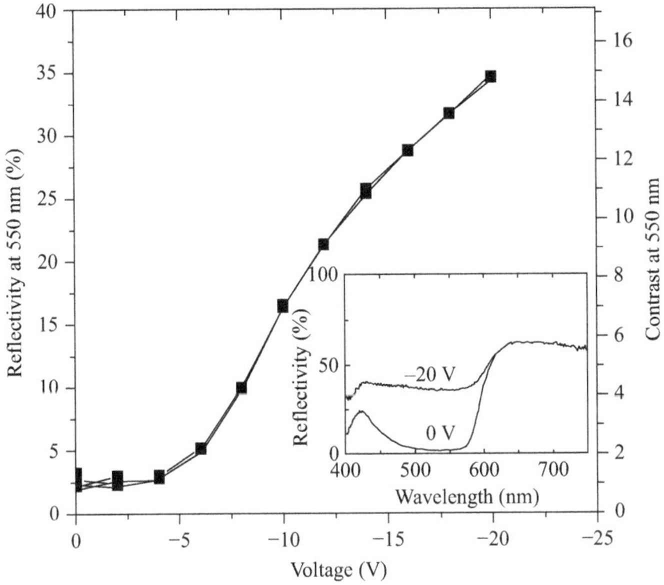

Figure 4.24 Reflectivity and contrast as a function of voltage for a pixel in electrowetting display. Reprinted with permission from Hayes et al. (2003) [20], copyright Nature Publishing Group

References

[1] F. Mugele and J.-C. Baret, "Electrowetting: from basics to applications," *Journal of Physics: Condensed Matter*, vol. 17, p. R705, 2005.

[2] X. Zeng and H. Jiang, "Polydimethylsiloxane microlens arrays fabrication through liquid-phase photopolymerization and molding," *Journal of Microelectromechanical Systems*, vol. 17, no. 5, pp. 1210–1217, 2008.

[3] X. Zeng, C. Li, D. Zhu, H.J. Cho, and H. Jiang, "Tunable microlens arrays actuated by various thermo-responsive hydrogel structures," *Journal of Micromechanics and Microengineering*, vol. 20, no. 11, 115035, 2010.

[4] M. Vallet, M. Vallade, and B. Berge, "Limiting phenomena for the spreading of water on polymer films by electrowetting," *The European Physical Journal B*, vol. 11, pp. 583–591, 1999.

[5] H.J.J. Verheijen and M.W.J. Prins, "Reversible electrowetting and trapping of charge: model and experiments," *Langmuir*, vol. 15, pp. 6616–6620, 1999.

[6] C.B. Gorman, H.A. Biebuyck, and G.M. Whitesides, "Control of the shape of liquid lenses on a modified gold surface using an applied electrical potential across a self-assembled monolayer," *Langmuir*, vol. 11, pp. 2242–2246, 1995.

[7] B. Berge and J. Peseux, "Variable focal lens controlled by an external voltage: an application of electrowetting," *The European Physical Journal B*, vol. 3, pp. 159–163, 2000.
[8] P.G. de Gennes, "Wetting: statics and dynamics," *Reviews of Modern Physics*, vol. 57, no. 3, pp. 827–863, 1985.
[9] T. Krupenkin, S. Yang, and P. Mach, "Tunable liquid microlens," *Applied Physics Letters*, vol. 82, pp. 316–318, 2003.
[10] S. Yang, T. Krupenkin, P. Mach, and E. A. Chandross, "Tunable and latchable liquid microlens with photopolymerizable components," *Advanced Materials*, vol. 15, pp. 940–943, 2003.
[11] S. Kuiper and B.H.W. Hendriks, "Variable-focus liquid lens for miniature cameras," *Applied Physics Letters*, vol. 85, pp. 1128–1130, 2004.
[12] F. Krogmann, W. Monch, and H. Zappe, "A MEMS-based variable microlens system," *Journal of Optics A: Pure and Applied Optics*, vol. 8, no. 7, pp. 330–336, 2006.
[13] X. Zeng, C.T. Smith, J.C. Gould, C.P. Heise, and H. Jiang, "Fiber endoscopes utilizing liquid tunable-focus microlenses actuated through infrared light," *IEEE/ASME Journal of Microelectromechanical Systems*, vol. 20, no. 3, pp. 583–593, June 2011.
[14] B.N. Ozbay, J.T. Losacco, R. Cormack, *et al.*, "Miniaturized fiber-coupled confocal fluorescence microscope with an electrowetting variable focus lens using no moving parts," *Optics Letters*, vol. 40, no. 11, pp. 2553–2556, 2015.
[15] C. Li and H. Jiang, "Electrowetting-driven variable-focus microlens on flexible surfaces," *Applied Physics Letters*, vol. 100, p. 251114, 2012.
[16] N. Smith, D. Abeysinghe, J. Haus, and J. Heikenfeld, "Agile wide-angle beam steering with electrowetting microprism," *Optical Express*, vol. 14, pp. 6557–6563, 2006.
[17] N. Smith, L. Hou, J. Zhang, and J. Heikenfeld, "Fabrication and demonstration of electrowetting liquid lens arrays," *Journal of Display Technology*, vol. 5, pp. 411–413, 2009.
[18] M.A. Bucaro, P.R. Kolodner, J.A. Taylor, A. Sidorenko, J. Aizenberg, and T. N. Krupenkin, "Tunable liquid optics: electrowetting-controlled liquid mirrors based on self-assembled Janus tiles," *Langmuir*, vol. 25, no. 6, pp. 3876–3879, 2008.
[19] C.U. Murade, J.M. Oh, D. van den Ende, and F. Mugele, "Electrowetting driven optical switch and tunable aperture," *Optics Express*, vol. 19, no. 16, pp. 15525–15531, 2011.
[20] R.A. Hayes and B.J. Feenstra, "Video-speed electronic paper based on electrowetting," *Nature*, vol. 425, no. 6956, pp. 383–385, 2003.

Chapter 5
Microcameras

Bader Aldalali[a] *and Hongrui Jiang*[b]

5.1 Introduction

There has been tremendous effort in miniaturizing cameras over the course of the last decade to meet specific application requirements a lot of which is in the medical field including but not limited to endoscopic and laparoscopic applications. These miniaturization efforts always took into account that certain camera metrics cannot be degraded during miniaturization including minimal aberrations of camera optics, high field of view (FOV), tunable focus, and high image quality in terms of detection and resolution. These challenges of maintaining the camera metrics while further miniaturizing the camera have produced a variety of novel and innovative microcameras. This chapter will discuss a number of these latest research level microcameras and their novelty in terms of the optics, the image detection, the image processing, or the camera system as a whole. This chapter includes works from research groups from all around the world.

In terms of aberrations, the conventional method of correcting aberrations used in standard cameras is by using more than one lens to correct for the aberration. This method cannot be implemented at the microscale because it increases the size of the camera as well as adding unreasonable complexity to the camera fabrication at the microscale. Fortunately, aberrations are lower for a smaller lens than that of a larger lens with the same *f*-number (focal length divided by aperture) and wavelength (1). Having said that, aberrations can be further miniaturized by using novel lens designs including a reflective-based lens. Aberrations can also be minimized not by adding more lenses but by curving the image detector to resemble that of a mammalian eye. Both of these aberration-limiting mechanisms are discussed in this chapter.

For the FOV, nature has inspired many of the works in this chapter to realize cameras with approximately hemispherical fields of view. Specifically, the compound eye of insects has been the common source of inspiration for all the high FOV microcameras in this chapter. Cameras that are capable of varying the FOV

[a]Kuwait University, Kuwait City, Kuwait
[b]University of Wisconsin, Madison, Madison, WI, USA

are also included. Hybrid cameras are also included that utilize the advantages of both the compound eye of insects and the mammalian eyes.

Microcameras are not limited to fixed focal length lenses. Cameras with tunable-focus lenses are also included that provide zoom capability using innovative lens fabrication techniques. In addition to capturing conventional images, microcameras are also included in this chapter that can provide multiple viewpoint imaging, refocusing, polarization-dependent imaging, image stitching, and stereoscopic display. Finally, application specific microcameras are included in the chapter to showcase how these microcameras can be applied in the medical field. Specifically, microcameras tailored to enhancing either endoscopic or laparoscopic procedures are included.

5.2 Microlens

Many novel microlens fabrication processes have been realized due to the limitation of the conventional lens fabrication process of grinding and polishing. Among these fabrication processes are mature technologies including photoresist thermal reflow, ink-jet printing, and grayscale lithography (2–4). These fabrication processes are typically used for fixed focal length microlenses. To tune the microlens focal length at the microscale, the typical method of mechanical translation of the lens cannot be used due to the difficulty of implementing mechanical translation at the microscale. Novel non-mechanical fabrication methods have been realized to provide tunable microlenses including electrowetting, electrophoretic lenses (5), solid tunable lenses (6), or the use of "smart" materials such as hydrogels (7). For electrowetting microlenses, refer the respective chapter for more details. For electrophoretic lenses and many other microlens fabrication techniques that are not included in this chapter, refer to the book: *Microlenses: Properties, Fabrication, and Liquid Lenses* by Hongrui Jiang and Xeufeng Zeng. As for hydrogels, their smartness comes from the fact that hydrogels are stimuli responsive water-based polymers that shrink and expand by absorbing or releasing water based on the presence of certain stimuli (8). Throughout this chapter, we will come across a number of microcameras that utilize hydrogel-based microlenses that are actuated by different stimuli but first we will see how hydrogels can be used to produce different focal length microlenses with no stimulus present.

5.2.1 Hydrogel microlenses

Aldalali *et al.* from the University of Wisconsin–Madison realized hydrogel-based microlens arrays utilizing the natural shrinkage of hydrogels in air (9). Hydrogels tend to naturally shrink in air due to the high concentration of water within the hydrogel that evaporates over time. The shrinkage of the hydrogel was carefully controlled and conformed such that the final shape of the shrunken hydrogel resembled that of a concave microlens. Figure 5.1 shows the fabrication process of the hydrogel-based microlens array. To initiate the fabrication process, a technique known as liquid phase photopolymerization is utilized using isobornyl acrylate (IBA)

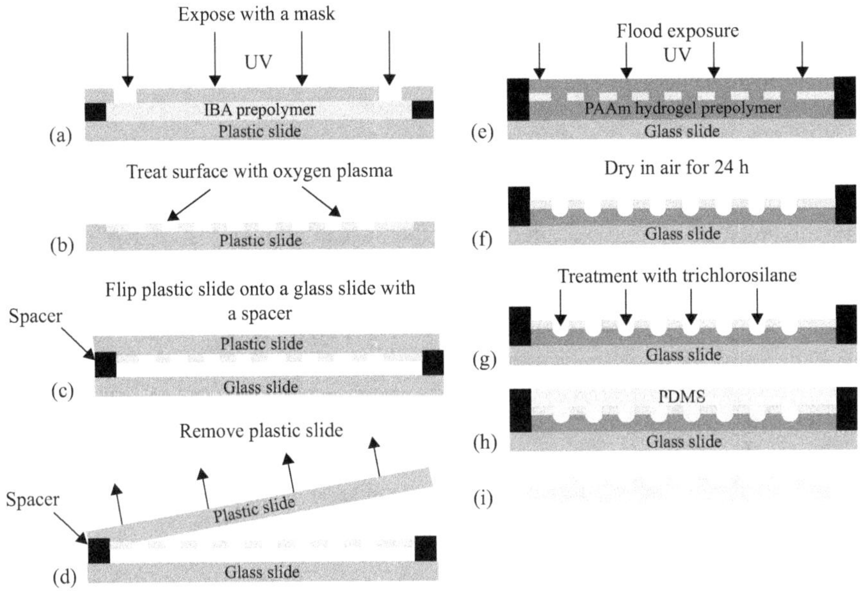

Figure 5.1 Fabrication process of the hydrogel based microlenses. (a) Isobornyl acrylate (IBA) prepolymer was pipetted into the plastic chamber through one of the cartridge holes until it filled the chamber. The IBA was photopatterened using ultraviolet light with an exposure of 10 mW/cm^2 for 25 seconds. (b) Top plastic surface is peeled off revealing the polymerized IBA aperture array. The remaining unpolymerized IBA was rinsed off with ethanol. The exposed IBA aperture array was treated with oxygen plasma. (c) An adhesive spacer was placed on top of a glass slide and the IBA aperture array was flipped on top of the spacer. (d) Photopolymerizable polyacrylamide (PAAm)-based pre-hydrogel is pipetted into the chamber filling the whole aperture array as well as the top surface. (e) The PAAm is then dried in air for one day. After drying, the PAAm will be pinned at the hydrophobic–hydrophilic boundary. (f) The microlens mold is rendered hydrophobic by treatment with octadecyltrichlorosilane. (g) Uncured PDMS is then poured on top of the mold and left to cure. (h) Cured PDMS is then peeled off. (i) The peeled off PDMS microlens array. Reprinted with permission from Aldalali (2013) [9], copyright Aldalali (et al.)

to define an array of circular apertures (10). The top surface of the aperture array and the sidewalls of each aperture are then rendered hydrophilic by oxygen plasma. The aperture array is then flipped and used as a lid of a chamber as shown in Figure 5.1(c). The bottom substrate of the aperture array is then peeled off exposing the bottom surface of the aperture array which is naturally hydrophobic.

This created a hydrophobic–hydrophilic boundary at the sidewalls/surface of the aperture array. The chamber as well as the surface of the aperture array is then filled with a photopolymerizable polyacrylamide (PAAm) as shown in Figure 5.1(e). PAAm is then photopolymerized by ultraviolet exposure which results in a hydrogel. The hydrogel is left to dry for 24 h and shrinks due to the decrease in its water content. However, the shrunken hydrogel will be pinned at the hydrophobic–hydrophilic boundary of the aperture array, thereby defining the curvature of a concave microlens as shown in Figure 5.1(f). The concave-shaped hydrogel aperture array is used as a mold and polydimethylsiloxane (PDMS) is cast on the mold, cured, and later peeled off to define the resulting PDMS convex microlens array as shown in Figure 5.1(i). After peeling off the PDMS, the hydrogel microlens mold can be reused.

The resulting microlenses had a 0.7 mm aperture and the process was also repeated for 1 mm microlenses. The significance of this fabrication process is that a single molding step has been used in comparison to the conventional photolithographic methods that require two molding steps to achieve transparent PDMS microlenses (11). Another benefit of this fabrication process is that the focal length of the microlens array can be tuned by tuning the time the hydrogel is left to dry. Increasing the drying time implies that the hydrogel will shrink further down the surface of the aperture array while still being pinned at the hydrophobic–hydrophilic boundary resulting in a lower radius of curvature and thereby a shorter focal length. The results show that for varying the shrinkage time between 9 h and 24 h the focal length drops from 39.57 mm to 5.63 mm.

This low-cost fabrication process can be utilized for batch processing of microcamera lenses instead of fabricating individual lenses. When compared to the traditional fabrication process such as photoresist thermal reflow, this fabrication process is useful at larger apertures while reflow is still simpler and more reproducible at smaller apertures. The main challenge with this fabrication process would be the requirement of high precision of the amount of hydrogel in order to control the resulting focal length of the lens.

5.2.2 *Tunable microlenses*

As previously mentioned, hydrogels can respond to certain stimuli to absorb or release water. This property can be utilized in tunable microlenses where the hydrogel can be used as the actuator that tunes the lens. Zhu *et al.* realized an array of tunable-focus microlenses on a hemispherical surface that are actuated by a fast thermo-responsive hydrogel (12). The microlens array also employs thermoelectric modules based on the Peltier effect for a more efficient heat transfer process between the heating and cooling down of the hydrogel.

Figure 5.2(a) and (b) are schematics of the tunable lens at short and long focal lengths, respectively. For the fabrication of the microlens array, two IBA molds were fabricated through photolithography and then PDMS was cast and cured on these molds. The first mold, shown as dark grey in Figure 5.2(a), defines the individual chamber that will house the microlens and the connecting bridges between the microlenses. The second mold defines the aperture of the microlenses and is

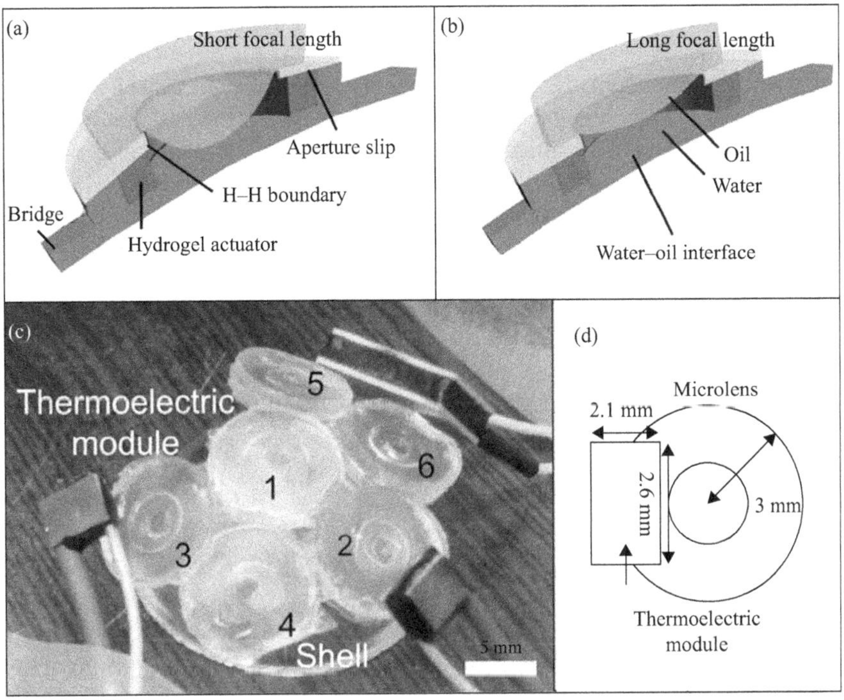

Figure 5.2 *(a) and (b) Models of a tunable microlens with short and long focal lengths, respectively. A PDMS aperture slip on top of a PDMS substrate form a cavity. GO–GMA hydrogel is patterned in the cavity as the actuator. The sidewalls of the aperture are rendered hydrophilic, while the top surfaces are naturally hydrophobic. A liquid-based microlens is formed by the curved water–oil interface which is pinned at the hydrophobic–hydrophilic (H H) boundary. Expanded hydrogel increases the meniscus curve which corresponds to a shorter focal length. Contracted hydrogel reduces meniscus curve which increases focal length. (c) A six-element array of 18 mm microlenses on a shell. The temperature of the microlenses is regulated by the thermoelectric modules placed around the lenses. Each microlens is labeled with a number. (d) Schematic of thermoelectric module position on the microlens. Reprinted with permission from Zhu (2012) [12], copyright IEEE*

shown as light grey in Figure 5.2(a). The hydrogel is poured and photopatterned to define a ring along the inner walls of the individual chambers. The sidewalls of the aperture slip are treated hydrophilic which results in a hydrophilic–hydrophobic boundary between the aperture surface and sidewall. Water is then poured into the chamber followed by oil thereby producing a water–oil meniscus. This meniscus will serve as a lens. As shown in Figure 5.2(a) and (b), as the temperature is increased, the hydrogel ring releases water, thereby shrinking in size. When the hydrogel shrinks, the water retains the volume lost by the hydrogel and the volume of water expands horizontally. This expansion forces the oil–water menisci to have a larger radius of curvature, thereby increasing the focal length of the lens. The microlens array contains six microlenses that are fabricated simultaneously on top of a thin PDMS substrate and are interconnected by thin PDMS bridges. The bridges absorb the stress that is induced when the microlens array is placed on a curved substrate. *N*-isopropyl acrylamide (NIPAAm) hydrogel is used and is embedded with glycidyl methacrylate (GMA) functionalized graphene oxide (GO), also known as a GO–GMA hydrogel. The benefit of a GO–GMA incorporated NIPAAm hydrogel compared with a NIPAAm hydrogel is that the GO–GMA has a much higher thermal response and hence will change volume faster than a plain NIPAAm hydrogel. Another reason the GO–GMA hydrogel-actuated microlenses boast a faster response time is due to the utilization of the thermoelectric modules. Without the presence of the thermoelectric module, heat reaching and dispersing from the hydrogel for actuation would be a natural slow process. To expedite the process, the Peltier effect based thermoelectric modules were placed in the vicinity of the hydrogel posts and increased the response time of both the heating and cooling processes of the hydrogel posts. In the cooling process, the connections of the thermoelectric module are reversed and the thermoelectric module acts as a cooler rather than a heater. After the fabrication process, the microlens array is placed on a hemispherical plastic dome structure. Figure 5.2(c) shows the microlens array along with the thermoelectric modules.

Each microlens in the array can be individually tuned by the use of the thermoelectric module. The microlenses had a diameter of 1.8 mm and can provide a focal length range between 7 mm and 120 mm. The microlenses can go through a forward cycle followed by a reverse cycle in approximately 11 s. The GO–GMA hydrogel-actuated tunable microlens array can image resolutions up to 25.3 line pairs per mm (lp/mm). In addition, the microlenses can also be actuated remotely by infrared (IR) light due to the presence of GO in the GO–GMA hydrogel which has a high IR light absorption. Without applying any heat and only IR light, the focal length of the microlenses can be tuned from 10 mm to 70 mm in 10 s. When the IR light source is turned off, the microlenses can be tuned back to 10 mm in an additional 6 s in the presence of the thermoelectric module as a cooler.

This fabrication process provides a proof of concept of an array of tunable lenses together with heating elements that can be individually tuned. As a concept, these lenses show great potential in terms of possible applications. To be applicable, further work would require the heating elements to be integrated within the lens' fabrication process as well as reducing the time of tuning in both the forward and reverse cycles.

5.2.3 Reflective cylindrical lens

Cylindrical lenses are typically used in beam shaping applications. A cylindrical lens focuses light onto a line rather than a point since curvature is only along one dimension. A cylindrical mirror also focuses light onto a line through reflection; however, the line is located at the same side of the light source rather than on the opposite side, similar to a lens. The benefit of reflection-based focusing rather than the typical refraction-based focusing is that aberrations are reduced since light does not interact within the mirror material and hence there are no absorption losses. Since the light does not interact with the mirror material this means that light of a wide range of spectrum can be used as a source. This is a perfect solution for x-ray imaging since x-rays cannot be refracted but can be reflected. However, the disadvantage of the mirror is that for certain applications, the image is required to be on the opposite side of the light source, such as in a camera. Huang and Jiang developed and realized a novel reflection-based cylindrical lens that accepts a wide spectrum of light and provides an extended FOV (13).

Schematic of the operation of the lens is shown in Figure 5.3(a). The idea is that the lens is not made up of a single optical element rather an array of optical structures. The lens includes an array of mirrored pillars placed on a curved transparent surface. Incoming light will be reflected off of the walls of the pillars but due to the tilt of the pillar, the reflected light from all of the pillars will focus onto a single line as shown in Figure 5.3. Fabrication of the high aspect ratio pillars was realized by lithography followed by deep reactive ion etching of a p-type silicon-on-insulator (SOI) wafer to realize the silicon pillars. Since the sidewalls are required to act as reflectors, the sidewalls were smoothed by a wet etching process and the SOI was then submerged in a buffered oxide etch (BOE) to etch the underlying oxide and allow the structures to be transferred to a flexible

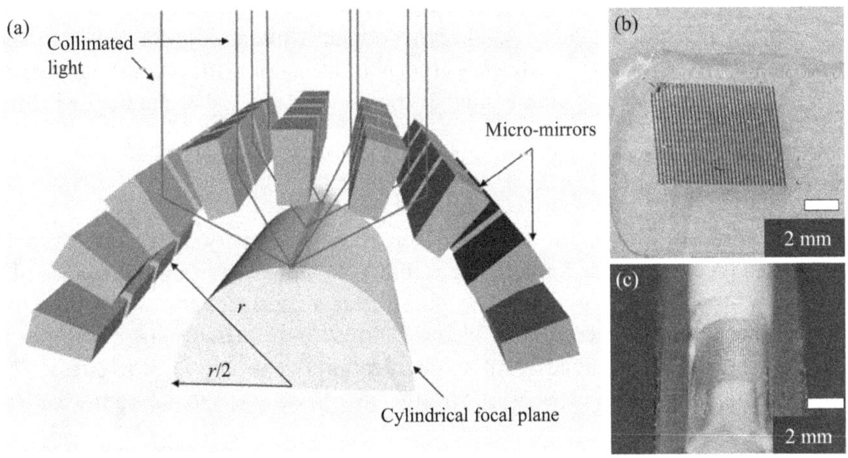

Figure 5.3 (a) Three-dimensional model of the reflective cylinder lens to illustrate the beam shaping and one-dimensional focusing capabilities. (b) High aspect ratio silicon pillars on a thin PDMS substrate. (c) PDMS substrate along with pillar array are curved into a cylindrical layout. Reprinted with permission from Huang (2014) [13], copyright Huang (et al.)

substrate. The BOE step is carefully timed such that the silicon structures are lightly attached to the substrate. Highly reflective aluminum was sputtered onto the silicon pillars to complete the fabrication of the pillars. To provide the wide FOV, a PDMS membrane was placed on the surface of the pillars and with a minor application of stress, the pillars contacted the PDMS membrane. The PDMS membrane along with the silicon pillars is then peeled off. The pillars had a height of 80 μm, length of 120 μm, and a width of 40 μm. The aspect ratio has been optimized to ensure that light is reflected off of one pillar only to contribute to the line focusing. Figure 5.3(a) clearly shows the reflected rays and their contribution to the focus. Figure 5.3(b) and (c) shows the images of the cylindrical lens before and after curving the PDMS substrate.

When comparing the reflective cylindrical lens with a commercial lens, imaging experiments conclude that the reflective cylindrical lens has shown superiority in terms of minimal chromatic aberrations (13). In terms of applications, the reflective cylindrical lens can be useful in scanning applications more than imaging applications due to the one-dimensional focusing nature of the cylindrical lens.

5.3 Electronic eye with curved image detector

Human eyes along with other single aperture eyes consist of a lens, a light adapting pupil, and an image-detecting retina (1). This retina can be considered as a curved photodetector, which provides an increased FOV and low aberrations. The reason behind the low aberrations is that the focal plane is actually a curved plane rather than a flat plane. On the other hand, conventional cameras do not use a curved detector but instead utilize a silicon–based photodetector with a planar geometry due to the planar nature of semiconductor processing and the technological difficulty of processing curved substrates.

5.3.1 Electronic eye camera with fixed focal length

Ko *et al.* from Professor John Roger's group at the University of Illinois, Urbana-Champaign realized a microcamera that mimics the human eye through the use of a curvilinear image detector (14). The curvilinear image detector is realized by combining standard semiconductor planar fabrication processes along with novel elastomeric transfer techniques with no degradation due to any form of stress. The novel transfer technique is what allows the image detector to be conformed to non-planar shapes.

Figure 5.4 shows the fabrication process of the curvilinear electronic eye camera image detector. The initial step involves fabrication of the elastomeric transfer element. The transfer element is realized by molding, using PDMS, the void created by a pair of opposing convex and concave lenses with equal radius of curvature. This results in a PDMS hemispherical mold along with a rim that surrounds the lens as shown in Figure 5.4. Using a specific mechanical setup, the hemispherical mold is then radially stretched from each direction into a planar

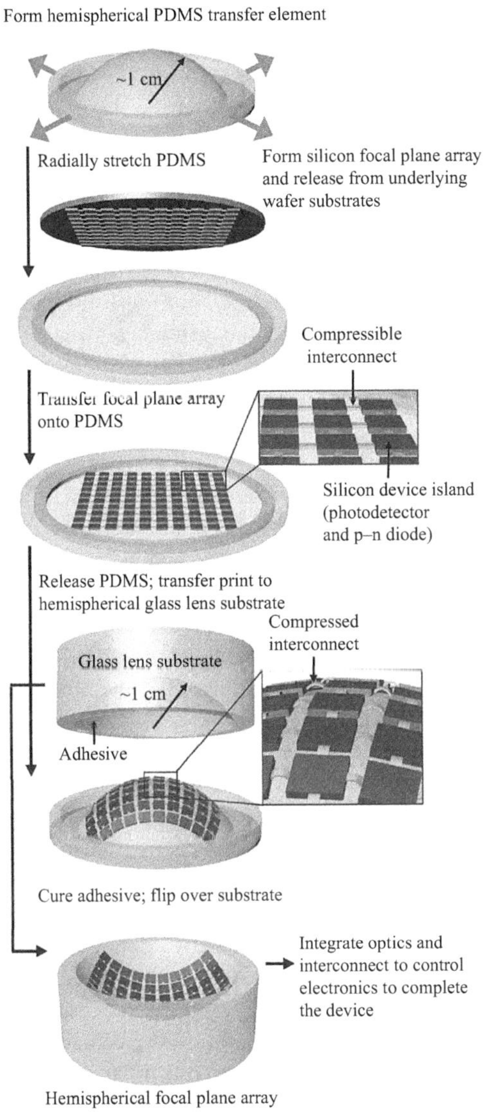

Figure 5.4 Fabrication process of a curvilinear electronic eye camera image detector. Reprinted with permission from Ko (2008) [14], copyright Nature Publishing Group

surface. The now flattened planar PDMS surface will be used later as a substrate for the curvilinear image detector. To fabricate the detector, standard semiconductor processing techniques are used on SOI wafers to realize a pixel array with each pixel containing a silicon photodetector, a current blocking diode, metal interconnects, and polyimide to provide support and act as a protective layer. The pixels are not touching each other but instead are connected to each other by very thin lines made of metal (360 μm long and 50 μm wide) on top of a thin layer of polyimide (100 μm wide). The buried oxide layer in the SOI wafer is then removed by hydrofluoric acid leaving the pixel array supported by a number of polymer posts on top of the underlying wafer.

The image detector is then brought into contact and bonded with the elastomeric planar PDMS surface as shown in Figure 5.4. The loosely connected underlying wafer is peeled away leaving behind the image detector on top of the planar-stretched PDMS surface. As described in Figure 5.4, the tensile stress placed on the PDMS surface is removed and the surface returned to its original hemispherical shape with approximately the same radius of curvature. Since the image detector is connected to the PDMS surface, the image detector will also follow the same curvature; however, the strains induced by the curvature will be absorbed by the metal/polyimide connecting lines which will compress and arc as shown in the inset in Figure 5.4. In other words, when transferring from a planar to a hemispherical curvature, the deformation of the image detector is limited to the connecting narrow lines which deform into an arc while all the pixels maintain their original shape and structure and with limited induced strain. To quantify the strains on the image detector when transferring from a planar geometry to a hemispherical one, the length between the pixels decreases by 20%. The maximum strain induced on the silicon pixels is approximately 0.01% while the maximum strain on the metal in the narrow lines is 0.3% (14).

The fabricated image detector was then aligned and connected to a printed circuit board which includes the readout circuitry and from there to a computer to output the images from the detector. To complete the electronic eye camera, a hemispherical enclosure that includes a single lens is placed on top of the image detector. In terms of results, the realized electronic eye camera includes a 16×16 pixel image detector and a 25.4 mm diameter plano-convex lens with a 35 mm focal length. The electronic camera was used to acquire images of a few letters from a typical eye chart. Figure 5.5 shows the completed camera as well as the images acquired by the camera. The image on the left of Figure 5.5(d) is of low spatial resolution due to the limited resolution of the camera (16×16). The spatial resolution can be improved by taking a number of images as the camera is rotated. A higher resolution image is then recreated from these images. The image on the right in Figure 5.5(d) clearly shows the higher resolution due to this technique. The authors also compared the images of their hemispherical detector with that of a planar detector and found that the hemispherical detector exceeds in the following: uniform focus across the focal plane, more uniform light intensity, and minimized geometric distortions (14).

These findings would be highly appealing for microcameras in biological imaging applications since with low aberrations the microcamera can function without the need of investing in multi-stage aberration correcting lenses which would augment the optical track of the microcamera and increase its cost. In our opinion, we believe the main limitation of the camera is the low spatial resolution.

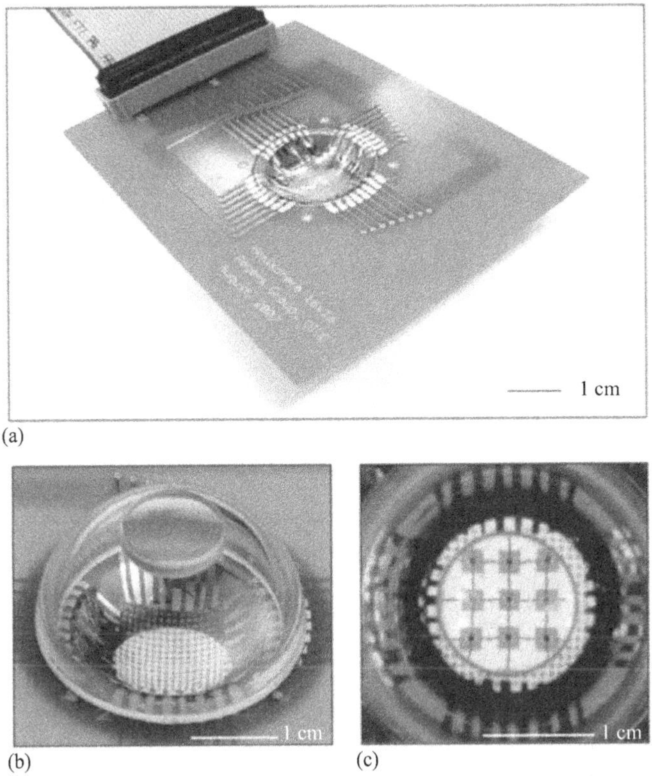

Figure 5.5 *Photographs of a hemispherical electronic eye camera and representative output images. (a) Photograph of a hemispherical image detector (center) fixed on a printed circuit board, along with a ribbon cable (upper left) to provide connection to a computer. (b) Photograph of the electronic eye camera after attaching a transparent hemispherical cap along with a single imaging lens (top). (c) Close-up photograph of the electronic eye camera as viewed directly through the imaging lens. For the specifications used here, this lens magnifies the image detector to show a small, 3 × 3 array of pixels. (d) Grayscale images of the first two rows in an eye chart captured using the electronic eye camera with a 16 × 16 pixel array. The rows are displayed on a hemispherical surface matching the detector surface (top) and projected onto a plane (bottom). The images on the left and right were acquired without and with scanning, respectively. Reprinted with permission from Ko (2008) [14], copyright Nature Publishing Group*

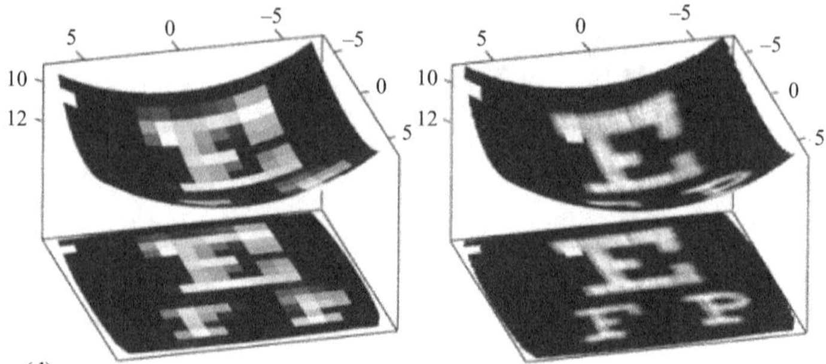

(d)

Figure 5.5 (Continued)

In the future, the main challenge facing the electronic eye camera is increasing the number of pixels while maintaining the hemispherical curvature of the camera. Overcoming this challenge would open up the door to an abundance of exciting biological, medical, and even military applications for the electronic eye camera.

5.3.2 Electronic eye zoom camera

Jung *et al.* also from Professor Roger's group expanded the work on the hemispherical electronic eye camera by adding zoom capability to the camera (15). The zoom capability was added by replacing the fixed single plano-convex lens with a single tunable liquid lens. Added to that, the image detector was made more versatile in the sense that the curvature of the image detector was made to change dynamically with the use of hydraulics under the image detector. The need to have a dynamically changing curvature of the image detector is due to the added zoom capability. When an image is captured by a convex lens, the true image sits on a non-planar surface called a Petzval surface. The curvature of this surface changes as the magnification is changed. Therefore, to correctly capture a minimally aberrated and high-quality image, the radius of curvature of the image detector has to change simultaneously with the tunable lens. The tunable lens is realized by a chamber filled with water and covered with a thin PDMS membrane. Adding more water into the chamber changes the curvature of the PDMS membrane, due to the added pressure, thereby changing the focal length. A very similar method was adopted for tuning the radius of curvature of the image detector. In this case, the image detector acted as the membrane on top of a water-filled chamber and the curvature changes by adding or removing water through the use of external pumps. Results of image testing of the electronic eye zoom camera with dynamically curved detector show superiority in terms of image quality when compared with a flat detector. This is due to the accurate dynamic conforming of the image detector curvature to the Petzval surface.

The ability of the electronic eye camera to provide zoom with only a single lens can potentially be very useful for medical imaging especially when fine details of different organs or tissues need to be imaged. Packaging specific parts of this electronic eye zoom camera might be a challenge in the future including the tunable liquid lens as well as the tunable image detector (external pumps, water chamber, etc.). That being said, the challenge is minimal compared to the superior image quality that the camera can deliver.

5.4 Compound eye cameras

The following chapter has a lot of information on artificial compound eyes on both flat and curved surfaces. Here, we will simply introduce a few examples on how can artificial compound eyes be applicable in medical imaging applications.

Compound eyes of insects have been the main inspiration behind a number of artificial compound eye structures. Compound eyes consist of an array of lenses on a curved surface and are considered the smallest known natural vision system. Compound eyes offer several advantages over single aperture eyes including a large FOV, polarization and high motion detection at the expense of lower spatial resolution. There are two main types of compound eyes: apposition and superposition. The main difference between apposition and superposition compound eyes lies in the arrangement of the lenses. In apposition compound eyes, each lens is connected to a distinct photoreceptor through an optical waveguide called the rhabdom. The rhabdom optically isolates the light from neighboring lenses. The lens, rhabdom, and photoreceptor combination is referred to as an ommatidium. In analogy, an ommatidium is similar to a camera structure. The number of ommatidia in an insect's eye can range from a few hundred to tens of thousands. The apposition compound eye therefore consists of an array of cameras optically isolated from one another. On the other hand, the superposition compound eye is composed of an array of lenses that share the same photoreceptor (1, 16). More discussion on the compound eyes and their mimicry can be found in Chapter 6.

5.4.1 Lobster eye camera

Huang and Jiang expanded on the work of the reflection-based cylindrical lens (13) and realized an artificial reflecting superposition compound eye (ARSCE) to provide, in addition to high-speed motion detection and wide FOV provided by refraction-based compound eyes, minimal chromatic aberration, and increased sensitivity to light (17). In nature, superposition compound eyes can be found in some decapods including the eyes of lobster and shrimp (14). As opposed to compound eyes of the dragonfly, where each ommatidium contains a lens, the ommatidium of these decapods includes a hollow "box" with highly reflecting sidewalls where incoming light is reflected off of two adjacent sidewalls of the box and onto a common photoreceptor. Figure 5.6 shows a schematic of the ARSCE which mimics the RSCE found in nature. As evident from Figure 5.6, boxes are packed closely and are organized onto a hemispherical surface which provides the high FOV. The photoreceptor as well sits on a hemispherical surface.

Fabrication of the ARSCE is very similar to the fabrication process of the cylindrical lens array with the major difference being that the pillars are replaced by hollow boxes and the boxes cover a hemispherical surface. Each box had a

(a)

(b)

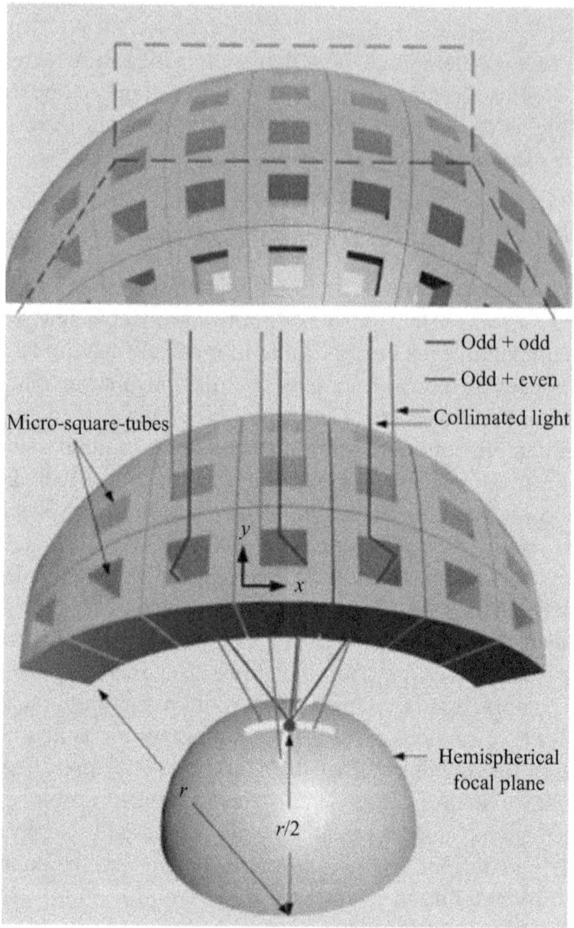

Figure 5.6 *Operating principles of a three-dimensional, artificial reflecting superposition compound eye (ARSCE) inspired by lobster eyes. (a) Model of the anatomical microstructures of a natural RSCE showing an array of closely packed, mirrored ommatidia with square cross-sections omnidirectionally arranged on a hemispherical surface. (b) A close up schematic representing the imaging features of a ARSCE consisting of an array of mirrored micro-hollow-boxes radially curved into a hemisphere (radius of curvature r) under collimated light. Even number of reflections on the sidewalls in either the X or Y coordinates within the hollow boxes forms the cross on the hemispherical focal plane halfway, namely r/2, to the center of the sphere. Odd number of reflections in both coordinates converge the reflected light toward a central focused point on the same focal plane. A real image is formed at the hemispherical focal plane with radius of curvature r/2 on the opposite side of the ARSCE from the source, acting like a refractive lens. Reprinted with permission from Huang (2014) [17], copyright Wiley*

height of 60 μm, an inner spacing of 20 μm with the sidewalls of the box having a thickness of 24 μm. The boxes are separated by a distance of 15 μm. These dimensions are very carefully chosen to resemble the dimensions of the natural optical structures found in a shrimp's eye. Since the sidewalls are required to act as reflectors, the sidewalls were smoothed by a wet etching process. Highly reflective aluminum was sputtered onto the silicon boxes to complete the fabrication of an array of 410×410 optical microstructures.

To provide a high FOV and as evident in Figure 5.6, the array of boxes is placed on a hemispherical surface. The hemispherical surface is realized by a process very similar to the transfer process of the electronic eye camera where the silicon microstructures were transferred onto a PDMS hemispherical surface. The FOV provided by the ARSCE is 165° making it ideal for medical imaging applications. The focusing capability of the ARSCE depends on the number of reflections the light goes through within the box. An odd number of reflections will focus the light into the center of the hemispherical plane, while an even number of reflections will not thereby producing this uniformly illuminated cross-patterned background. Results of imaging with the ARSCE show that the ARSCE does not suffer from chromatic distortion due to its reflection-based imaging.

The absence of chromatic aberration opens the door for the ARSCE to be used in applications that cover a wide range of the electromagnetic spectrum. The ARSCE can potentially be used in either medical x-ray imaging or in low-light applications where the absorption of light by the optics becomes an issue. A potential challenge would be overcoming the monochromatic aberration faced by the nature of the ARSCE (the background cross present in the image plane).

5.4.2 TOMBO compound eye camera

An early version of the compound eye camera based on planar optics was realized by Tanida *et al.* (18). The camera was named TOMBO which is an acronym for thin observation module by bound optics. TOMBO is also the name given to the *dragonfly* in Japanese which also happens to have an apposition compound eye. The camera is considered an apposition compound eye in the sense that it includes a microlens array situated on top of a charge-coupled device (CCD) or a complementary metal oxide semiconductor (CMOS) chip. As opposed to having a single photoreceptor or a pixel under each microlens, the TOMBO has multiple pixels under each microlens. The TOMBO can be visualized as an array of microcameras where each camera unit consists of a microlens on top of a small array of pixels. To prevent crosstalk, the microlenses are separated from the array of pixels under them by means of a metallic wall that cover all the sidewalls between adjacent camera units. The relation between the total number of pixels and the pixels under each microlens can be summarized by the following equations:

$$N = \mu v \tag{5.1}$$

$$s = \frac{d}{v} \tag{5.2}$$

where N is the total number of pixels in the photodetector, μ is the number of microlenses, v is the number of pixels under each microlens, s is the pixel width, and d is the width of each microlens.

TOMBO was initially realized using an experimental setup rather than a full-fledged camera. The setup utilizes a glass microlens array with a total of 32 microlenses. Each microlens has a diameter of 250 μm and a focal length of 650 μm. The metallic light-blocking separation layer is formed by utilizing two 120 μm thick metallic sheets. Each sheet is laser etched to form an array of square hollow tubes and the two sheets are brought together to form a 240 μm separation layer. The microlens and the separation layer are placed in front of a CCD sensor which is covered with an imaging fiber plate for better light coupling efficiency. The CCD sensor has 739 × 575 pixels. The number of pixels under each microlens is 22.7 pixels.

Each camera unit in TOMBO takes an image of the same scene from slightly different angles. The final image is rendered through image processing by the use of two methods: the sampling method and the backprojection method. The sampling method selects certain pixels under each unit and renders the final image based on the selected pixels. The final image can be altered based on which pixels are selected. For example, selecting pixels close to the optical axis of each unit will produce an erect image while different pixels might produce magnified or minified images. In the backprojection method, all the pixels are utilized for the rendering of the final image rather than selecting certain pixels.

In terms of medical imaging applications, the TOMBO can provide additional functionality to a conventional endoscope in terms of multiple viewpoint imaging or post processing of captured images. As will be seen later in the chapter, a more extensive version of the TOMBO was realized as an endoscope.

5.4.3 Compound eye zoom camera

Compound eye microcameras have also been realized that are capable of providing zoom. Typically, in conventional cameras, zoom capability is provided by the use of two sets of variable focus lenses. The first set of lenses provides the zoom by changing the effective focal length, while the second set of lenses provides the focus. The requirement of two sets of lenses thus increases the optical path and complexity which might be hard to implement in microcameras. Duparré *et al.* realized an artificial compound eye zoom camera that utilizes a single variable focus liquid lens along with a fixed focus microlens array (19). In their work, they were able to use a single variable focus lens rather than two due to the short focal length of the microlens array. The short focal length of the microlens arrays implies that no focus readjustment is necessary since the depth of focus is inherently large.

A schematic of the camera is shown in Figure 5.7. The camera includes a commercial variable focus electrowetting liquid lens, a polymer microlens array, black matrix polymers to define the apertures, and a CMOS image sensor. The electrowetting lens has a diameter of 4 mm and a response time on the order of 10 ms. The microlens array includes 80 × 60 microlenses with the microlenses

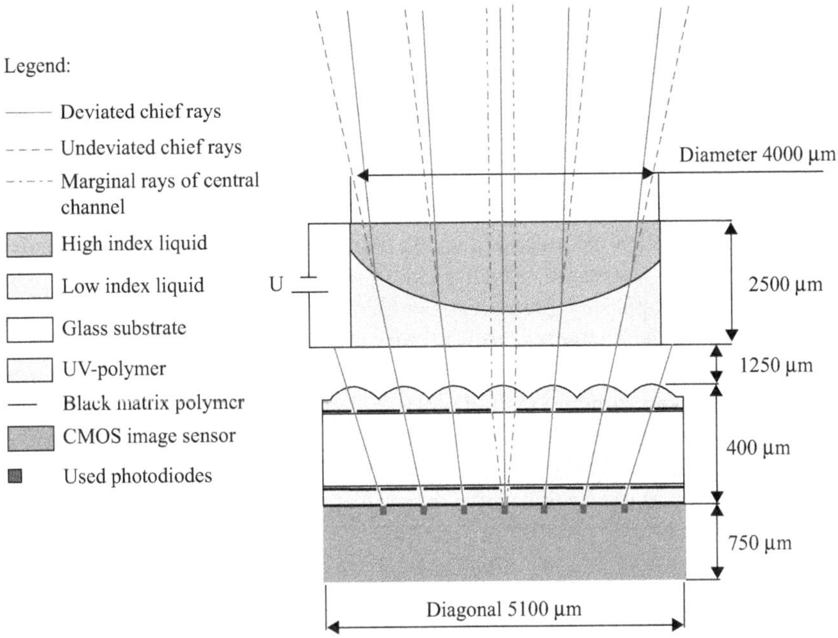

Figure 5.7 Schematic of artificial compound eye zoom camera (not to scale). A tunable liquid lens is arranged on top of an ultra-thin artificial apposition compound eye. Reprinted and reproduced with permission from Duparré (2008) [19], copyright IOP Publishing

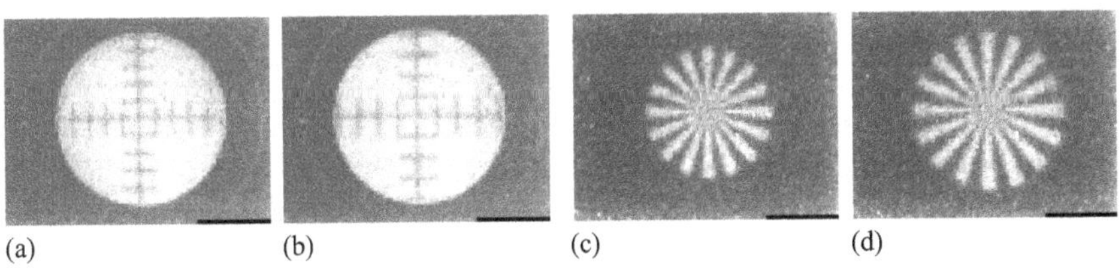

Figure 5.8 (a) Captured image. Maximum FOV, five scale lines visible. (b) Captured image. Minimum FOV, four scale four visible. (c) Radial star target, maximum FOV. (d) Radial star target, minimum FOV. Reprinted and reproduced with permission from Duparré (2008) [19], copyright IOP Publishing

having a focal length of 0.252 mm. The resulting FOV of the camera is 49° × 37°. The total thickness of the camera is 6.9 mm. In terms of zoom, the camera provides a zoom factor of 1.17 and has a defocus of 0.013 mm. Figure 5.8 shows images of test targets captured by the camera at maximum and minimum FOV, respectively. The 1.17 zoom, however, is well below standard zoom factors of at least three in conventional consumer products. The limited zoom factor is due to the limited optical power change of the liquid lens. Using a liquid lens that can provide a higher optical change, by using a higher refractive index liquid, can potentially increase the zoom factor to between three and four (19).

The artificial compound eye zoom camera can potentially be used as a microcamera for medical imaging due to its minimal thickness and zooming capability. Similar to the electronic eye zoom camera by Roger's group, this camera provides optical zoom with a single tunable lens. The tuning mechanism is different however, whereas in Roger's electronic eye camera the lens was tuned pneumatically by applying pressure on a thin PDMS membrane. In Duparré's zoom camera, the lens is an electrowetting lens. Another difference between the two cameras is that the curvature of the detector in the electronic eye camera changes through pneumatic pressure to account for the changing zoom. In Duparré's camera, the presence of the short focal length microlens array reduces the requirement of changing the curvature of the detector. However, we believe that the Duparré camera would suffer from aberrations due to the planar nature of the detector.

5.5 Multiple viewpoint camera

In a number of camera applications, the lack of space prevents the use of conventional cameras to achieve the desired viewing angle. An example of that is in endoscopy and specifically in procedures that involve body cavities. Maneuvering the endoscope within body cavities elevates the risk of damaging tissue, nerves, or internal organs (20). An ideal solution would be an endoscope that can provide images from different viewpoints of the same scene without maneuvering the endoscope. This solution can be realized by introducing the concept of plenoptic cameras to form microscale cameras that can work in such environments.

The difference between an ordinary camera and a plenoptic camera is the addition of a microlens array in a specified location between the camera lens and image sensor. Rather than capturing a single image, the plenoptic camera captures data about the surrounding scene (21). This data, called the lightfield, can then through image processing produce images of the same scene but as if viewed from a different perspective. The lightfield, a four-dimensional function, provides information on the light's direction and position (22).

Lightfield photography is a century old notion that has its origins in Gabriel Lippmann's 1908 concept of integral photography (21). Lippmann came up with the idea that if an array of miniature lenses was placed on top of a photographic plate then the three-dimensional scene can be captured onto a two-dimensional surface (23). However, due to the fact that lightfield capture requires digital post processing the concept did not proceed further because of the lack of digital photography at that time. With the advent of digital photography, the idea of capturing the lightfield was revived and many approaches were realized that captured the lightfield including using camera arrays, encoded mask, pinhole array, and a plenoptic camera (24–27). Aldalali *et al.* used the same

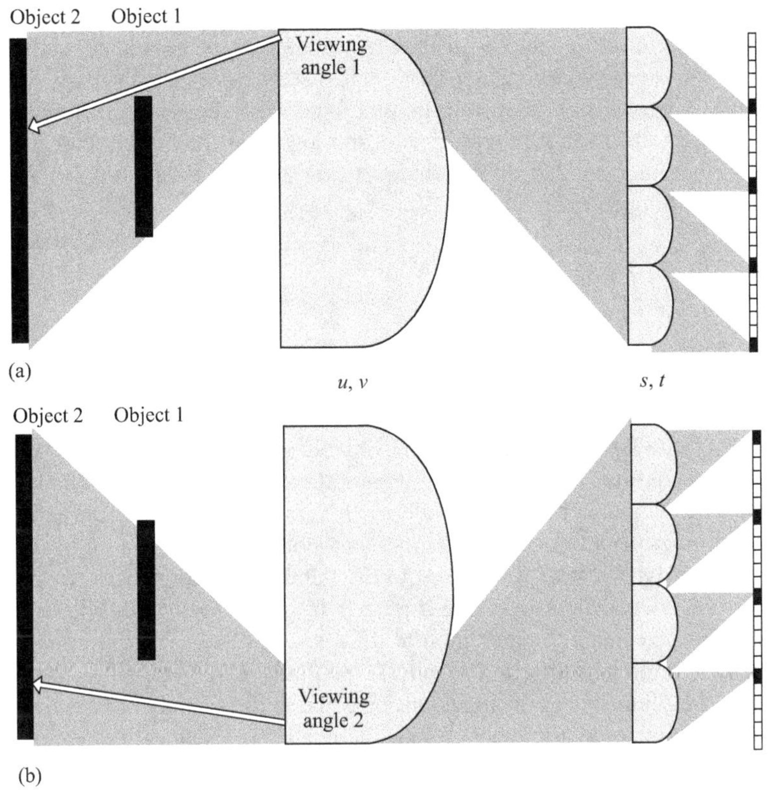

Figure 5.9 Schematic of a micro plenoptic camera. Camera includes a main lens, microlens array, and image sensor. (a) and (b) are two different viewpoints of object 2 imaged by the main lens onto a distinct pixel under each microlens. In (a) light is focused onto the bottom pixel under each microlens, while in (b) light is focused onto the top pixel under each microlens

concepts behind the plenoptic camera to realize the multiple viewpoint camera at the microscale (28).

Let us first describe the concepts behind lightfield capture and consider the lightfield as a four-dimensional function $L(u,v,s,t)$ and parameterized as a two-plane Cartesian coordinate parameterization the (u,v) plane and the (s,t) plane. Figure 5.9 is a diagram of the multiple viewpoint camera. The camera is comprised of three components: a single main lens on top of a microlens array which is on top of an image sensor. The key behind the camera is the precise location of the different components. The image sensor is positioned at a focal length away from the microlens array while the microlens array is positioned at the image plane of the main single lens. As seen in the schematic above, the (u,v) plane is the plane of the camera lens while the (s,t) plane is the plane of the microlens array. In terms of operation of the camera, incident light from a distinct viewpoint is focused by a portion of the main lens onto the image plane (microlens array) and is then focused by the microlens array onto a distinct pixel of the image sensor under each microlens. As seen in Figure 5.9, both schematics correspond to two different viewpoints where at each viewpoint light is focused on the same pixel under each microlens (bottom pixel for Figure 5.9(a) and top pixel for Figure 5.9(b)). Quantitatively, the equation that directs the exposure at each pixel assuming paraxial rays and a fully open main lens is (27):

$$I(s,t) = \iint L(u,v,s,t)dudv \tag{5.3}$$

This equation shows that extracting an image pertaining to a specific viewpoint corresponds to keeping u and v constant (choosing a specific region in the lens plane) and extracting all s and t (microlens array). The number of pixels under each microlens is equivalent to the number of available viewpoints that can be extracted. This also means that the number of regions the main lens can be divided to is equivalent to the number of pixels under each microlens. Choosing the same pixel under each microlens to extract a distinct viewpoint will result in an image from that viewpoint with a resolution equal to the number of microlenses in the array. Visually this is equivalent to taking a picture from that viewpoint with a camera having a lens size equal to one region of the main lens.

Another important criterion to ensure accurate image extraction is to have the image side *f*-number of the main lens be equal to the *f*-number of the microlens array. The image side *f*-number of the main lens is equal to the ratio of the distance between the main lens and microlens array and the main lens aperture. Matching the *f*-numbers ensures optimal utilization of the pixels under each microlens.

For Aldalali *et al.*'s multi-viewpoint microcamera, the microlens array was fabricated using photoresist thermal reflow process while the main lens and image sensor were both commercial (3). Figure 5.10(a) and (b) shows the simple fabrication process along with the microlenses before and after reflow. In the reflow process, blocks of the array apertures are first patterned on a glass substrate by lithography. For example, for circular microlenses, a cylinder is patterned while for a square microlens a square block is patterned. The only difference between a conventional photolithography process and a reflow process is that baking is done at a much higher temperature (165°C in this case) such that the temperature exceeds the glass transition temperature of the photoresist and it starts to melt, or reflow. This reflow will result in a spherical cap structure due to surface tension amid the photoresist and the glass substrate. The focal length of the resulting lens depends on the thickness of the pre-baked photoresist, baking time, and temperature. The focal length of the fabricated microlenses is determined by the following lens equation:

$$f = \frac{R}{n-1} \tag{5.4}$$

where R is the radius of curvature and n is the refractive index.

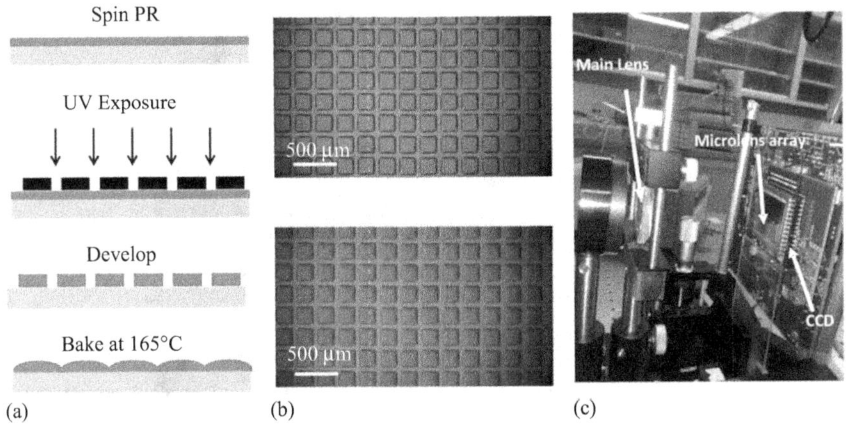

Figure 5.10 *The fabrication process. (a) The focal length of the microlenses is governed by the initial thickness of the photoresist cylinders, pitch, reflow time, and temperature. (b) The microlens array before and after photoresist reflow. (c) Experimental setup of the microcamera showing a main lens (aperture 15.2 mm; f/2.5), microlens array (97 × 97; pitch size: 230 μm; focal length: ~1.2 mm), and a 1-megapixel image sensor. Reprinted with permission from Aldalali (2012) [28], copyright IEEE*

There were two experimental cameras realized by Aldalali *et al.* The first camera has a main lens aperture of 15.2 mm, a 97 × 97 microlens array with a pitch of 230 μm, and a 1-megapixel image sensor which translates into approximately 100 different viewpoints. Figure 5.10(c) shows the experimental setup of the first camera where the lenses and image sensor are clearly identified. The second camera has a 9 mm main lens aperture, an 80 μm pitch 56 × 42 microlens array, and a higher resolution 5-megapixel image sensor. The second camera can provide approximately 1000 different viewpoints. Results from the second camera are shown in Figure 5.11. The camera was used to image a scene of two wires placed on the same axis. For an ordinary camera, the wire at the back would have only been captured if the camera was placed off-axis and aiming at the back wire. However, with the multi-viewpoint camera placed on axis and without changing its position, simply digitally choosing the region of the main lens toward the edge would suffice since at this region incoming light reflected off the back wire is not blocked by the front wire. Figure 5.11(a) shows the lightfield data captured by the camera. Figure 5.11(b) shows the 47 different extracted viewpoints and it can be noticed that at certain viewpoints the back wire is clearly visible.

As stated previously, the camera captures the lightfield image and the final viewpoint-specific images are synthetically produced via image processing. The image processing of the captured lightfield is a very simple algorithm done in MATLAB. The algorithm starts by resizing the captured lightfield to ensure that

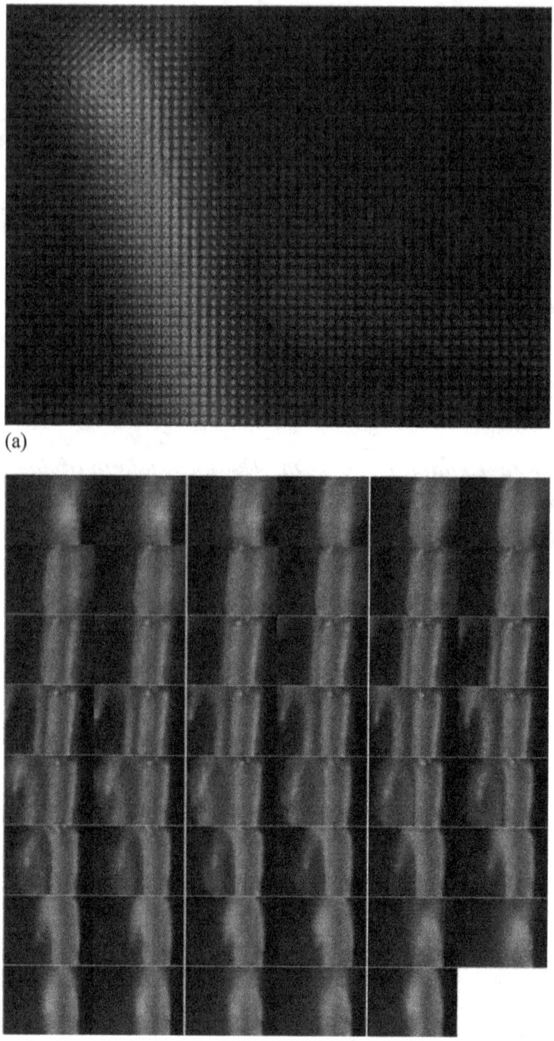

Figure 5.11 Captured lightfield and synthetic images from the second camera. (a) The captured lightfield of two wires placed along the same axis. (b) Digitally post-processed images with 47 different perspectives of the wires. Center rows (3–6) are viewpoints where both wires are visible. The remaining rows are viewpoints where only the wire closer to the camera is visible. Reprinted with permission from Aldalali (2012) [28], copyright IEEE

there are an integer number of pixels under each microlens. The two-dimensional captured lightfield is then divided into a four-dimensional array of (u,v,s,t) in which u and v refer to the two-dimensional pixel array under each microlens and s and t refer to the microlens array. Since each pixel under a microlens pertains to a specific viewpoint, rendering a viewpoint specific image is equivalent to choosing the same pixel under each microlens resulting in a synthetic image of resolution $s \times t$. If another viewpoint is to be chosen, then a different value of u and/or v has to be used. It is clear that there has to be a compromise between the number of viewpoints available and the resolution of the synthesized image. The resolution of the synthesized image is solely determined by the number of microlenses. Yet if more microlenses are added while maintaining overall size and resolution of the image sensor, then the pixels would be distributed over more microlenses and hence the number of viewpoints would decrease.

Keeping in mind the low resolution of the synthesized images from the microcameras, the multi-viewpoint cameras can still be of potential use in medical imaging specifically in endoscopic procedures where movement of the endoscope is very limited. Having a stationary microcamera that allows multiple viewpoint imaging will also limit the risks of tissue or nerve damage as well as minimize the handling requirements of the scope. The key going forward with this multi-viewpoint camera is to find the optimal compromise between the number of available viewpoints and the resolution of the synthesized image.

5.6 Camera arrays

In certain applications, it is highly beneficial to have a hybrid camera system that has the advantage of the high FOV integral to compound eye cameras and high-resolution inherent in "eye" cameras or single aperture cameras. Aldalali *et al.* from the University of Wisconsin–Madison realized a camera array that is inspired by both compound eyes and mammalian eyes (29). The camera array includes three cameras with each camera consisting of an *f*/10.9 mm microlens on top of a miniaturized 1 mm^2 image sensor. The microlens is fabricated by a liquid-phase photopolymerization of droplets of an optical adhesive on top of a Teflon-coated substrate. Figure 5.12(c) shows a profile of the lens using a goniometer. The three cameras are connected together by a very flexible 3 mm long elastic polymer called Ecoflex. The added flexibility allows for the instantaneous reconfiguration of the camera to change the FOV or the viewing angle. Figure 5.12(a) and (b) shows the camera array at two different configurations. In terms of image acquisition, the images from the three cameras are digitally stitched together with the requirement that there be sufficient overlap between neighboring cameras for successful and seamless stitching. The microcamera array of only three cameras can achieve a FOV of 130° which is mainly the result of the flexibility of the array.

As opposed to the multi-viewpoint camera, which synthesizes lower resolution viewpoint specific images, the camera array does not lose resolution but instead increases the FOV when its images are stitched. An ideal application for the camera array would be to increase the number of cameras in the array to provide a spherical FOV. Based on the results of the three camera arrays, a spherical FOV could be achieved by using approximately nine cameras. A spherical FOV is of high benefit to doctors since it provides a view of the whole working space from a single image capture of the array.

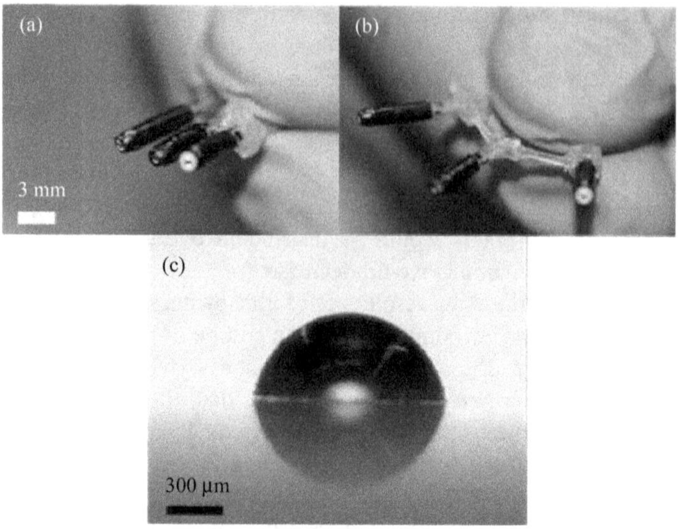

Figure 5.12 Flexible miniaturized camera array. (a) Photograph of the camera array in the compact state. Each camera is facing approximately the same scene. (b) Photograph of the camera array in the flexed state. Each camera is facing a different scene. (c) Profile of the optical adhesive microlens taken with a goniometer. The lens has a diameter of 0.9 mm and a focal length of 0.8 mm. Reprinted with permission from Aldalali (2013) [29], copyright IEEE

5.7 Applications

Among the main limitations of current endoscopes and laparoscopes are a limited FOV as well as lack of depth perception (30). The following are application-specific microcameras which were developed specifically to be used as endoscopes or laparoscopes to tackle their limitations. These are only examples of ongoing research into "smarter" endoscopes focusing mainly on tunable liquid lenses and electrically tuned lenses. There is existing research into endoscopes with tunable liquid crystal lenses (31) which have not been included. Also there is numerous published work on capsule endoscopy which is beyond the scope of this chapter.

5.7.1 Endoscopes

The previously mentioned work of the TOMBO compound eye camera was recently further developed by Kagawa *et al.* to be used as a multifunctional endoscope (32). The TOMBO endoscope can provide an extended depth of field, depth estimation, a varying FOV, different viewpoint images, and polarization imaging. Figure 5.13(a) is a schematic of the TOMBO endoscope. The extended depth of field is provided by the wavefront coding technique (33). As shown in Figure 5.13, the depth estimation is provided by pattern projection via the combination of

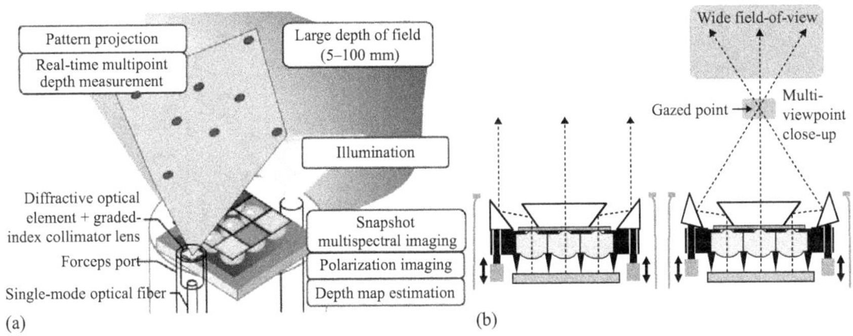

Figure 5.13 Schematic of the TOMBO endoscope. (b) Variable FOV TOMBO endoscope in the parallel and cross configuration. Reprinted with permission from Kagawa (2012) [32], copyright IEEE

optical elements as well as the optical fiber, placed at the periphery of the TOMBO compound eye, followed by some software processing. As previously described, the TOMBO compound eye camera produces a number of images from different angles. The TOMBO endoscope utilizes these images and after some image processing techniques realize full stereoscopic display (32).

Limited FOV is one of the main limitations of current endoscopes. The TOMBO compound eye camera inherently has a limited FOV due to the planar nature of the microlens array as well as the limited size of the microlens array. To increase the FOV, combinations of fixed and movable mirrors were embedded into the endoscope as shown in Figure 5.13(b). The mirrors provide the endoscope with two configurations: the parallel configuration and the cross configuration. The parallel configuration, as shown in Figure 5.13(b) on the left, while the FOV is still limited, is optimized to provide three-dimensional contour data. In the cross configuration, as shown in Figure 5.13(b) on the right, the mirrors are tilted to provide a much wider FOV (>100°) as well as providing images from different viewpoints for nearby objects. The TOMBO endoscope also allows for polarization-dependent imaging of tissues. Depending on the polarization of the light, objects on the surface of the tissue or deep inside the tissue can be imaged. To achieve that, a grid of metal wires was placed on the light sources in the TOMBO endoscope (the rectangular block on the thin fiber at the edges of the endoscope in Figure 5.13(b)). Polarization imaging is only possible in the cross configuration since the tissue being imaged has to be placed within the gazed point as shown in Figure 5.13. Kawahito and Kagawa also recently added refocusing capability to the TOMBO endoscope through the addition of an array of band pass filters on top of the compound eye camera (34).

The TOMBO endoscope has demonstrated versatility in providing multiple functions to the endoscope in terms of extended FOV and focus, stereoscopic display, and refocusing capability, as well as polarization-dependent imaging. These added functionalities could potentially be used to further simplify endoscopy to both the user and the patient and simultaneously adding more diagnostic capabilities within the endoscope.

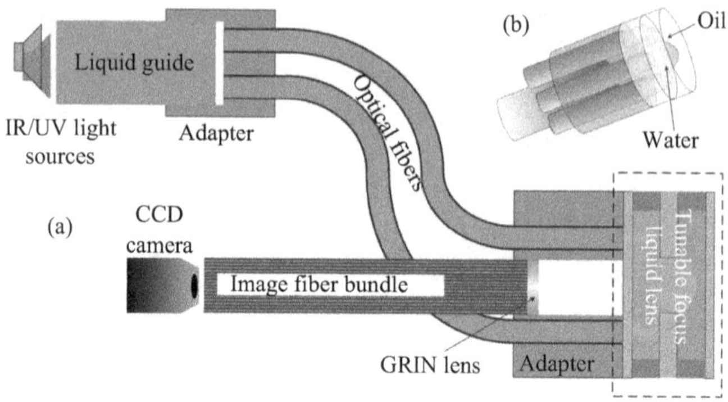

Figure 5.14 (a) Schematic and (b) three-dimensional model of the tunable focus IR actuated endoscope. IR light is transmitted via a set of optical fibers to tune the microlens focal length. Images captured by the microlens are sent to a camera via another set of optical fibers (image acquisition fibers). Images from the microlens are transferred to a camera via the image acquisition fiber bundle. Reprinted with permission from Zeng (2011) [35], copyright IEEE

Zeng *et al.* realized a fiber endoscope which utilizes a liquid lens that can tune its focus through the use of IR light (35). A schematic of the endoscope is shown in Figure 5.14. The endoscope includes the liquid lens, IR-light responsive NIPAAm hydrogel posts to provide lens tuning, an optical fiber bundle that transfers the focused light from the lens to the image sensor, and optical fibers that actuate the hydrogel under the liquid lens. As seen from the schematic, the fiber bundle is situated in the center directly behind the lens, while the actuating optical fibers are locating in the periphery directly behind each hydrogel post.

The liquid lens is fabricated using the same methods as described above in the work of Zhu *et al.* The only major difference is that the hydrogel is embedded with gold nanoparticles, instead of GO–GMA, which are used for the actuation of the lens. When the IR light illuminates the hydrogel post, the gold nanoparticles within the hydrogel absorb the light and in turn produce heat which leads to contraction of the hydrogel. When the IR light source is removed, the gold nanoparticles no longer generate heat and the hydrogel swells back to its original volume. During this contraction and swelling cycle of the hydrogel, the curvature of the microlens (oil–water meniscus) changes from divergent to convergent.

The tunable-focus fiber endoscope demonstrated a number of realized capabilities. The lens can be tuned from a focal length of 52 mm to 8 mm. The minimum response time of the lens to IR actuation is 20 s. The endoscope can realize a FOV between 77° and 128°. To demonstrate the capabilities of the fiber endoscope in a typical endoscopic environment, a plastic model of the human colon was constructed with 3–10 mm colonic polyps made from simulated tissue. Figure 5.15 is a frame-by-frame sequence of focused images of the model colon from one cycle scan of the fiber endoscope's liquid lens. As can be clearly seen from the sequence of images, at 1.8 s,

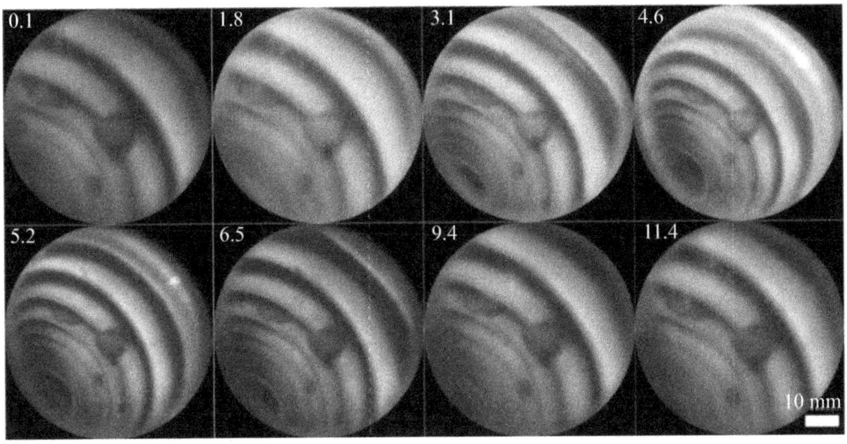

Figure 5.15 *Frame sequence of the focused images of the simulated human colon and polyps captured from the fiber endoscope in a single scanning cycle of the microlens. Initially, three polyps could be observed on the colon. From 1.8 s, the IR light was turned on, and the field of view began to increase. At 4.6 s, an extra polyp and the other end of the simulated colon lumen could be observed. From 5.2 s, the IR light was turned off, and the field of view of the microlens began to decrease. At 11.4 s, the microlens went back to the initial status. Reprinted with permission from Zeng (2011) [35], copyright IEEE*

the IR light is activated and from then onwards the FOV of the endoscope starts to increase. This is evident by the visibility of an extra polyp at 4.6 s compared to 0.1 s. At 5.2 s, the IR light is off and the FOV starts to decrease back to its initial value.

The IR tunable fiber endoscope exhibits great promise as a future endoscope. The optical tuning mechanism instead of the more popular electrically tuned lenses gets rid of the requirement of a high-voltage source for the lens. On the other hand, the response time of the IR endoscope is slow compared to the response time of electrowetting lenses.

Most of the existing tunable lenses have been mainly liquid based regardless of whether they are actuated by electricity (electrowetting) or smart actuators (hydrogels). Zou *et al.* from the University of Singapore have realized a miniature endoscope with a tunable solid lens (6). The tunability of the lens is based on the Alvarez principle. Alvarez-based lenses are a pair of lenses that provide a change in optical power whenever one of the lenses is shifted laterally from the other lens (36). A schematic of the realized endoscope is shown in Figure 5.16. The endoscope consists of a pair of solid lenses that are made from an optical adhesive. The lenses have an optically flat surface on one side and a freeform surface on the other. When attached together, the lenses provide no focusing capability and do not refract any light. Once one of the lenses is laterally displaced relative to the other lens, light is refracted from the overall lens pair. The amount of light refracting/focusing depends on the extent of lateral displacement. As can be seen from Figure 5.16, the endoscope also includes a pair of piezoelectric arms connected to one of the lenses on each side by mechanical springs to provide the displacement. The arms are actuated separately by two external voltage sources. For imaging, an image fiber bundle is used that connects the captured light to external cameras, as well as illumination fibers that are situated around the periphery of the endoscope as shown.

The solid lenses are fabricated by a precision diamond turning process to define the freeform surface. The defined surface is then molded onto PDMS and finally a second molding onto the optical adhesive. The profile of the freeform surfaces was based on a sixth-order polynomial and optimized using optical design software. In terms of results, the endoscope can provide a focal length tuning between 4.9 mm and 7.4 mm when the voltage actuating the piezoelectric arms is varied between −90 V and 90 V. The dimensions of the endoscope probe were 5 mm × 5 mm.

The proposed endoscope with solid lenses shows great promise for medical imaging applications. Specifically, the endoscope tackles the challenge faced by other tunable lens technologies including packaging issues of liquid lenses, as well as difficult mechanical translation of optical lenses at the microscale. We believe the challenges that face this endoscope lie in the high-level optimization required

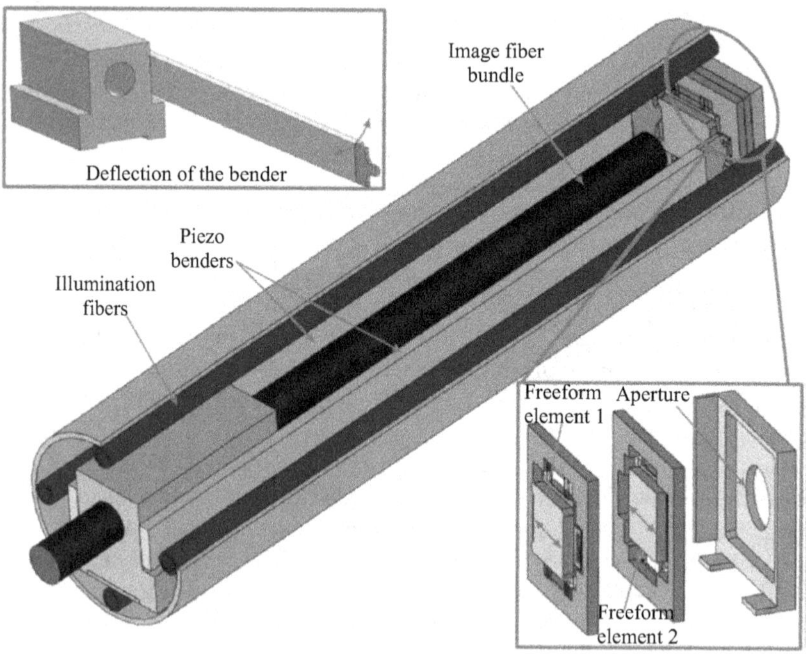

Figure 5.16 Schematic of the miniature endoscope with a solid tunable lens endoscope. The top and bottom insets illustrate the deflection of the piezoelectric bender and the close-up view of the solid tunable lens configuration, respectively. Reprinted with permission from Zou (2015) [6], copyright the Optical Society

for the freeform surfaces as well as the high precision fabrication requirement. Another challenge would be the need for two voltage sources to actuate both arms as opposed to only one used in other tunable lens technologies.

5.7.2 Laparoscopes

Kanhere *et al.* from the University of Wisconsin–Madison realized a reconfigurable microcamera array that was developed specifically for laparoscopic procedures (37). Laparoscopy, a minimally invasive surgical procedure, relies on the use of multiple small incisions over the abdomen. Each of these incisions will be used to house different surgical instruments or a camera, through the use of surgical ports. As shown in Figure 5.17, these ports safeguard the abdomen walls from the different surgical instruments. In terms of the camera, current typical laparoscopic procedures rely on fixed focus cameras which suffer from a number of deficiencies including but not limited to: narrow FOV, limited depth perception, glare when in the presence of reflective organ surfaces. Another limitation is that if the surgeon requires a different viewing angle, the camera is shifted into a different port that provides the needed viewing angle (37). To overcome all of the preceding limitations, Kanhere *et al.* developed a system of microcameras (four total) that when combined provide a large FOV. In the context of laparoscopy, this system of microcameras will be connected

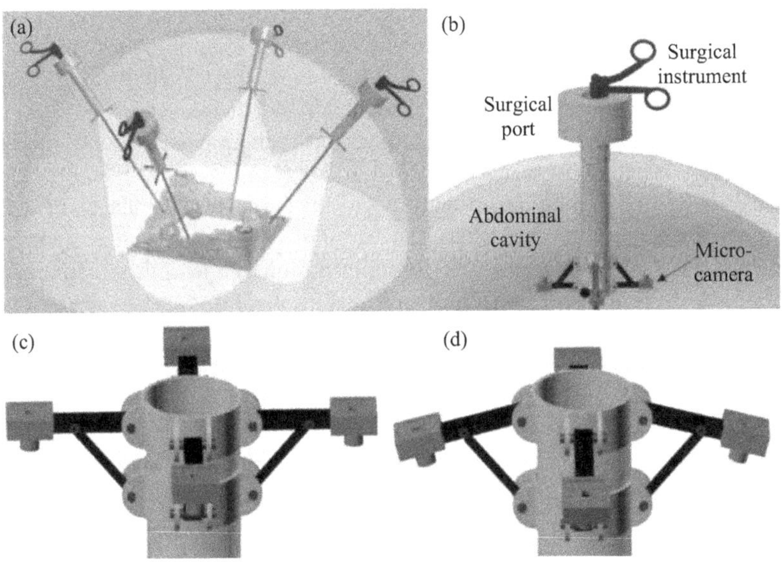

Figure 5.17 Schematic representation of (a) microcamera array, as applied in a laparascopic surgery setting. (b) A close-up view of an integrated surgical port microcamera assembly after insertion within an abdominal cavity. (c) and (d) The reconfigurability of camera viewpoints using an umbrella-like mechanical actuator. Reprinted with permission from Kanhere (2014) [37], copyright IEEE

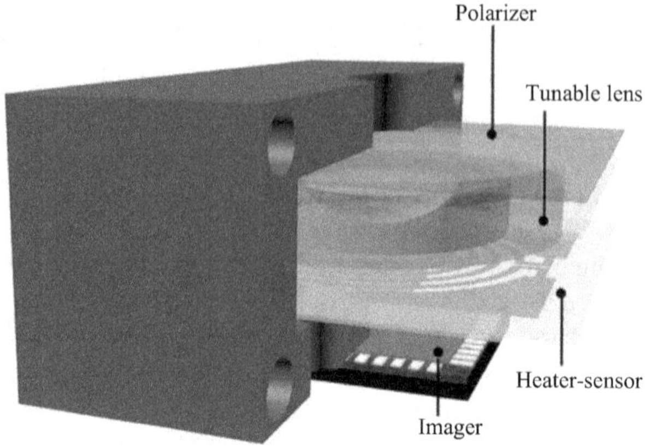

Figure 5.18 Three-dimensional representation of the microcamera unit. Each camera module consists of a commercial image sensor (1 mm × 1 mm), a tunable thermo-responsive liquid lens on top of a heater-sensor, and a polarizer. Reprinted with permission from Kanhere (2014) [37], copyright IEEE

to the end of each port as shown in Figure 5.17(b)–(d). The microcameras can be simultaneously opened or closed mechanically very similar to how an umbrella operates. The large FOV can be achieved by stitching the images provided by the four individual microcameras. In their first prototype, the camera array used commercial cameras that included the same miniaturized image sensor as in Aldalali *et al.*'s work (1 mm × 1 mm) as well as fixed focal length lenses (5 mm), and each camera was connected together by a PDMS bridge (38).

In their second prototype, in addition to the large FOV, each camera itself consisted of a tunable lens to provide a varying focal length and a polarizer to reduce the effect of glare. Figure 5.18 is an illustration of one of these cameras. The camera consists of a micro image sensor (1 mm × 1 mm), a heater-sensor to provide lens tunability, a tunable water–oil-based microlens, and a polarizer. The tunability of the lens was achieved by utilizing a thermo-responsive hydrogel using a similar fabrication method as that of Zhu *et al.* [12].

To expedite the heating process, a heater-sensor-actuator was utilized in their design. The heater-sensor is an aluminum coil that is fabricated using a lift-off process. The temperature is increased by increasing the resistance of the conductive coil since resistance is related to temperature by the following equation:

$$R_T = R(1 + \alpha\Delta T) \tag{5.5}$$

where R_{T} is the resistance at the desired temperature, R is the resistance at ambient temperature, α is the temperature coefficient at the ambient temperature, and ΔT is the difference between the desired temperature and the ambient temperature. The polarizer is fabricated by curing polyvinyl alcohol (PVA) on top of a pre-etched glass substrate to define the orientation of the polarizer. The cured PVA film was then removed and sandwiched between two thin PDMS substrates to serve as a flexible polarizer. In terms of results, the tunable lenses can be tuned between 5 mm and 50 mm. The time it takes to tune the lens between 5 mm and 50 mm is approximately 20 s. The FOV achieved by the camera array is 128°. The temperature of the heater-sensor was varied between 42°C and 65°C.

There are several advantages of the microcamera array laparoscope by Kanhere *et al.* The first is the fast tunable lenses which allow the surgeon for an increased working space since the cameras need not be close to the scene being imaged. The second advantage is the increased perception of depth for the surgeon due to the camera array and the stitching of the individual images. The third advantage is the placement of the array which is on the periphery of the surgical port. This positioning neither interferes with any of the other surgical tools nor requires a separate surgical port. The main limitation at this stage would be the relatively slow response time of the tunable lens when compared with other tunable lens technologies. Another issue that might arise would be the versatility of the image stitching under low-light conditions which is inherent in laparoscopic procedures.

Within laparoscopy, Tsai *et al.* from the University of California, San Diego, realized a laparoscopic zoom camera utilizing two fluidic lenses and performed experimental surgeries with the camera (39). As opposed to Kanhere *et al.*'s work where the camera array was envisioned to be mounted around the periphery of the surgical port, the camera by Tsai *et al.* is tailored to be fixed at the surface of the abdomen's internal wall. The fluidic lenses are made by encapsulating an optically clear fluid in a chamber with a thin optically clear membrane. Applying pressure to the chamber will change the curvature of the membrane which will change the focal length (40). The camera can provide 4× optical zoom, can work under very low-light conditions down to 160 lux of a light emitting diode (LED) light source while maintaining a total length of 17 mm. The camera can also provide multiple viewpoint capability through the insertion of a spherical mirror a small distance away from the camera. Another viewpoint is produced when the spherical mirror is rotated.

The laparoscopic zoom camera by Tsai *et al.* offers a lot of advantages in terms of enhanced light sensivity, tunable focus, as well as high optical zoom. A possible limitation of the laparoscopic zoom camera would be the added requirement of fixing the camera to the abdominal wall. While this requirement maximizes working space for the surgeon, it limits the viewpoint of the camera as well as adding the step of fixing the camera to the internal wall in each procedure.

Tsai *et al.* also compared their laparoscope with a commercial laparoscope and concluded that the zoom camera provides better functionality under low-light conditions. Another advantage was the added working space when utilizing the zoom camera compared to the commercial laparoscope. Using a commercial laparoscope with a fixed focal length lens, the lens has to be close to the intended scene to clearly visualize it. Their zoom camera was also used in an experimental gall bladder removal surgery where the zoom capability enhanced the details of the imaged organs (41).

The aforementioned microcameras for both endoscopy and laparoscopy have shown great promise in the plethora of advantages including but not limited to zoom capability, wide varying FOV, polarization imaging, and stereoscopic display. These advantages add benefit to both the surgeon in terms of ease of use, added functionality, and improved diagnostics as well as the patient in terms of limiting the time of the surgery and minimizing discomfort.

5.8 Conclusion

This chapter provided the latest research grade microcameras that can be used for medical imaging. The optics of the camera was discussed along with the latest innovative solutions for simple fabrication processes, focus tunability, zoom capability, and wide FOV. In addition to the optics, a novel design and implementation for the curvilinear image detector was also realized to minimize aberrations. Insect compound eye inspired microcameras with large FOV were also described in terms of functionality and different fabrication techniques on both curved and planar substrates. In terms of added functionality, a microcamera with multiple viewpoint imaging capability by capturing a single image was discussed. Details of large FOV multi-camera systems with either fixed focal length or tunable lenses were also included in this chapter. Finally, research level endoscopes and laparoscopes were discussed. The different scopes had integrated some of the mentioned innovative techniques and some have shown better performance than current commercial scopes.

References

1. Völkel R, Eisner M, Weible KJ. Miniaturized imaging systems. Microelectronic Engineering. 2003;67–68(1):461–72.
2. Beihl S, Danzebrink R, Oliveira P, Aegerter MA. Refractive microlens fabrication by ink jet process. Journal of Sol-Gel Science and Technology. 1998;13(1):177–82.
3. Yang H, Chao C-K, Wei M-K, Lin C-P. High fill-factor microlens array mold insert fabrication using a thermal reflow process. Journal of Micromechanics and Microengineering. 2004;14(8):1197–204.
4. Rogers JD, Karkkainen AHO, Tkaczyk T, Rantala JT, Descour MR. Realization of refractive microoptics through grayscale lithographic patterning of photosensitive hybrid glass. Optics Express. 2004;12(7):1294–303.
5. Cheng C-C, Chang CA, Yeh JA. Variable focus dielectric liquid droplet lens. Optics Express. 2006;14(9):4101–6.
6. Zou Y, Zhang W, Chau FS, Zhou G. Miniature adjustable-focus endoscope with a solid electrically tunable lens. Optics Express. 2015;23(16):20582–92.
7. Jiang H, editor. Bio-inspired microlenses and their biomedical applications; Proceedings SPIE 8598, April 12, 2013.
8. Dong L, Agarwal AK, Beebe DJ, Jiang H. Adaptive liquid microlenses activated by stimuli-responsive hydrogels. Nature. 2006;442(7102):551–4.
9. Aldalali B, Kanhere A, Fernandes J, Huang C-C, Jiang H. Fabrication of polydimethylsiloxane microlenses utilizing hydrogel shrinkage and a single molding step. Micromachines. 2014;5(2):275–88.

10. Agarwal AK, Beebe DJ, Jiang H. Integration of polymer and metal microstructures using liquid-phase photopolymerization. Journal of Micromechanics and Microengineering. 2006;16(2):332–340.
11. Daly D. *Microlens Arrays*; Taylor and Francis: New York, NY, USA, 2001.
12. Zhu D, Lo C-W, Li C, Jiang H. Hydrogel-based tunable-focus liquid microlens array with fast response time. Journal of Microelectromechanical Systems. 2012;21(5):1146–55.
13. Huang C-C, Jiang H. Bio-inspired wide-angle broad-spectrum cylindrical lens based on reflections from micro-mirror array on a cylindrical elastomeric membrane. Micromachines. 2014;5(2):373–84.
14. Ko HC, Stoykovich MP, Song J, *et al.* A hemispherical electronic eye camera based on compressible silicon optoelectronics. Nature. 2008;454 (7205):748–53.
15. Jung I, Xiao J, Malyarchuk V, *et al.* Dynamically tunable hemispherical electronic eye camera system with adjustable zoom capability. Proceedings of the National Academy of Sciences. 2011;108(5):1788–93.
16. Keum D, Jung H, Jeong K. Planar emulation of natural compound eyes. Small. 2012;8(14):2169–73.
17. Huang C-C, Wu X, Liu H, Aldalali B, Rogers JA, Jiang H. Large-field-of-view wide-spectrum artificial reflecting superposition compound eyes. Small. 2014;10(15):3050–7.
18. Tanida J, Kumagai T, Yamada K, *et al.* Thin observation module by bound optics (TOMBO): concept and experimental verification. Applied Optics. 2001;40(11):1806–13.
19. Duparré J, Wippermann F, Dannberg P, Bräuer A. Artificial compound eye zoom camera. Bioinspiration & Biomimetics. 2008;3(4):046008.
20. Kim K, Kamiuchi H, Masamune K, Dohi T. A new, safer, controllable field-of-view endoscope avoiding movement inside body cavities. Medical Engineering and Physics. 2011;33(2):174–9.
21. Adelson EH, Wang JYA. Single lens stereo with a plenoptic camera. IEEE Transactions on Pattern Analysis and Machine Intelligence. 1992;14(2): 99–106.
22. Levoy M, Hanrahan P. Light field rendering. Proceedings of the 23rd Annual Conference on Computer Graphics and Interactive Techniques. 237199: New York: ACM; 1996. p. 31–42.
23. Lippman G. La photographie integrale. Comptes Rendus de l'Académie des Sciences. 1908;146:6.
24. Wilburn B, Joshi N, Vaish V, *et al.* High performance imaging using large camera arrays. ACM SIGGRAPH 2005 Papers; Los Angeles, California. 1073259: New York: ACM; 2005. p. 765–76.
25. Veeraraghavan A, Raskar R, Agrawal A, Mohan A, Tumblin J. Dappled photography: mask enhanced cameras for heterodyned light fields and coded aperture refocusing. ACM Transactions on Graphics. 2007;26(3):69.
26. Chih-Chieh C, Yi-Chang L, Ming-Shing S, editors. Light field based digital refocusing using a DSLR camera with a pinhole array mask. 2010 IEEE

International Conference on Acoustics Speech and Signal Processing (ICASSP); Mar. 14–19, 2010.

27. Ng R, Levoy M, Brédif M, Duval G, Horowitz M, Hanrahan P. Light field photography with a hand-held plenoptic camera. Stanford University Computer Science Technical Report. 2005; 2005–02.
28. Aldalali B, Li C, Zhang L, Jiang H. Micro cameras capable of multiple viewpoint imaging utilizing photoresist microlens arrays. Journal of Microelectromechanical Systems. 2012;21(4):945–52.
29. Aldalali B, Fernandes J, Almoallem Y, Jiang H. Flexible miniaturized camera array inspired by natural visual systems. Journal of Microelectromechanical Systems. 2013;22(6):1254–6.
30. Szema R, Rastegar J, Lee L. An artificial compound eye for stereoendoscopy. Journal of Medical Engineering & Technology. 2004;28(3):117–24.
31. Chen H-S, Lin Y-H. An endoscopic system adopting a liquid crystal lens with an electrically tunable depth-of-field. Optics Express. 2013;21(15):18079–88.
32. Kagawa K, Shogenji R, Tanaka E, Yamada K, Kawahito S, Tanida J, editors. Variable field-of-view visible and near-infrared polarization compound-eye endoscope. Engineering in Medicine and Biology Society (EMBC), 2012 Annual International Conference of the IEEE; Aug. 28–Sep. 1, 2012.
33. Dowski ER, Cathey WT. Extended depth of field through wave-front coding. Applied Optics. 1995;34(11):1859–66.
34. Kawahito S, Kagawa K. Endoscopic application of a compact compound-eye camera. Makara Journal of Technology. 2014; 18(3):128–32.
35. Zeng X, Smith CT, Gould JC, Heise CP, Jiang H. Fiber endoscopes utilizing Liquid tunable-focus microlenses actuated through infrared light. Journal of Microelectromechanical Systems. 2011;20(3):583–93.
36. Barbero S, Rubinstein J. Adjustable-focus lenses based on the Alvarez principle. Journal of Optics. 2011;13(12):125705.
37. Kanhere A, Van Grinsven KL, Huang C-C, *et al.* Multicamera laparoscopic imaging with tunable focusing capability. Journal of Microelectromechanical Systems. 2014;23(6):1290–9.
38. Kanhere A, Aldalali B, Greenberg JA, Heise CP, Zhang L, Jiang H. Reconfigurable micro-camera array with panoramic vision for surgical imaging. Journal of Microelectromechanical Systems. 2013;22(5):989–91.
39. Tsai FS, Johnson D, Francis CS, *et al.* Fluidic lens laparoscopic zoom camera for minimally invasive surgery. Journal of Biomedical Optics. 2010;15(3):030504.
40. Tsai FS, Johnson D, Sung Hwan C, Qiao W, Arianpour A, Yu-Hwa L, editors. Bio-inspired fluidic lens surgical camera for MIS. Engineering in Medicine and Biology Society, 2009 EMBC 2009 Annual International Conference of the IEEE; Sep. 3–6, 2009.
41. Tsai FS, Johnson D, Francis CS, *et al.*, editors. Fluidic lens laparoscopic zoom camera for minimally invasive surgery. 2010 Conference on Lasers and Electro-Optics (CLEO) and Quantum Electronics and Laser Science Conference (QELS); May 16–21, 2010.

Chapter 6

Biologically inspired optical surfaces for miniaturized optical systems

Ki-Hun Jeong, Jae-Jun Kim, Dongmin Keum, Hyukjin Jung, Minhee Kang and Sung-Pyo Yang

6.1 Introduction

Mining smartness from vision organs found in nature becomes of much interest in optical applications such as imaging, display, or lighting. Unlike conventional bulk optics, miscellaneous hierarchical structures at micro- or nanoscale deliver highly efficient light management with a small form factor. For example, natural species have evolved their eyes to obtain all necessary visual information from surrounding environment. Natural imaging schemes can be chiefly classified by three different types of pinholes, camera, and compound eyes. Pinhole eyes found in clam are sensitive enough to allow the animals to protect themselves from dangerous environment but not so sensitive to collect all visual inputs. Unlike other types, the pinhole eye as one of natural eyes with the simplest and thinnest optical configurations is well known for infinite depth-of-field, i.e., no blurred imaging depending on object distance and thus these unique features have been implemented on early-stage simple camera imaging systems. Besides, advanced pinhole eyes are also found in viper snakes of Crotalinae and some python of Boidae, which combine both pinhole eye and ordinary camera eye in order to confer infrared (IR) as well as visible imaging for warm-blooded preys. Compound eyes found in arthropods exhibit many intriguing features for wide field-of-view (FOV), fast motion detection, polarization sensing, color imaging, or high-resolution imaging with compact optical configuration unlike other types. They comprise arrays of integrated optical units called ommatidia. The individual component consists of a facet lens, a crystalline cone, a light-guiding rhabdom, and photoreceptor cells. Furthermore, nature exhibits ten different optical schemes of compound eyes, which have some attractive figures-of-merits for sustainable life style in visual acuity, photon collection efficiency, and polarization or spectral sensitivity. Such biological inspiration recently and actively provides new opportunities for

Department of Bio and Brain Engineering, Korea Advanced Institute of Science and Technology (KAIST), 291 Daehak-ro (373-1 Guseong-dong), Yuseong-gu, Daejeon 305-701, Republic of Korea improving optical capability of conventional imaging systems by incorporating nano- and microfabrication methods and furthermore it delivers technical solutions for miniaturized optical systems in medical, industrial, and military fields. In this chapter, we will review engineering approach inspired from diverse biological organs, which can be utilized for miniaturized optical systems in diverse optical applications.

6.2 Biological inspiration from index gradient

6.2.1 Natural gradient index

Animal eyes have been evolved for optimal survival by using index gradients to enhance focusing power, correct aberrations, and simplify optical configuration. The index gradient can be mainly found in either human eyes or fish eyes. The human eye has only two lens elements of a cornea and a crystalline lens but produces nearly aberration-free imaging. The crystalline human lens serves as an aspheric compensator for correcting the corneal induced-spherical aberrations while avoiding any major off-axis coma generation. This unique optical aberration correction is caused by a dual compensation mechanism of an aspheric lens surface shape and an internal gradient index distribution. Consequently, the human eyes can offer relatively aberration-free images with only two optical components of the cornea and a gradient refractive index (GRIN) crystalline lens. The refractive index distribution of the human eye lens is intrinsically created by approximately 22,000 non-planar layers of proteins whose index difference is about 0.03, radially changing from a maximum refractive index, $n = 1.42$, at the lens core to a minimum refractive index, $n = 1.37$, at the lens surface. Some fish eyes with a spherical lens typically provide the gradient index in order to almost perfectly correct spherical aberration.

The refractive index of a lens as well as its surface shape serves as the key parameters that determine the refractive power; the refractive index of the surrounding media also contributes to the lens refraction [1]. In a homogeneous index medium, refraction takes place at the lens surface and the ray continues without deviation until it is refracted again at the surface point from which it exits the lens. In the case of a gradient index lens, such as in the eye, refraction continues inside the lens medium. The index gradient determines the lens power and thus the image quality on the retina by minimizing aberrations (deviations from a perfect point focus that deteriorates the image quality), which vary across species with different gradient forms. This variation depends on the medium constituents, the lens proteins, their concentrations and magnitudes, and how much each protein class contributes to the refractive index. Variations, most evident across species, have been carefully designed so that the refractive power and the aberration control was implemented on the cornea optics as well as with the biometry of the eye such as axial length or position of the photoreceptor layer to meet the functional visual demands.

The eyes of aquatic animals typically have higher optical power than those of terrestrial animals because the lens power of the cornea is significantly reduced in water. Piscine lenses have all been found to be spherical or nearly spherical. Spherical aberration obviously expected from such lenses with homogenous index is significantly reduced by the gradient index profile. A comparison of studies using non-invasive techniques (mostly ray tracing) demonstrates that the piscine lens gradients are relatively steep and have maximal values at the center, which are among the highest values found in nature. Magnitudes reach from 1.540 in the African Cichlid fish *Haplochromis burtoni* to 1.55–1.57 in a goldfish *Carassius auratus auratus* for a wavelength of 633 nm. The lowest refractive index found at the lens edge varies from 1.35 for the goldfish *C. auratus auratus* to 1.408 for the Blue Eye Trevally lens *Hyperoglyphe antarctica*. For other wavelengths, the edge index was 1.3716 in the lens of the trout *Oncorhynchus mykiss* for 550 nm and 1.368 in the lens of the Black Dory *Allocyttus niger* for 590 nm. Most fishes have similar values at the edges and centers of their lenses with the exception of the Blue Eye Trevally, which has a very large eye, e.g., over 9.5 g as well as lens relative to its size. With such a high refractive index at the lens surface, a substantial refraction occurs at the interface between the lens and surrounding media. In terms of visual demands based on habitat, it has not been well explained why this lens should be different from other fish species.

6.2.2 Mimicking index gradients

New paradigms for bioinspired optical designs that could not be implemented by using homogeneous lens materials have been extensively explored by using active spatial modification, layer-by-layer methods, and subwavelength structures to control the refractive index distribution. The Index gradients with diverse functional forms can be spatially modulated in both the radial and axial directions and thus enable the fabrication of GRIN lenses. For instance, the refractive index of optical materials can be spatially modified by varying the neutron radiation dose over selected areas of a lens to correct for different types of optical aberrations [2]. Depending on the irradiation times, neutron irradiation leads to index modifications of the required order of magnitude with the stability of the index modifications as well as the absence of residual radioactivity to make an irradiated lens suitable for optical applications. A parabolic-index profile can be realized on gradient-index rod lenses by using a double Na–Ag ion exchange process [3]. In particular, the high diffusing rate for Ag ions allows the implementation of large-sized (larger than 10 mm) rod lenses for photographic uses. The index gradient of Maxwell fish-eye spherical lenses has been also demonstrated by using ion exchanging in optical glasses [4]. Phosphate glasses serve as a material for GRIN-lens production by silver–alkaline ion exchange, where phosphate glasses transform in this sequence from one-dimensional (1D) to three-dimensional (3D) space arrangement of the glass-forming network and realize the maximum index variation of ~0.17 [5]. In addition, ultraviolet (UV) energy-controlled exposure process can be employed for the fabrication of GRIN lenses [6]. A V-shaped gel zone was formed in the reaction tube to prevent the formation of bubbles in the polymer matrix after photo-induced polymerization. Gel effects on the fabrication and properties of GRIN polymer rods and the effect of concentrations of diphenyl sulfide, poly(methyl methacrylate) (PMMA), and photo initiator on the optical properties of the rod lenses were well investigated depending on UV exposure dose. Recently, a maskless lithography tool and optically initiated diffusive photopolymer enable arbitrary two-dimensional (2D) gradient index (GRIN) polymer lens profiles such as quadratic, millimeter aperture GRIN lenses, Zernike polynomials, and GRIN Fresnel lenses [7]. The lithography tool uses a pulse-width modulated deformable mirror device to control the 8-bit grayscale intensity pattern on the material. A custom polymer responds with a self-developing refractive index profile that is nonlinear with exposure dose. This nonlinear material response can be pre-compensated with the intensity pattern to yield high fidelity, optically induced index profiles.

A symmetrical spherical index distribution can be achieved in a plastic spherical lens by using suspension polymerization in a single operation without grinding or polishing [8]. The index profile of the spherical GRIN lens was measured by shearing interferometry. Embedded in the transparent polymer solid, the GRIN sphere with almost parabolic index distribution remarkably reduces the spherical aberration compared with that of a homogeneous sphere. Besides, GRIN spheres with quadratic- and linear-index distributions were further obtained by two different methods to confirm the effect of the GRIN profile on the focusing property of the sphere lens [9]. An interfacial gel-copolymerization technique can also be utilized for the monolithic fabrication of radial-, axial-, or spherical-GRIN polymer materials, which demonstrated steric radial-GRIN optical circuit components and 2D GRIN lens arrays with quadratic index distribution against the distance from the center axis [10]. Novel fiber-optic gradient-index (GRIN) lenses can also be fabricated by using a modified chemical vapor deposition process [11]. Fusing GRIN fiber lenses onto a single-mode fiber yields automatically aligned all-fiber probes that insert into tissue through hypodermic needles as small as 31-gauge (inner diameter, 127 μm) to measure microscopic brain motions in vivo by using low-coherence interferometry (LCI). Nanolayer polymer extrusion also allows the fabrication of practical GRIN polymer optical materials with an index contrast of ~0.17 for flat GRIN lenses and shaped lenses with specified internal index gradients [12]. The nanolayered films were stacked to create a material with an index gradient depending on the stacking order of different nanolayered films and therefore these materials were molded into the GRIN lenses or a shaped lens. The biological layer motif of human eye inspires a new class of polymer optics with index gradients such as *f*/2.25 GRIN singlet based on a nanolayered polymer composite material in order to correct spherical aberration [13]. Compared to a homogeneous BK7 glass singlet with a similar *f*-number, the modulation transfer function of GRIN singlet lens is better at all spatial frequencies and even its weight is approximately an order of magnitude less than the homogeneous glass lens. A nanolayered polymer films approach for fabricating GRIN lenses with controlled refractive index distribution profiles and an independently prescribed lens surface geometry have been demonstrated to reduce the overall wavefront error [14]. This approach utilized nanolayered polymer materials, which constructed with PMMA and a styrene-co-acrylonitrile copolymer with a tailored refractive index intermediate to bulk materials. The fabrication of nanolayered polymer GRIN lens was achieved through thermoforming and finishing steps. This method has been utilized for a *f*/#2.25 spherical GRIN plano-convex singlet with one quarter of the weight of a similar BK7 lens and a bioinspired aspheric human eye GRIN lens.

Nano-structured graded index micro-optical components have been recently demonstrated by using the stack-and-draw method used for photonic crystal fibers [15]. Spherical nano-structured microlenses with a flat facet were fabricated with this method and designed using an algorithm based on the Maxwell Garnett mixing formula that describes the discrete structures with an effective refractive index distribution. Fabrication at subwavelength scales can discover novel artificial media with anomalous optical features called metamaterials. All-dielectric metamaterials possess high future potential for practical applications over the whole electromagnetic spectrum because of low loss and optical isotropy. Some metamaterials have been actively implemented to harness the flexibility of gradient index materials for imaging applications [16]. For example, a broadband lens with a FOV of ~180° and zero *f*-number has been successfully demonstrated by incorporating transformation optics.

Gradient metasurfaces are 2D optical devices which can manipulate light by imparting local, space-variant phase changes on an incident electromagnetic wave. These optical surfaces have thus far been constructed from nanometallic optical antennas, and high diffraction efficiencies have been limited for only reflection modes. Dielectric gradient metasurface optical elements with high efficiencies in transmission mode in the visible spectrum have been very recently demonstrated in a visible range [17]. Ultrathin gratings, lenses, and axicons have been fabricated by patterning a 100-nm-thick silicon layer into a dense arrangement of Si nanobeam antennas. In addition, subwavelength silicon through-hole arrays can serve as an all-dielectric gradient index metamaterial with broadband terahertz operation [18]. The unit cell consists of a single subwavelength through-hole on highly resistive monocrystalline silicon. Depending on the fill-factor and period, the effective index was linearly modulated at 0.3–1.6 THz. The experimental results also indicate Si-GRIN lenses with parabolic index profiles through the spatial modification of a single unit cell along the radial direction. Si-GRIN lenses either focus 0.4–1.6 THz beam down to the diffraction-limit size or serve as a flat and thin solid immersion lens on the backside of terahertz photoconductive antenna for highly efficient pulse extraction. This all-dielectric GRIN metamaterial opens up great opportunities for integrated terahertz GRIN optics. The use of semiconductors can broaden the general applicability of gradient metasurfaces, as they offer facile integration with electronics and can hasten the realization by mature semiconductor fabrication technologies.

6.2.3 Summary

Bioinspired GRIN optics has actively demonstrated distinctive advantages for singlet size and mass reduction of traditional lens systems. For decades, polymer materials have extensively derived but most previous methods are still struggling in demonstrating significant GRIN advantages. Material selections, precise control of index, and mass production have served as bottlenecks for real applications. However, gradient metasurface incorporating with transformation optics recently becomes of much interest in taking full benefits for practical GRIN optics.

6.3 Biological inspiration from focal tunability

6.3.1 Tunable focus found in nature

Optical focalization is a fundamental feature of diverse devices for different fields of application, such as consumer electronics (e.g., cameras, mobile phones, surveillance systems), medical diagnostics (e.g., endoscopes and lab-on-a-chip units), and optical communications (e.g., optical fiber components). In conventional active optical systems, focal length tuning is usually obtained by displacing one or more constant-focus lenses. However, miniaturization of such systems is typically complex and expensive owing to the need for moving parts and thus the resulting structures are prone to be bulky and limited by friction.

An imaging system with a focal length tunability enables an animal to capture accurate and clean images of target objects at different distances. For instance, the human eye serves as the most familiar tunable imaging system found in nature. Target objects at different distances are focused by a flexible and controllable crystalline lens. The ciliary muscles control tension on the lens to change its curvature and thus its focal length. Including only human eyes, the animal kingdom exhibits diverse types of tunable eyes. The bird eye has the added control of reshaping and deforming the cornea as well [19]. Brucke's muscles attached to bony ossicles in reptiles and birds actively change the lens thickness. Birds have an additional muscle, Crampton's muscle, which can alter the shape of the cornea. In contrast, the whale eye uses hydraulics to move the lens itself closer or farther from the retina; a chamber behind the lens is filled or emptied with fluid depending on the focal length needed [20]. This design allows for good vision in and out of the water, and compensates for increased pressure in deeper aquatic environments. The protractor lentis in some amphibian eyes moves a fixed-shape lens closer or farther from the retina for accommodation.

6.3.2 Biomimicry

While conventional focusing systems are reliable and have a good focusing ability, their use of glass lenses and an electric motor to control the distance between the glass lenses make them difficult to include in a compact lens system. Besides, precise and expensive mechanisms are required to position the lenses along an optical axis. To overcome such drawbacks, tunable-focus optical lenses with no moving parts are intensively studied. Different types of technologies are being explored, using "smart" materials driven by a variety of stimuli, such as hydraulic/pneumatic. Examples based on electrical driving include lenses integrated with shape-memory alloys (SMAs), liquid-crystal lenses, and liquid lenses. In hydraulically or pneumatically driven lenses, an external pump pressurizes a fluid into a lens-shaped flexible chamber. Recently, fluidic actuation was used in an electronic eye camera system consisting of a tunable lens and an array of photodetectors on an elastomeric membrane; hydraulics allowed for changes of curvature not only of the lens but also of the detector-supporting membrane, yielding to improve image quality. Lenses with electromagnetic activation include examples filled with ferrofluids that are displaced by electromagnetic fields, as well as magnetic-responsive membranes that pump inert fluids into flexible chambers. Photo-activated lenses can be made with photosensitive polymer membranes that bend under irradiation of light. Thermally or pH driven lenses have been demonstrated by exploiting volume changes of either a liquid in a chamber, or a hydrogel changing the volume and shape of a liquid droplet. Electro-thermal activation has also been reported with SMA-driven lenses. In electrochemically tunable lenses, the shape of a liquid droplet is varied by modifying its surface tension through redox processes. Focal length changes are obtained in liquid-crystal lenses by varying the refractive index with an applied electric field. Electrowetting-based liquid lenses exploit electro-capillarity to change the contact angle between two immiscible liquids.

6.3.2.1 Pneumatic membrane deformation

An intriguing example for the tunable lens, external pressure manipulates the shape of a liquid droplet and thereby adjusts its optical properties. Chronis *et al.* demonstrated an elastomer-based tunable liquid-filled microlens array integrated on top of a microfluidic network using soft lithographic techniques [21]. The simultaneous control of the focal length of all the microlenses composing the elastomeric array was accomplished by pneumatically regulating the pressure of the microfluidic network. A focal length tuning range of hundreds of microns to several millimeters was achieved. Such an array can be used potentially in dynamic imaging systems and adaptive optics. Jeong *et al.* demonstrated a tunable micro-doublet lens capable of creating dual modes of biconvex or meniscus lens (Figure 6.1) [22]. The microdoublet lens consists of a tunable liquid-filled lens and a solid negative lens. It can be tuned either by changing the shape of the liquid-filled lens into bi-convex or meniscus or by changing a filling media with different refractive indices. The microfabrication is based on photopolymer microdispensing and elastomer micromolding methods. The microdoublet lens can provide a solution for minimizing optical aberrations and maximizing the tunability of focal length or FOV by controlling variable and fixed lens curvatures. More recently, Zhang *et al.* demonstrated a self-contained, liquid tunable microlens has been developed by using polyacrylate membranes integrated with compact on-chip thermo-pneumatic actuation fabricated using full-wafer processing [23]. Silicone oil as the optical liquid is pushed or pulled into the lens cavity via an extended microfluidic channel structure without any pumps, valves, or other mechanical means. The heat load generated by the thermal actuator is physically isolated from the lens chamber. The back focal length may be tuned from infinity to 4 mm with a maximum power consumption of 300 mW. The principal application is fine tuning of the back focal length, for which tuning time constants as small as 100 ms are suitable. Iimura *et al.* also demonstrated a liquid-filled tunable lenticular lens for

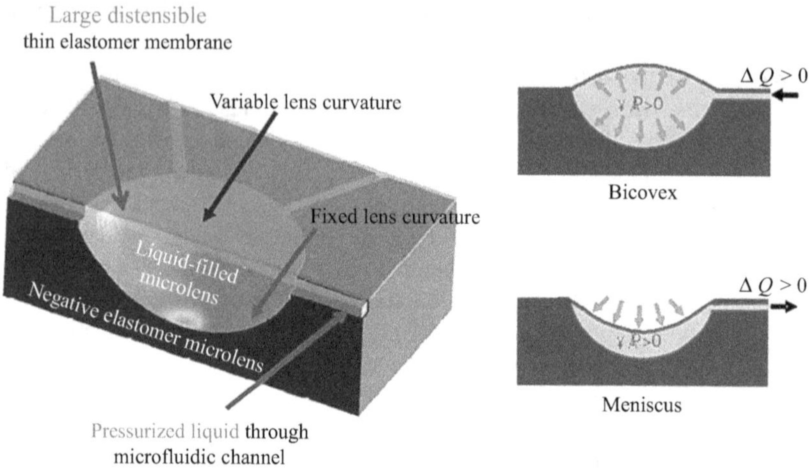

Figure 6.1 Pneumatic tunable lens. The basic configurations of a tunable microdoublet lens consisting of a tunable liquid-filled microlens and a solid negative lens of different refractive indices ($n_{elastomer}$ and n_{liquid}) acting in combination. Reprinted with permission from Jeong (2003) [[22]], copyright the Optical Society

switching between 2D and 3D images in naked-eye 3D displays [24]. Compared with previous 2D/3D switchable displays, this tunable lenticular lens that is directly attached to a smartphone display can project both a 2D image with the original resolution of the smartphone display and a 3D image with high brightness. This lens is simply composed of transparent polydimethylsiloxane (PDMS) microchannels. While the thin top membrane on the microchannels is normally flat to transmit light without deflection for displaying 2D images, applying pressure to the microchannel deforms the membrane to acquire characteristics of lenticular lenses for 3D images.

6.3.2.2 pH sensitive volume change

Dong *et al.* first demonstrated a stimuli–responsive liquid lens system that allows for autonomous focusing, i.e., with a pH-sensitive hydrogel integrated into a microfluidic system and serving as the container for a liquid droplet, with the hydrogel simultaneously sensing the presence of stimuli and actuating adjustments to the shape—and hence focal length—of the droplet (Figure 6.2) [25]. By working at the micrometer scale where ionic diffusion and surface tension scale favorably 14, this work can use pinned liquid–liquid interfaces to obtain stable devices and realize response times of ten to a few tens of seconds. The microlenses, which can have a focal length ranging from 23 to −12 mm (convergent and divergent). Ding and Ziaie further demonstrated a two-step casting process to fabricate a bifunctional hydrogel-based microlens array, which responds to both temperatures, becomes opaque above certain temperature, and pH, changes its focal length at different pH levels, and can be operated in air for an extended period of time [26]. Each lens in the array is 1 mm

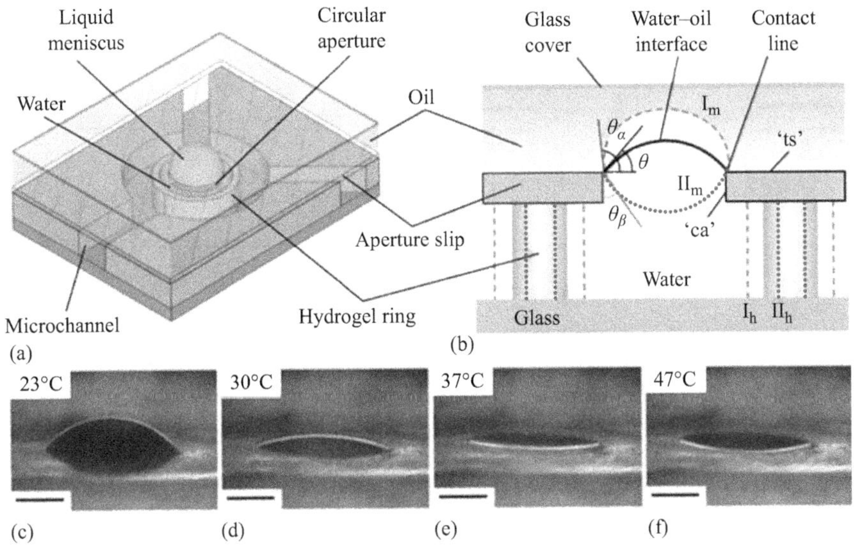

Figure 6.2 *pH sensitive tunable lens. Smart microlens using a pinned liquid–liquid interface. (a) The water–oil interface forms the liquid microlens. The microchannels allow the flow of fluids to the microlens structure. (b) Smart variable-focus mechanism. The hydrophilic sidewall and bottom surface ("ca") and hydrophobic top surface ("ts") of the aperture pin the water–oil meniscus along the contact line "ca–ts." The expansion and contraction of the hydrogel regulates the shape of the liquid meniscus by changing the angle θ of the pinned water–oil interface. The dashed lines show the expanded state of the hydrogel ring ("I_h") and the corresponding divergent microlens ("I_m") at $\theta = \theta_\alpha$. The dotted lines show the contracted state of the hydrogel ring ("II_h") and the corresponding convergent microlens ("II_m") at $\theta = -(90° - \theta_\beta)$. (c)–(f) The shape of the liquid microlens varies with local environmental temperature. Scale bar: 1.0 mm. Reprinted with permission from Dong (2006) [25], copyright Nature Publishing Group*

in diameter and its focal length changes from 4.5 to 55 mm when the environmental pH is varied between 2.0 and 5.0. The light-switching capability is measured to be 92% when temperature increases from 25 to 35°C.

6.3.2.3 Artificial muscles

Inspired by the biomechanics of the human eye, Carpi *et al.* first demonstrated an electrically tunable lens made of dielectric elastomers (DEs), an emerging class of "artificial muscle" materials for actuation as shown in Figure 6.3 [27]. The optical device is inspired by the architecture of the crystalline lens and ciliary muscle of the human eye. It consists of a fluid-filled elastomeric lens integrated with an annular elastomeric actuator working as an artificial muscle. Upon electrical activation, the

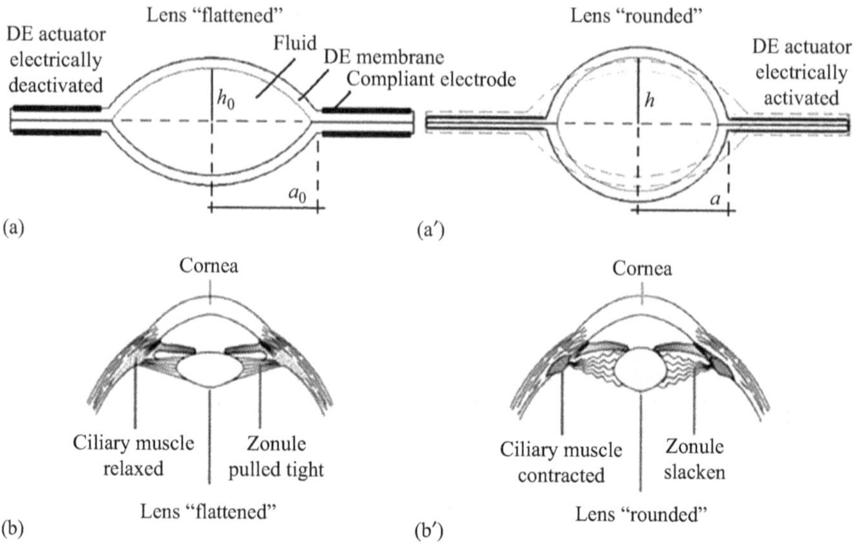

Figure 6.3 Biomimetics of muscle. Functional analogy between the bioinspired lens and the human lens. (a) and (b) Schematic sectional views (not in scale) of the two systems in the rest state. (a′) and (b′) Corresponding views in an activation state. The annular DE actuator works as an artificial muscle, functionally analogous to the combined ciliary muscle and zonule; it radially stretches and relaxes the lens, so as to change the radius of curvature. Reprinted with permission from Carpi (2011) [27], copyright John Wiley and Sons

artificial muscle deforms the lens, so that a relative variation of focal length comparable to that of the human lens is demonstrated. The device combined optical performance with compact size, low weight, fast and silent operation, shock tolerance, no overheating, low power consumption, and possibility of implementation with inexpensive off-the-shelf elastomers. Results show that combining bioinspired design with the unique properties of DEs as artificial muscle transducers has the potential to open new perspectives on tunable optics. Choi *et al.* also demonstrated the variable focus lens system consists of a PDMS lens, winding-type SMA actuator, and load arms as a counterpart of crystalline lens, ciliary muscles, and zonular fibers of the human eye [28]. The PDMS lens is encircled and stretched by the load arms joined to an outer ring that is rotated by the winding-type SMA actuator, thereby changing the focal length of the lens. In contrast to other single tunable-focus lenses, the proposed system uses a gel-type PDMS lens, which is insensitive to gravity and external vibration, and can be made into a biconvex or aspheric shape.

6.3.3 Summary

In summary, active visual organs of animal kingdom clearly show the potential solution to enable a new generation of adaptive optical lenses, which are suitable with small form factor (i.e., compact size and low weight), fast operation, and low power consumption. Moreover, these biomimetic tunable lenses can be readily integrated into arrays on common optical devices that may find use in applications such as optical sensing, 2D/3D imaging [24, 29], beam steering, medical diagnostics, and lab-on-a-chip technologies.

6.4 Biological inspiration from wide field of view

6.4.1 Compound eyes found in nature

Compound eyes consist of an array of optical units called ommatidia whose individual optical element contains a facet lens, a crystalline cone, a rhabdom, and photoreceptor cells (Figure 6.4(a) and (b)). From several hundred up to tens of thousands of ommatidia along a curved surface receive visual signal from slightly different direction and thus independently contribute to the formation of the final image. Each ommatidium is surrounded by screening pigment cells that prevent an incoming light with a large incident angle [30].

In nature, around a dozen different optical designs of compound eyes exist depending on the light focusing mechanisms of ommatidia. The schemes can be divided by seven apposition eyes, three superposition eyes and one unusual eye. The apposition eyes cover simple apposition, open rhabdom, neural superposition, afocal apposition, transparent light guiding, transparent axial gradient, and transparent radial gradient eye. The superposition eyes can be further classified into refracting, reflecting, and parabolic eye (Figure 6.5) [31]. The simple apposition eye is the most basic design of compound eyes, where rhabdom is connected at the focal plane of a corneal lens. The open rhabdom eye holds the unique feature of an ommatidium with separated rhabdomeres with photoreceptor cells, where each rhabdomere directs to slightly different orientations. The open rhabdom can be evolved with the neural superposition by merging the outermost rhabdomeres within a single ommatidium with those of neighbor ommatidia to enhance photon collection efficiency without loss of angular

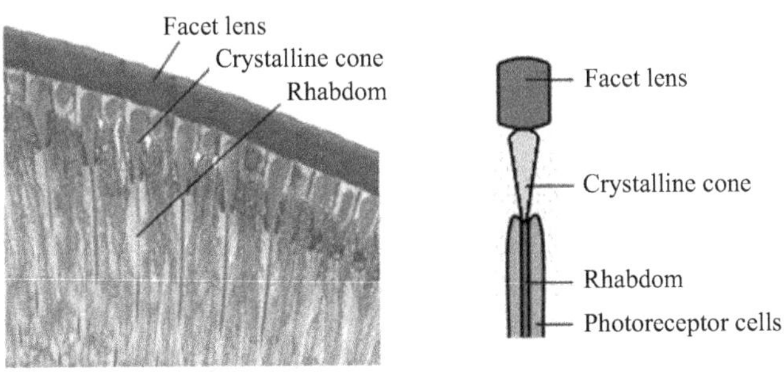

(a) Natural compound eye (b) Natural ommatidium

Figure 6.4 *(a) A histology section of a honeybee's eye. (b) Each ommatidium consists of a facet lens, a crystalline cone, a rhabdom, and photoreceptor cells. Reprinted with permission from Jeong (2006) [49], copyright the American Association for the Advancement of Science*

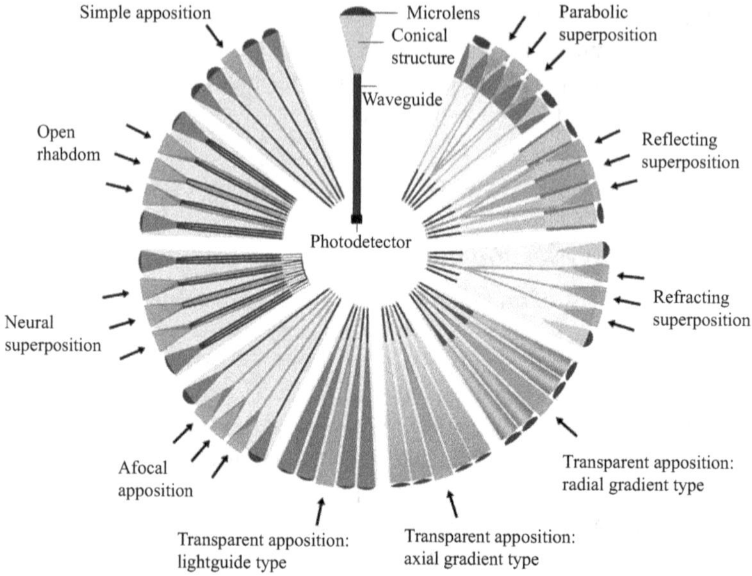

Figure 6.5 Schematic illustration of the optical types of compound eyes. The optical designs can be divided into two groups: apposition and superposition eyes. The apposition eyes include simple apposition, open rhabdom apposition, neural superposition, afocal apposition, transparent apposition lightguide, transparent apposition axial gradient, and transparent apposition radial gradient eyes. The superposition eyes consist of refracting, reflecting, and parabolic superposition eyes. The terms of a mirolens, a conical structure, and a waveguide are used for referring to a corneal facet lens, a crystalline cone, and a rhabdom, respectively. Reprinted with permission from Keum (2012) [36], copyright John Wiley and Sons

resolution. The afocal apposition eye has an additional optical component in the crystalline cone. The proximal part of the crystalline cone includes a graded-index region, behaves as an additional proximal lens, and the corneal facet lens and the proximal lens together contribute to the image formation. This unique type provides narrow angular sensitivity relatively compared to the simple apposition eye. The transparent apposition eyes of a lightguide type, axial gradient type, and radial gradient type have little pigmentation for camouflage. The ommatidia can be optically isolated by extremely long crystalline cone or additional graded index lens in the crystalline cone. In contrast, the superposition eyes have clear zone where the crystalline cone and the rhabdom are physically separated to superpose a bundle of light from neighboring corneal lenses into a single rhabdom. The crystalline cone serves as light re-collimation by implementing diverse methods such as refracting, reflecting, or parabolic mirror. The optical schemes can improve photon sensitivity compared to apposition eyes because one rhabdom collects light from several corneal lenses

[31, 32]. Interestingly, nature also provides an unusual visual system, which is not included in apposition or superposition eyes. For example, *Xenos peckii*, an endoparasite of paper wasps, has only about 50 lenses per eye, an order of magnitude fewer than conventional insect eyes. Each lens has their own retina, containing more than 100 photoreceptors. This unit is called an eyelet instead of an ommatidium, and the eyelet captures image of different part within total field of view. Large diameter of facet lens and multiple photoreceptors under each lens result in higher resolution and better sensitivity compared to conventional compound eyes [33–35].

6.4.2 Biomimicry

Compound eyes have attracted great research interest in advanced photonic systems because the optical schemes exhibit distinguished benefits in wide FOV, fast motion detection, and low aberration. The physiological features and the visual information processing mechanism may provide novel technical solutions for miniaturized optical systems in military, industrial, and medical fields.

6.4.2.1 Planar emulation of natural compound eyes

Keum *et al.* reported a novel approach for understanding the optical designs and functions of natural compound eyes, based on planar micro-optics as shown in Figure 6.6(a). Natural ommatidium can be emulated by a cylindrical microlens, a conical structure, a rib waveguide, and a photodetector. Among ten different types in nature, two representative types of the simple apposition and the reflecting superposition were emulated with 2D polymer structures by photolithography. Light propagation inside both types was directly visualized by patterning the structures with fluorescence dye (Rhodamine 6G) doped photosensitive polymer resin (SU-8). This optical scheme can be utilized not only for understanding the visual principles of natural compound eyes but also for designing the artificial compound eyes (ACEs) [36].

6.4.2.2 Artificial compound eyes with flat substrates

A number of ACEs were implemented on a flat substrate due to the fabrication technology limitations and conventional image sensors. The optical designs are suitable for combining with the commercialized image sensor to function as camera modules, but the FOV is limited by the inherent flatness of the substrate. Tanida *et al.* developed a compact image-capturing system called thin observation module by bound optics (TOMBO), which is composed of multiple sets of optical units (Figure 6.6(b)). Each unit consists of a microlens, a separation layer to prevent crosstalk, and photodetector arrays. TOMBO captures a set of images with low spatial resolution simultaneously, and then the images are combined to the single image with high resolution by various image restoration algorithms, which correspond to the “superresolution”. They mostly focused on the reconstructing of high-resolution image rather than achieving wide FOV [37–41]. Fraunhofer Institute has reported various ACEs on a planar substrate. The ACE obtained extended FOV by adjusting the pitch difference of optical components in each layer. In their first version of the ACE, each channel detects optical signal from its own direction

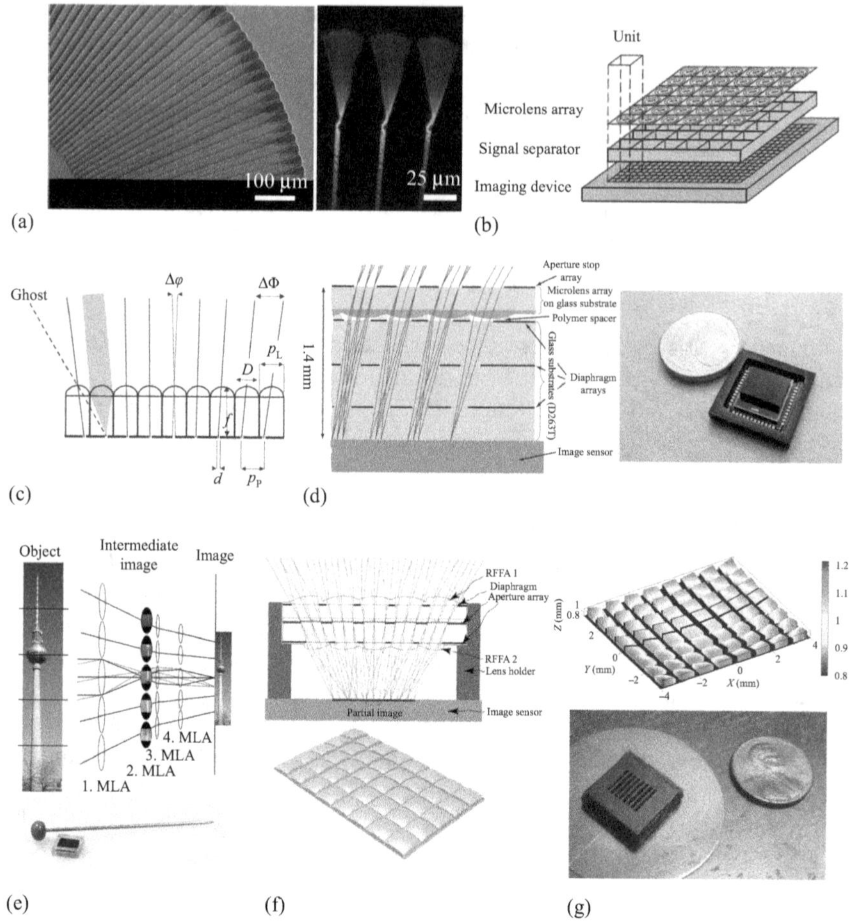

Figure 6.6 ACEs with flat substrates. (a) Planar emulation of natural compound eyes. (b) Thin observation module by bound optics (TOMBO). (c) Artificial apposition compound eye. (d) Schematic illustration of the electronic cluster eye and the optical image of the prototype. (e) Working principle of the optical cluster eye and the assembled system. (f) ACE with refractive freeform microlens arrays. (g) Surface profile of the ACE with freeform microlens array and assembly of the camera. Reprinted with permission from Keum (2012) [36], copyright John Wiley and Sons; from Tanida (2015) [37], copyright Springer; from Duparré (2005) [43], copyright the Optical Society; from Brückner (2010) [44], copyright the Optical Society, from Meyer (2011) [45], copyright the Optical Society; from Brückner (2014) [47], copyright Brückner (et al.); from Li (2012) [48], copyright the Optical Society

as a point, and the final image was obtained by combining those points like that of a natural apposition compound eye (Figure 6.6(c)) [42, 43]. After, they integrated the microlens with multiple diaphragms to reduce optical aberration and crosstalk. Each channel receives the visual information as a partial image, and all partial images are stitched together to reconstruct total final image. The demonstrated prototype achieves video graphics array resolution with a reduced thickness compared to single-aperture equivalents (Figure 6.6(d) and (e)) [44, 45]. Recently, they are also developing multi-aperture camera module using refractive freeform microlens arrays (RFFAs) to enhance image quality as shown in Figure 6.6(f), [46, 47]. Li and Yi proposed a unique freeform microlens array for a compact compound eye camera (Figure 6.6(g)). Freeform microlens arrays were manufactured by ultra-precision diamond machining and used to steer and focus incident light simultaneously. A micro aperture array was assembled with the freeform microlens arrays to prevent crosstalk between neighboring channels. The assembled module was directly mounted on a complementary metal-oxide-semiconductor (CMOS) image sensor to form a compact compound eye camera system, and the camera has a FOV of 48° × 48° [48].

6.4.2.3 Artificial compound eyes with curved surfaces

ACEs with curved surface are more analogous to the natural compound eyes. Imaging optics on a curved surface facilitates receiving light over wide range of view; however, commercially available photodetectors are not compatible with these schemes. Jeong *et al.* reported a biomimetic configuration of ACE using the polymer replication with the deformed elastomer membrane and self-written waveguides as shown in Figure 6.7(a). The hemispherical shape of the ACE with 8370 honeycomb-packed hexagonal microlenses was achieved by a reconfigurable PDMS membrane with microlens patterns. And then the artificial ommatidia were accomplished by a microlens-assisted self-written waveguides. The results show that both the physical dimensions and the optical properties are very comparable to natural compound eyes [49]. Song *et al.* presented an arthropod-inspired digital camera with stretchable hemispherical shapes of photodetector arrays (Figure 6.7(b)). The ommatidium of the camera consists of elastomeric convex microlens on the post-structure and one pixel of thin silicon photodetector. Each ommatidium is surrounded by black perforated sheet to eliminate stray light. The camera has totally 180 numbers of ommatidia, which means 180 resolvable points. Although the resolution is not enough for practical applications yet, the camera has wide FOV (160°), which is similar to natural compound eyes [50]. Floreano *et al.* also proposed an ACE integrated with curved photodetector (Figure 6.7(c)). The artificial ommatidia include three layers, optical layer composed of microlens arrays and aperture arrays, photodetector (CMOS chip) layer, and interconnection layer (printed circuit board). Three layers were assembled in planar configuration, and then diced down to the flexible interconnection layer, which remains intact. The ommatidial arrays were curved along the bendable direction and finally attached to a curved rigid semicylindrical substrate. The prototype has 180° × 60° of FOV, and motion extraction capability was

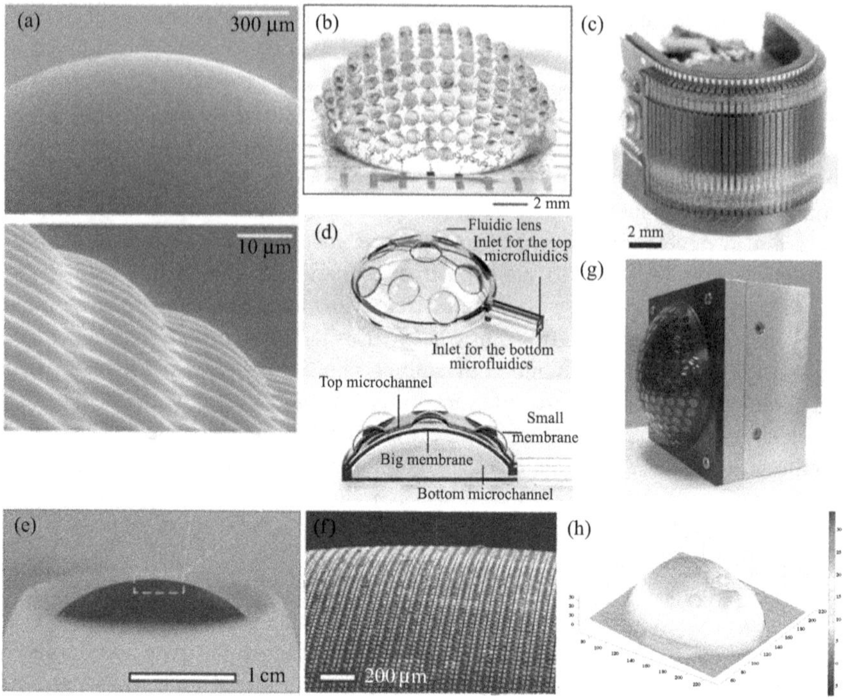

Figure 6.7 *ACEs with curved surfaces. (a) ACE with artificial ommatidium consists of a refractive polymer microlens and a self-written waveguide. (b) Digital camera with hemispherical photodetectors. (c) Miniature curved ACE. (d) Insect-human hybrid eye consists of bi-layered microfluidics. (e) and (f) Artificial reflecting superposition eye. (g) and (h) ACE for 3D object detection and the measured result. Reprinted with permission from Jeong (2006) [49], copyright the American Association for the Advancement of Science; from Song (2013) [50], copyright Nature Publishing Group; from Wei (2014) [52], copyright Royal Society of Chemistry; from Huang (2014) [53], copyright John Wiley and Sons; from Ma (2014) [54], copyright the Optical Society*

demonstrated [51]. An adaptive optofluidic lens combining the properties of inset and human eyes were developed by Wei *et al.* as given in Figure 6.7(d). They proposed a reconfigurable polymeric optofluidic device that combines the merits of both vision mechanisms, large FOV of compound eye, and adaptive focusing capabilities of human eye. The device consists of bi-layered microfluidics: an array of fluidic lenses integrated into the top layer and elastomeric big membrane embedded within the bottom layer. Fluidic lenses in the top layer serves to tune the focusing power, while the elastomeric membrane in the bottom layer controls the overall FOV. Focal length of each single lens and total FOV of the device were adjusted by alternating the fluid volume in each membrane [52]. Huang *et al.* reported an artificial reflecting superposition eye using silicon micro-square-tube arrays on a stretchable hemispherical membrane (Figure 6.7(e) and (f)). Natural reflecting superposition eye has less chromatic aberration, because the dispersion by the optical materials is minimized in reflection-based optical systems. Rays passing through the channels are superimposed at the hemispherical focal plane located halfway to the geometric center of the sphere. The device demonstrated exceptional FOV up to 165° without distortion and modest aberrations [53]. A compound eye can also be applicable for 3D object detection because multiple lenses are working together as a cluster. Ma *et al.* presented an ACE for 3D object detection, consisting of lenslet arrays and a light-steering meniscus lens (Figure 6.7(g)). A multi-position calibration method was developed for establishment of the relationship between the incident light rays and the image points. Measurement results show that the system can detect a 3D object with angular accuracy of better than 1 mrad (Figure 6.7(h)) [54].

6.4.3 Summary

In summary, the anatomical features of compound eyes and their inspired optical systems are reviewed in this chapter. Each type of compound eyes has its own figures-of-merits for obtaining visual information of surrounding environment. Spherically arranged imaging units facilitate wide FOV, high sensitivity of motion, and reduced optical aberration. Moreover, multi-channel nature of the architecture has great advantages in development of miniaturized imaging devices. Researchers have developed various configurations of ACEs, which have huge potential for advanced optical systems in military, industrial, and medical fields.

6.5 Biological inspiration from antireflection

6.5.1 Antireflection found in nature

Some insects benefit from antireflective (AR) structures, either on their eyes to see under low-light conditions or on their wings to reduce Fresnel reflections in transparent areas for camouflage. Therefore, antireflection occurs on the corneas of moth and butterfly eyes [55, 56] and on the transparent wings of moths and cicadas [57–59]. These consist of nipples, with rounded tips, in a hexagonal arrangement with a periodicity of around 240 nm (Figure 6.8(a) and (b)). Effectively they introduce a gradual refractive index profile at an interface between chitin (a polysaccharide, often embedded in a proteinaceous matrix; refractive index of 1.54) and air, and hence suppress reflectance by a factor of 10. Because the AR structures have dimensions less than the wavelength, i.e., located in the sub-wavelength scale or nanoscale, light is insensitive to the AR structures and tend to refract progressively as if the AR surface has a GRIN (Figure 6.8(c) and (d)). Even though the angle of incidence is changed, the AR surface still exhibits a relatively smooth change of refractive index toward the incident direction of light, thus suppressing the light reflection for a broad range of wavelength. Besides, the reflection of light with either s- or p-polarization can be suppressed to a very low level, because the transmission of light with different polarizations is insensitive to the media with extremely low disparity of refractive index [60, 61].

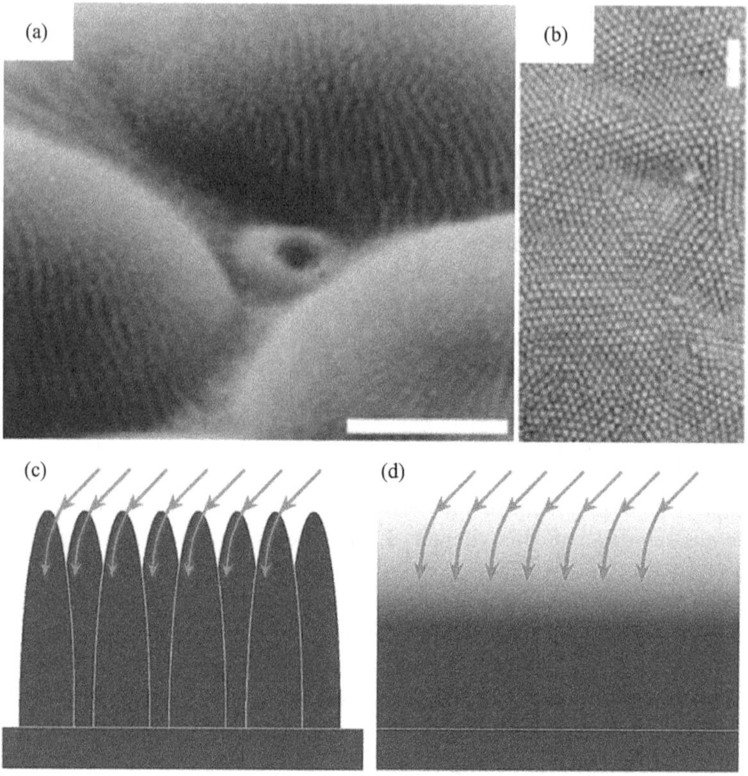

Figure 6.8 Antireflective nanostructures in nature. Scanning electron micrographs of AR surfaces. (a) Ridges on four facets on a 45-million-year-old dolichopodid fly's eye. Scale bar: 3 μm. (b) Moth-eye AR surfaces. Scale bar: 1 μm. (c) Interaction of incident light with the subwavelength-size nanoarray. (d) Schematic illustration of the refractive index change corresponding to (c). Reprinted with permission from Chattopadhyay (2010) [60], copyright Elsevier; from Cai (2015) [61], copyright Royal Society of Chemistry

6.5.2 Biomimicry

Since the "moth-eye structure" was first reproduced at its correct scale by crossing three gratings at 120° using lithographic techniques [62], the diverse methods of AR structure fabrication have been extensively applied in a biomimetic fashion on planar substrates such as silicon [63], III–V semiconductor [64], glass [65], or polymer [56].

Among many effective fabrication techniques for high-performance AR structures with various materials, dry etching by reactive ion etching (RIE) can produce AR surfaces based on both aperiodic and periodic nanostructure arrays (NSA) of Si, III–V compounds, and silica with or without an etching mask. Zhu *et al.* reported the fabrication of a-Si:H nanowire and nanocone arrays, through a

RIE process using close-packed silica monolayers as masks [66]. These a-Si:H nanostructures show great absorption over a broadband optical region and wide angles of incidence, due to suppressed reflection. The great absorption effect is particularly enhanced in shape of nanocone arrays, which offer nearly perfect impedance matching between a-Si:H substrate and air through an effective index gradient. Besides silica colloidal crystals, self-assembled polymer sphere monolayers were also used as etching masks for generating AR Si nanopillar arrays combining RIE methods. For example, Park *et al.* proposed a novel method to extend the antireflection spectral range shorter than the lattice constant of the nanostructure by combining AR coatings and the moth's eye structure without a complicated process to fabricate a feature size of sub-300 nm [67]. Hexagonal close-packed monolayer polystyrene nanosphere crystals were used as etching masks to construct novel graded-index nanostructures integrating AR nanoisland coating arrays on top of silicon nanoconical frustum arrays. These complex structures not only exhibit good antireflection properties in the visible wavelength, but also decrease the average reflectance in the near-UV spectral range (300–400 nm) from 9.2% for sharp-tipped nanocone structures to 3.8%. Second, an electroless wet chemical etching method is very useful and simple for fabricating AR surfaces of single-crystalline, polycrystalline, and amorphous silicon with or without patterned metal catalysts. Chan *et al.* implemented density-graded surface nanostructures on ultrathin silicon solar microcells by silver-nanoparticle-catalyzed wet chemical etching [68]. Compared to the devices of the bare silicon, the black silicon surface can enhance the energy conversion efficiency of the thin film microcells by 148% and 50% with and without a diffuse backside reflector, respectively, while the combination of the bare silicon and a conventional one-layer AR coating can only lead to a smaller increase in the energy conversion efficiency. Besides, SiO_2 and ZnO nanoarray structures were also directly fabricated on the surface of solid solar cells as effective AR coatings [69, 70]. Third, solution growth has been widely employed to construct AR surfaces of metal oxides, such as ZnO and TiO_2. Tapered ZnO nanorod arrays fabricated by Lee *et al.* using a solution-growth method on seeded substrates display broadband reflection suppression from 400 to 1200 nm [71]. Nanorod parameters, such as nanorod tip diameter, diameter of the non-tapered region, thickness of the fused base layer, overall nanorod length, and length of the tapered region, have a great effect on the macroscopic AR performance. ZnO pyramidal array structures on zinc substrates with a gradient change of refractive index, which were prepared through a hydrothermal method, can effectively suppress the reflection of light at a range of wavelength from UV through the visible part to the near-IR region, with reflectivity <2.5%, resulting in its black color [72]. The alignments of ZnO nanoarray structures provide different surface profiles, resulting in different behaviors of scattered light. In this regard, Chao *et al.* fabricated three types of ZnO nanoarray structures with different alignments (well-aligned, quasi-aligned, and flower-like) using the hydrothermal process, and investigated their AR performance. The results demonstrated that the nanorod arrays with large alignment variation (i.e., flower-like ZnO nanoarrays) exhibit broadband and omnidirectional AR properties due to the gradual index profile [73].

ZnO NSAs are always used as AR coatings in thin film solar cells due to their easy anisotropic growth via either the solution method or vapor deposition. Recently, several NSAs including nanorod, nanotree, dandelion-shaped, branched nanorod, and branched nanoflower arrays have been hydrothermally fabricated as AR coatings on Cu(In,Ga)Se_2 or Cu_2ZnSnS_4 solar cells, all of which exhibit the efficient AR performance, thus largely enhancing the short-circuit current density [70, 74]. Fourth, nanoimprinting is a scalable and low-cost method to fabricate polymer AR surfaces using nanostructure array templates. Template imprinting is the most widely used technique in preparing polymer AR coatings based on nanostructure arrays. In a recent report, polyhedral oligomeric silsesquioxane-based AR moth's eye nanostructures on glass were developed by a double-side nanoimprint lithography (NIL) using a Ni mold [75]. These AR nanostructures exhibited excellent broadband and quasi-omnidirectional AR properties, improving the transmittance of the resulting glass to 98.2%. Daglar *et al.* fabricated large-area nanostructured polymer films using a drop casting technique on the tapered silicon molds and the polymer film with optimized single-side nanostructures can reach 92% transmission [76]. A polyethylene terephthalate film with similar double-side nanopores was fabricated by a so lithographic approach with a modified polyurethane acrylate mold, and its average transmission efficiency can be enhanced to 98.7% at normal incidence and 92.5% at an incident angle of 60 over a range of 400–800 nm. Lee *et al.* fabricated conical sub-wavelength gratings of an SU-8 polymer at one- and both-sides of the sapphires by a similar so lithography technique with PDMS as the mold, enhancing the transparency of sapphires over a wide wavelength region of 400–1000 nm [77]. Self-assembled porous films were used as templates to fabricate PDMS replica molding, which were covered by a compact distribution of hemispherical nanodomes about 250 nm in diameter and about 100 nm in height [78]. The transmission can be enhanced largely toward wavelengths ranging from visible to near-IR regions when these nanostructures are applied on a glass surface. Wu *et al.* fabricated an interesting 3D moth-eye structure via a one-step imprinting process, which exhibited better AR performance than 2D arrays over most wavelengths from 400 to 2400 nm, with the reflectance lower than 5.7% at all of the wavelengths [79]. A finely controlled anodized aluminium oxide (AAO) template structure could be transferred with exactitude to the polymer surface, and the resulting five-level film exhibited the best AR properties, with an average reflectance of 0.64% at a wavelength range of 400–800 nm [80]. Then, they developed a roll-to-roll thermal NIL to fabricate nanoporous honeycomb structures with high throughput and AR performance [81]. Recently, Li *et al.* fabricated moth-eye nanostructures of PMMA via the hot-press imprinting method using alumina as templates, and optimal broadband wide-angle AR performance (average specular reflectance of 0.5% at 8–45 for wavelengths of 400–2000 nm) was obtained [82]. The AR films after depositing a thin layer of hydrophilic silica nanoparticles showed superhydrophilic antifogging function. Fifth, chemical vapor deposition (CVD) growth is frequently used to fabricate vertically aligned carbon nanostructures including carbon nanotubes and nanowalls. Diedenhofen *et al.* prepared GaP nanorod arrays by CVD using the vapor–liquid–solid (VLS) method, and the produced surfaces exhibited enhanced transmission and reduced reflection properties over a broad wavelength and angular range, owing to a graded refractive index in the nanorod layer [64]. Then, base-tapered InP nanowires on the top of an InP substrate were prepared by combining patterned gold catalyst and VLS growth by the same group [83]. The specific geometry of the nanowires endowed them with a broadband and omnidirectional absorbing medium, and more than 97% of the light was absorbed in the system. Afterwards, they fabricated tapered GaP nanowires on the top of AlInP/GaAs substrates using the VLS method, which could reduce the reflection over a broad spectral range, still due to the graded refractive index of the layer [84]. An interesting randomly branched vertical array of InSb nanowires was obtained by Mohammad *et al.* via electrodeposition within branched porous anodic alumina membranes, which exhibited low reflectance over the visible and IR regions as well as wavelength-dependent absorbance in the IR region [85]. GaAs nanopyramids prepared by using a combination of nanosphere lithography, nanopyramid metal organic chemical vapor deposition growth, and gas-phase substrate removal processes showed excellent optical absorption over a broad range of wavelengths, at various incident angles and at large curvature bending [86].

The AR surfaces can effectively suppress the reflection and improve the absorption or transmission of light. Therefore, they have many important applications, covering light-emitting diode (LED), solar cells, high-quality optical elements, etc. Kim *et al.* fabricated an ARS arrays on an LED device [87]. The ARS arrays effectively draw out the emitted light to outer space and improve the efficiency of LED devices. Recently, Li *et al.* fabricated a white light organic light-emitting diode using biomimetic AR surfaces as anode, the luminance efficiency of the device in the normal direction is increased a factor of 1.4 [88]. In the field of solar cells, the AR surfaces can significantly improve their efficiency. For silicon-based solar cells, the refractive index in the visible and near-IR region of silicon is large, so that the surface reflection losses are very high (about 30%). These losses can be substantially reduced by the use of suitable AR surfaces on solar cells. Therefore, the efficiency of the solar cell can be dramatically improved. Zhu *et al.* fabricated a novel nanodome silicon-based solar cells [89]. They used AR surfaces to reduce the reflection and enhance absorption over a broad spectral range. The efficiency of such solar cells is higher than that of the flat film silicon solar cells. Although the applications of biomimetic AR surfaces have been tried by several groups, the preparation cost is the bottleneck for putting them into practical applications.

6.5.3 Summary

Diverse fabrication methods have been developed to form high-performance bioinspired AR structures of various materials. First, dry etching by RIE can produce AR surfaces based on both aperiodic and periodic nanostructure arrays of Si, III–V compounds, and SiO_2 with or without an etching mask. Second, an electroless wet chemical etching method is very useful and simple for fabricating AR surfaces of single-crystalline, polycrystalline, and amorphous silicon with or without patterned metal catalysts. Third, solution growth has been widely employed to construct AR surfaces of metal oxides, such as ZnO and TiO_2. Fourth, nanoimprinting is a scalable and low-cost method to fabricate polymer AR surfaces using nanostructure array templates. Fifth, CVD growth is frequently used to fabricate vertically aligned carbon nanostructures including carbon nanotubes and nanowalls. Additionally, two or more methods can be combined to construct hierarchical AR structures. These synthesized AR surfaces have exhibited excellent AR properties (e.g., broadband, omnidirectional, and polarization-insensitive antireflection) together with multifunctions, and they have shown promising applications in a wide range of fields including solar cells, LEDs, detection, and optical imaging.

6.6 Biological inspiration from color

6.6.1 Structural color

Most colors in nature come from the inherent colors of materials, i.e., pigments or dyes, but it can also arise from physical interaction of light with biological structures. These color-creating structures called *structural colors* have attracted much attention in a wide variety of research fields. The structural colors in nature are considered to originate from the following fundamental optical processes or combinations of these processes [90]; thin-film interference (Figure 6.9(a)) [91, 92], multilayer interference (Figure 6.9(b)), a diffraction grating effect (Figure 6.9(c)) [93], photonic crystals, and light scattering. In animals include on the feathers of birds and the scales of butterflies, interference is originated by multilayer interference, diffraction gratings, and photonic crystals.

Among the many examples of structural color found in nature, the brilliant blue color of a blue *Morpho* butterfly is due to interference and diffraction from the periodic structures and the irregular heights of ridges. In fact, the structural analysis of the scales of butterflies was concluded after the invention of the electron microscope. Anderson and Richards Jr. first observed the structure of the *Morpho cypris*, giving rise to the brilliant blue color and reported that it consisted essentially of hundreds of vanes on each wing scale, the vanes possessing linear thickenings 0.2 μm apart which reinforce the reflection of blue light [94]. Vukusic *et al.* invested the absolute reflectivity and transmissivity from the single-scale microstructures for various wavelengths of two species of *Morpho* butterfly, i.e., *Morpho rhetenor* and *Morpho didius* and explained the mechanisms behind their wide-angle visibility (Figure 6.10(a)) [95]. Meanwhile, Vukusic *et al.* also demonstrated the origin of the bright green coloration of the wing scales of the Indonesian male *Papilio palinurus* butterfly (Figure 6.10(b)). The microstructure generates an extraordinary combination of both yellow and blue iridescence [96]. The dual color arises from a modulation imposed on the multilayer. On the other hand, some fishes have different structural colors from those of the scales of butterflies. Especially, the structural color of the neon tetra is distinguishable because the colors can change in response to the light intensity of surroundings. The lateral stripe of the neon tetra shows brilliant blue in the daytime, while the color changes to deep violet at night. Nagaishi *et al.* reported the presence of regular layer of thin guanine crystals in the iridescent cells of neon tetra, which contains stacks of periodically

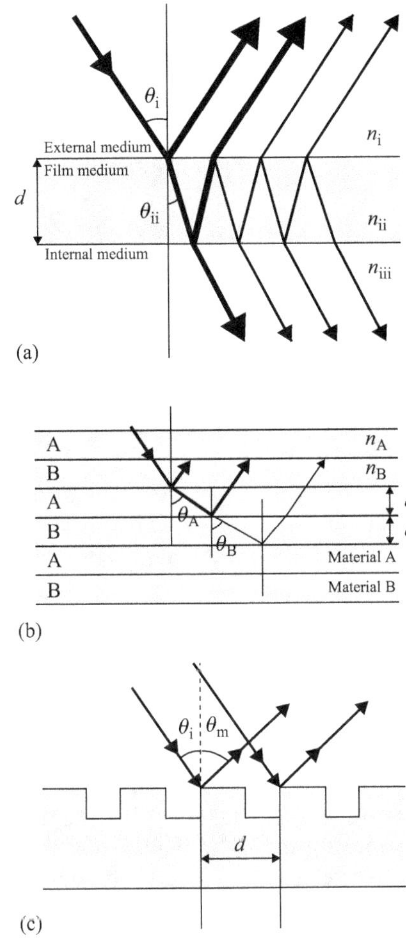

Figure 6.9 Fundamental optical processes related to structural colors. (a) Configuration of thin-film interference. (b) Configuration of the interference from a multilayer consisting of layers A and B with a thickness of d_A and d_B and a refractive index of n_A and n_B, respectively. (c) Configuration of diffraction grating and the light path

arranged light-reflecting platelets (Figure 6.10(c)) [97]. Two rows of parallel platelets, which can cause multilayer optical interference, simultaneously change their inclinations result in an alteration of the distance between the platelets. Among the butterfly species in Danaidae and Nymphalidae, there exist species showing surprisingly metallic pupae bearing gold or silver colors. These pupae were described by Poulton. He reported that the metal-like appearance of pupae of *Vanessa urticae* and these metal-like glitters need moisture [98]. This kind of metallic reflection is also found in beetles. The elytron of beetles is the most well-known case [99]. Parker *et al.* described a beetle, *Pachyrhynchus argus*, which has a metallic coloration that is visible from any direction owing to a photonics crystal

Figure 6.10 An example of iridescence through structural coloration in nature. (a) Morpho didius, *(b) Green Buddha Swallowtail butterfly (*Papilio blumei*), (c) neon tetra, (d)* Pachyrhynchus argus, *(e)* Chrysochroa vittata, *(f)* Chrysina gloriosa, *(g)* Betta splendens. *Photos by (a) Didier Descouens, (b) Siga, (c) H. Krisp, (d) and (e) Hectonichus, (f) Poul Beckmann, and (g) Anandarajkumar. Wikipedia, Creative Commons*

structure analogous to that of opal (Figure 6.10(d)) [100]. Vigneron *et al.* investigated cuticle of the multilayer structure from Buprestidae *Chrysochroa vittata* by scanning electron microscopy (SEM) (Figure 6.10(e)) [101]. Sharma *et al.* have been examining the structure on the exocuticle of the beetle *Chrysina gloriosa* (or *Plusiotis gloriosa*), which selectively reflects left circularly polarized light and possesses a brilliant metallic appearance (Figure 6.10(f)) [102]. Khoo *et al.* reported ultrastructural morphology of highly valued metallic iridescence in the Siamese fighting fish (*Betta splendens*) by transmission electron microscopy (Figure 6.10(g)) [103]. Metallic color on the body of *B. splendens* is attributed to refraction, reflection, and thin-film interferences from reflecting platelets in iridophores that are closely associated with other chromatophore types.

Like animals, plants produce structural color by both multilayer interference and diffraction gratings, but very little studied in plants. Blue iridescence appears of a few plants from the shady environments of humid tropical forests. The most common example of blue iridescence in plants is *Selaginella*. Two species of *Selaginella*, *Selaginella willdenowii* and *Selaginella uncinata*, produce a vivid blue–green iridescence on their leaves when growing in shade. Hébant and Lee described the ultrastructural basis of iridescence in *S. willdenowii* and *S. uncinata* [104]. They found that *Selaginella* leaves had two layers in the outer cell wall within epidermal cells. These layers, visible under transmission electron microscopy, were each approximately 80 nm thick, the predicted thickness to cause multilayer interference that would result in the observed iridescence. Interestingly, these two layers were not found in ordinary green *Selaginella* leaves grown under higher light conditions and lacking iridescence. Multilayers generating iridescence are also found in the fruits of *Elaeocarpus angustifolius* and *Delarbrea michieana*, in this case arising from a structure called an "iridosome." This is secreted to the region outside the cell membrane of fruit epidermal cells, and consists of layers of cellulose that are of the predicted thickness to cause interference coloration [105]. Diffraction gratings were identified in plants more recently, with the first report of their presence on the petals of species including *Tulipa* sp., *Hibiscus trionum*, and *Mentzelia lindleyi* [106]. In these species, the petal epidermal cells are elongated and flat and the overlying cuticle produces a series of long, ordered ridges with a periodicity that acts as a diffraction grating and splits the light reflecting from the surface into component wavelengths.

6.6.2 Biomimicry

6.6.2.1 Structural color

Structural colors serve as fundamental sources of inspiration of the development of artificially colored devices, and photonic materials. During the last decade, micro and nanofabrication methods have enabled investigation and functional mimicry of structural colors in nature. Chung *et al.* have fabricated a *Morpho*-mimetic thin film by depositing a dielectric multilayer on top of size-distributed silica microsphere layer without complicated fabrication methods (Figure 6.11(a)) [107]. Jewel beetles, also known as tiger beetles, exhibit beautiful iridescent structural colors that are thought to arise from several alternating electron-lucent and electron-dense layers on the epicuticle of the elytra. Crne *et al.* used breath figure templated assembly with atomic layer deposition for mimicking the green color on the wings of a butterfly *P. palinurus*, which is a method of using hexagonally packed water microspheres as templates for the fabrication of microporous polymer films (Figure 6.11(b)) [108]. Yabu *et al.* have demonstrated multicolor light reflectors by mimicking the surface structures of tiger beetle shells (Figure 6.11(b)) [109]. They prepared dimpled and multilayered films containing thin black layers, which had multi-angle reflectivity, by embossing a silicone microlens array. The surface structuring caused the reflected light to be scattered at multiple angles and thus visible at different viewing angles. This scattering system increases the visibility of the biological surface colors and blends the reflected colors of tiger beetles. These results demonstrate that combining multilayered structures with surface structuring produces unique optical properties. Kolle *et al.* combined colloidal self-assembly, sputtering, and atomic layer deposition to fabricate photonic structures. With this combination of layer deposition, they mimic the color mixing effect of the Indonesian butterfly *P. blumei* wings (Figure 6.11(c)) [110]. Khudiyev *et al.* investigated the iridescent neck feathers of *Anas platyrhynchos* drakes, shows that they feature an unusual arrangement of 2D photonic crystals and further exhibit a superhydrophobic surface, and mimicked this multifunctional structure using a nanostructure composite fabricated by top-down iterative size reduction method, which avoids the above-mentioned fabrication challenges, provides macroscale control, and enhances hydrophobicity through the surface structure [111].

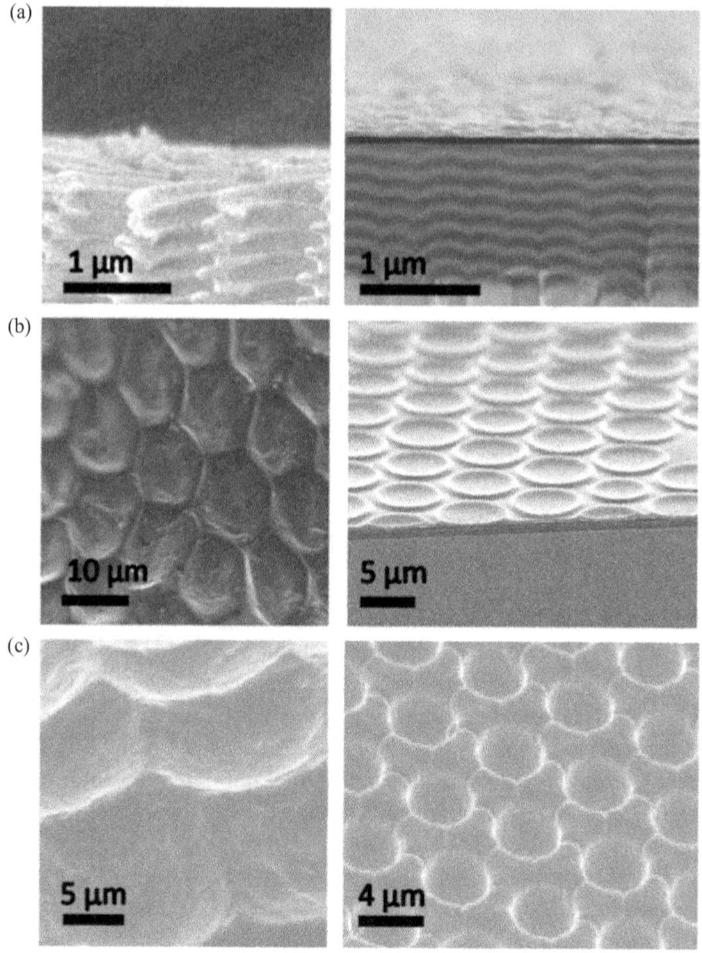

Figure 6.11 *(a) Cross-sectional SEM image of the multilayered ridges on the dorsal ground scale of a* M. didius *butterfly and a Morpho-mimetic thin film by depositing a dielectric multilayer on top of a silica microsphere base layer (right). (b) SEM image of the tiger beetle elytra and mimetic dimpled and multilayered film by embossing a PDMS microlens array (right). (c) SEM image of the surface of the wing scale from the* P. blumei *butterfly (left) and an artificial mimicry concavities (right). Reprinted with permission from Chung (2012) [107], copyright John Wiley and Sons; from Yabu (2014) [109], copyright Nature Publishing Group, from Kolle (2010) [110], copyright Nature Publishing Group*

Colors of nature cannot be expressed using a simple fabrication method, but we can mimic the appearance of color in nature by using metallic nanostructures, which is well known as plasmonics. When the metallic structures become small enough, strong optical resonance associated with plasmon frequencies of the conduction electron in the metals can read to qualitatively new phenomena. One of the most well-known example of this is the red color of a nanosized dispersion of gold colloid. The plasmonic nanostructures, i.e., nanohole and nanoslit arrays, can control the conversion between photons and plasmons. Recently, such effects have been reported using a metallic nanohole array to filter color by tuning the resonant transmission peak at the visible range [112]. Other attempts such as nanoslits combined with period grooves [113] or inserted into a metal–insulator–metal waveguide also show color filtering effect. In summary, plasmonic colors hold immense practical and commercial potential in device miniaturization, integration, and active color control, thus providing new frameworks and avenues for a host of exciting and sophisticated applications.

6.6.3 Summary

Butterflies, beetles, and other groups of insects have long been the focus of iridescence research and industrial biomimicry due to their structural colors and color filtering effect. Structural colors differ from pigments or dyes. The structural colors arise from a light interaction with periodic structure and cannot be mimicked by dyes or pigments. Moreover, various colors can be displayed corresponding to the periodicity of the periodic nanostructures or environmental conditions, just using a single material. This chapter presents a brief summary of structural colors and recent progress on their biomimicry. Structural coloring shows promise in a variety of applications including sensors, optical filters, reflectors, metamaterials, and photonic devices.

6.7 Illumination

6.7.1 Bioluminescence found in nature

Bioluminescence is an effective source of light in dim or dark environment and has various functions such as attracting mates or prey [114–116], detecting prey [117], or camouflage [118, 119]. Luminescent organisms have evolved complex photonic structures, which offer effective light modulation by using micro- and nanostructures for enhancing light emission effectiveness. Light organs of terrestrial and aquatic animals such as fireflies, squids, and fishes have optical components surrounding a photogenic part where the light is generated. The optical components of light organs consist of a reflector and a transmitted part. The photonic structures of reflector enhance light output power by effectively reflecting bioluminescent light. The reflector of fireflies has uric acid nanospheres and that of squids and fishes has multilayered platelets. The transmitted part of fireflies is cuticle layer and that of squids and fishes is lens. The micro- and nanostructures of cuticle layer improve light extraction for efficient optical communications. The lens of aquatic animals modulates light distributions for improving performance of camouflage. The photonic structures of optical components can effectively manage bioluminescent light for the survival of organisms.

6.7.1.1 Fireflies

Fireflies have a distinctive light organ (also called a "lantern") to produce optical signal in sexual communication. Each firefly species has particular light pulse patterns. Males send specific courtship signals to females, which selectively respond with their own flash pattern [114, 115, 120]. The highly conspicuous light signal is basically required for successful sexual communication. The light organ is composed of three layers: the cuticle, photogenic, and dorsal layer [121, 122]. These layers are obviously associated with highly efficient emission of bioluminescence. The light organs of adult fireflies have intriguing photonic structures to effectively manage bioluminescence.

Nanostructures and microstructures were found on the cuticle layer and the reflector layer of firefly light organs [87, 123, 124]. The lantern (L) cuticle of a firefly (*Luciola lateralis* Motschulsky) has highly ordered longitudinal nanostructures, which is parallel with the sagittal plane of a firefly body and the normal (N) cuticle has amorphous structures (Figure 6.12(a)–(c)) [87]. The width, height, and period of longitudinal nanostructures are about 150 nm, 110 nm, and 250 nm, respectively. The nanostructures with these physical dimensions show high transmission of bioluminescence by reducing the index mismatch between air and the cuticle layer. The lantern cuticle of a firefly (*Photuris* sp.) has microstructures with an asymmetric triangular profile (Figure 6.12(d) and (e)) [123]. The physical dimensions of tilted microstructures are 10 μm in width and 3 μm in height. The asymmetric microstructures can improve light extraction of bioluminescence by suppressing the TIR at air–cuticle interface. Both the nanostructures and microstructures on the cuticle layer can help highly efficient emission of bioluminescence by reducing the index mismatch and TIR. The reflector layer of a firefly (*L. cerata*) has spherical uric acid granules (Figure 6.12(f)–(h)) [124]. Most of the granules are highly regular in size about 700 nm in diameter. In the edge of the reflector layer, unusual large granules were found and its diameter was up to 4 μm in diameter. The densely packed uric acid granules can improve the bioluminescence intensity by reflection.

6.7.1.2 Squids

A squid (*Euprymna scolopes*) has unique optical components in its light organ such as a reflector and a lens (Figure 6.13(a)) [125]. The squid uses bioluminescent light for counter-illumination, which is light production for camouflaging by matching down-welling light [119]. The optical components of squid light organs can help efficient counter-illumination by reflect and focus the bioluminescent light.

The reflector of squid light organs has multilayered platelets of high refractive index and spacing layers of low refractive index (Figure 6.13(b)). These

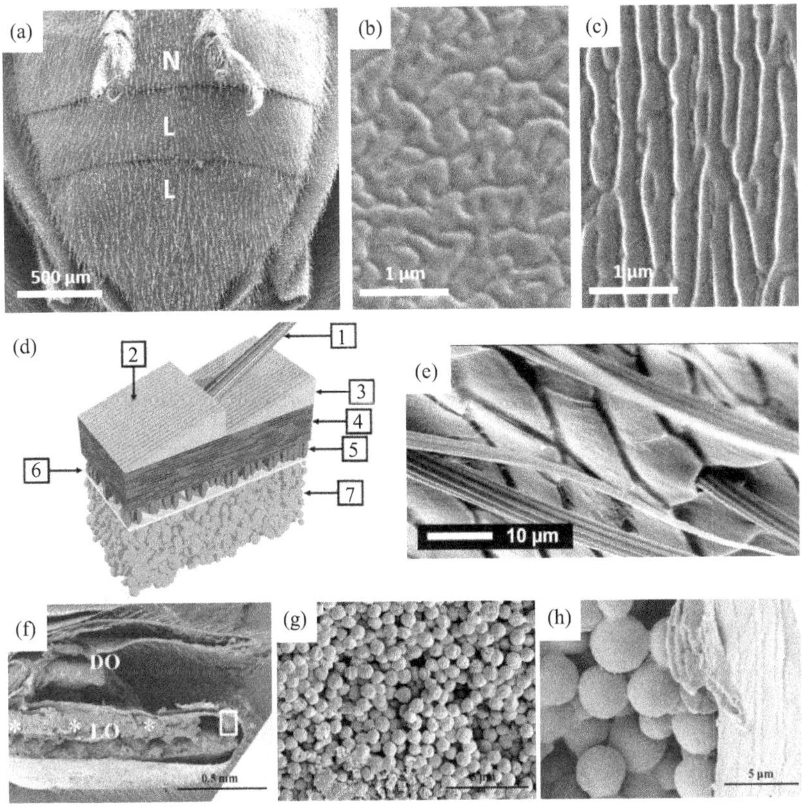

*Figure 6.12 Photonic structures of a firefly light organ. (a) Normal (N) and lantern (L) cuticles of a firefly (*L. lateralis *Motschulsky) in male. (b) Amorphous patterns on the normal cuticle. (c) Highly ordered longitudinal nanostructures on the lantern cuticle. (d) The photonic structures of a firefly (*Photuris *sp.). (e) 10 μm width and 3 μm height of the asymmetric microstructures. (f) Size investigation in different regions (indicated with * and □) of the seventh segment of* L. cerata. *(g) The highly regular-sized granules about 700 nm in diameter in the * region. (h) The unusually large granules at the edged region of reflector layer in the □ region. Reprinted with permission from Bay (2013) [123], copyright the Optical Society; from Goh (2013) [124], copyright Goh (et al.)*

organized layers generate thin-film interference, which help effective reflection of bioluminescence. The platelets of squids are composed of protein called reflectin [126]. The reflectin film, which is coated by using a flow-coating method, shows multiple reflectance peaks in visible wavelengths depend on the film thickness [127, 128].

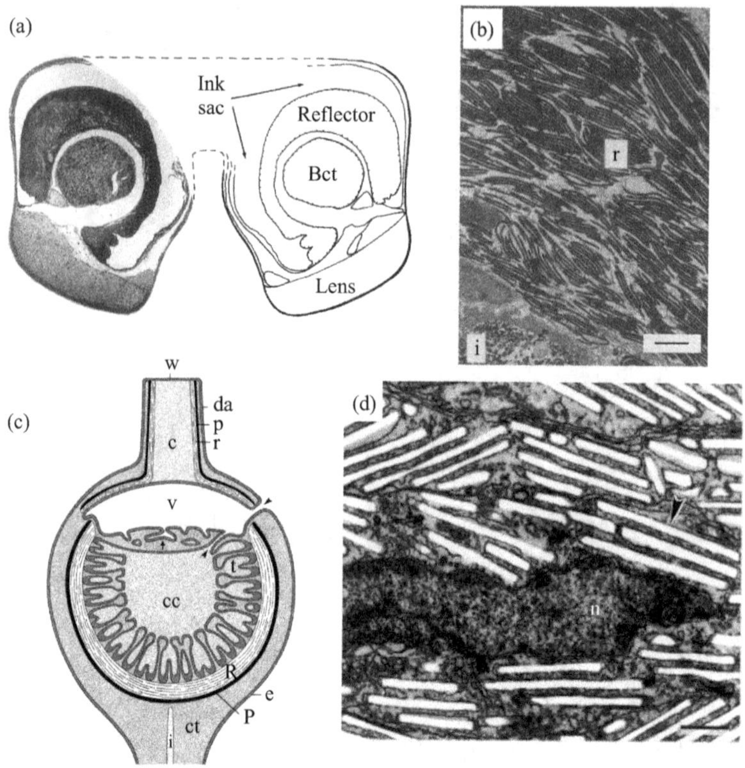

*Figure 6.13 Photonic structures of aquatic animals. (a) Optical micrograph and scheme of the squid (*Е. scolopes*) light organ. (b) Transmission electron micrograph of the reflector. Multilayered reflector platelets, r, and ink granules, i, in the ink sac. Scale bar: 2 μm. (c) Schematic diagram of an anglerfish lure. cc: central cavity, v: vestibule, e: epidermal, t: tubules, r: guanine reflector, p: pigment layer, da: distal appendage, c: light pipe core, w: clear window, i: illicium, and ct: connective tissue. (d) Electron micrograph of guanine crystals in the reflector of Oneirodes eschrichti. Scale bar: 0.5 μm. Reprinted with permission from McFall-Ngai (1990) [125], copyright the Marine Biological Laboratory; from Munk (1999) [129], copyright John Wiley and Sons*

6.7.1.3 Fishes

Fishes have various types of photophores, which have several functions such as attracting prey [129], detecting prey [117], and camouflage [130, 131]. The deep-sea anglerfishes have light organ in upper part of head and they use their bioluminescence for attracting prey. The flashlight fishes have a large photophore below the eyes to assist detecting prey. The hatchetfish *Argyropelecus* generates bioluminescent light for camouflaging by matching the down-welling light.

Most of the fish light organs have a reflector for improving emission efficiency of bioluminescent light (Figure 6.13(c)). Guanine crystals are the well-known reflective material in the photophores of fish. The reflector has multilayered platelets of high refractive index, guanine crystals, and spacing layers of low refractive index, cytoplasm (Figure 6.13(d)). The multilayers of reflector show effective reflection of bioluminescence by thin-film interference.

6.7.2 Biomimicry

Photonic structures found on the cuticle of firefly light organs have recently been investigated and provide inspirations for highly efficient illumination systems [87, 132]. Cuticular nanostructures and microstructures were found on the light organ cuticle of *L. lateralis* Motschulsky and *Photuris* sp., respectively. The cuticular nanostructures were mimicked by fabricating nanostructures on an LED lens surface (Figure 6.14(a)–(d)) [87]. The bioinspired LED lens offers highly efficient LED illumination by reducing the index mismatch between air and lens materials. The cuticular microstructures were mimicked by constructing factory-roof microstructures on a GaN LED (Figure 6.14(e) and (f)) [132]. The bioinspired microstructures increase light extraction by suppressing TIR. These biological inspirations may provide new opportunities in display and illumination applications.

6.7.2.1 Bioinspired LED lens

The bioinspired LED lens was fabricated by using a nanotemplating and a nanomolding method. A large-area nanotemplate was fabricated by using colloidal lithography and RIE. The period, width, and height of nanopillars were determined by diameter of nanospheres, isotropic etching of nanospheres, and anisotropic etching of substrates, respectively. The nanopillars were replicated by PDMS with heat-induced shear thinning, which enhance replication fidelity of the dense nanostructures. The LED lens with nanostructures was constructed by using a curved PDMS mold with nanostructures and a UV curable resin. Figure 6.14(c) and (d) show the SEM images of the LED lens with highly ordered nanostructures. The bioinspired LED lens considerably increases light transmission over visible ranges, compared with a smooth surface lens. This improvement is comparable to conventional antireflection coating.

6.7.2.2 Bioinspired overlayer on a GaN LED

The bioinspired overlayer, factory-roof microstructure, was fabricated by using direct laser writing. The factory-roof microstructures were constructed by using AZ 9245 photoresist on a GaN LED. Figure 6.14(f) shows that the factory-roof microstructures are well defined on the GaN LED. The geometrical parameters of bioinspired overlayer differ from those found on the firefly light organ due to the difference of refractive index of incident medium and emission wavelength. The optimal dimensions of microstructures are 5 μm in period and 5 μm in height. The bioinspired overlayer substantially improves light extraction up to 55% compared with the plane surface.

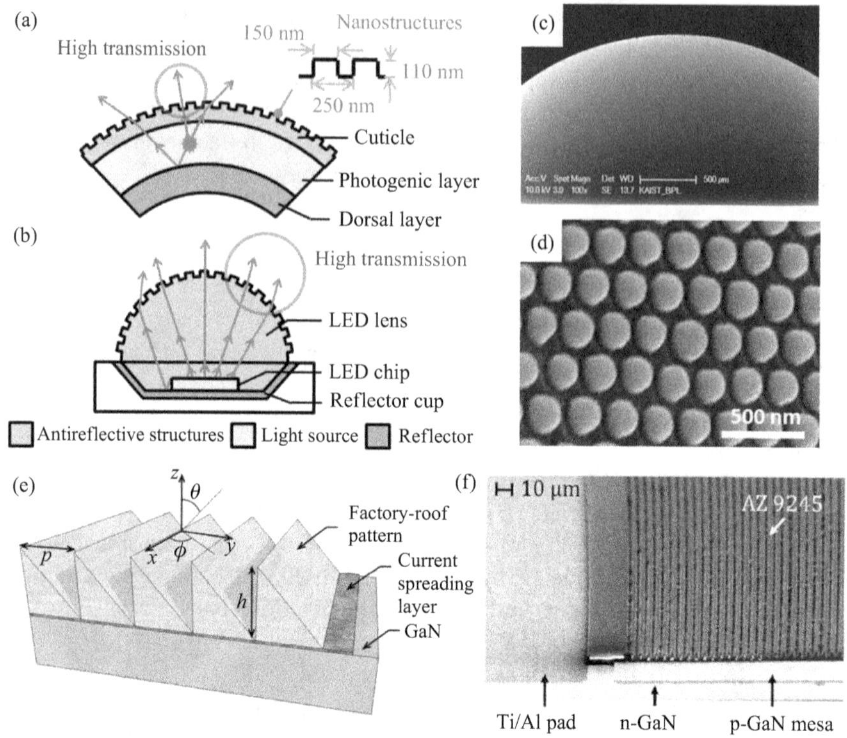

Figure 6.14 Engineering inspiration from photonic structures of light organs. Schematic diagrams of (a) a firefly lantern and (b) a high-power LED package. The firefly lantern consists of a cuticle, a photogenic, and a dorsal layer. The high-power LED package has similar components such as a reflector cup, an LED chip, and a nanostructured lens. SEM images of (c) a bioinspired LED lens and (d) highly ordered nanostructures on curved lens surface. (e) A schematic diagram of bioinspired overlayer on a GaN LED. The factory-roof microstructures have optimized period p and height h. (f) A SEM image of factory-roof microstructures fabricated on a GaN LED by direct laser writing. Reprinted with permission from Bay (2013) [132], copyright the Optical Society

6.7.3 Summary

Bioluminescent light organs of terrestrial and aquatic organisms have intriguing photonic structures for effective illumination. Fireflies have micro- and nanostructures in a cuticle and a reflective layer. The asymmetric microstructures of a cuticle layer improve light extraction by reducing TIR. The longitudinal nanostructures of a cuticle layer show high transmission by reducing the index mismatch between air and the cuticle material. The spherical uric acid granules of a reflective layer effectively reflect bioluminescent light. Squids and fishes have dielectric mirror, which consist of reflectin or guanine. Multilayered platelets in reflector enhance light output power by effectively reflecting bioluminescent light. The photonic structures of light organs offer efficient management of bioluminescent light and provide biological inspirations for highly efficient illumination systems. Cuticular nanostructures of *L. lateralis* Motschulsky were mimicked by fabricating nanostructures on the LED lens and cuticular microstructures of *Photuris* sp. were mimicked by constructing factory-roof microstructures on a GaN LED. These biological inspirations can offer new insights for illumination and display applications.

6.8 Conclusion

Diverse biological organs have intriguing photonic structures, which offer highly efficient light modulation by using micro- and nanoscale structures. Smartness of biological organs gives inspirations for various optical applications such as imaging, display, or lighting. Many engineering approaches have been developed to improve performance of optical systems. Bioinspired GRIN optics has proved unique advantages for singlet size and mass reduction of traditional lens systems. Gradient metasurface with micro- or nanostructures become of much interest in taking full benefits for practical GRIN optics. Bioinspired tunable lenses have been developed for small form factor, fast operation, and low power consumption. Diverse tunable lens systems were developed such as pneumatically, thermally, pH, or electrochemically driven tunable lenses. Compound eyes have extensive research interest in advanced photonics due to the notable benefits in wide FOV, fast motion detection, and low aberration. Researchers have developed various optical configurations of ACEs. Planar compound eye is a new approach for emulating the optical schemes of natural compound eyes and their functions. The simple apposition and reflecting superposition were emulated with 2D polymer structures by photolithography. ACEs were implemented on a flat substrate and a curved surface. ACEs on a flat substrate can combine conventional image sensors. ACEs with a curved surface have comparable FOV to natural compound eyes. The ACEs can offer technical solutions for advanced miniaturized optical systems in military, industrial, and medical fields. High-performance bioinspired AR structures have been developed by diverse fabrication methods. Dry etching by RIE can produce AR surfaces based on both aperiodic and periodic nanostructures. Chemical etching method is very useful and simple for fabricating AR surfaces of single-crystalline, polycrystalline, and amorphous silicon with or without patterned metal catalysts. Solution growth has been widely employed to construct AR surfaces of metal oxides. Nanoimprinting is a scalable and low-cost method to fabricate polymer AR surfaces using nanostructure array templates. CVD growth is frequently used to fabricate vertically aligned carbon nanostructures. Moreover, these methods can be combined to construct hierarchical AR structures. These AR structures have shown promising applications in a wide range including solar cells, LEDs, detection, and optical imaging. Insects, such as butterflies and beetles, have been the focus of iridescence research and industrial biomimicry due to their conspicuous structural colors. Various colors can be displayed corresponding to the periodicity of the periodic nanostructures or environmental conditions, just using a single material. Structural coloring shows promise in a variety of applications including sensors, functional films, optical filters, color reflectors, metamaterials, and photonic devices. Not only the superior optical properties but also the multi-functions such as robust mechanical strength and good wettability are also exploited in various applications. Bioluminescent light organs have intriguing photonic structures for effective illumination. Fireflies have micro- and nanostructures in a cuticle and a reflective layer. The photonic structures offer efficient light extraction for successful sexual communication. Squids and fishes have multilayered platelets in reflector. The platelets enhance light output power by effectively reflecting bioluminescent light. The photonic structures provide biological inspirations for highly efficient illumination systems. Cuticular nanostructures and microstructures were mimicked by fabricating nanostructures on the LED lens and factory-roof microstructures on a GaN LED, respectively. These biological inspirations can offer new insights for illumination and display applications.

References

[1] Pierscionek B. K. and Regini J. W. "The gradient index lens of the eye: An opto-biological synchrony". *Prog. Ret. Eye. Res.* 2012, vol. 31, pp. 332–349.

[2] Sinai P. "Correction of optical aberrations by neutron irradiation". *Appl. Opt.* 1971, vol. 10(1), pp. 99–104.

[3] Ohmi S., Sakai H., Asahara Y., Nakayama S., Yoneda Y., and Izumitani T. "Gradient-index rod lens made by a double ion-exchange process". *Appl. Opt.* 1988, vol. 27(3), pp. 496–499.

[4] Lv H., Shi B., Guo L., and Liu A. "Fabrication of Maxwell fish-eye spherical lenses and research on distribution profiles of gradient refractive index". *J. Opt. Soc. Am. A.* 2008, vol. 25(3), pp. 609–611.

[5] Tagantsev D. K., Lipovskii A. A., Schultz P. C., and Tatarintsev B. V. "Phosphate glasses for GRIN structures by ion exchange". *J. Non-Cryst. Solids.* 2008, vol. 354, pp. 1142–1145.

[6] Jahromi M. J. N. and Liu J.-H. "Gel effects on the fabrication of gradient refractive index plastic rods via energy-controlled polymerization". *J. Taiwan Inst. Chem. Eng.* 2012, vol. 43, pp. 301–305.

[7] Urness A. C., Anderson K., Ye C., Wilson W. L., and McLeod R. R. "Arbitrary GRIN component fabrication in optically driven diffusive photopolymers". *Opt. Express.* 2015, vol. 23(1), pp. 264–273.

[8] Koike Y., Sumi Y., and Ohtsuka Y. "Spherical gradient-index sphere lens". *Appl. Opt.* 1986, vol. 25(19), pp. 3356–3363.

[9] Koike Y., Kanemitsu A., Shioda Y., Nihei E., and Ohtsuka Y. "Spherical gradient-index polymer lens with low spherical aberration". *Appl. Opt.* 1994, vol. 33(16), pp. 3394–3400.

[10] Koike Y., Takezawa Y., and Ohtsuka Y. "New interfacial-gel copolymerization technique for steric GRIN polymer optical waveguides and lens arrays". *Appl. Opt.* 1988, vol. 27(3), pp. 486–491.

[11] Reed W. A., Yan M. F., and Schnitzer M. J. "Gradient-index fiber-optic microprobes for minimally invasive *in vivo* low-coherence interferometry". *Opt. Lett.* 2002, vol. 27(20), pp. 1794–1796.

[12] Shirk J. S., Sandrock M., Scribner D., *et al.* "Biomimetic gradient index (GRIN) lenses". *NRL Rev.* 2006, pp. 53–61.

[13] Beadie G., Shirk J. S., Rosenberg A., *et al.* "Optical properties of a bio-inspired gradient refractive index polymer lens". *Opt. Express.* 2008, vol. 16(15), pp. 11540–11547.

[14] Ji S., Yin K., Mackey M., Brister A., Ponting M., and Baer E. "Polymeric nanolayered gradient refractive index lenses: Technology review and introduction of spherical gradient refractive index ball lenses". *Opt. Eng.* 2013, vol. 52(11), p. 112105.

[15] Hudelist F., Buczynski R., Waddie A. J., and Taghizadeh M. R. "Design and fabrication of nano-structured gradient index microlenses". *Opt. Express.* 2009, vol. 17(5), pp. 3255–3263.

[16] Kundtz N. and Smith D. R. "Extreme-angle broadband metamaterial lens". *Nature Mater.* 2010, vol. 9, pp. 129–132.

[17] Lin D., Fan P., Hasman E., and Brongersma M. L. "Dielectric gradient metasurface optical elements". *Science.* 2014, vol. 345(6194), pp. 298–302.

[18] Park S. G., Lee K., Han D., Ahn J., and Jeong K. H. "Subwavelength silicon through-hole arrays as an all-dielectric broadband terahertz gradient index metamaterial". *Appl. Phys. Lett.* 2014, vol. 105, p. 091101.

[19] Land M. F. and Nilsson D. E. *Animal Eyes*. Oxford: Oxford University Press; 2002.

[20] Smythe R. H. *Vision in the Animal World*. London: Macmillan Press; 1975.

[21] Chronis N., Liu G. L., Jeong K. H., and Lee L. P. "Tunable liquid-filled microlens array integrated with microfluidic network". *Opt. Express.* 2003, vol. 11(19), pp. 2370–2378.

[22] Jeong K. H., Liu G. L., Chronis N., and Lee L. P. "Tunable microdoublet lens array". *Opt. Express.* 2004, vol. 12(11), pp. 2494–2500.

[23] Zhang W., Zappe H., and Seifert A. "Wafer-scale fabricated thermo-pneumatically tunable microlenses". *Light Sci. Appl.* 2014, vol. 3, p. e145.

[24] Iimura Y., Onoe H., Teshima T., *et al.* "Liquid-filled tunable lenticular lens". *J. Micromech. Microeng.* 2015, vol. 25(3), p. 035030.

[25] Dong L., Agarwal A. K., Beebe D. J., and Jiang H. "Adaptive liquid microlenses activated by stimuli–responsive hydrogels". *Nature.* 2006, vol. 442(7102), pp. 551–554.

[26] Ding Z. and Ziaie B. "A pH-tunable hydrogel microlens array with temperature-actuated light-switching capability". *Appl. Phys. Lett.* 2009, vol. 94(8), p. 081111.

[27] Carpi F., Frediani G., Turco S., and Rossi D. D. "Bioinspired tunable lens with muscle-like electroactive elastomers". *Adv. Funct. Mater.* 2011, vol. 21(21), pp. 4152–4158.

[28] Choi J., Son H., and Lee Y. "Biomimetic variable-focus lens system controlled by winding-type SMA actuator". *Opt. Express.* 2009, vol. 17(10), pp. 8152–8164.

[29] Donner J. S., Morales-Dalmau J., Alda I., Marty R., and Quidant R. "Fast and transparent adaptive lens based on plasmonic heating". *ACS Photonics.* 2015, vol. 2(3), pp. 355–360.

[30] Stavenga D. G. "Angular and spectral sensitivity of fly photoreceptors. I. Integrated facet lens and rhabdomere optics". *J. Comp. Physiol. A.* 2003, vol. 189(1), pp. 1–17.

[31] Stavenga D. G. and Hardie R. C. (eds.). *Facets of Vision*. Berlin: Springer-Verlag; 1989. p. 30.

[32] Land M. F. and Nilsson D. E. *Animal Eyes*. Oxford: Oxford University Press; 2002.

[33] Buschbeck E., Ehmer B., and Hoy R. "Chunk versus point sampling: Visual imaging in a small insect". *Science.* 1999, vol. 286(5442), pp. 1178–1180.

[34] Buschbeck E. K., Ehmer B., and Hoy R. R. "The unusual visual system of the Strepsiptera: External eye and neuropils". *J. Comp. Physiol. A.* 2003. vol. 189(8), pp. 617–630.

[35] Maksimovic S., Layne J. E., and Buschbeck E. K. "Behavioral evidence for within-eyelet resolution in twisted-winged insects (Strepsiptera)". *J. Exp. Biol.* 2007, vol. 210(16), pp. 2819–2828.

[36] Keum D., Jung H., and Jeong K. H. "Planar emulation of natural compound eyes". *Small.* 2012, vol. 8(14), pp. 2169–2173.

[37] Tanida J., Mima H., Kagawa K., Ogata C., and Umeda M. "Application of a compound imaging system to odontotherapy". *Opt. Rev.* 2015, vol. 22(2), pp. 322–328.

[38] Tanida J., Kumagai T., Yamada K., *et al.* "Thin observation module by bound optics (TOMBO): Concept and experimental verification". *Appl. Opt.* 2001, vol. 40(11), pp. 1806–1813.

[39] Kitamura Y., Shogenji R., Yamada K., *et al.* "Reconstruction of a high-resolution image on a compound-eye image-capturing system". *Appl. Opt.* 2004, vol. 43(8), pp. 1719–1727.

[40] Nitta K., Shogenji R., Miyatake S., and Tanida J. "Image reconstruction for thin observation module by bound optics by using the

iterative backprojection method". *Appl. Opt.* 2006, vol. 45(13), pp. 2893–2900.

[41] Mendelowitz S., Klapp I., and Mendlovic D. "Design of an image restoration algorithm for the TOMBO imaging system". *J. Opt. Soc. Am. A.* 2013, vol. 30(6), pp. 1193–1204.

[42] Duparré J., Dannberg P., Schreiber P., Bräuer A., and Tünnermann A. "Artificial apposition compound eye fabricated by micro-optics technology". *Appl. Opt.* 2004, vol. 43(22), pp. 4303–4310.

[43] Duparré J., Dannberg P., Schreiber P., Bräuer A., and Tünnermann A. "Thin compound-eye camera", *Appl. Opt.* 2005, vol. 44(15), pp. 2949–2956.

[44] Brückner A., Duparré J., Leitel R., Dannberg P., Bräuer A., and Tünnermann A. "Thin wafer-level camera lenses inspired by insect compound eyes". *Opt. Express.* 2010, vol. 18(24), pp. 24379–24394.

[45] Meyer J., Brückner A., Leitel R., Dannberg P., Bräuer A., and Tünnermann A. "Optical cluster eye fabricated on wafer-level". *Opt. Express.* 2011, vol. 19(18), pp. 17506–17519.

[46] Dunkel J., Wippermann F., Brückner A., Reimann A., Müller M., and Bräuer A. "Fabrication of refractive freeform array masters for artificial compound eye cameras". *Proceedings of SPIE;* Brussels, Belgium, April 2014. pp. 91300P-1–91300P-6.

[47] Brückner A., Oberdörster A., Dunkel J., Reimann A., Müller M., and Wippermann F. "Ultra-thin wafer-level camera with 720p resolution using microoptics". *Proceedings of SPIE*; San Diego, United States, Aug 2014. pp. 91930W-1–91930W-8.

[48] Li L. and Yi A. Y. "Design and fabrication of a freeform microlens array for a compact large-field-of-view compound-eye camera". *Appl. Opt.* 2012, vol. 51(12), pp. 1843–1852.

[49] Jeong K. H., Kim J., and Lee L. P. "Biologically inspired artificial compound eyes". *Science.* 2006, vol. 312(5773), pp. 557–561.

[50] Song Y. M., Xie Y., Malyarchuk V., *et al.* "Digital cameras with designs inspired by the arthropod eye". *Nature.* 2013, vol. 497(7447), pp. 95–99.

[51] Floreano D., Pericet-camara R., Viollet S., *et al.* "Miniature curved artificial compound eyes". *Proc. Natl. Acad. Sci. USA.* 2013, vol. 110(23), pp. 9267–9272.

[52] Wei K., Zeng H., and Zhao Y. "Insect–human hybrid eye (IHHE): An adaptive optofluidic lens combining the structural characteristics of insect and human eyes". *Lab Chip.* 2014, vol. 14(18), pp. 3594–3602.

[53] Huang C.-C., Wu X., Liu H., Aldalali B., Rogers J. A., and Jiang H. "Large-field-of-view wide-spectrum artificial reflecting superposition compound eyes". *Small.* 2014, vol. 10(15), pp. 3050–3057.

[54] Ma M., Guo F., Cao Z., and Wang K. "Development of an artificial compound eye system for three-dimensional object detection". *Appl. Opt.* 2014, vol. 53(6), pp. 1166–1172.

[55] Bernhard C. G. "Structural and functional adaptation in a visual system". *Endeavour.* 1967, vol. 26(98), pp. 79–84.

[56] Jung H. and Jeong K. H. "Monolithic polymer microlens arrays with antireflective nanostructures". *Appl. Phys. Lett.* 2012, vol. 101(20), p. 203102.

[57] Yoshida A., Motoyama M., Kosaku A., and Miyamoto K. "Antireflective nanoprotuberance array in the transparent wing of a hawk moth *Cephonodes hylas*". *Zool. Sci.* 1997, vol. 14(5), pp. 737–741.

[58] Deparis O., Khuzayim N., Parker A., and Vigneron J. P. "Assessment of the antireflection property of moth wings by three-dimensional transfer-matrix optical simulations". *Phys. Rev. E.* 2009, vol. 79(4), p. 041910.

[59] Sun M., Liang A., Zheng Y., Watson G. S., and Watson J. A. "A study of the anti-reflection efficiency of natural nano-arrays of varying sizes". *Bioinspir. Biomim.* 2011, vol. 6(2), p. 026003.

[60] Chattopadhyay S., Huang Y. F., Jen Y. J., Ganguly A., Chen K. H., and Chen L. C. "Anti-reflecting and photonic nanostructures". *Mater. Sci. Eng. R.* 2010, vol. 69(1–3), pp. 1–35.

[61] Cai J. G. and Qu L. M., "Recent advances in antireflective surfaces based on nanostructure arrays". *Mater. Horiz.* 2015, vol. 2(1), pp. 37–53.

[62] Gale M. "Diffraction, beauty and commerce". *Phys. World.* 1989, vol. 2(10), pp. 24.

[63] Huang Y. F., Chattopadhyay S., Jen Y. J., *et al.* "Improved broadband and quasi-omnidirectional anti-reflection properties with biomimetic silicon nanostructures". *Nat. Nanotechnol.* 2007, vol. 2(12), pp. 770–774.

[64] Diedenhofen S. L., Vecchi G., Algra R. E., *et al.* "Broad-band and omnidirectional antireflection coatings based on semiconductor nanorods". *Adv. Mater.* 2009, vol. 21(9), pp. 973–978.

[65] Oh Y. J., Kim J. J., and Jeong K. H. "Biologically inspired biophotonic surfaces with self-antireflection". *Small.* 2014, vol. 10(13), pp. 2558–2563.

[66] Zhu J., Yu Z., Burkhard G. F., *et al.* "Optical absorption enhancement in amorphous silicon nanowire and nanocone arrays". *Nano Lett.* 2009, vol. 9(1), pp. 279–282.

[67] Park H., Shin D., Kang G., Baek S., Kim K., and Padilla W. J. "Broadband optical antireflection enhancement by integrating antireflective nanoislands with silicon nanoconical-frustum arrays". *Adv. Mater.* 2011, vol. 23(48), pp. 5796–5800.

[68] Chan L., Kang D., Lee S. M., Li W., Hunter H., and Yoon J. "Broadband antireflection and absorption enhancement of ultrathin silicon solar microcells enabled with density-graded surface nanostructures". *Appl. Phys. Lett.* 2014, vol. 104(22), p. 223905.

[69] Ho C. H., Lin G. J., Fu P. H., *et al.* "An efficient light harvesting scheme using SiO_2 nanorods for InGaN multiple quantum well solar cells". *Sol. Energ. Mat. Sol. Cells.* 2012, vol. 103, pp. 194–198.

[70] Hsieh M. Y., Kuo S. Y., Han H. V., *et al.* "Enhanced broadband and omnidirectional performance of Cu(In,Ga)Se_2 solar cells with ZnO functional nanotree arrays". *Nanoscale.* 2013, vol. 5(9), pp. 3841–3846.

[71] Lee Y. J., Ruby D. S., Peters D. W., McKenzie B. B., and Hsu J. W. P. "ZnO nanostructures as efficient antireflection layers in solar cells". *Nano Lett.* 2008, vol. 8(5), pp. 1501–1505.

[72] Tian Y. S., Hu C. G., Xiong Y. F., Wan B. Y., He X. S., and Liu H. "ZnO pyramidal arrays: Novel functionality in antireflection". *J. Phys. Chem. C.* 2010, vol. 114(22), pp. 10265–10269.

[73] Chao Y. C., Chen C. Y., Lin C. A., and He J. H. "Light scattering by nanostructured anti-reflection coatings". *Energy Environ. Sci.* 2011, vol. 4(9), pp. 3436–3441.

[74] Wang Y. C., Lin B. Y., Liu P. T., and Shieh H. P. D. "Photovoltaic electrical properties of aqueous grown ZnO antireflective nanostructure on Cu(In,Ga)Se_2 thin film solar cells". *Opt. Express.* 2014, vol. 22(101), pp. A13–A20.

[75] Raut H. K., Dinachali S. S., He A. Y., *et al.* "Robust and durable polyhedral oligomeric silsesquioxane-based anti-reflective nanostructures with broadband quasi-omnidirectional properties". *Energy Environ. Sci.* 2013, vol. 6(6), pp. 1929–1937.

[76] Daglar B., Khudiyev T., Demirel G. B., Buyukserin F., and Bayindir M. "Soft biomimetic tapered nanostructures for large-area antireflective surfaces and SERS sensing". *J. Mater. Chem. C.* 2013, vol. 1(47), pp. 7842–7848.

[77] Lee S. H., Leem J. W., and Yu J. S. "Transmittance enhancement of sapphires with antireflective subwavelength grating patterned UV polymer surface structures by soft lithography". *Opt. Express.* 2013, vol. 21(24), pp. 29298–29303.

[78] Galeotti F., Trespidi F., Timò G., and Pasini M. "Broadband and crack-free antireflection coatings by self-assembled moth eye patterns". *ACS Appl. Mater. Interfaces.* 2014, vol. 6(8), pp. 5827–5834.

[79] Wu F., Shi G., Xu H., *et al.* "Fabrication of antireflective compound eyes by imprinting". *ACS Appl. Mater. Interfaces.* 2013, vol. 5(24), pp. 12799–12803.

[80] Choi K., Park S. H., Song Y. M., *et al.* "Nano-tailoring the surface structure for the monolithic high-performance antireflection polymer film". *Adv. Mater.* 2010, vol. 22(33), pp. 3713–3718.

[81] Choi K., Park S. H., Song Y. M., Cho C., and Lee H. S. "Robustly nano-tailored honeycomb structure for high-throughput antireflection polymer films". *J. Mater. Chem.* 2012, vol. 22(33), pp. 17037–17043.

[82] Li J., Zhu J., and Gao X. "Bio-inspired high-performance antireflection and antifogging polymer films". *Small.* 2014, vol. 10(13), pp. 2578–2582.

[83] Diedenhofen S. L., Janssen O. T. A., Grzela G., Bakkers E. P. A. M., and Rivas J. G. "Strong geometrical dependence of the absorption of light in arrays of semiconductor nanowires". *ACS Nano.* 2011, vol. 5(3), pp. 2316–2323.

[84] Diedenhofen S. L., Grzela G., Haverkamp E., Bauhuis G., Schermer J., and Rivas J. G. "Broadband and omnidirectional anti-reflection layer for III/V multi-junction solar cells". *Sol. Energ. Mat. Sol. Cells.* 2012, vol. 101, pp. 308–314.

[85] Mohammad A., Das S. R., Khan M. R., Alam M. A., and Janes D. B. "Wavelength-dependent absorption in structurally tailored randomly branched vertical arrays of InSb nanowires". *Nano Lett.* 2012, vol. 12(12), pp. 6112–6118.

[86] Liang D., Huo Y., Kang Y., *et al.* "Optical absorption enhancement in freestanding GaAs thin film nanopyramid arrays". *Adv. Energy Mater.* 2012, vol. 2(10), pp. 1254–1260.

[87] Kim J. J., Lee Y., Kim H. G., *et al.* "Biologically inspired LED lens from cuticular nanostructures of firefly lantern". *Proc. Natl. Acad. Sci. USA.* 2012, vol. 109(46), pp. 18674–18678.

[88] Li Y., Li F., Zhang J., *et al.* "Improved light extraction efficiency of white organic light-emitting devices by biomimetic antireflective surfaces". *Appl. Phys. Lett.* 2010, vol. 96(15), p. 153305.

[89] Zhu J., Hsu C. M., Yu Z., Fan S., and Cui Y. "Nanodome solar cells with efficient light management and self-cleaning". *Nano Lett.* 2010, vol. 10(6), pp. 1979–1984.

[90] Kinoshita S. and Yoshioka S. "Structural colors in nature: The role of regularity and irregularity in the structure". *Chem. Phys. Chem.* 2005, vol. 6, pp. 1442–1459.

[91] Stavenga D. G. "Thin film and multilayer optics cause structural colors of many insects and birds". *Mater. Today Proc.* 2014, IS, pp. 109–121.

[92] Kooi C. J., Wilts B. D., Leertouwer H. L., Staal M., Elzenga J. T M., and Stavenga D. G. "Iridescent flowers? Contribution of surface structures to optical signaling". *New Phytologist.* 2014, vol. 203, pp. 667–673.

[93] Seago A. E., Brady P., Vigneron J.-P., and Schultz T. D. "Gold bugs and beyond: A review of iridescence and structural colour mechanisms in beetles (Coleoptera)". *J. R. Soc. Interface.* 2009, vol. 6, pp. S165–S184.

[94] Anderson T. F. and Richards Jr. A. G. "An electron microscope study of some structural colors of insects". *J. Appl. Phys.* 1942, vol. 13, pp. 748–758.

[95] Vukusic P., Sambles J. R., Lawrence C. R., and Wootton R. J. "Quantified interference and diffraction in single *Morpho* butterfly scales". *Proc. R. Soc. Lond. B.* 1999, vol. 266, pp. 1403–1411.

[96] Vukusic P., Sambles J. R., and Lawrence C. R. "Color mixing in wing scales of a butterfly". *Nature,* 2000, vol. 404, pp. 457.

[97] Nagaishi H., Oshima N., and Fujii R. "Light-reflecting properties of the iridophores of thin neon tetra, *Paracheirodon innesi*". *Comp. Biochem. Physiol.* 1990, vol. 95A(3), pp. 337–341.

[98] Poulton E. B. "XV. Notes in 1887 upon lepidopterous larvæ, &c, including a complete account of the life-history of the larvæ of *Sphinx convolvuli* and *Aglia tau*". *Trans. Ent. Soc. Lond.* 1888, vol. 36(4), pp. 515–606.
[99] Neville A. C. "Metallic gold and silver colours in some insect cuticles". *J. Insect Physiol.* 1977, vol. 23, pp. 1264–1274.
[100] Parker A. R., Welch V. L., Driver D., and Martini N. "Structural colour: opal analogue discovered in a weevil". *Nature*, 2003, vol. 426, pp. 786–787.
[101] Vigneron J. P., Rassart M., Vandenbem C., *et al.* "Spectral filtering of visible light by the cuticle of metallic woodboring beetles and microfabrication of a matching bioinspired material". *Phys. Rev. E.* 2006, vol. 73, pp. 041905–041908.
[102] Sharma V., Crne M., Park J. O., and Srinivasarao M. "Structural origin of circularly polarized iridescence in Jeweled beetles". *Science.* 2009, vol. 325, pp. 449–451.
[103] Khoo G., Lim T. M., and Phang V. P. E. "Cellular basis of metallic iridescence in the Siamese Fighting fish, Betta splendens". *IJA.* 2014, vol. 66, pp. 988–997.
[104] Hébant C. and Lee D. W. "Ultrastructural basis and developmental control of blue iridescence in *Selaginella* leaves". *Amer. J. Bot.* 1984, vol. 71(2), pp. 216–219.
[105] Lee D. W., Taylor G. T., and Irvine A. K. "Structural fruit coloration in *Delarbrea michieana* (Araliaceae)". *Int. J. Plant Sci.* 2000, vol. 161(2), pp. 297–300.
[106] Whitney H. M., Kolle M., Andrew P., Chittka L., Steiner U., and Glover B. J. "Floral iridescence, produced by diffractive optics, acts as a cue for animal pollinators". *Science.* 2009, vol. 323, pp. 130–133.
[107] Chung K., Yu S., Heo C.-J., *et al.* "Flexible, angle-independent, structural color reflectors inspired by *Morpho* butterfly wings". *Adv. Mater.* 2012, vol. 24, pp. 2375–2379.
[108] Crne M., Sharma V., Blair J., Park J. O., Summers C. J., and Srinivasarao M. "Biomimicry of optical microstructures of *Papilio palinurus*". *EPL.* 2011, vol. 93, 14001-p.1-14001-p.4.
[109] Yabu H., Saito Y., and Shimomura M. "Unique light reflectors that mimic the structural colors of tiger beetles". *Polymer Journal.* 2014, vol. 46, pp. 212–215.
[110] Kolle M., Salgard-Cunha P. M., Scherer M. R. J., *et al.* "Mimicking the colourful wing scale structure of the *Papilio blumei* butterfly". *Nat. Nanotechnol.* 2010, vol. 5, pp. 511–515.
[111] Khudiyev T., Dogan T., and Bayindir M. "Biomimicry of multifunctional nanostructures in the neck feathers of mallard (*Anas platyrhynchos L.*) drakes". *Sci. Rep.* 2014, vol. 4.
[112] Lee H.-S., Yoon Y.-T., Lee S.-S., Kim S.-H., and Lee K.-D. "Color filter based on a subwavelength patterned metal grating". *Opt. Express.* 2007, vol. 15(23), pp. 15457–15463.

[113] Laux E., Genet C., Skauli T., and Ebbesen T. W. "Plasmonic photon sorters for spectral and polarimetric imaging". *Nat. Photonics*, 2008, vol. 2, pp. 161–164.

[114] Ohba N. "Flash communication systems of Japanese fireflies". *Integr. Comp. Biol.* 2004, vol. 44(3), pp. 225–233.

[115] Lewis S. M. and Cratsley C. K. "Flash signal evolution, mate choice, and predation in fireflies". *Annu. Rev. Entomol.* 2008, vol. 53, pp. 293–321.

[116] Herring P. J. "Sex with the lights on? A review of bioluminescent sexual dimorphism in the sea". *J. Mar. Biol. Ass. UK.* 2007, vol. 87(4), pp. 829–842.

[117] Morin J. G., Harrington A., Nealson K., Krieger N., Baldwin T. O., and Hastings J. W. "Light for all reasons: Versatility in the behavioral repertoire of the flashlight fish". *Science.* 1975, vol. 190(4209), pp. 74–76.

[118] Young R. E. and Roper C. F. "Bioluminescent countershading in midwater animals: Evidence from living squid". *Science.* 1976, vol. 191(4231), pp. 1046–1048.

[119] Jones B. W. and Nishiguchi M. K. "Counterillumination in the Hawaiian bobtail squid, *Euprymna scolopes* Berry (Mullusca: Cephalopoda)". *Mar. Biol.* 2004, vol. 144(6), pp. 1151–1155.

[120] Michaelidis C. I., Demary K. C., and Lewis S. M. "Male courtship signals and female signal assessment in *Photinus greeni* fireflies". *Behav. Ecol.* 2006, vol. 17(3), pp. 329–335.

[121] Beams H. W. and Anderson E. "Light and electron microscope studies on the light organ of the firefly (*Photinus pyralis*)". *Biol. Bull.* 1955, vol. 109 (3), pp. 375–393.

[122] Smith D. S. "The organization and innervation of the luminescent organ in a firefly, *Photuris pennsylvanica* (Coleoptera)". *J. Cell Biol.* 1963, vol. 16(2), pp. 323–359.

[123] Bay A., Cloetens P., Suhonen H., and Vigneron J. P. "Improved light extraction in the bioluminescent lantern of a *Photuris* firefly (Lampyridae)". *Opt. Express.* 2013, vol. 21(1), pp. 764–780.

[124] Goh K. S., Sheu H. S., Hua T. E., Kang M. H., and Li C. W. "Uric acid spherulites in the reflector layer of firefly light organ". *PLoS One.* 2013, vol. 8(2), p. e56406.

[125] Mcfallngai M. and Montgomery M. K. "The anatomy and morphology of the adult bacterial light organ of *Euprymna scolopes* berry (Cephalopoda, Sepiolidae)". *Biol. Bull.* 1990, vol. 179(3), pp. 332–339.

[126] Crookes W. J., Ding L. L., Huang Q. L., Kimbell J. R., Horwitz J., and McFall-Ngai M. J. "Reflectins: The unusual proteins of squid reflective tissues". *Science.* 2004, vol. 303(5655), pp. 235–238.

[127] Kramer R. M., Crookes-Goodson W. J., and Naik R. R. "The self-organizing properties of squid reflectin protein". *Nat. Mater.* 2007, vol. 6(7), pp. 533–538.

[128] Qin G., Dennis P. B., Zhang Y., *et al.* "Recombinant reflectin-based optical materials". *J. Polym. Sci., Part B: Polym. Phys.* 2013, vol. 51(4), pp. 254–264.

[129] Munk O. "The escal photophore of ceratioids (Pisces; Ceratioidei): A review of structure and function". *Acta Zool.* 1999, vol. 80(4), pp. 265–284.

[130] Denton E. J. "On the organization of reflecting surfaces in some marine animals". *Phil. Trans. R. Soc. B*. 1970, vol. 258(824), pp. 285–313.

[131] Denton E. J., Gilpin-brown J. B., and Wright P. G. "The angular distribution of the light produced by some mesopelagic fish in relation to their camouflage". *Proc. R. Soc. Lond. B*. 1972, vol. 182(1067), pp. 145–158.

[132] Bay A., André N., Sarrazin M., *et al.* "Optimal overlayer inspired by *Photuris* firefly improves light-extraction efficiency of existing light-emitting diodes". *Opt. Express.* 2013, vol. 21(S1), pp. A179–A189.

Chapter 7

Tuning nanophotonic cavities with nanoelectromechanical systems

Guangya Zhou, Feng Tian and Xiongyeu Chew

7.1 Introduction

Ever since the realization of optical fibers and integrated optical components in the 60s and 70s in the last century, optics has revolutionized the communication systems of the digital era. Presently, optical communications are only vastly used in long-haul communications and still rely heavily on high-speed electronics for signal processing. The explosion of internet resulted in an exponential growth of internet subscribers ever hungrily downloading and uploading music, videos, films, photos etc., thus demanding more bandwidth and speed. Meanwhile, the switching speed of high-speed electronics is reaching its limits, due to the challenging issues in the time-delay responses, heat dissipations, crosstalk interferences, etc. It seems that the development of an all-optical digital communication is the rational way to proceed to the next level. Indeed today, many major semiconductor/technology corporations such as Intel and IBM are developing integrated photonic devices that are compatible with existing communication hardware, capable of higher speed and larger bandwidth, and yet CMOS compatible with low fabrication costs.

The concept of integrated photonic devices is analogous to the electronic integrated circuits (ICs), where multiple photonic devices such as waveguides, optical buffers, filters, and modulators are integrated as a multi-functional system. Among those devices, silicon nanophotonic devices have attracted much attention in recent years [1]. Silicon nanophotonics refers to the utilization of engineered silicon nanostructures to achieve desired photonic functions. Here, silicon is used as medium for light propagation and interaction. The uniqueness and impact of this area is that it can leverage on the current existing semiconductor manufacturing capabilities and has the potential to form hybrid devices consisting of both electronic and photonic components due to their common silicon-based substrate.

Photonic micro/nano resonators are among the most important fundamental building blocks in integrated photonic devices [2, 3]. They strongly confine light in

Micro and Nano Systems Initiative, Department of Mechanical Engineering, National University of Singapore, 9 Engineering Drive 1, Singapore 117576, Singapore small mode volumes at resonant frequencies thus enhancing light–matter interactions and have already found many applications in science and engineering, for example, efficient and compact lasers, optical wavelength filters, and wavelength-selective add/drop multiplexers. In recent years, photonic resonators have also been used in label-free highly sensitive bio and chemical sensors, which detect the optical resonance shift induced by the minute refractive index changes inside or near the resonators [4]. Tunable photonic resonators are attractive as they can provide greater flexibility in a dynamic photonic system and offer post-process compensation capability for fabrication imperfections.

Thermal-optic tuning [5] and carrier-injection modulation [6] are two widely used silicon compatible approaches to implement photonic resonator tuning. Thermal-optic tuning approach utilizes localized electric current to heat up the resonator to induce a refractive index change thus shifting the optical resonance. Carrier-injection modulation method however changes the refractive index of the photonic resonator by introducing charge carriers into it. Both of the two methods lack the ability to provide a large change in the effective index, hence resulting in an insufficient wavelength tuning range. In addition, they also consume significant amount of power. In recent years, a new class of interesting photonic resonator tuning mechanism has attracted much attention [7, 8]. This mechanism tunes photonic resonance mechanically using on-chip integrated microelectromechanical systems (MEMS) or nanoelectromechanical systems (NEMS) [9]. Traditionally, MEMS devices have an advantage of performing accurate and dynamic mechanical actuation induced by electrostatic forces. These devices have outstanding advantages including low power consumption, small size, high positioning accuracy and good controllability, absence of exotic materials, and compatible with silicon micro/nano-fabrication processes. A number of MEMS/NEMS devices have already excelled in the optical communication applications [10], such as micromirrors for large array cross-connect switches [11], variable optical attenuators [12], add-drop multiplexers, and tunable external cavity lasers [13]. Hence, it is not surprised that researchers are beginning to place much focus in integrating MEMS/NEMS for tunable nanophotonic devices.

In this chapter, after short introductions of some basic concepts including MEMS/NEMS, photonic crystals (PhCs), and photonic nanocavities, we present three approaches for dynamic and reversible tuning of nanophotonic resonators or cavities using on-chip integrated NEMS actuators. They are: (1) resonance tuning through cavity evanescent field perturbation using NEMS-driven dielectric nanoprobes, (2) resonance wavelength splitting/shifting/tuning of coupled nanophotonic resonators through NEMS-induced coupling strength variations, and (3) resonance tuning by resonator's deformation induced by NEMS actuators.

7.2 PhC nanocavity designs

Photonic resonators are wavelength selective, and their optical responses are often measured by their quality factors (Q-factor) and mode volumes. In photonics, a

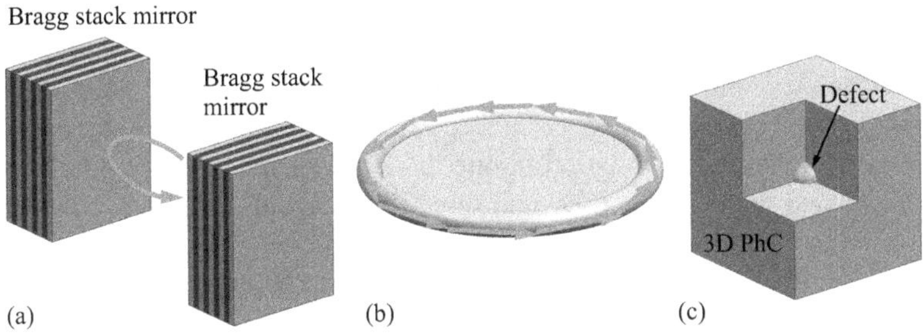

Figure 7.1 Schematic illustrations showing (a) an F–P cavity constructed by two dielectric mirrors, (b) a whispering gallery mode resonator, and (c) a photonic cavity resonator constructed by embedding a defect into a 3D PhC structure

desirable resonator should posses a high Q-factor and small mode volume to provide large enhancement of light–matter interactions. Perhaps the most commonly known photonic resonators are Fabry–Pérot (F–P) cavities [14], as shown in Figure 7.1(a), that trap light within two reflecting mirrors for applications such as lasers. Conventional bench-top F–P cavities have very low energy loss rates and therefore high finesses (or Q-factors). However, when the size of an F–P cavity decreases, the leakage of light energy increases causing a reduction in its Q-factor. This effect grows proportionally larger when F–P cavity scales into the micro regime, effectively limiting the miniaturization of F–P cavities. As shown in Figure 7.1(b), another concept of photonic resonators is the whispering gallery mode resonators that utilize continuous total-internal-reflection (TIR) as a method to confine light waves [15, 16] in a circular path. Small mode volumes with ultra high Q-factors as high as 8×10^9 have been reported utilizing thermally reflowed microspheres [17] and toroidal micro rings [18], which exploit surface tension to provide extremely low surface roughness to reduce TIR scattering losses. For some applications, the mode volumes for these whispering gallery mode resonators may still not small enough. The difficulties associated with the miniaturization of these resonators are quite similar to those of the F–P cavities. In other words, the light energy loss increases when the radius of the circular optical path (i.e. the size of the microsphere or ring) reduces. A new class of photonic resonators is then proposed as shown schematically in Figure 7.1(c), boasting ultra small mode volume and high Q. These are now known as PhC micro/nano cavities [19]. PhC micro/nano cavities are formed by introducing defects such as missing holes, modified radius holes, or even chirped-holes in PhC structures, where light are localized in the defect region and cannot escape due to the photonic bandgap structures thus forming photonic resonance modes.

PhCs are dielectric or metallic structures that have a certain structural periodicity that is analogous to the periodic arrangement of atoms in a crystalline medium. First proposed by Yablonovitch [20] and John [21] in 1987, a "photonic bandgap" is a structure whereby photons with frequencies lying within certain bands are not allowed to propagate thus acting as perfectly reflecting mirror or an excellent confinement medium. The types of PhCs are categorized by the number of coordinates or directions needed to specify a periodically varying dielectric constant (or reflective index) over space. Notably, a one-dimensional (1D) PhC refers to a periodic index variation in one direction such as a Bragg-stack mirror. Likewise, a two-dimensional (2D) and three-dimensional (3D) PhCs are periodic index variations in two and three directions, respectively. A good example of a successful commercial application of 2D PhCs is the PhC-based fibers [22], where reflective index variation is arranged periodically in a 2D plane perpendicular to the light propagation direction along the length of the fiber. 3D PhCs such as the woodpile structure [23] were also experimentally demonstrated by micro assembly of multiple 2D PhC slabs. Clearly, comparing with their 1D and 2D counterparts, they are significantly more difficult to fabricate. When properly engineered, PhC structures exhibit unique optical properties, such as photonic bandgap [24], negative refractive index [25], slow light [26], superprisms [27], and self-collimation [28] just to name a few. Among these interesting phenomena, PhCs are now being extensively used to confine light forming resonant micro or nano cavities for various applications due to their prominent photonic bandgap feature.

Although photonic micro/nano cavities formed by embedding defects in 3D PhCs, as shown schematically in Figure 7.1(c), have been reported [29], the technical difficulties in micro and nano fabrication processes for these devices still remain. In the past decade, researchers and engineers have also realized that a combination of light confinement methods using TIR, and photonic bandgap could be exploited for constructing photonic cavities that are easy to fabricate and yet have the ability to confine light in all three directions [30].

2D PhC micro and nano cavities are one type of such devices utilizing PhC slabs. Due to the refractive discontinuity in the vertical direction, 2D PhC micro and nano cavities utilize TIR method (usually the index of the slab material in the middle is much larger than those of the upper and lower media) for vertical confinement and photonic bandgap for the in-plane confinement. An example is shown schematically in Figure 7.2(a), where a line defect (a few missing holes in a row) is

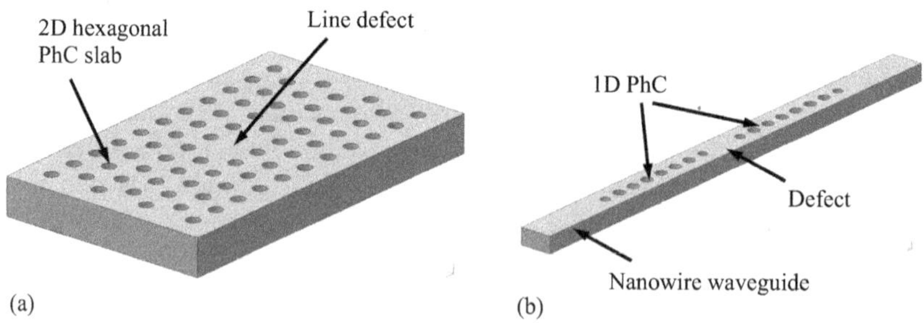

Figure 7.2 Schematic illustrations of (a) a 2D line-defect PhC cavity using a slab and (b) a 1D PhC cavity using a nano-beam waveguide

embedded in a hexagonal PhC lattice slab. In the defect region, guided waves can be established as the waves are bonded to the defect vertically due to the index guiding mechanism and horizontally due to the photonic bandgap of the crystal. The resonant modes of such cavity can be intuitively understood as standing waves developed by counter propagating guided waves in the line defect. The types of 2D PhC cavities developed can be further categorized as point defect cavities [31], line defect cavities [32], and heterostructure cavities [33]. In addition to 2D PhC cavities, 1D PhC cavities use a similar combination of TIR and photonic bandgap methods for light confinement [34]. As shown schematically in Figure 7.2(b), light wave is localized at the defect region using photonic bandgap confinement along the nanowire's length direction with the 1D PhCs and index guiding (or TIR confinement) along the width and thickness directions of the nanowire. Again, a standing wave will be established in the defect resulting in a photonic resonant mode. In this chapter, we will concentrate on the 1D PhC cavities for demonstration of the proposed tuning methods, as they are relatively easier for integration with on-chip MEMS/NEMS structures. It is should be noted that these tuning methods are certainly applicable to other photonic resonators.

A layout of a typical air-bridged 1D PhC nanocavity is shown schematically in Figure 7.3(a). A series of periodic nano-sized holes are patterned on a free-suspending silicon nanowire waveguide to function as a 1D PhC. When the light frequency falls within its photonic bandgap, it functions as a Bragg-stack mirror. The 1D PhC nanocavity is formed by an F–P-like resonator that is composed of two such 1D PhCs. TCM theory [35, 36] can be used to gain physical insight of the 1D

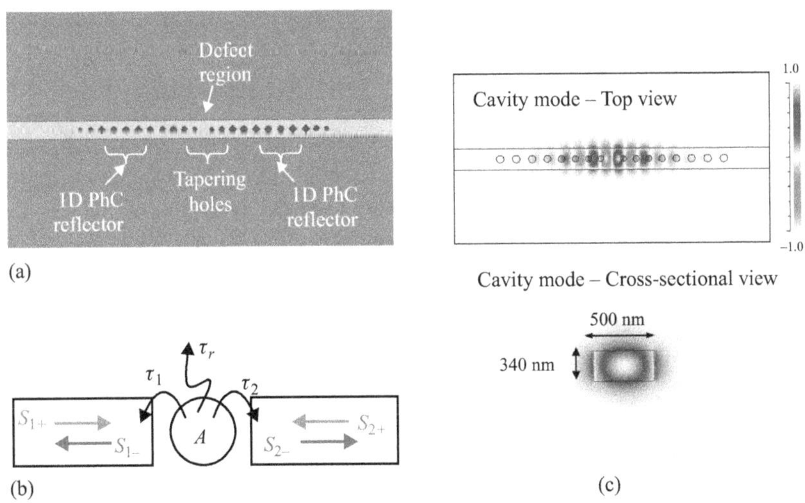

Figure 7.3 (a) Typical layout of an air-bridged 1D PhC nanocavity, (b) schematic used in the temporal coupled mode (TCM) theory of such a 1D PhC nanocavity, and (c) a typical finite-difference time-domain (FDTD) simulation result showing the cavity mode field

PhC nanocavity design. As shown schematically in Figure 7.3(b), assuming only one input port (i.e. $s_{2+} = 0$), the TCM equations are [35]:

$$\frac{da}{dt} = i\omega_0 a - \left(\frac{1}{\tau_r} + \frac{1}{\tau_1} + \frac{1}{\tau_2}\right) a + \sqrt{\frac{2}{\tau_1}} s_{1+}, \tag{7.1}$$

$$s_{1-} = -s_{1+} + \sqrt{\frac{2}{\tau_1}} a, \tag{7.2}$$

$$s_{2-} = \sqrt{\frac{2}{\tau_2}} a, \tag{7.3}$$

where a and ω_0 are, respectively, the amplitude and resonant frequency of the cavity mode, $1/\tau_r$ is the mode amplitude decay rate into the surrounding air (intrinsic radiation loss of the cavity), $1/\tau_1$ and $1/\tau_2$ are the mode decay rate into the input and output waveguides, s_{1+} is the amplitude of the incident wave field, and s_{1-} and s_{2-} are amplitudes of the reflected and transmitted wave fields, respectively. When $s_{1+} = S[\exp(i\omega t)]$, i.e. the incident light has a frequency of ω, then cavity mode also oscillates at ω, i.e. $a = A[\exp(i\omega t)]$ and $da/dt = i\omega A[\exp(i\omega t)]$. Plugging these into above equations, the transmission spectrum can be easily obtained:

$$T(\omega) = \frac{|s_{2-}|^2}{|s_{1+}|^2} = \frac{\frac{4}{\tau_1\tau_2}}{(\omega - \omega_0)^2 + \left(\frac{1}{\tau_r} + \frac{1}{\tau_1} + \frac{1}{\tau_2}\right)^2}. \tag{7.4}$$

This equation shows the transmission spectrum has a Lorentzian peak centered at $\omega = \omega_0$. For a symmetric waveguide–cavity–waveguide system, i.e. $\tau_1 = \tau_2$, we define overall decay rate into the input and output waveguides as $1/\tau_w = 1/\tau_1 + 1/\tau_2 = 2/\tau_1$, and accordingly, the quality factor due to the coupling of waveguides is then $Q_w = \omega_0\tau_w/2$. Similarly, the radiative quality factor is $Q_r = \omega_0\tau_r/2$ for radiation decay rate $1/\tau_r$, and the total quality factor considering both radiation loss and waveguide coupling is then $Q = \omega_0\tau/2$, where total decay rate is defined as $1/\tau = 1/\tau_r + 1/\tau_w$. Clearly, $1/Q = 1/Q_r + 1/Q_w$. Replacing the decay rates in (7.4) with their respective Q factors, the transmission spectrum can then be rewritten as:

$$T(\omega) = \frac{\frac{1}{4Q_w^2}}{\left(\frac{\omega-\omega_0}{\omega_0}\right)^2 + \frac{1}{4Q^2}}. \tag{7.5}$$

The maximum transmission is then at $\omega = \omega_0$, with the peak value:

$$T(\omega_0) = \left(\frac{Q}{Q_w}\right)^2 = \left(\frac{Q_r}{Q_w + Q_r}\right)^2 \tag{7.6}$$

It is obvious then from this equation that due to the presence of the radiation loss (finite value of Q_r) the peak transmission of the 1D PhC nanocavity is always less than 100%, which is different from the conventional F–P cavity whose peak transmission at resonance is close to 100%. In order for such 1D PhC cavities to achieve high transmission, one has to ensure that Q_r is much larger than Q_w. In other words, the cavity mode must decay more quickly into the waveguides than the surrounding air through radiation.

To suppress the radiation loss and achieve high Q_r, the air holes in the 1D PhC mirrors near the defect region have to be modified [37]. Without this modification, the mode field inside the cavity terminates abruptly at the 1D PhC boundary, and this leads to a large amount of spatial frequency components inside the leaky region thus resulting in a large radiation loss from the cavity [32]. As shown in Figure 7.3(a), one way to reduce this radiation loss is to taper the sizes and lattice periods of the PhC holes into the defect region such that the cavity mode profile has a Gaussian-like attenuation when terminating at the PhC boundary [34, 38].

1D PhC nanocavities are usually designed through a series of FDTD numerical simulations, as there are generally no analytical solutions available for such complex geometries at the nanoscale. FDTD is a numerical analysis technique used for solving Maxwell's equations in complex geometries. Using a fully vectorial method discretizing both space and time, FDTD can solve 1D, 2D, and 3D Maxwell's equations in dispersive and some simple nonlinear media. The FDTD method is a grid-based differential numerical modeling method. The time-dependent Maxwell's equations are discretized on a discrete mesh made up of so-called Yee cells [39]. Discretized electromagnetic fields are solved with FDTD method as a function of time. By performing Fourier transforms, the result of FDTD simulation can provide the electromagnetic fields as functions of frequency or wavelength.

FDTD uses a rectangular Cartesian style mesh. The electric and magnetic fields are calculated at each mesh point. In spatial domain, each electric field component is located midway between a pair of magnetic field components, and similarly each magnetic field component is located midway between a pair of electric field components. This spatially staggered configuration facilitates numerical computation of curls in Maxwell's equations. In time domain, the electric field components are calculated at time $t = n\Delta t$ (n is an integer), while the magnetic components are calculated at $t = (n + 1/2)\Delta t$, which is shifted by half a time step Δt. Obviously, smaller meshes allow for a more accurate representation of the device under simulation, but at a substantial cost of simulation time and memory requirements. In addition, to guarantee the numerical stability of an FDTD simulation, the upper limit of the time step must be bounded by the well-known Courant–Friedrichs–Lewy criterion that restricts the electromagnetic field propagation from cell to cell in an update cycle from being faster than the light velocity c. Consequently, the criterion for a general Yee FDTD grid is as follows [40]:

$$c\Delta t \leq \frac{1}{\sqrt{\left(\frac{1}{\Delta x}\right)^2 + \left(\frac{1}{\Delta y}\right)^2 + \left(\frac{1}{\Delta z}\right)^2}}, \tag{7.7}$$

where Δx, Δy, and Δz are cell grid sizes along different axes.

Electromagnetic sources are also very important for a FDTD simulation, which act as excitations for a structure under investigation. Many types of sources are available and well-developed, such as plane waves and broadband pulses (e.g. Gaussian pulses). When a pulsed source is activated, the simulation runs usually until there are essentially no electromagnetic fields left in the simulation region. A commonly used excitation pulse is a Gaussian pulse defined as:

$$E(t) \sim \exp\left[-\frac{4\pi(t-t_0)^2}{\tau^2}\right], \qquad (7.8)$$

where the time constant τ decides the width of Gaussian pulse and its frequency band. The response of the system over a wide range of frequencies can then be obtained by performing Fourier transforms on the FDTD simulation results. FDTD simulation using broadband pulses is useful in applications where resonant frequencies of the photonic devices are not exactly known, such as the case here for PhC cavity designs.

A typical design result is illustrated in Figure 7.3(c) showing the electric field profile of the cavity mode. The silicon nanowire waveguide is 500-nm wide and 340-nm thick surrounded by air. We found that, through a series of FDTD simulations, five decreasing tapered holes yielded the best performances with a radiative Q_r of 4×10^6. The waveguide-loaded Q is dependent on the number of holes in the Bragg mirrors patterned. We used a two-hole Bragg mirrors in our design to achieve a waveguide-loaded Q of 4×10^3. Further increase in the number of holes in Bragg mirrors results in an increased overall waveguide-loaded Q; however, the maximum transmission is decreased as indicated in (7.6). We will mainly use this design for our demonstration of MEMS tunable nanophotonic cavities.

7.3 MEMS and NEMS

MEMS technology [41] is evolved from semiconductor manufacturing technology and uses various deposition, lithography, etching, and release processes to integrate microelectronic and micromechanical components on a chip. Typical dimensions of the MEMS devices are in the micrometer range. This technology has a number of outstanding advantages including low cost, fast operation speed, light weight, and enhanced signal-to-noise ratio due to on-chip integrated electronics. MEMS technology has already found a wide range of applications in various areas, for example, micro accelerometers, micro gyroscopes, and miniature pressure sensors. More recently, as the semiconductor nanofabrication capabilities have made substantial progresses, there is a trend of research toward scaling both mechanical and electrical elements further down into the nano regime with an aim to create more powerful functionalities and devices featuring enhanced performances, faster operational speed, greater bandwidth, and higher level of integration. This area is now also known as NEMS [9]. Typical dimensions for NEMS devices are in the nanometer range, which leads to ultra low mass, very high mechanical resonant frequencies, and extremely large surface-to-volume ratio that is highly desirable for surface-based sensors.

In the past decades, MEMS technology has also combined with optics generating a tremendously useful technology (now termed Optical MEMS) that is still rapidly developing presently, and some optical MEMS devices have already been successfully incorporated into various commercial applications ranging from telecommunication and biomedical healthcare systems to defense and homeland security [10]. Optical MEMS not only enables the miniaturization of conventional optics thus making the optical systems more compact and lower cost through the IC-like batch fabrication processes but also unfolds new dimensions for innovations in optical and photonic systems. Examples include micromirror-based endoscope probes [42], laser scanning portable projectors [43], Texas Instruments' digital micromirror device (DMD) [44, 45], micro-optical switches and filters in fiber communication networks [10], and micro-cantilever-based resonant chemical sensors with optical readouts [46]. More recently, similar to the scaling from MEMS to NEMS, there is also a trend in research to scale optical MEMS further down to the nano region. This new dynamic nanophotonic devices and applications are emerging at a fast pace. Examples include NEMS tunable PhC nanocavities [8], NEMS-driven nanophotonic switches [47], and nanophotonic accelerometers [48].

7.3.1 MEMS/NEMS comb-drive actuator design

Comb-drive actuators [49], which utilize the electrostatic attraction forces to generate mechanical movements, are most commonly used actuators in MEMS and NEMS. They are very attractive due to their outstanding advantages compared with other types of micro/nano actuators, for example, very low power consumption, compatible with silicon processes, and linear actuation (force is independent of displacement). In fact, these are also the main reasons why we choose NEMS electrostatic comb-drive as a nanoscale actuation mechanism for our tunable PhC nanocavities.

As shown schematically in Figure 7.4(a), a standard electrostatic comb-drive actuator consists of interdigitated fingers as driving electrodes and a folded-beam suspension. The electrostatic force can be evaluated using [49]:

$$F_e = \frac{n\varepsilon t V^2}{g}, \tag{7.9}$$

where n is the number of movable fingers, ε is the permittivity of the medium where comb-drive is operated ($\varepsilon = 8.85 \times 10^{-12}$ F/m for air), t and g are, respectively, the finger thickness and gap spacing between fixed and movable fingers, and V is the applied voltage. The displacement output of a comb-drive actuator can be easily computed by balancing the electrostatic force of the comb-drive with the mechanical restoring force of the folded-beam suspension, and this leads to:

$$\delta_x = \frac{F_e}{k_x} = \frac{n\varepsilon t V^2}{k_x g}, \tag{7.10}$$

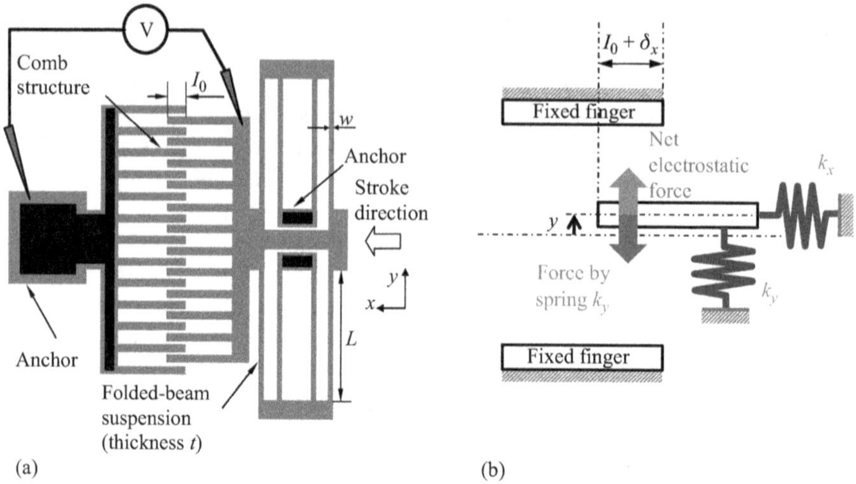

Figure 7.4 (a) Schematic showing a comb-drive actuator with a standard folded-beam suspension, and (b) forces exerting on a movable finger after a small disturbance y

where δ_x is the displacement output and k_x is spring constant along the x-direction, which can be determined using the following equation for a folded-beam suspension:

$$k_x = \frac{2Etw^3}{L^3}, \tag{7.11}$$

where E is the Young's modulus of the beam material; w and L are, respectively, the width and length of the flexure beams; and t is the beam thickness (usually comb-drive fingers and folded-beams have the same thickness, since they are fabricated in the same layer. We hence use the same symbol t here).

The maximum travel range of an electrostatic comb-drive actuator is typically limited by its side instability pull-in, i.e. a state where movable fingers move perpendicularly along the y-direction and cause an electrical short circuit. To analyze the condition for this pull-in, one must consider that the movable fingers are perturbed by an external disturbance and shifted slightly along the y-direction from its original equilibrium state as shown schematically in Figure 7.4(b). The net electrostatic force along the y-direction is then:

$$F_y = \frac{n\varepsilon t(l_0+\delta_x)V^2}{2(g-y)^2} - \frac{n\varepsilon t(l_0+\delta_x)V^2}{2(g+y)^2} \approx \frac{2n\varepsilon t(l_0+\delta_x)V^2}{g^3}y, \tag{7.12}$$

where electrostatic forces for parallel plate actuators are used, and l_0 is the initial finger overlap without voltage applied. In the above approximation, Taylor expansion is employed, and only the first-order term of the disturbance y is kept. Clearly, the net electrostatic force along the y-direction will not restore the movable finger structure to its original position after a disturbance but cause it to move further away instead. Hence, the comb-drive actuator is not stable without a spring suspension strong enough to hold each movable finger electrode in the middle of two fixed fingers. The stable operation condition without pull-in is then:

$$k_y > \frac{2n\varepsilon t(l_0 + \delta_x)V^2}{g^3}, \tag{7.13}$$

where k_y is the spring constant of the suspension along the y-direction as shown in Figure 7.4(b). Combining (7.10) and (7.13) and eliminating $n\varepsilon tV^2$ yields

$$k_y > \frac{2k_x\delta_x(l_0 + \delta_x)}{g^2}. \tag{7.14}$$

This inequality can be easily solved for δ_x and thus leading to the maximum travel range:

$$\delta_{x,max} = \frac{1}{2}\sqrt{l_0^2 + 2\frac{k_y}{k_x}g^2} - \frac{l_0}{2}. \tag{7.15}$$

It can be observed from the above equation that the maximum travel of a comb-drive actuator without pull-in is proportional to the ratio of k_y over k_x and finger gap g. A good suspension design that is rigid along the y-direction and at the same time flexible along the x-direction is desirable for large displacement comb-drives. In addition, increasing the gap g also helps to stabilize the actuator and increases its travel range, but at the expense of the increased driving voltage.

For a standard folded-beam suspension, while k_x can be easily estimated using (7.11), the spring constant k_y however is complicated. It reduces significantly as the beams bent when the displacement δ_x increases. An approximate relation between k_y and δ_x can be found [50]:

$$k_y = \frac{200Etw^3}{36L\delta_x^2}. \tag{7.16}$$

Inserting (7.11) and (7.16) into (7.15), and assuming the initial finger overlap is small and negligible (i.e. $l_0 = 0$), we obtain

$$\delta_{x,max} = \left(\frac{25}{18}\right)^{\frac{1}{4}}\sqrt{gL}. \tag{7.17}$$

The above equation can be used to roughly estimate the stable travel range of an electrostatic comb-drive actuator. Further enhancement of the stability and thus travel range is possible through suspension designs, for example, the tilted fold-beam suspension [51] whose flexural beams are pre-tilted against their movement directions. In fact, for a single-sided travel, a tilted fold-beam suspension is always better than a standard fold-beam suspension.

The design of tilted folded-beam suspension originates from the observation of (7.14). The two sides of the inequality are computed individually and are plotted in Figure 7.5(b), both as functions of the comb-drive displacement δ_x. Key parameters

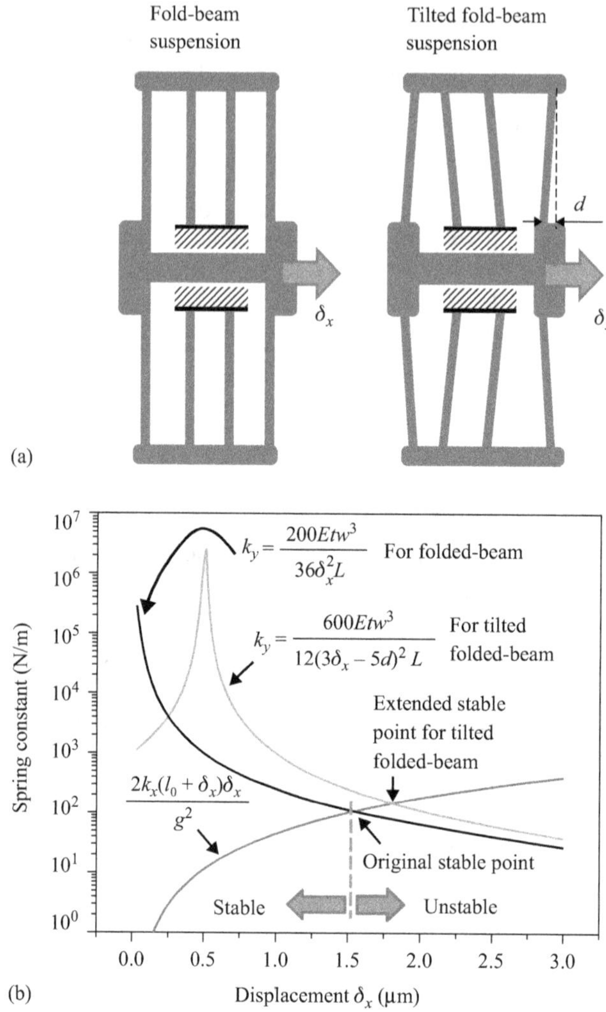

Figure 7.5 (a) Schematics show the traditional folded-beam and tilted folded-beam suspensions. The folded-beam suspension has straight flexural beams, while the tilted folded-beam suspension has flexural beams pre-tilted against the movement direction. (b) Spring constant k_y *of the two suspension designs as functions of displacement* δ_x. *The enhancement of comb-drive's stability is clearly visible in the graph*

used are $E = 165$ GPa, $t = 340$ nm, $w = 200$ nm, $L = 10$ μm, $g = 200$ nm, and the initial finger overlap is negligible $l_0 = 0$. As shown in the figure, the stable travel range is predicted at around 1.5 μm, which is consistent with the estimation from (7.17). From this figure, it can be seen that when δ_x is small the spring k_y is very strong (in fact it peaks at $\delta_x = 0$), which is unnecessary because the term

$2k_x\delta_x(l_0+\delta_x)/g^2$ is also small at small displacement. If one can shift the peak location of the k_y to some positive δ_x, the stability of the comb-drive actuator can be improved greatly and the stable travel range is enhanced at the same time. Indeed, this is achieved through a tilted folded-beam suspension design, where flexural beams are pre-tilted against the movement direction so that the beams bent from originally tilted to straight positions and subsequently further from straight to tilted positions. The spring constant k_y is related to δ_x according to the following equation [51]:

$$k_y = \frac{600Etw^3}{12(3\delta_x - 5d)^2L}, \tag{7.18}$$

where d is the amount of pre-tilt as indicated in Figure 7.5(a). This is a more general expression for the spring constant k_y. Clearly, if there is no pre-tilt of the beams ($d=0$), (7.18) is reduced to (7.16), which agrees well with the fact that a traditional fold-beam suspension is a special case for a tilted folded-beam suspension with $d=0$. The spring constant k_y for a tilted fold-beam suspension with a pre-tilt amount d setting at 300 nm is also plotted in Figure 7.5(b) for a comparison. The rest of the parameters here are the same as what we used for the fold-beam suspension calculations. It is very clear that the stability of the comb-drive is enhanced and the stable travel range is enlarged.

In our work here, we have used the above theory to design our comb-drives at the nanoscale to act as actuators for tuning PhC cavity resonance. As can be seen in Figure 7.3(c), for a PhC nanocavity, the resonant mode is tightly confined in silicon with a very short evanescent "tail" extending into the air region. The length of such "tail" is typically a few hundred of nanometers, which can be accessed for perturbation tuning of the cavity resonance. It is thus clear that the travel range of the NEMS comb drive actuator required for tuning is less than a micron. The typical key parameters for NEMS comb-drives here are then $g \sim 200$ nm for finger gap and $L \sim 10$ μm for folded-beam length, which translate into a more than one micron stable range according to (7.17). In some devices, for extra stability, tilted folded-beams are used.

7.3.2 NEMS fabrication processes

We now describe the micro/nano fabrication process steps for the tunable PhC nano cavities with on-chip integrated NEMS actuators. We have successfully developed the process utilizing a combination of electron-beam (E-beam) lithography and conventional silicon-on-insulator (SOI) micromachining methods to achieve on-chip integration of both nanophotonic devices and nanomechanical structures. As shown schematically in Figure 7.6, this process consists of two E-beam lithography processes to define the silicon nano patterns and two additional optical lithography processes to define micro-sized electrical pads and routing wires.

In our process, we first pattern the 1D PhC nanocavities, NEMS comb-drive structures, nano flexural suspensions, and photonic nanowire waveguides through 100 keV electron beam lithography using ELS-7000 (Elionix, Inc.) on a SOI wafer (SOITEC, Inc.). The SOI wafer has a 260-nm or 340-nm silicon device layer on a

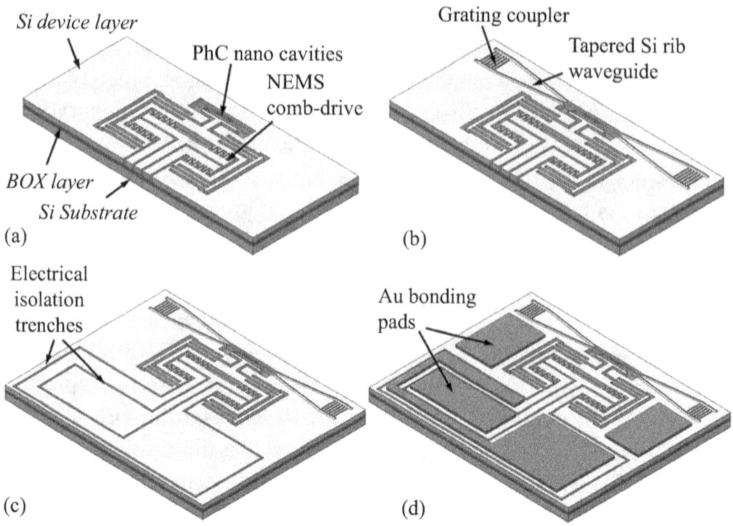

Figure 7.6 Schematic illustration of the micro/nano fabrication process for the nanophotonic resonators and their on-chip integrated NEMS. (a) Device after the first E-beam lithography and ICP-RIE etching, (b) device after the second E-beam lithography and ICP-RIE etching, (c) device after a UV photolithography process and RIE etching, and (d) device after lift-off patterning of gold pads

1.0-μm buried SiO_2 (BOX) layer. The silicon device layer is p-doped to a resistivity of 14–22 Ω cm, which is sufficiently conductive for electrostatic actuation yet the light absorption coefficient is low enough for optical wave guiding at the NIR fiber communication bands around 1550 nm. First, a layer of electron beam resist ZEP520 is spin-coated onto the SOI wafer at a speed of 5000 pm forming a 350-nm-thick layer which acts as a mask for subsequent etching process. An electron beam with a registered dose of 320 $\mu C/cm^2$ is optimized to pattern the PhC cavities, waveguides, and NEMS structures. The top silicon device layer is then selectively etched with inductive-coupled-plasma reactive ion etcher (ICP-RIE) (Oxford Instruments Ltd.) using SF_6 and C_4F_8 chemistries to attain straight sidewall profiles. The parameters used for this silicon etching process are listed as follows: 450 W ICP power, 30 W RF power, gas pressure at 15 mTorr, and etching gas ratio SF_6:C_4F_8 at around 1:2. The etch rate of silicon is determined to be approximately at 200 nm/min, which is about three times faster than that of the electron beam resist. This etching completely removes the exposed silicon device layer stopping on the BOX layer. After etching, the electron beam resist is removed from the SOI wafer by a solvent followed by a typical piranha (a mixture of sulfuric acid H_2SO_4 and hydrogen peroxide H_2O_2) etch. Sample cleaning is critical to remove polymeric residues for low-loss waveguides. A schematic illustration showing the device at this stage is given in Figure 7.6(a). The electrical pads and routing wire parts of the device are purposely removed to show the depth of the etching in the cross-section area.

Next, a similar step of E-beam lithography and ICP-RIE etching is carried out, and the pattern is precisely aligned to that of the previous step. In this step, as shown schematically in Figure 7.6(b), the silicon device layer is shallowly etched to a depth of 80 nm to define the grating couplers and tapered rib waveguides that are used for launching laser light into and out of the device. This etch depth is optimized for maximizing the light coupling efficiency of the grating coupler, which has a size of 12 μm by 12 μm to match the input-mode profile of a single-mode (SM) fiber. Tapered waveguides are used to connect the grating couplers and the nanowire waveguides with 1D PhC nanocavities. Their sizes are gradually reduced to minimize the insertion losses. Additionally, since the electric signals are necessary to drive the on-chip integrated NEMS actuators, electrodes and isolation trenches are necessary and they are fabricated by a conventional UV photolithography and ICP-RIE etching stopping on the BOX layer, as shown in Figure 7.6(c). Next in step (d), another UV photolithography and metal electron beam evaporating process are used for lift-off patterning of the gold (Au) pads, which are needed for bonding Au wires in our device package later.

The final step of the fabrication process is the etching of BOX layer to release the movable NEMS structures. The nanowire waveguides and nanocavity structures in our devices are mechanically extremely weak and susceptible to surface stiction as the liquid dries during the release process [52]. Typical MEMS structures are relatively strong with a thickness ranging from 2 um to 100 um. Hence, carefully dipping the MEMS device in a buffered oxide etcher (BOE) or hydrofluoric acid (HF) solution, rinsing it with DI water, and dry with CO_2 critical drying will not result in surface tension-induced stiction failure. However, as our current nanomechanical designs possess nano-sized beams with a thickness of 260–340 nm and a width ranging from 100 nm to 500 nm. The lengths of these nano-beams are at tens of micrometers. Hence, the release of these free-suspending nano features utilizing the conventional release techniques is deemed to be extremely challenging. A scanning electron microscopy (SEM) image showing the collapsed nano-beam waveguide due to stiction is given in Figure 7.7(a). In view of the major obstacles in the wet release processes, we used another alternative release method namely the HF vapor release technique [53, 54]. With the oxide etching carried out in the vapor phase, the surface tension-induced stiction problem is overcome. This etching mechanism involves the volatile nature of concentrated HF (49%). The device is mounted on top of the HF solution and heated with a small heater to push the dew point away in the duration of sacrificial release to avoid the formation of water droplets. In our last fabrication step, the sacrificial BOX layer is then removed using the HF vapor etch system having an etch rate of approximately 100 nm/min. The samples are finally cleaned again in an oxygen plasma RIE (Trion Technology, Inc.) dry etcher to remove any remnants of non-volatile residues left from sacrificial oxide etch. The successfully released nanomechanical structures are shown in Figure 7.7(b)–(d).

7.3.3 NEMS and nanophotonic devices testing

As the movement of the NEMS comb-drive for resonance tuning is typically less than a micron, the characterization of the actuator is done under a SEM with special

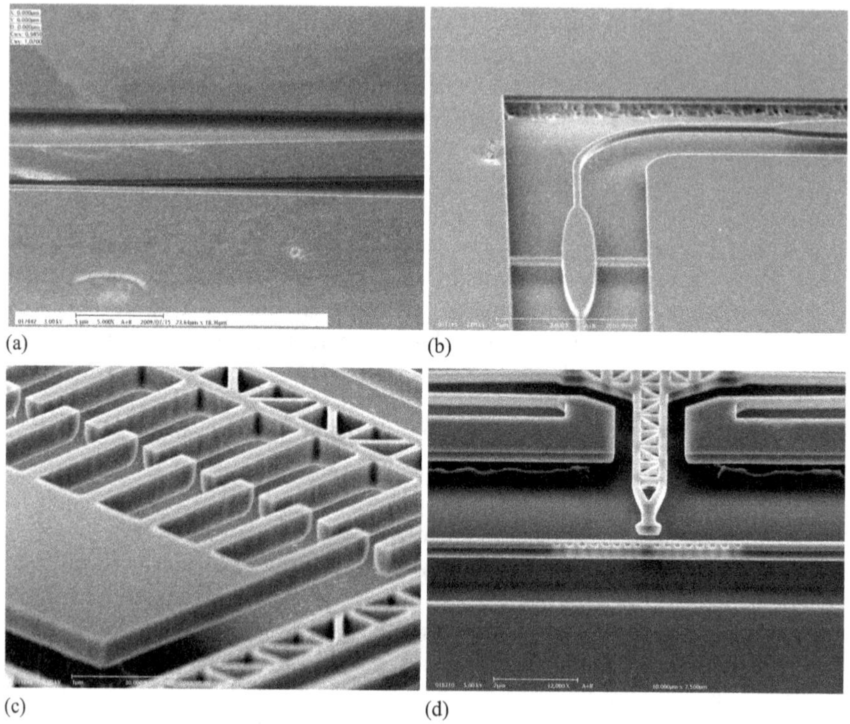

Figure 7.7 (a) Collapsed nanowire waveguide due to stiction in the wet release process. (b–d) Successfully released free-standing nanomechanical structures using an HF vapor release setup

electrical feed-throughs. This allows real-time precision displacement monitoring and measurement in nanometer accuracy when actuating the device at various voltages. The static displacement and voltage relationship of the device is thus calibrated. The device is then placed in a setup as shown schematically in Figure 7.8(a) for optical resonance characterization.

A tunable laser (ANDO AQ4321D) having a spectral range from 1520 nm to 1620 nm is used as the light source. It is controlled and synchronized with an optical spectrum analyzer (OSA) using General Purpose Interface Bus (GPIB) connection. After passing through a fiber polarization controller, the laser light from tunable laser is launched into an SM fiber and coupled into the device with a grating coupler. The DUT and optical fibers are individually mounted on precision alignment stages, which allow submicron fine adjustment for minimizing coupling losses. This alignment is performed under an optical microscope, which is installed on the top of stages. A photo showing the alignment stages and coupling fibers is provided in Figure 7.8(b). A voltage source is connected to the electrodes on the device chip to drive the NEMS actuator according to the displacement and voltage relationship calibrated in the previous SEM measurement. The fiber polarization controller is employed to maximize the TE-like modes in the device to excite the cavity resonance. The laser light transmits

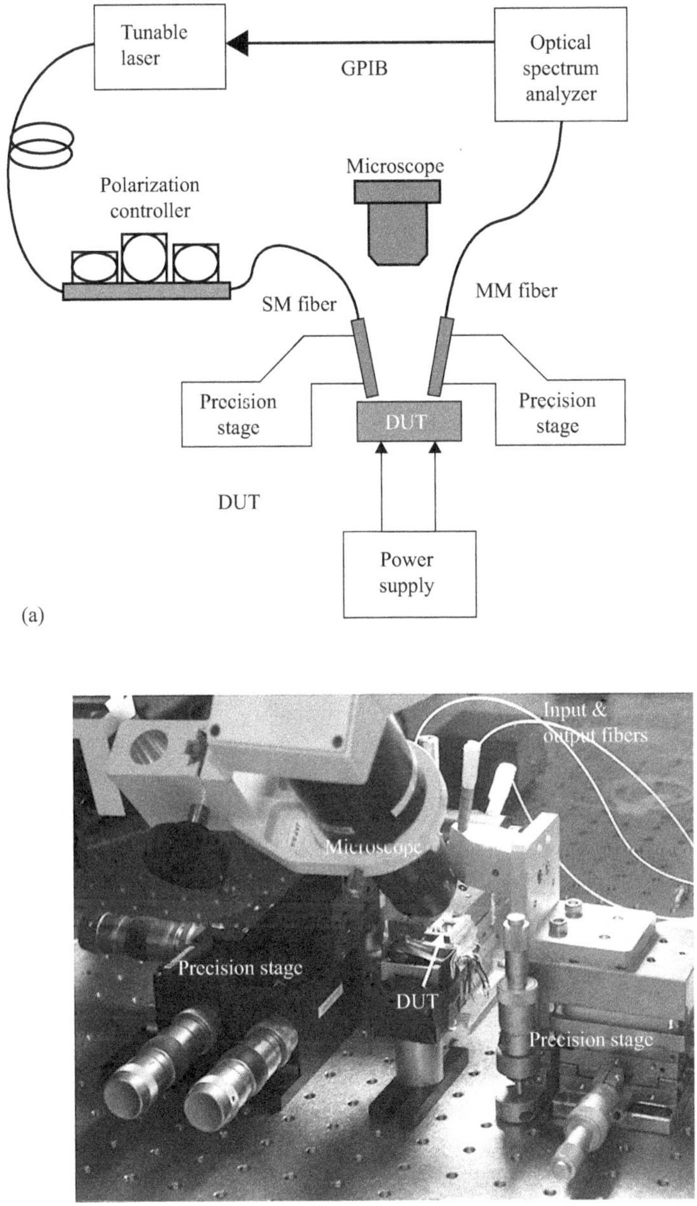

Figure 7.8 *(a) Schematic of the setup to test the MEMS/NEMS tunable photonic cavities, and (b) photo showing the precision manual alignment stages, coupling fibers, and the device under test (DUT).*

through the PhC cavity under test and couples to an output waveguide that is connected to an output grating coupler. Finally, a multi-mode fiber is aligned to the grating coupler to collect the signal, which in turn is sent into an OSA or a photodetector and an oscilloscope if the dynamic testing is required. The OSA (ANDO AQ6317C) synchronized with the tunable laser source is used for static spectrum characterization where the incident laser wavelength is scanned from 1520 nm to 1620 nm.

7.4 Tuning of PhC nanocavities with NEMS-driven dielectric probes

In 2003, Kanamori *et al.* proposed the idea of tuning the transmission characteristics of PhCs by inserting nano-rod defects [55]. By utilizing FDTD method, he numerically investigated a mechanically tunable nanophotonic device with sufficient extinction ratio at communication transmission frequencies. In 2005, Koenderink *et al.* theoretically demonstrated that through precisely displacing a dielectric nanoprobe into the near-field regime of the tightly confined optical field, the optical resonance of the high Q nanocavity could be tuned mechanically [56]. Soon in 2006, Märki *et al.* experimentally proved the concept using a silicon Atomic Force Microscope (AFM) tip to tune the resonant properties of a 2D PhC microcavity by inserting the tip into one of the Bragg mirror holes, where a few nanometer red-shift in cavity resonant wavelength is observed upon insertion of tip [57]. The tip also induces a damping effect and reduction of Q-factor to the cavity. This is due to coupling of light into the tip and strong scattering of light generated due to the altered refractive indices close to the cavity. Similar work was also reported in that period [58, 59].

The theory of resonance tuning of a photonic resonator with perturbing objects near or inside the resonator is well developed for microwave cavities [60] and photonic cavities at communication bands [56]. Generally speaking, resonant modes of a photonic resonator are eigen-solutions of a generalized Hermitian eigenvalue problem [35]:

$$\nabla \times \nabla \times \boldsymbol{E}(\boldsymbol{r}) = \left(\frac{\omega}{c}\right)^2 \varepsilon(\boldsymbol{r})\boldsymbol{E}(\boldsymbol{r}), \tag{7.19}$$

for real (lossless) and positive $\varepsilon(\mathbf{r})$. Using the standard time-independent perturbation theory and treating the perturbing object as a small perturbation to the Hermitian operator, the resonance frequency shift of the photonic resonator can be found [35]:

$$\frac{\Delta\omega}{\omega} \approx -\frac{1}{2}\frac{\int d^3\mathbf{r}\Delta\varepsilon(\mathbf{r})|\mathbf{E}(\mathbf{r})|^2}{\int d^3\mathbf{r}\varepsilon(\mathbf{r})|\mathbf{E}(\mathbf{r})|^2}, \tag{7.20}$$

where $\mathbf{E}$ and ω are, respectively, the electric mode profile and frequency of an interested resonance mode for the unperturbed perfectly linear and lossless dielectric function $\varepsilon(\mathbf{r})$, and $\Delta\varepsilon(\mathbf{r})$ is the small change in the dielectric function due to the perturbing object. From the above equation, a few conclusions can be drawn for the perturbation-induced frequency shift. First, an increase in the refractive index due to the perturbing object results a red shift in resonance wavelength (lower mode frequency). Second, the amount of frequency shift is proportional to the overall change in the index multiplied by the electric field energy inside the perturbation object integrated over the perturbed region. Clearly, taking into consideration that there exists a large electric field gradient near the photonic resonator, the resonant wavelength of the resonator will progressively red shift as a dielectric perturbation probe is moved toward the photonic resonator due to the increasingly larger electric field energy inside the probe. Third, the amount of frequency shift is inversely proportional to the mode volume of the resonance mode. Smaller mode volume results in larger frequency shift.

7.4.1 Tuning by single-tip dielectric probes

We first investigate a simple rectangular-shaped dielectric (silicon) nano-probe [8] shown schematically in Figure 7.9(a). A series of nano-sized holes are patterned to function as 1D PhCs (or Bragg-stack mirrors/reflectors) on a free-suspending 500-nm wide and 340-nm-thick silicon nanowire waveguide. These holes are periodic with a radius of 91 nm and a period of 370 nm. Similar to an F–P cavity, a nanophotonic resonator is formed between two 1D PhCs. As discussed in Section 7.2, closer to the nanocavity, the air-hole size and periodicities are tapered down to ensure that the cavity mode attenuates gradually and terminates at each PhC boundary for strong light localization and high radiative Q-factor. As shown in the figure, the rectangular nano-tip has a width of 400 nm and length of 1.5 μm. The purpose of this simple rectangular design is to gain a physical insight to understand the in-plane perturbing characteristics of the 1D PhC nanocavity. The nano-probe is attached to an NEMS comb-drive actuator having interdigitated fingers with a

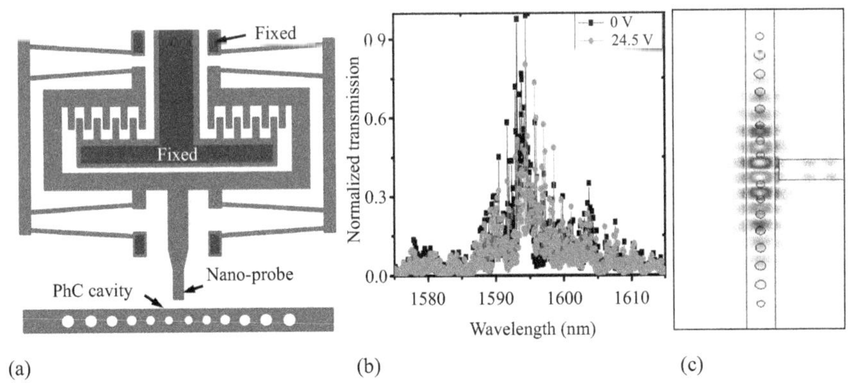

Figure 7.9 (a) Schematic of a 1D PhC cavity tuned by a simple rectangular nano probe, (b) measured transmission spectra of the device with and without the driving voltage [8], and (c) FDTD simulation results showing the light energy leakage through the probe [8]. Reprinted with permission from OSA © 2010

finger gap spacing of 200 nm. Each comb finger is 200-nm wide and 1.5-μm long. The tilted folded-beam suspension has beams of 10-μm length and 150-nm width. Through applying a driving voltage across two comb-like electrodes, the induced electrostatic force displaces the nano-probe into the near-field regime of the cavity to perturb the resonance mode thereby changing the resonance wavelength.

Let us consider the case where multiple laser beams with different wavelengths are coupled into the waveguide and fed to the 1D PhC nanocavity. The frequencies of these laser beams are all within the photonic band gap of the 1D PhC. As the PhC cavity resonance is designed to be located within the PhC photonic bandgap as well, only the light whose wavelength matches the resonance mode of the cavity is able to transmit through the structure with a high transmission. Laser beams at other wavelengths are reflected by the PhC. Here, the device can hence function as a tunable wavelength filter if the probe can perturb the cavity and change its resonance.

The device is fabricated using the process described in Section 7.3. A SEM calibration of the device is carried out first. The NEMS comb-drive works as expected and can achieve a displacement about 700 nm at a driving voltage about 23 V, which closes the gap between the rectangular probe and cavity from originally 750 nm to less than 100 nm. The calibrated displacement to voltage relationship is then utilized in the subsequent optical testing, and the transmission spectra recorded are provided in Figure 7.9(b). The 1D PhC cavity has a central resonant wavelength of 1597 nm with an experimental waveguide-loaded Q of 700 before nano-probe perturbation. After applying a driving voltage and displacing the probe, the resonance wavelength shift of the cavity is detectable, which proves the working principle of the perturbation tuning. However, an issue is also identified. As shown in Figure 7.9(b), the tuning is accompanied with large degradation of Q factor. When the probe moves to a gap less than 100 nm, the Q is reduced to less 300. Further displacing the probe yields a hardly distinguishable resonance peak along with a large noise background. The degradation of Q here is mainly due to the fact that the high refractive index silicon nano-probe is indeed a waveguide that effectively causes energy leakage of the cavity resonant mode. A FDTD simulation result shown in Figure 7.9(c) clearly shows the noticeable energy leakage through the nano-probe.

To address this energy leakage problem, we investigated a set of nano-probe designs in an attempt to reduce the energy coupling to the guided modes into the probe itself. Figure 7.10 shows one of the successful designs with a lunate-shaped nano-probe. The FDTD simulation results in Figure 7.10(b) clearly confirm the low energy leakage through the meniscus-like nano-probe. A SEM showing the fabricated device is provided in Figure 7.10(a). The comb-drive NEMS actuator is first calibrated under the SEM, and the results showing the probe to cavity gap as a function of driving voltage are given in Figure 7.11(a), where the movement of the probe is clearly observable. Optical characterization of the device is subsequently carried out. Experimental results are shown in Figure 7.11(b). With zero driving voltage, the probe is not effectively perturbing the PhC nanocavity. The recorded unperturbed cavity resonance and waveguide-loaded Q are about 1602 nm and 1300, respectively. After a driving voltage of 24.75 V is applied, the nano-probe is displaced to a gap spacing of less than 100 nm from the edge of the nanocavity causing

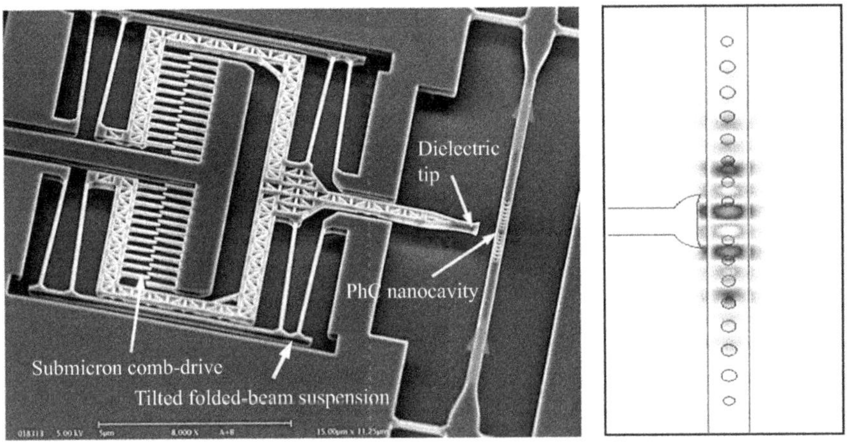

(a) (b)

Figure 7.10 *(a) Schematic of a 1D PhC cavity tuned by a lunate-shaped nano-probe [8] and (b) FDTD simulation results showing that the light energy leakage through the probe is greatly suppressed [8]. Reprinted with permission from OSA © 2010*

a resonance wavelength shift of 2.25 nm, as shown in the transmission spectra in the figure. The measured Q factor of the perturbed resonance is about 1000, which indicates very little energy losses caused by the perturbing nano-probe.

7.4.2 Tuning by multi-tip dielectric probes

To take the probe perturbation tuning method to the next step to achieve larger wavelength shift with minimal degradation of cavity Q factor, we have to consider the light coupling to a rectangular waveguide. As shown in Figure 7.12(a), the coupling efficient of an incident light wave to a waveguide eigenmode is determined by eigenmode expansion of the field [62]:

$$\eta_m = \frac{\left|\iint G(x,y,0)E_m^*(x,y)dxdy\right|^2}{\iint |G(x,y,0)|^2 dxdy \iint |E_m^*(x,y)|^2 dxdy}, \tag{7.21}$$

where $G(x,y,0)$ is the incident wavefield at the end-fire boundary $z = 0$, and $E_m(x,y)$ is the field distribution of mth-order eigenmode of the waveguide, and the asterisk denotes complex conjugate of the field. The above overlap integral measures the similarity of the incident wave field and the waveguide eigenmode. As shown in Figure 7.12(a), $E(x,y)$ denotes a fundamental order quasi-TE mode, which has a mirror symmetry about the $x = 0$ plane. If the incident wave field $G(x,y)$ has a similar symmetry, as shown with a solid line in the figure, the coupling efficiency is maximized and thus the probe effectively couples the light energy out, hence the Q of the cavity is degraded severely. However, if the incident field $G'(x,y)$ has an anti-symmetric about the $x = 0$ plane instead (shown as a dashed-line in the figure),

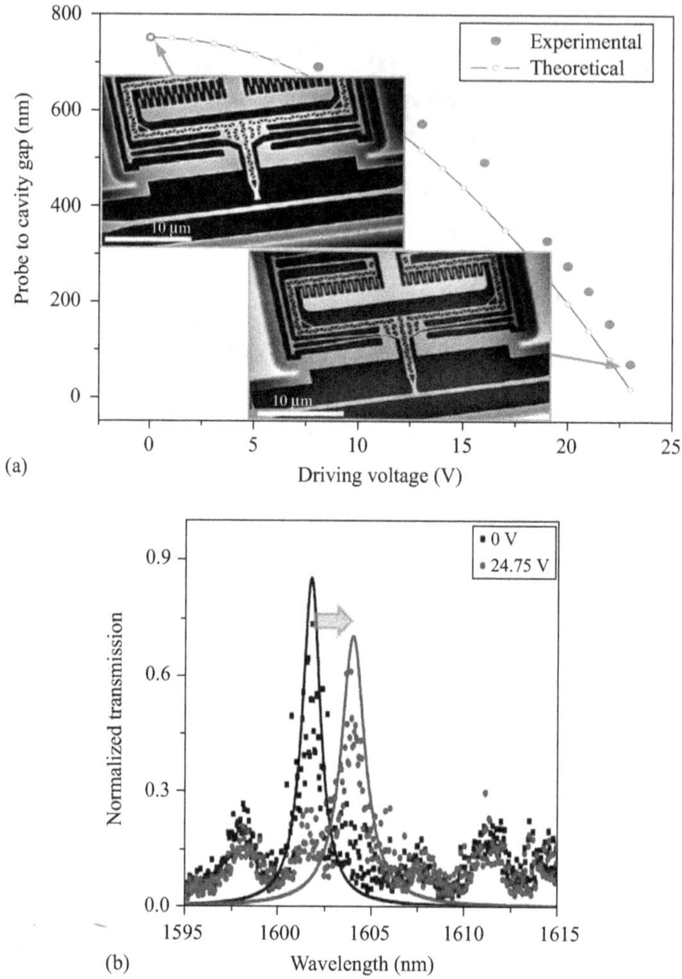

Figure 7.11 (a) SEM measured probe to cavity gap as a function of comb-drive voltage and (b) measured optical resonance shift at different driving voltages [8]. Reprinted with permission from OSA © 2010

the coupling efficient is zero according to the (7.21), and thus the energy leakage through this fundamental mode is minimized.

Based on the above analysis, we can now outline a few important design considerations to achieve an optimal dielectric probe for the perturbation tuning of a photonic cavity [61]. First, the probe must be strategically placed close to the node point of the cavity mode where the field energy is minimum. In this way, the mode field has anti-symmetry with respect to the center of the probe (i.e. waveguide), thus effectively suppress the coupling from the cavity mode to the fundamental mode of the probe. Second, the width of the rectangular probe must be small so that it does not

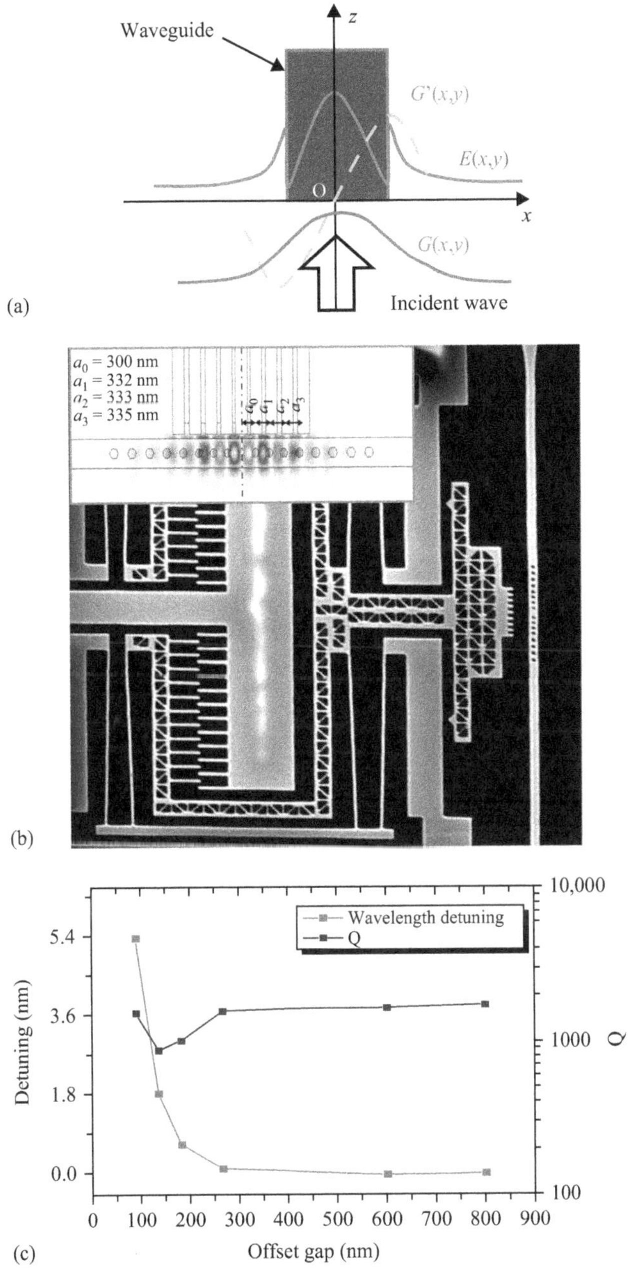

Figure 7.12 (a) Coupling of an incident wave to a rectangular dielectric waveguide, (b) SEM image showing the design of a 1D PhC cavity tuned by a multi-tip dielectric probe [61], and (c) summary of the resonant wavelength shift and Q factor of the cavity at different probe to cavity gap [61]. Reprinted with permission from SPIE © 2011

support higher order mode. It is noted that some higher order modes have anti-symmetry about the center of the waveguide. If these modes exist, the cavity mode will couple to them thus causing energy leakage and Q degradation when the probe is placed at the node point of the cavity mode. In fact, the first-generation rectangular probe we developed is a good example of this leakage. As shown in Figure 7.9(c), the FDTD result shows that the cavity mode is coupled to a higher order mode because the width of the probe is too large. Last but not the least, since the cavity mode field may have multiple node points, multiple probes (or tips) can be used to increase the overall volume of the perturbed region such that large resonance wavelength shift can be achieved.

Following the above design considerations, we demonstrate experimentally that the nano-probe tuning approach can be enhanced by introducing multiple dielectric nano-tips. Figure 7.12(b) shows a SEM image of the device fabricated using the same SOI microfabrication process. While there is no essential difference in the waveguides, PhC nanocavity, and driving submicron comb-drive actuator designs compared with the previous devices, the design of the nano-probe is however significantly improved. It consists of a total of nine nano-tips; each has a width of 250 nm and is placed at a node point of the cavity mode field, as shown in the FDTD simulation result in the inset of Figure 7.12(b). It is also observed from this simulation that the multi-tip design can effectively suppress the light energy leakage through the perturbation probe. The device is calibrated first under a SEM and then tested optically. Experimental transmission spectra at different tip-to-cavity gaps are measured, and a summary is shown in Figure 7.12(c). The unperturbed cavity resonance is located at around 1610 nm with a waveguide-loaded Q factor of 1700. The resonance changes minimally when the probe moves and closes the gap from the original 800 nm down to 250 nm. The resonance wavelength, however, red-shifts exponentially as the gap further reduces, which indicates that the multi-tip probe is in the near-field regime and perturbs effectively the cavity mode field. Finally, when the multi-tip probe is displaced to an offset gap of about 88 nm from the cavity, a maximum wavelength shift about 5.4 nm is recorded. The Q factor of the cavity is also monitored and shown as a solid blue line in Figure 7.12(c). It is well above 1000 most of the time throughout the tuning range, which indicates that the multi-tip design is very effective in reducing the energy leakage through the probe itself.

7.5 Tuning of PhC nanocavities with NEMS-driven coupled cavities

Compared with the probe tuning of the PhC nano-beam cavity, the coupled PhC nano-beam cavities can be mechanically tuned over a larger wavelength range and with further reduced degradation of Q factor [63].

7.5.1 Tuning with dual-cavity coupling

The physics of coupled resonators is well studied, and the phenomenon of resonance splitting into symmetric and anti-symmetric modes through coupling of

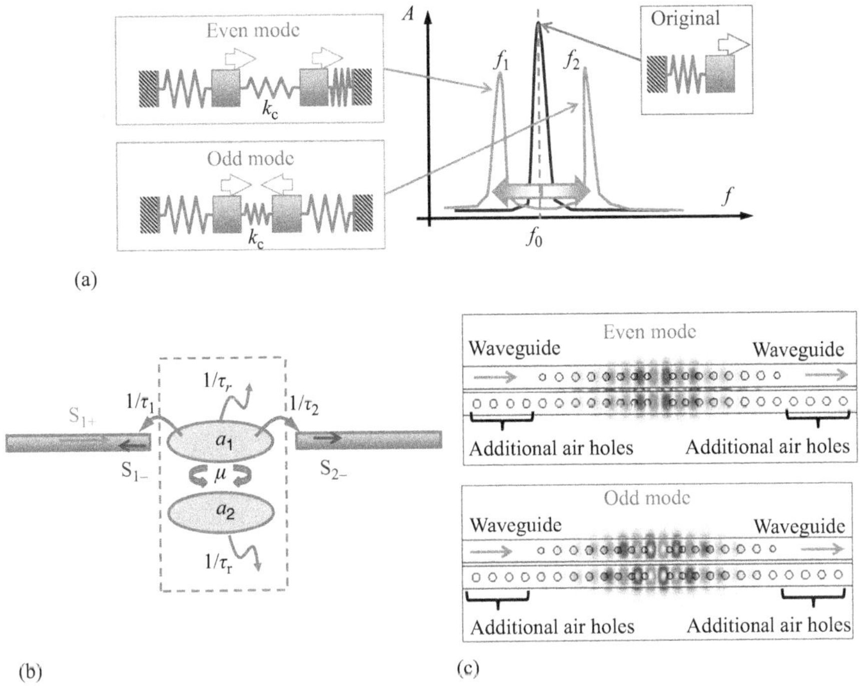

Figure 7.13 *(a) Vibration modes and mode splitting of double coupled mechanical spring-mass resonators, (b) schematic showing coupled photonic resonators and the input and output waveguides, and (c) possible implementation of the schematic (b) using 1D PhC nanobeam cavities with electric profiles showing the even and odd modes due to the coupling [8]. Reprinted with permission from OSA © 2010*

resonators is ubiquitous in nature. For example, as shown in Figure 7.13(a), two identical spring-mass resonators have the same resonant frequency f_0 without coupling. If an additional spring k_c connects the two masses, the two resonators are then coupled and this coupling produces symmetric and anti-symmetric vibration modes at f_1 and f_2, respectively, as shown in the figure and the original resonance at f_0 disappears. This phenomenon is well known in mechanics.

The same phenomenon occurs when two photonic resonators are coupled, as shown in Figure 7.13(b). We start by assuming that the two photonic resonators in the dashed box are weakly coupled and lossless (i.e. $1/\tau_1$, $1/\tau_2$, and $1/\tau_r$ are all zero). The rate changes of mode amplitudes a_1 and a_2 of the two resonators with respect to time t due to the coupling μ can be expressed using TCM theory [36]:

$$\frac{da_1}{dt} = i\omega_0 a_1 - i\mu a_2, \tag{7.22}$$

$$\frac{da_2}{dt} = i\omega_0 a_2 - i\mu a_1, \tag{7.23}$$

where ω_0 is the resonant frequency of the two identical cavities, and μ is the coupling coefficient between them. If the coupled resonators are illuminated with a laser with frequency ω, then everywhere in this system is oscillating at this frequency. Hence, assuming a dependence $\exp(i\omega t)$ of the mode amplitudes, i.e. $a_1 = A_1\exp(i\omega t)$ and $a_2 = A_2\exp(i\omega t)$. The above equation can be rewritten into a matrix form:

$$\begin{bmatrix} i(\omega - \omega_0) & i\mu \\ i\mu & i(\omega - \omega_0) \end{bmatrix} \begin{bmatrix} A_1 \\ A_2 \end{bmatrix} = \mathbf{MA} = 0. \tag{7.24}$$

In order to obtain the non-trivial solutions of A_1 and A_2, the determinant of the matrix $\mathbf{M}$ must be zero. This results in two resonant frequencies of the dual coupled cavities:

$$\omega = \omega_0 + \mu, \tag{7.25}$$

$$\omega = \omega_0 - \mu. \tag{7.26}$$

It is clear from above that the coupling results in the splitting of the original resonance at ω_0 to a blue-shifted odd mode (anti-symmetric) at $\omega_0 + \mu$ and a red-shifted even mode (symmetric) at $\omega_0 - \mu$. Hence, the resonance frequencies or wavelengths of the coupled cavities can be tuned through varying the coupling coefficient μ, in other words, changing the gap spacing between two cavities.

Next, we consider only one photonic cavity is connected with input and output waveguides with decay rates $1/\tau_1$ and $1/\tau_2$, respectively, and both cavities have radiation decay to free space $1/\tau_r$, as shown in Figure 7.13(b). In this case, we can treat the two coupled cavities as a single multi-mode photonic resonator with symmetric and anti-symmetric super modes, as indicated inside the dashed box. The system is then identical to the photonic resonator shown in Figure 7.3(b), and the theory we discussed before is also applicable to this coupled cavities. To implement this physically, as shown in Figure 7.13(c), the cavity a_1 has to be strongly coupled to the waveguides with fewer numbers of air holes in its 1D PhC Bragg stack mirrors, and the other cavity a_2 must have more numbers of air holes in its stack mirrors to prevent the light coupling out.

Mode splitting using double coupled 1D PhC cavities was demonstrated experimentally and showed the static tuning of the resonant modes by fabricating various coupled cavities with different separations [64] and subsequently using electrostatically actuated nano-beams with limited displacements [65]. Here, we introduce the real-time on-chip tuning by integrating a NEMS actuator into one of the coupled cavities [8, 63]. A SEM image of a fabricated device to demonstrate the coupled-cavity tuning approach is shown in Figure 7.14(a). The designs of the two 1D PhC cavities, with one fixed and coupled to the input and output waveguides (suspended nano silicon wires of 500-nm width and 340-nm thickness) and the other movable and connected to a NEMS comb-drive, are identical except the difference of the air-hole numbers in their stack mirrors. Thus, they have the same resonant wavelength when decoupled. The 1D PhC used has a lattice constant of 370 nm and air-hole radius of 96 nm, which acts as a Bragg mirror in the communication bands near wavelength 1550 nm. Inside the nanocavity, the periodicity and radius of the five tapered air holes are gradually reduced to suppress the

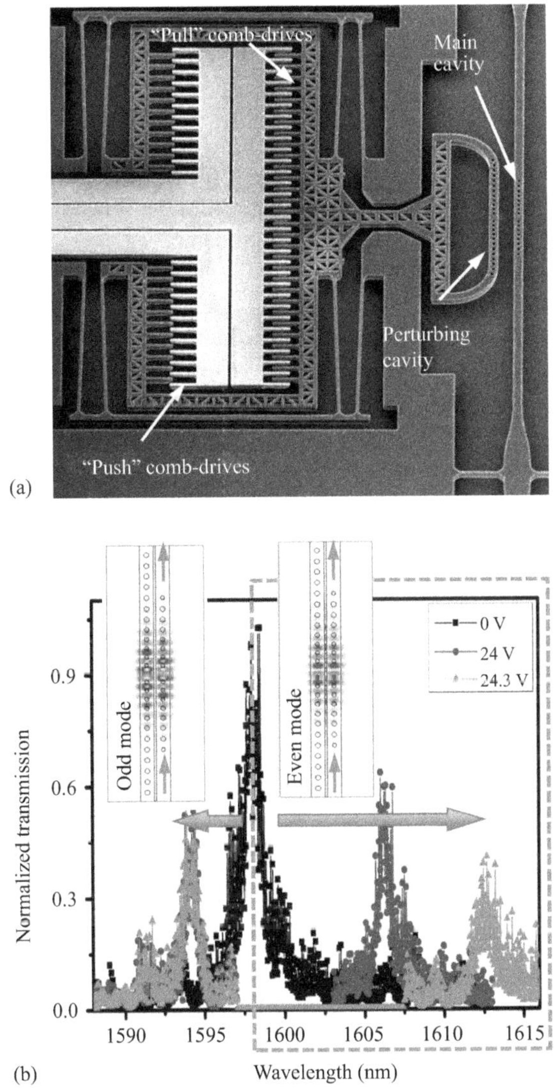

Figure 7.14 *(a) SEM image of the fabricated coupled cavities [8]. The main cavity is fixed and attached to input and output waveguides, and the perturbing cavity is attached to a NEMS comb-drive actuator. (b) Experimentally recorded spectra showing the resonance splitting into even and odd modes [8], whose wavelengths can be controlled through applying voltages to the NEMS actuator. Reprinted with permission from OSA © 2010*

radiation loss to free space and enhance the radiative Q factor. By optimizing the five-hole taper design, simulated radiative Q-factor of a single nano-beam cavity greater than 10^6 can be achieved. Here, the length of the cavity is designed at 440 nm as shown in the figure, which corresponds to a resonant wavelength within

1520–1620 nm suitable for our test setup. The FDTD-simulated electric field profiles of the even and odd modes of the two-cavity system when they are coupled are shown in the inset of Figure 7.14(b).

Similar NEMS comb-drive actuator is used as shown in the figure. The "Push" comb-drives are able to displace the perturbing cavity toward the 1D PhC main nanocavity and close the coupling gap from originally 750 nm down to less than 100 nm. In our previous experiments, we noticed that when the perturbing object was displaced nearer (within 100 nm) to the vicinity of the main nanocavity, sudden fluctuations in driving voltage or external vibrations might cause the perturbing object to adhere to the main PhC cavity surface. Hence, we design a second set of "Pull" comb-drives that are able to move the perturbing cavity in the opposite direction to enable retractable motion to recover the device. SEM testing results show that this NEMS actuator again can move more than 650 nm with a driving voltage of 24 V, which closes the cavities' coupling gap to less than 100 nm.

The measured transmission spectra are plotted in Figure 7.14(b). At 0 V driving voltage, the two cavities are separated with a large gap and are effectively decoupled. The main cavity demonstrates a resonance wavelength of 1598 nm and Q factor of 1.1×10^3. When a sufficiently large driving voltage is applied to the submicron comb-drive microactuator, the offset gap between two PhC nanocavities closes resulting in mode coupling through overlapping of the evanescent tails of the two tightly confined optical modes. We then observe the resonance splitting/shifting phenomenon due to this mode coupling. At a driving voltage of 23 V, which corresponds to a 225-nm coupling gap, it is sufficient to achieve ± 3 nm shift respectively for both even and odd modes from initial resonant wavelength of 1598 nm. Increasing the driving voltages to 23.5 V, which corresponds to 180-nm gap, will initiate the unequal spectral split between both even mode (4 nm) and odd mode (-3.5 nm). Further increasing the driving voltage results in a phenomenon where the resonance wavelength of the odd mode changes little with the nano-beam separation, whereas the even mode disperses rapidly to longer wavelength with the separation decreases. This may be due to the strong coupling of the cavities at such gaps that the linear equations (7.22) and (7.23) fail to predict the mode splitting accurately. Finally, a driving voltage of 24.4 V, which corresponds a 70-nm gap distance, induces an 18-nm wavelength shift for the even mode, while the odd mode remains at -3.9-nm wavelength shift. The results are in good agreement with the FDTD-simulated results.

It should be noted that unlike the previously described nano-probe tuning approach which shifts a single resonance over a spectral range, coupled-cavity tuning will generally produce two resonances, with one blue-shifting (odd mode) and the other red-shifting (even mode) while varying the coupling strength between the two coupling cavities. This produces significantly larger spectral tuning range with higher Q-factor as both nanocavity modes are tightly confined and thus can eliminate the energy leakage problem through the perturbing object. For applications including tunable wavelength filters that typically require shifting of a single resonance over a certain wavelength band, one has to design the initial resonance of the decoupled resonators to be just outside of the spectral band of interest, which is indicated as the brown dashed box in Figure 7.14(b). Tuning of a single resonance within the band can then be achieved through controlling the cavity coupling strength. Usually, in this case, the even mode should be employed as the filter's pass band because it has a much larger spectral tuning range compared with that of the odd mode.

7.5.2 Tuning with triple-cavity coupling

Enhanced resonance splitting/shifting is further demonstrated by a simple modification utilizing a triple-beam coupled-nanocavity configuration driven by two identically designed NEMS actuators [66]. It is also well known that the coupling of three SM resonators will generally result in three resonance modes, whose spectral shifts as functions of coupling coefficients are complicated. Such systems may not be useful in tunable resonance devices. However, by elaborating the physical aspects using TCM theory, we demonstrate that by proper design and arrangement of three cavities, the system can indeed be simplified to a dual resonance system similar to that of a doubly coupled nanocavities system.

The physical phenomenon of the triple-cavity coupled nano-resonators can also be explained using TCM theory. As shown schematically in Figure 7.15(a), again we treat this system as a photonic resonator (indicated as a dashed box) with multiple super modes and it is connected with the input and output waveguides. To get a physical insight of the super modes, we assume weak coupling among three ideal lossless ($1/\tau_1$, $1/\tau_2$, $1/\tau_R$, $1/\tau_L$, and $1/\tau_C$ are all zero) nanocavities. In addition, since the three cavities are arranged in a plane and there is a large distance between

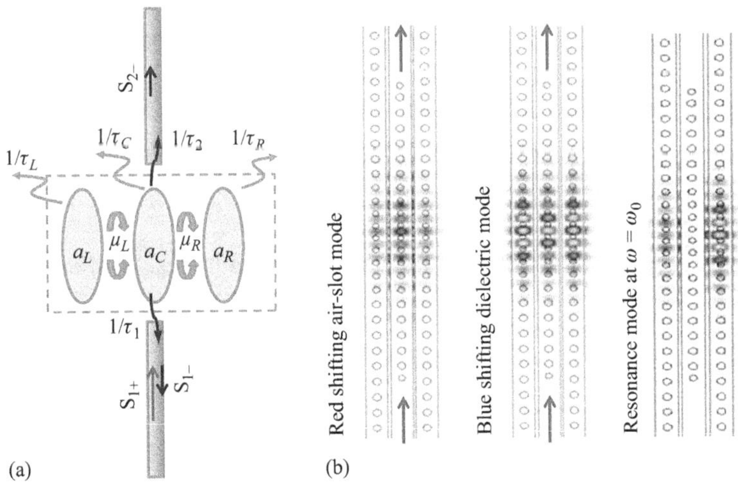

Figure 7.15 (a) Schematic of triple coupled cavities with the center cavity coupled to the input and output waveguides, and (b) a possible implementation of the schematic (a) using 1D PhC nano-beam cavities. The electric profiles show the three modes resulting from the cavities' coupling [66]. Reprinted with permission from IEEE © 2011

the right and left perturbing cavities a_R and a_L, we also assume that their coupling is negligible. The mode amplitudes a_L, a_R, and a_C can then be expressed as:

$$\frac{da_L}{dt} = i\omega_0 a_L - i\mu_L a_C, \tag{7.27}$$

$$\frac{da_C}{dt} = i\omega_0 a_C - i\mu_L a_L - i\mu_R a_R, \tag{7.28}$$

$$\frac{da_R}{dt} = i\omega_0 a_R - i\mu_R a_C, \tag{7.29}$$

where ω_0 is the natural frequency of the three identical cavities, μ_L and μ_R are the coupling coefficients. Similarly, assuming a dependence exp($i\omega t$) of the mode amplitudes, i.e. $a_L = A_L[\exp(i\omega t)]$, $a_C = A_C[\exp(i\omega t)]$, and $a_R = A_R[\exp(i\omega t)]$. We solve for the nature frequencies of the coupled resonators and obtain

$$\begin{bmatrix} i(\omega-\omega_0) & i\mu_L & 0 \\ i\mu_L & i(\omega-\omega_0) & i\mu_R \\ 0 & i\mu_R & i(\omega-\omega_0) \end{bmatrix} \begin{bmatrix} A_L \\ A_C \\ A_R \end{bmatrix} = \mathbf{MA} = 0, \tag{7.30}$$

Again in order to obtain the non-trivial solutions of A_L, A_C, and A_R, the determinant of the matrix **M** must be zero:

$$(\omega-\omega_0)^3 - \left(\mu_L^2 + \mu_R^2\right)(\omega-\omega_0) = 0, \tag{7.31}$$

which results in three resonance modes with frequencies centered at:

$$\omega = \omega_0, \tag{7.32}$$

$$\omega = \omega_0 + \sqrt{\mu_L^2 + \mu_R^2}, \tag{7.33}$$

$$\omega = \omega_0 - \sqrt{\mu_L^2 + \mu_R^2}. \tag{7.34}$$

Clearly, the in-plane coupling of the three nanocavities results in three modes, with one located at the original resonance without the coupling, one blue-shifted, and another red-shifted. FDTD simulation results further reveal the mode shapes (electric field profiles) of these three modes as shown in Figure 7.15(b). Here in our implementation of the triple-cavity coupling, we again use additional air holes in the left and right perturbing cavities to prevent light leaking out through the perturbing structures. As shown in the figure, the red-shifted resonance mode ($\omega = \omega_0 -$ sqrt ($\mu_L^2 + \mu_R^2$)) has electrical field maxima located in the air gap regions. Hence, we name it "air-slot mode." Comparing to the even mode shown in Figure 7.13(c), we see that this air-slot mode in fact originates from the even mode previously observed in the dual coupled cavities. On the other hand, the blue-shifted resonance mode ($\omega = \omega_0 +$ sqrt($\mu_L^2 + \mu_R^2$)) has electrical field maxima located in silicon regions, and therefore we call it "dielectric mode." Again, it is quite clear that this dielectric mode originates from the odd mode of the dual coupled cavities. Both the air-slot and dielectric modes are affected by the coupling and shift accordingly in response to the overall coupling coefficients arising from the coupling to the left and right cavities. Last but not the least, the third resonance mode located at the original frequency ω_0 is not affected by the coupling and does not shift at all. Its mode shape reveals that it has an anti-symmetry about the center cavity, and the electric field at the center is minimal. Consider a symmetric waveguide mode (e.g. fundamental mode) launching from the input waveguide, this third resonance mode will not be excited due to the mode mismatch (near zero overlapping integral). Hence, the resonance characteristic of this triple-beam coupled nanocavity system is found to be quite similar to that of the previously reported dual-beam coupled-cavity system, except its larger splitting/shifting of resonant frequencies/wavelengths according to (7.33) and (7.34).

The fabricated device of NEMS-tuned triple-coupled PhC nano-beam cavities is shown in Figure 7.16. The three cavities have identical designs except their PhC mirror strengths, and hence they have the same original resonance without coupling. All the three cavities have PhC periodicity of $a = 370$ nm and air-hole radius of $r = 0.23a$ etched into a nano silicon beam with width of 450 nm and thickness of 340 nm. At the center of the cavity, there is a defect region, which could locate the first-order resonance at 1543 nm. Tapering of five holes at both sides adjacent to defect region is utilized to achieve high Q-factor due to the gradual mode termination at the PhC mirror boundary. For center cavity, there are only five periodic holes on each side forming the PhC Bragg mirror region and one aperiodic out-tapered hole to enhance the waveguide coupling. However, right and left cavities have nine periodic holes for Bragg mirrors to strengthen the PhC mirrors and prevent the light from leaking to the perturbing structures. As shown in the figure, light is launched into and out of the center cavity using input and output waveguides, respectively.

Similarly, triple-coupled cavities driven by NEMS actuators are measured. The cavity coupling gap changes with respect to the driving voltages applied to the

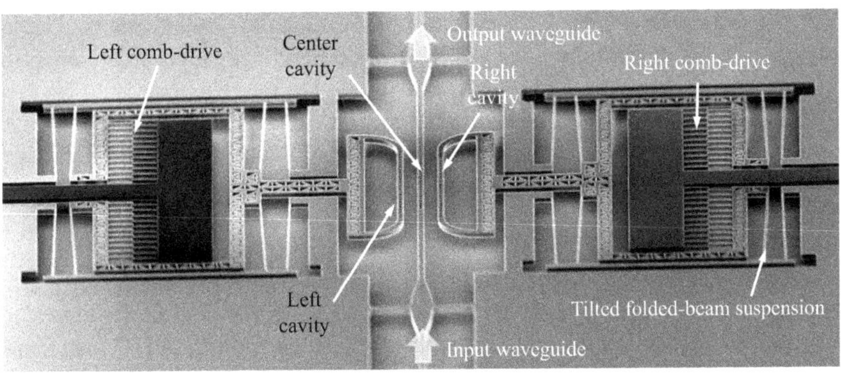

Figure 7.16 SEM image of the triple-coupled nano-beam cavities integrated with NEMS comb-drive actuators

left and right NEMS comb-drives are calibrated individually under a SEM. Both of the comb-drives can move more than 700 nm thus closing their respective gaps with the center cavity from the original 800 nm down to less than 100 nm with a driving voltage about 14 V. There is, however, a slight difference between their displacement–voltage relationships due to the fabrication imperfections. Next, the triple-cavity coupling device is tested using our established testing platform. We drive simultaneously both the left and right cavities to achieve a triple-beam coupled-nanocavity system. The perturbing nanocavities are first kept at identical or near-identical offset gaps to demonstrate the enhanced resonance tuning of such a triple-nanobeam coupled-cavity system. To achieve this, we used the SEM measured gaps to estimate the required actuating voltages so that both the left and the right cavities can be displaced at almost equal offset gaps.

Figure 7.17(a) shows the recorded resonance spectra of the device. From this result, we observe that the original resonance of the center cavity is located at the wavelength of 1544 nm with an experimentally measured waveguide-loaded Q factor of 3.8×10^3. As the offset gaps of both the left and the right cavities with respect to the center cavity are identically reduced to 200 nm, the resonance splits into the dielectric and air-slot modes, respectively, due to the cavities' coupling as predicted from (7.33) and (7.34). As shown the light-grey curve in Figure 7.17(a), the air-slot mode red-shifts to wavelength 1549 nm and the Q is measured to be 2.7×10^3, while the dielectric mode blue-shifts to wavelength 1542 nm with the recorded Q around 7.9×10^3. Further decreasing the coupling gaps of the left and right cavities with respect to the center cavity equally to 100 nm, we record that the air-slot mode resonance further red-shifts to wavelength 1559 nm, while the dielectric mode resonance blue-shifts to 1539 nm, as shown the dark-grey curve in Figure 7.17(a). The Q factors of the air-slot and dielectric modes resonances are, respectively, about 5.6×10^3 and 7×10^3, which show no obvious degradation.

Next, the resonance splitting and shifting effect is also examined when both perturbing cavities are unequally displaced. For simplicity, only the resonance of the air-slot mode is shown in Figure 7.17(b). First, the right cavity is gradually displaced, while keeping the left cavity fully decoupled at its original position. The results of resonance wavelength shift are then identical to a dual-cavity coupled system as shown in the figure. An 8.45-nm wavelength shift is recorded when the coupling gap of the right cavity is reduced to about 110 nm. Next, the right cavity is then kept stationary maintaining a 110-nm gap, and at the same time the left cavity is gradually displaced. As shown in the upper part of Figure 7.17(b), the resonance wavelength continues to red-shift. A total of 16.7 nm resonance shift is measured, when the left cavity is also displaced to a gap of 110 nm.

Throughout both equal and unequal gap tuning experiments, only the air-slot and dielectric modes are found as demonstrated above. We do not find the third mode predicted by the TCM theory, which is located at the original frequency ω_0 when the cavities are coupled. This is indeed expected from the FDTD-simulated mode shapes shown in Figure 7.15(b), as it does not match with the symmetric shape of the incident waveguide mode. This exercise demonstrates the capability of achieving resonance tuning using two perturbing cavities. More importantly, it

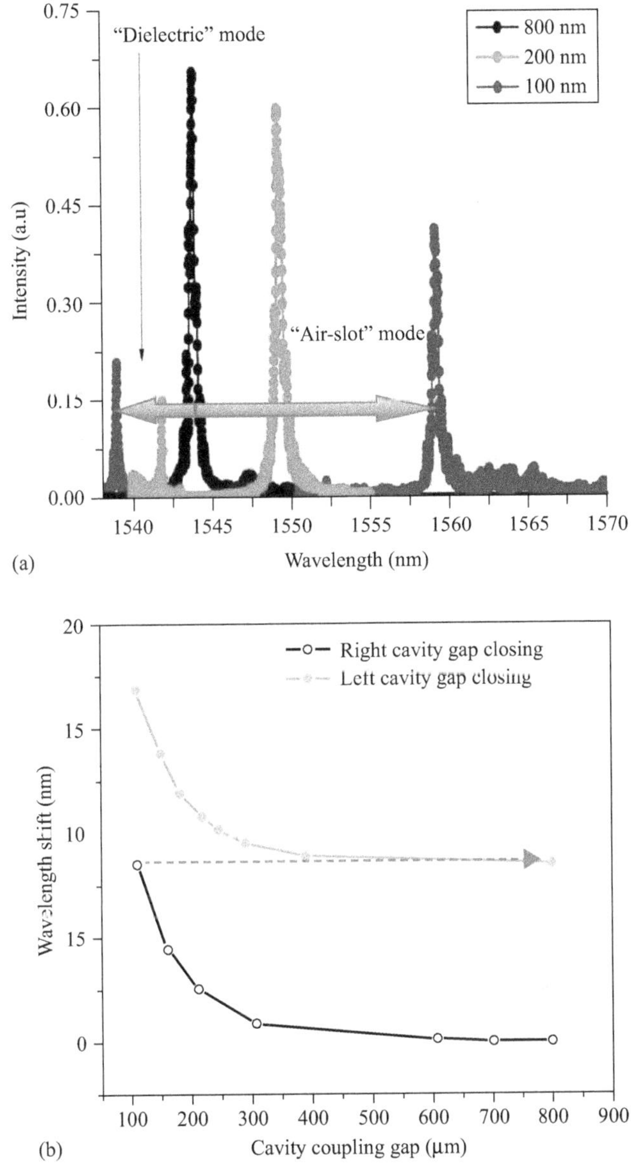

Figure 7.17 *(a) Measured spectra of the triple-coupled PhC nano-beam cavities [66]. The two perturbing cavities are driven to move synchronously to maintain an equal coupling gap with respect to the center cavity. (b) Experimental wavelength shift of the air-slot mode in response to the unequal coupling gap. The right cavity is first driven to close the gap, while the left cavity is kept at its original decoupled position. Next, the right cavity is held at 110-nm gap, and the left cavity is displaced to close its gap. Reprinted with permission from IEEE © 2011*

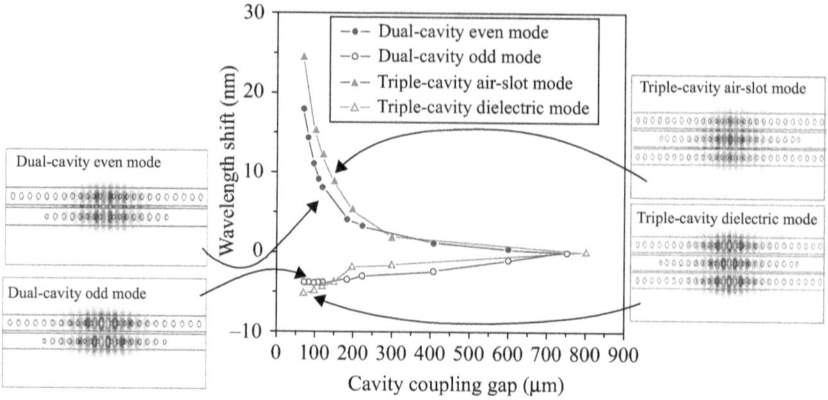

Figure 7.18 A summary of the experimental resonance tuning results for both triple-cavity and dual-cavity coupling systems. Mode shapes are also provided for comparison

shows that the requirement of having high-precision voltage sources for synchronized equal-gap tuning is not necessary, as this triple-cavity coupled system is reduced to a dual-cavity coupled system.

The experimental data for triple-cavity tuning are further summarized, together with those for dual-cavity tuning, in Figure 7.18. The relevant mode shapes are also provided in the figure for the convenience of comparing. Clearly, when the cavities are fully coupled within a coupling gap below 200 nm, the triple-cavity coupling system demonstrates a significantly enhanced tuning range. We show experimentally that at an equal offset gap of 70 nm to both left and right cavities, we obtain air-slot mode resonance red-shift about 24.36 nm, while the measured Q is still maintained at around 3×10^3. At the same time, the dielectric mode blue-shifts about 5 nm with a measured Q of 5×10^3. This demonstrates approximately a 30% improvement in resonance shift for both dielectric and air-slot modes as compared to those of the respective modes of a dual-cavity coupled system.

7.5.3 Ultrafine tuning of double-coupled multi-mode cavities

In previous sections, the real-time tuning of coupled cavities by on-chip integrated NEMS actuators is demonstrated. We show that a nanoscale displacement can result in a drastic shift of resonant wavelength. In our experiments, we also find that the displacement outputs of NEMS comb-drives (a typical result is shown in Figure 7.11) are not accurate enough for precision control of the gaps for coupled cavities. For applications that require a fine-tuning of cavity resonance within a small wavelength range, such devices are not good enough and improvements are needed. In this section, we will show the designs to achieve ultrafine displacement control together with very high Q cavity resonances.

As shown in Figure 7.19(a), a displacement shrinkage mechanism is incorporated into the on-chip integrated NEMS driving structures. The electrostatic force

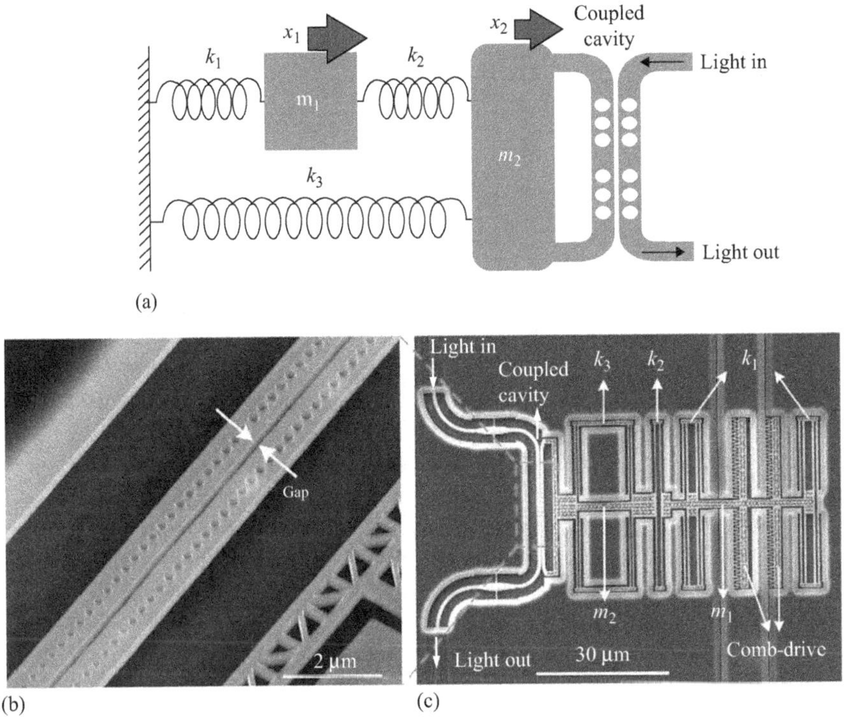

Figure 7.19 (a) Schematic of the device configuration with a displacement shrinkage spring mechanism. The NEMS comb-drive acts on the mass m_1. (b) Magnified SEM image of the double-coupled PhC nanobeam cavities with a separation gap of 144.2 nm [67]. (c) SEM image of the whole device, where folded-beam springs, rigid masses, and comb-drive is annotated [67]. Reprinted with permission from OSA © 2013

produced by the comb-drive acts on the mass m_1, and the perturbing cavity is attached to the mass m_2. The two masses are connected with a suspension network with springs k_1, k_2, and k_3. Assuming that the displacements of m_1 and m_2 are x_1 and x_2, respectively, at the equilibrium state under the influence of the electrostatic force F_e, we obtain the force equations:

$$k_3x_2 = k_2(x_1 - x_2), \tag{7.35}$$

$$F_e = k_1x_1 + k_2(x_1 - x_2), \tag{7.36}$$

for the two masses. Clearly, (7.35) indicates that the displacement output x_2 at m_2 is much smaller than x_1 when $k_2 << k_3$, with a displacement shrinkage ratio:

$$\frac{x_2}{x_1} = \frac{k_2}{k_2 + k_3}. \tag{7.37}$$

In addition, using (7.37) to eliminate x_1 in (7.36) leads to a relationship that allows us to control the movement x_2 of the perturbing cavity using the comb-drive electrostatic force F_e:

$$x_2 = \frac{F_e}{k_1 + k_3 + \frac{k_1 k_3}{k_2}}. \tag{7.38}$$

We now show the use of the displacement shrinkage mechanism in combination with the NEMS comb-drive for ultrafine tuning of photonic resonance through coupled cavities. Here, our target is to demonstrate the precision adjustment of cavity resonance within a small wavelength range. The 1D PhC cavity is hence designed to achieve a high Q-factor larger than 10^4. Figure 7.19(b) is the SEM image of the double-coupled PhC cavities with high Q and high transmission. The design of the cavity follows the rule of gradually tapered air-hole sizes in a constant PhC lattice [68]. Maintaining a constant lattice with tapered holes minimizes the light scattering losses inside the cavity as well as at the cavity–waveguide interfaces. There is no defect region in the cavity, but the tapered holes possess gradually varying mirror strengths from center to two sides. Multiple resonance modes can be formed in the cavity with very high Q-factor and high transmission through the cavity. The dimensions of the coupled cavities in Figure 7.19(b) are as follows: silicon nano-beam waveguide thickness and width are 260 nm and 670 nm, respectively, PhC lattice constant is 300 nm, diameters of the air holes are tapered from 190 nm at center to 40 nm at the two ends through a total number of 39 holes on each side, and the gap between the two coupled cavities is around 144 nm.

The overall view of the device is shown in Figure 7.19(c). One of the PhC cavities is fixed and connected with the input and output waveguides. The other cavity is movable and integrated with a displacement shrinkage mechanism that consists of three folded-beam springs. The springs and masses are annotated according to its schematic shown in Figure 7.19(a). The three suspension springs have spring constants of k_1, k_2, and k_3, respectively, and a NEMS comb-drive shown in Figure 7.19(c) applies an electrostatic force on mass m_1, thus causing the movable cavity (attached to the mass m_2) to displace toward the other fixed one. The flexible beams of the spring k_1 have length of 15 μm, thickness of 260 nm, and width of 374 nm, while those of the springs k_2 and k_3 have identical beam dimensions of 15-μm length, 260-nm thickness, and 165-nm width, thereby making k_1:k_2: $k_3 \approx 47{:}1{:}2$ from theoretical and FEM calculations. For NEMS comb-drive, there are 84 movable fingers with an initial finger overlap of 480 nm, finger width of 180 nm, and finger gap spacing of 230 nm. The displacement shrinkage ratio of m_1 and m_2 thus is determined to be $k_2{:}(k_2 + k_3) \approx 1{:}3$ using the (7.37).

The device at its original cavity coupling gap of 144 nm without actuating the NEMS comb-drive is then measured, and its spectrum is shown in Figure 7.20(a). It is noted that there are multiple resonance modes for a single decoupled 1D PhC cavity used here due to its unique design. Hence, it is expected that each single cavity resonance shall then split into even and odd modes due to the coupling effect. Resonance peaks of the second-order even ($TE_{e,2}$), the third-order odd ($TE_{o,3}$), the

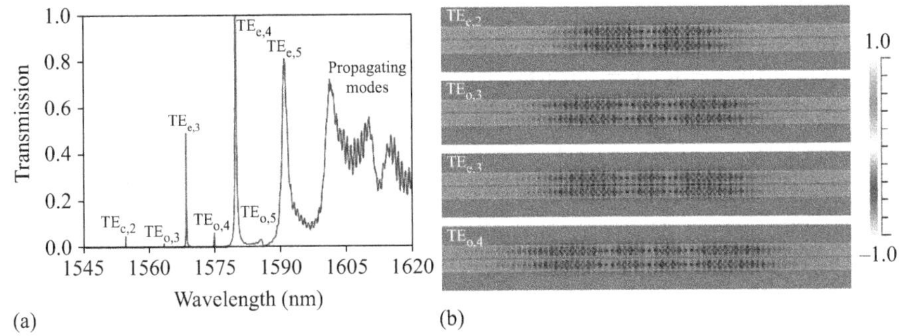

Figure 7.20 (a) Measured transmission spectrum of the coupled cavities shown in Figure 7.19(b), where various resonance modes and propagating modes are marked [67]. $TE_{e,2}$, second-order even mode; $TE_{o,3}$, third-order odd mode; $TE_{e,3}$, third-order even mode; $TE_{o,4}$, fourth-order odd mode; $TE_{e,4}$, fourth-order even mode; $TE_{o,5}$, fifth-order odd mode; $TE_{e,5}$, fifth-order even mode. The spectrum is normalized to $TE_{e,4}$ with the highest transmission. (b) FDTD-simulated electric field profiles of the $TE_{e,2}$, $TE_{o,3}$, $TE_{e,3}$, and $TE_{o,4}$ modes [67]. Reprinted with permission from OSA © 2013

third-order even ($TE_{e,3}$), the fourth-order odd ($TE_{o,4}$), the fourth-order even ($TE_{e,4}$), the fifth-order odd ($TE_{o,5}$), and the fifth-order even ($TE_{e,5}$) TE modes are recorded and marked as shown in the figure. These mode frequencies match well with the FDTD simulations. When the wavelength is longer than ~1602 nm, the cavities' Bragg mirrors are no longer functional, as the light frequency is now outside of their photonic bandgaps. In this case, instead of localized cavity resonance modes, the propagation modes appear. The cavity geometries after fabrication are measured under a SEM and then loaded into FDTD simulations, with which electric field profiles of $TE_{e,2}$, $TE_{o,3}$, $TE_{e,3}$, and $TE_{o,4}$ modes are obtained and plotted in Figure 7.20(b). Their corresponding wavelengths are simulated to be 1558.8 nm, 1571.9 nm, 1574.8 nm, and 1585.5 nm, respectively. As shown, the even modes have more electric field energies concentrated in the slot of the coupling gap than those of the odd modes, while the electric fields of the odd modes are near zero in the slot gap since the odd modes are anti-symmetric about this gap. These simulated resonance wavelengths are slightly deviated from the peaks in the measured spectrum as shown in Figure 7.20(a), and the main reason is due to the SEM cavity geometry measurement errors and simulation errors from the limited mesh size in FDTD.

The NEMS comb-drive actuator and the displacement shrinkage mechanism are calibrated first as usual when actuated under a SEM. Since the expected displacements of the perturbing cavity are now only a few tens of nanometers with the integrated displacement shrinkage mechanism, they are too small to be measured precisely by visually checking the two edges of the coupling gap even with a high SEM magnification. Hence, we develop an image processing technique to accurately measure the coupling gap, and this is based on the fact that the charges in

SEM tend to accumulate at the edges of the cavity beams thus causing the beam edges to have the highest local brightness values in the captured images. We first take a series of gap images at different voltages with the gap arranged vertically. We then compute the pixel distance between two peak brightness values in each row and then average the pixel distances for all rows in the record SEM image. And finally, we convert the average pixel distance to geometry distance by comparing it to a reference structure nearby with a known geometric dimension. In this way, the two cavities' coupling gaps at different driving voltages are precisely measured and plotted in Figure 7.21(e). Clearly, the coupling gap can be finely tuned by the NEMS comb-drive in conjunction with the displacement shrinkage mechanism.

The part labels (a)–(d) of Figure 7.21 are zoom-in views of the resonance peaks of $TE_{e,2}$, $TE_{o,3}$, $TE_{e,3}$, and $TE_{o,4}$ marked in Figure 7.20(a). The data are fitted

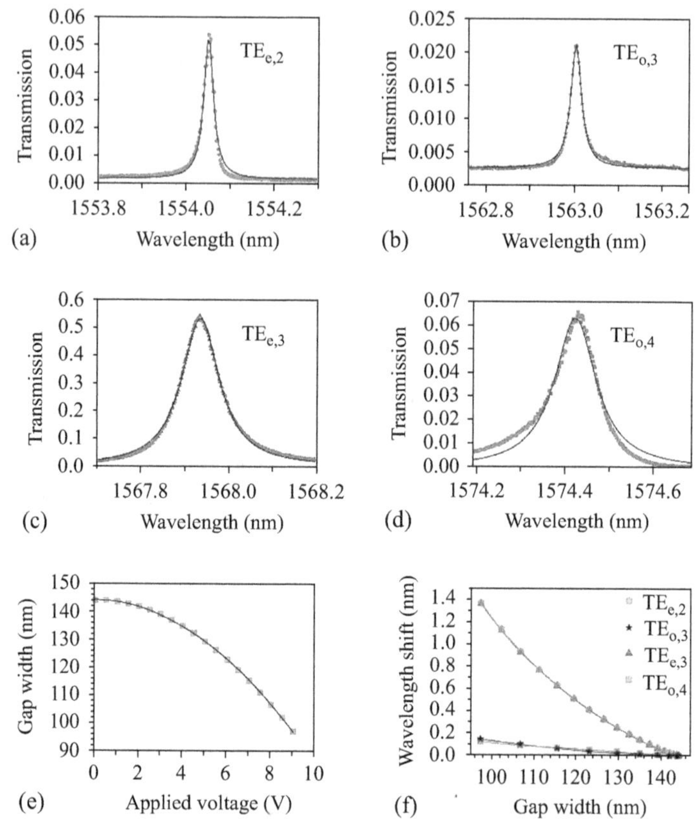

Figure 7.21 (a–d) detailed resonance peaks of $TE_{e,2}$ (c), $TE_{o,3}$ (d), $TE_{e,3}$ (e), and $TE_{o,4}$ (f) modes with respective Lorentz fits [67]. (e) The gap width between the coupled cavities changed with the voltage applied to NEMS comb-drive [67]. (f) The wavelength shifts of the $TE_{e,2}$, $TE_{o,3}$, $TE_{e,3}$, and $TE_{o,4}$ modes with respect to cavities' gap width [67]. Reprinted with permission from OSA © 2013

by Lorentz curves. As shown, the peaks of these curves are determined to be 1554.052 nm, 1563.003 nm, 1567.936 nm, and 1574.423 nm, respectively, and the *Q*-factors of the resonances calculated from the full width at half maximum of these Lorentz curves are 61,500, 61,900, 16,300, and 14,800 respectively. Lower order resonances cannot be experimentally detected in this device because of the resolution limitation of our OSA coupled with the cavities' low transmissions at those resonances. The device is then actuated according to the calibrated gap width versus voltage relationship as shown in Figure 7.21(e), and the wavelength shifts of the four marked resonance modes are recorded as functions of the coupling gap width. The measured data in Figure 7.21(f) are fitted to quartic curves which can be utilized in the accurate calculation of optomechanical coupling coefficients [69]. Clearly, these data demonstrate the ultrafine photonic resonance tuning capability of our device. It is also observed that even modes ($TE_{e,2}$ and $TE_{e,3}$) vary significantly with the gap change, while the odd modes ($TE_{o,3}$ and $TE_{o,4}$) are tuned slightly.

NEMS-driven ultrafine mechanical tuning of coupled PhC cavities can be important as well for the investigation of cavity optomechanics [70–72], which studies the interaction between light and mechanical objects at the nanoscale. For example, based on the results reported here, optomechanical coupling coefficients, characterizing the strength of the interaction between an optical field and a mechanical structure, can be derived, and subsequently the thermo-optic effect can be decoupled from the optomechanical effect in the bipolar optical force experiments [67].

7.6 Tuning of PhC nanocavities with NEMS-driven nano-deformation

We have previously discussed the tuning of PhC nano-cavities utilizing NEMS-driven probes and NEMS-driven coupled cavities. Perturbing the evanescent field of a cavity by inserting a nano-probe can change the resonant wavelength of the cavity, but the drawback of this method is that the tuning range is small. Although various probe shapes have been explored for extending the tuning range, the widest tuning range reported is still very limited (about several nanometers) [61]. In addition, the nano-probe also scatters the light of the cavity mode and thus degrades the cavity's *Q* factor. The other tuning scheme by changing the coupling of double-coupled cavities achieves a tuning range up to 18 nm [8]. Further enhancements such as utilizing triple-coupled cavities can improve the tuning range to 24 nm [66]. The *Q*-factor is not degraded during the tuning process, because the light modes in the perturbing cavities are also well confined. However, the resonance shift through coupled cavities is accompanied by a resonance splitting. For some applications, such as filters and lasers, this split resonance is undesirable. In this section, a new type of tunable cavity is introduced to overcome the drawbacks mentioned above.

Tuning the resonance of a photonic cavity through its own deformation is not new. Many ideas have been tested and reported, for example, tuning of 1D PhC nanocavity with direct strain induced by piezoelectric actuation [74] as shown schematically in Figure 7.22(a). However, the requirement of large forces and

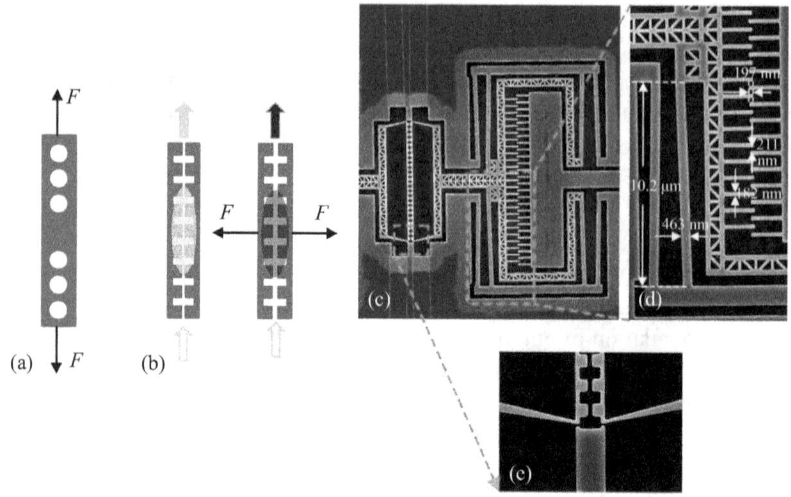

Figure 7.22 *(a) Schematic of a 1D PhC cavity tuned with direct strain induced by a piezoelectric actuator. (b) Schematic of NEMS comb-drive tuned split-ladder cavity. (c) A SEM overview of the NEMS tunable split-ladder cavity [73]. (d) Details of the driving comb-drive with critical dimensions indicated [73]. (e) Zoom-in view showing the coupling between the air-suspended waveguide and the split-ladder cavity [73]. Reprinted with permission from OSA © 2012*

small cavity deformation (hence narrow tuning range) are among the major limiting factors of such approaches. Here, we introduce an interesting cavity design termed "split-ladder cavity" [73], which has a ladder-like structure with an air-slot in its central axis. As shown schematically in Figure 7.22(b), this cavity can be tuned by changing its slot width, which is implemented by on-chip integrated NEMS comb-drives. Let us consider the case where multiple laser beams illuminate the cavity as shown in the figure. The device can essentially function as a tunable filter to selectively pass the signal beam whose wavelength matches the cavity resonance. Since the left and right parts of the split-ladder cavity are all suspended by their own springs and are not connected directly, a small electrostatic force can induce a large deformation of the cavity shape (i.e. changing the slot width) and thus inducing a large resonance shift.

A SEM image of the fabricated device is shown in Figure 7.22(c). A split-ladder cavity is fabricated using the micro and nano fabrication technology described in Section 7.3.2. As shown, both the left and right halves of the cavity are supported by tilted folded-beam springs. In addition, the left part is stationary without a driving actuator, while the right part is attached to a NEMS electrostatic comb-drive. When a driving voltage is applied to the comb-drive, the electrostatic force produced will pull the movable structure to the right and thus expending the slot. Major dimensions of the NEMS comb-drive structure are provided in

Figure 7.22(d). The relationship between displacement of the comb-drive and applied voltage can be approximately calculated by (7.10). Unlike the previous comb-drive designs with a large stable travel range around 700 nm for probe tuning and coupled cavities, here the required displacement of the comb-drive actuator is only a few tens of nanometers, which is sufficient for a large resonance tuning range. A magnified view showing the coupling region between cavity and waveguide is provided in Figure 7.22(e). The waveguide is fixed, while the cavity is deformable. Therefore, they must be disconnected to allow free movement of the cavity structures. The gap between the waveguide and the cavity, as shown in the figure, is 65 nm. Obviously, there is an insertion loss caused by this gap as well as the supporting structures in this coupling region. Calculated by FDTD simulations, the insertion loss caused by the gap and the supports is around 3 dB.

We first investigate the photonic band structure of a perfect split-ladder. All major dimensions except the thickness (t) of the device are annotated in Figure 7.23(a). Figure 7.23(b) shows its axial TE-like band structure, calculated by 3D RSoft BandSOLVE^{TM} software. The device dimensions used for band structure calculation are normalized to its lattice constant a as follows: $a/a = 610/610$, $w/a = 730/610$, $t/a = 260/610$, $g/a = 120/610$, $hz/a = 280/610$, and $hx/a = 520/610$. Thirty-two axial calculation steps are made for one lattice period, which is sufficiently fine to obtain the convergent band structure. As shown, a photonic band gap exists between the lowest two bands. Electric field component in the x-direction (E_x) is used to represent the TE-like light field in a single PhC lattice cell, and the E_x profile at the first band (lowest order band) edge is shown in the inset of Figure 7.23(b).

We now describe the design of the split-ladder cavity. Usually, photonic resonance of a PhC cavity is formed by introducing a defect into a perfect PhC

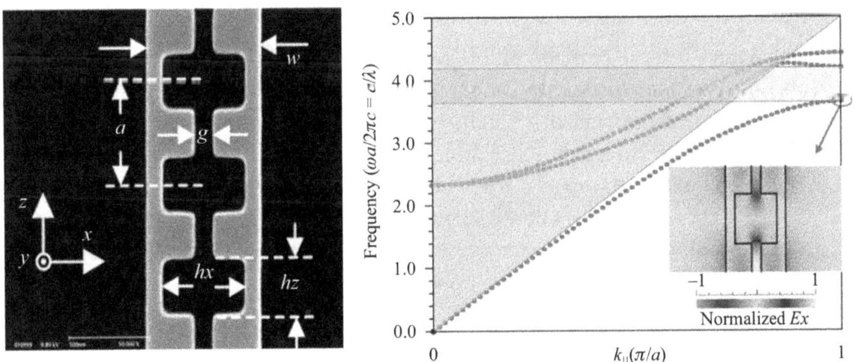

(a) (b)

Figure 7.23 *(a) Local SEM image of the split-ladder cavity [73].* a *is the PhC lattice constant;* w *is the cavity width;* g *is the slot width;* hz *is the axial air-hole lengths;* hx *is the transverse air-hole widths. (b) Axial TE-like band structure of a split-ladder lattice [73]. The inset shows the normalized electric field component in* x*-direction at the first(lowest)-order band edge. Reprinted with permission from OSA © 2012*

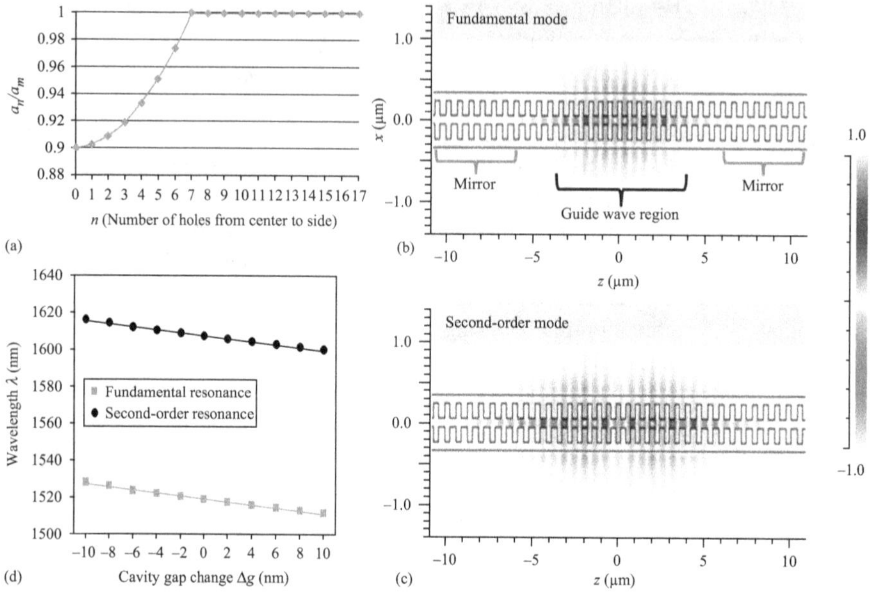

Figure 7.24 *(a) Normalized lattice periods of the half split-ladder cavity (from center n = 0 to one end n = 17) [73]. (b) The fundamental and (c) second-order resonance mode profiles of the split-ladder cavity consisting of 35 holes [73]. (d) FDTD-simulated resonance wavelength shifts of the fundamental and second-order modes as functions of the cavity gap change [73]. Reprinted with permission from OSA © 2012*

structure. Here, a total number of 35 periods of the split-ladder lattice is used to illustrate the working principle of our device. As illustrated in Figure 7.24, the lattice constants of the central 15 cells in the split-ladder cavity are modulated as the defect region, while keeping the dimensions of all air holes constant. We denote the lattice constant of the cavity's PhC Bragg mirror region by a_m and that of the defect region by a_n, where n is an integer varying from 0 to 7 corresponding to the lattice cell number from the center to either end of the cavity defect region. As shown in Figure 7.24(a), the lattice constants in the defect region are modulated according to:

$$a_n = \left(0.9 + \frac{n^2}{490}\right) \times a_m. \tag{7.39}$$

The lattice constant in the cavity's center is reduced to 90% of a_m. This gradual reduction of lattice periods shifts the first band up into the photonic band gap of the Bragg mirror region, hence certain light frequencies are confined to the defect region by the two Bragg mirrors and photonic resonances are formed. By using 3D FDTD simulations, the resonance modes of the designed split-ladder cavity are obtained, and their E_x mode profiles are shown in Figure 7.24(b) and (c) for fundamental and second-order resonances, respectively. It is quite clear from the simulated mode shapes that the light is concentrated in the air-slot region between the two ladder parts. We also investigate the changes of these two resonances as functions of the slot width numerically using FDTD. Results shown in Figure 7.24(d) demonstrate that the resonant wavelengths drastically shift when the slot width varies, which is mainly due to its extremely small mode volumes comparing with the previous cavity designs. This is the working principle of the tunable split-ladder cavity.

The split-ladder cavity designs are then fabricated and tested. The transmission spectra of a typical split-ladder cavity are shown in Figure 7.25(a,b) for the fundamental and second-order resonance modes, whose peaks are located at wavelengths of 1523.4 nm and 1592.2 nm, respectively. As shown, the fundamental resonance has a much lower transmission than the second-order resonance. The gap between the fundamental and second-order resonances is 68.8 nm, which is a large free spectral range and is advantageous for applications such as tunable filters. However, the measured Q-factors of the split-ladder cavity, 1500 and 930 for the fundamental and second-order resonances respectively, are not considered high for

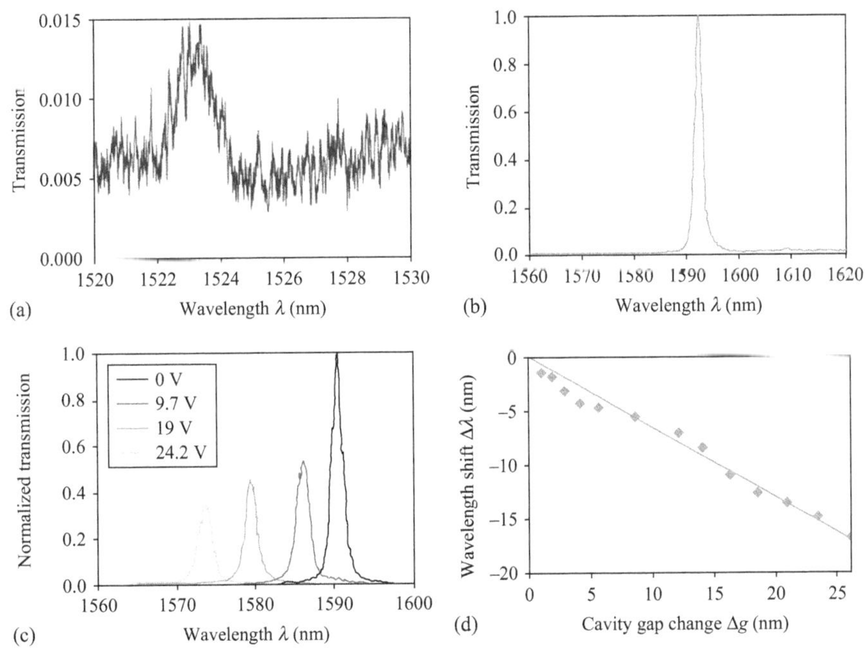

Figure 7.25 *(a,b) The measured spectra of the fundamental and second-order resonance modes of the split-ladder cavity, respectively [73]. (c) Second-order resonance wavelength varies accordingly with the applied comb-drive voltage [73]. (d) Second-order resonance wavelength shift $\Delta\lambda$ is plotted as a function of cavity gap change Δg [73]. Reprinted with permission from OSA © 2012*

such cavity designs. Considering the fact that the fundamental resonance is very close to the lower wavelength limit of our tunable laser (1520–1620 nm) and the blue-shifting effect when the cavity gap expands, we only investigate the second-order resonance tuned by the NEMS-driven slot broadening. Figure 7.25(c) shows the experimental results, where the second-order resonance shifts under various applied voltages to the comb-drive (higher voltage corresponds to wider slot gap). It can be observed that the applied voltages of 9.7 V, 19 V, and 24.2 V cause the resonance wavelength to blue-shift 4.3 nm, 10.9 nm, and 16.9 nm, respectively. As shown in the graph, the transmission peak broadens slightly along with the resonance shift indicating a somewhat reduction of cavity Q factor when tuning.

The slot width (or cavity gap) versus the voltage applied to the comb-drive is calibrated using a SEM with electrical feed-throughs. The wavelength shift of the second resonance as a function of the cavity gap change Δg is measured, calculated, and plotted in Figure 7.25(d). The recorded maximum resonance blue-shift in our experiment is 17 nm, which corresponds to a cavity gap expansion of 26 nm. The ratio of the resonance wavelength shift to the stroke of the NEMS actuator for this split-ladder cavity is much larger than the perturbative probe and coupled-cavity tuning approaches discussed in previous sections. This is mainly due to their comparatively smaller cavity mode volumes for split-ladder cavities. The experimentally obtained 17-nm wavelength shift is, however, not the tuning limit for our split-ladder cavities, and larger wavelength shift can be realized by simply further enlarge the stroke of the on-chip integrated NEMS actuator.

7.7 Conclusions

Photonic micro/nanocavity resonator is one of the most important fundamental building blocks in modern photonic systems with a myriad of applications include compact lasers, wavelength-selective add/drop multiplexers, optical filters, and high-sensitive label-free biochemical sensors. Tunable photonic cavities or resonators are attractive as they can provide a greater flexibility in dynamic photonic systems. In this chapter, we have investigated tuning of photonic resonance using near-field perturbation and cavity deformation approaches with on-chip integrated MEMS/NEMS actuators. For near-field perturbation method, we have further demonstrated resonance tuning utilizing both nano-probe perturbation and coupled-cavity approaches. Compared with the conventional thermal-optic tuning and carrier-injection modulation methods, these MEMS/NEMS tuning approaches have many outstanding advantages including low power consumption, large tuning range, absence of exotic materials, and compatible with silicon micro/nano-fabrication processes. We selected PhC-based nanocavities here as an experimental platform for the demonstration of resonance tuning approaches using MEMS/NEMS due to their excellent capability to confine light in very small mode volumes with high Q-factors. It is also noted that the reported MEMS/NEMS tuning approaches can certainly be applied to any other photonic micro/nano resonators.

References

[1] W. Bogaerts, R. Baets, P. Dumon, *et al.*, "Nanophotonic waveguides in silicon-on-insulator fabricated with CMOS technology," *Journal of Lightwave Technology*, vol. 23, pp. 401–412, Jan 2005.

[2] D. K. Armani, T. J. Kippenberg, S. M. Spillane, and K. J. Vahala, "Ultra-high-Q toroid microcavity on a chip," *Nature*, vol. 421, pp. 925–928, Feb 27, 2003.

[3] K. J. Vahala, "Optical microcavities," *Nature*, vol. 424, pp. 839–846, Aug 14, 2003.

[4] Y. Liu and H. W. M. Salemink, "Photonic crystal-based all-optical on-chip sensor," *Optics Express*, vol. 20, pp. 19912–19920, Aug 27, 2012.

[5] M. W. Pruessner, T. H. Stievater, M. S. Ferraro, and W. S. Rabinovich, "Thermo-optic tuning and switching in SOI waveguide Fabry–Perot microcavities," *Optics Express*, vol. 15, pp. 7557–7563, Jun 11, 2007.

[6] C. Li, L. J. Zhou, and A. W. Poon, "Silicon microring carrier-injection-based modulators/switches with tunable extinction ratios and OR-logic switching by using waveguide cross-coupling," *Optics Express*, vol. 15, pp. 5069–5076, Apr 16, 2007.

[7] P. T. Rakich, M. A. Popovic, M. R. Watts, T. Barwicz, H. I. Smith, and E. P. Ippen, "Ultrawide tuning of photonic microcavities via evanescent field perturbation," *Optics Letters*, vol. 31, pp. 1241–1243, May 1, 2006.

[8] X. Y. Chew, G. Y. Zhou, H. B. Yu, *et al.*, "An in-plane nano-mechanics approach to achieve reversible resonance control of photonic crystal nanocavities," *Optics Express*, vol. 18, pp. 22232–22244, Oct 11, 2010.

[9] H. G. Craighead, "Nanoelectromechanical systems," *Science*, vol. 290, pp. 1532–1535, Nov 24, 2000.

[10] M. C. Wu, A. Solgaard, and J. E. Ford, "Optical MEMS for lightwave communication," *Journal of Lightwave Technology*, vol. 24, pp. 4433–4454, Dec 2006.

[11] D. T. Neilson, R. Frahm, P. Kolodner, *et al.*, "256 × 256 Port optical cross-connect subsystem," *Journal of Lightwave Technology*, vol. 22, pp. 1499–1509, Jun 2004.

[12] K. Isamoto, K. Kato, A. Morosawa, C. H. Chong, H. Fujita, and H. Toshiyoshi, "A 5-V operated MEMS variable optical attenuator by SOI bulk micromachining," *IEEE Journal of Selected Topics in Quantum Electronics*, vol. 10, pp. 570–578, May–Jun 2004.

[13] A. Q. Liu and X. M. Zhang, "A review of MEMS external-cavity tunable lasers," *Journal of Micromechanics and Microengineering*, vol. 17, pp. R1–R13, Jan 2007.

[14] S. O. Kasap, *Optoelectronics and Photonics: Principles and Practices*. New Jersey: Prentice-Hall, Inc., 2001.

[15] A. B. Matsko and V. S. Ilchenko, "Optical resonators with whispering-gallery modes – Part I: Basics," *IEEE Journal of Selected Topics in Quantum Electronics*, vol. 12, pp. 3–14, Jan–Feb 2006.

[16] V. S. Ilchenko and A. B. Matsko, "Optical resonators with whispering-gallery modes – Part II: Applications," *IEEE Journal of Selected Topics in Quantum Electronics*, vol. 12, pp. 15–32, Jan–Feb 2006.

[17] S. M. Spillane, T. J. Kippenberg, and K. J. Vahala, "Ultralow-threshold Raman laser using a spherical dielectric microcavity," *Nature*, vol. 415, pp. 621–623, Feb 7, 2002.

[18] T. J. Kippenberg, S. M. Spillane, and K. J. Vahala, "Demonstration of ultra-high-Q small mode volume toroid microcavities on a chip," *Applied Physics Letters*, vol. 85, pp. 6113–6115, Dec 20, 2004.

[19] T. Asano, B. S. Song, Y. Akahane, and S. Noda, "Ultrahigh-Q nanocavities in two-dimensional photonic crystal slabs," *IEEE Journal of Selected Topics in Quantum Electronics*, vol. 12, pp. 1123–1134, Nov–Dec 2006.

[20] E. Yablonovitch, "Inhibited spontaneous emission in solid-state physics and electronics," *Physical Review Letters*, vol. 58, pp. 2059–2062, May 18, 1987.

[21] S. John, "Strong localization of photons in certain disordered dielectric superlattices," *Physical Review Letters*, vol. 58, pp. 2486–2489, Jun 8, 1987.

[22] J. C. Knight, J. Broeng, T. A. Birks, and P. S. J. Russel, "Photonic band cap guidance in optical fibers," *Science*, vol. 282, pp. 1476–1478, Nov 20, 1998.

[23] K. Aoki, H. T. Miyazaki, H. Hirayama, *et al.*, "Microassembly of semi-conductor three-dimensional photonic crystals," *Nature Materials*, vol. 2, pp. 117–121, Feb 2003.

[24] E. Yablonovitch, T. J. Gmitter, and K. M. Leung, "Photonic band-structure – the face-centered-cubic case employing nonspherical atoms," *Physical Review Letters*, vol. 67, pp. 2295–2298, Oct 21, 1991.

[25] C. Y. Luo, S. G. Johnson, J. D. Joannopoulos, and J. B. Pendry, "Sub-wavelength imaging in photonic crystals," *Physical Review B*, vol. 68, pp. 1–15, Jul 15, 2003.

[26] Y. A. Vlasov, M. O'Boyle, H. F. Hamann, and S. J. McNab, "Active control of slow light on a chip with photonic crystal waveguides," *Nature*, vol. 438, pp. 65–69, Nov 3, 2005.

[27] L. J. Wu, M. Mazilu, T. Karle, and T. F. Krauss, "Superprism phenomena in planar photonic crystals," *IEEE Journal of Quantum Electronics*, vol. 38, pp. 915–918, Jul 2002.

[28] H. Kosaka, T. Kawashima, A. Tomita, *et al.*, "Self-collimating phenomena in photonic crystals," *Applied Physics Letters*, vol. 74, pp. 1212–1214, Mar 1, 1999.

[29] M. H. Qi, E. Lidorikis, P. T. Rakich, *et al.*, "A three-dimensional optical photonic crystal with designed point defects," *Nature*, vol. 429, pp. 538–542, Jun 3, 2004.

[30] E. Chow, S. Y. Lin, S. G. Johnson, *et al.*, "Three-dimensional control of light in a two-dimensional photonic crystal slab," *Nature*, vol. 407, pp. 983–986, Oct 26, 2000.

[31] M. Loncar, T. Yoshie, A. Scherer, P. Gogna, and Y. M. Qiu, "Low-threshold photonic crystal laser," *Applied Physics Letters*, vol. 81, pp. 2680–2682, Oct 7, 2002.

[32] Y. Akahane, T. Asano, B. S. Song, and S. Noda, "High-Q photonic nanocavity in a two-dimensional photonic crystal," *Nature*, vol. 425, pp. 944–947, Oct 30, 2003.

[33] Y. Takahashi, H. Hagino, Y. Tanaka, B. S. Song, T. Asano, and S. Noda, "High-Q nanocavity with a 2-ns photon lifetime," *Optics Express*, vol. 15, pp. 17206–17213, Dec 10, 2007.

[34] P. B. Deotare, M. W. McCutcheon, I. W. Frank, M. Khan, and M. Loncar, "High quality factor photonic crystal nanobeam cavities," *Applied Physics Letters*, vol. 94, pp. 1–3, Mar 23, 2009.

[35] J. D. Joannopoulos, S. G. Johnson, J. N. Winn, and R. D. Meade, *Photonic Crystals: Molding the Flow of Light*, 2nd ed. New Jersey: Princeton University Press, 2008.

[36] H. A. Haus, *Waves and Fields in Optoelectronics*. New Jersey: Prentice-Hall, Inc., 1984.

[37] P. Lalanne, S. Mias, and J. P. Hugonin, "Two physical mechanisms for boosting the quality factor to cavity volume ratio of photonic crystal microcavities," *Optics Express*, vol. 12, pp. 458–467, Feb 9, 2004.

[38] A. R. M. Zain, N. P. Johnson, M. Sorel, and R. M. De la Rue, "Ultra high quality factor one dimensional photonic crystal/photonic wire micro-cavities in silicon-on-insulator (SOI)," *Optics Express*, vol. 16, pp. 12084–12089, Aug 4, 2008.

[39] K. S. Yee, "Numerical solution of initial boundary value problems involving Maxwells equations in isotropic media," *IEEE Transactions on Antennas and Propagation*, vol. Ap14, pp. 302–307, 1966.

[40] R. Courant, K. Friedrichs, and H. Lewy, "Partial differential equations of mathematical physics," *Mathematische Annalen*, vol. 100, pp. 32–74, 1928.

[41] K. E. Petersen, "Silicon as a mechanical material," *Proceedings of the IEEE*, vol. 70, pp. 420–457, 1982.

[42] Y. T. Pan, H. K. Xie, and G. K. Fedder, "Endoscopic optical coherence tomography based on a microelectromechanical mirror," *Optics Letters*, vol. 26, pp. 1966–1968, Dec 15, 2001.

[43] S. T. S. Holmstrom, U. Baran, and H. Urey, "MEMS laser scanners: A review," *Journal of Microelectromechanical Systems*, vol. 23, pp. 259–275, Apr 2014.

[44] J. Y. Liang, Y. Zhou, A. W. Winkler, *et al.*, "Random-access optical-resolution photoacoustic microscopy using a digital micromirror device," *Optics Letters*, vol. 38, pp. 2683–2686, Aug 1, 2013.

[45] J. Y. Zheng, R. M. Pasternack, and N. N. Boustany, "Optical scatter imaging with a digital micromirror device," *Optics Express*, vol. 17, pp. 20401–20414, Oct 26, 2009.

[46] S. T. Koev, R. Fernandes, W. E. Bentley, and R. Ghodssi, "A cantilever sensor with an integrated optical readout for detection of enzymatically produced homocysteine," *IEEE Transactions on Biomedical Circuits and Systems*, vol. 3, pp. 415–423, Dec 2009.

[47] K. I. Umemori, Y. Kanamori, and K. Hane, "Photonic crystal waveguide switch with a microelectromechanical actuator," *Applied Physics Letters*, vol. 89, pp. 1–3, Jul 10, 2006.

[48] A. G. Krause, M. Winger, T. D. Blasius, Q. Lin, and O. Painter, "A high-resolution microchip optomechanical accelerometer," *Nature Photonics*, vol. 6, pp. 768–772, Nov 2012.

[49] W. C. Tang, T. C. H. Nguyen, M. W. Judy, and R. T. Howe, "Electrostatic-comb drive of lateral polysilicon resonators," *Sensors and Actuators A – Physical*, vol. 21, pp. 328–331, Feb 1990.

[50] R. Legtenberg, A. W. Groeneveld, and M. Elwenspoek, "Comb-drive actuators for large displacements," *Journal of Micromechanics and Micro-engineering*, vol. 6, pp. 320–329, Sep 1996.

[51] G. Y. Zhou and P. Dowd, "Tilted folded-beam suspension for extending the stable travel range of comb-drive actuators," *Journal of Micromechanics and Microengineering*, vol. 13, pp. 178–183, Mar 2003.

[52] R. T. Howe, "Surface micromachining for microsensors and micro-actuators," *Journal of Vacuum Science & Technology B*, vol. 6, pp. 1809–1813, Nov–Dec 1988.

[53] D. F. Weston and R. J. Mattox, "Hf vapor-phase etching (Hf-Vpe) – Production viability for semiconductor manufacturing and reaction model," *Journal of Vacuum Science & Technology*, vol. 17, pp. 466–468, 1980.

[54] Y. I. Lee, K. H. Park, J. Lee, *et al.*, "Dry release for surface micromachining with HF vapor-phase etching," *Journal of Microelectromechanical Systems*, vol. 6, pp. 226–233, Sep 1997.

[55] Y. Kanamori, K. Inoue, K. Horie, and K. Hane, "Photonic crystal switch by inserting nano-crystal defects using MEMS actuator," *2003 IEEE/Leos International Conference on Optical Mems,* pp. 107–108, 2003.

[56] A. F. Koenderink, M. Kafesaki, B. C. Buchler, and V. Sandoghdar, "Controlling the resonance of a photonic crystal microcavity by a near-field probe," *Physical Review Letters*, vol. 95, pp. 1–4, Oct 7, 2005.

[57] I. Marki, M. Salt, and H. P. Herzig, "Tuning the resonance of a photonic crystal microcavity with an AFM probe," *Optics Express*, vol. 14, pp. 2969–2978, Apr 3, 2006.

[58] W. C. L. Hopman, A. J. F. Hollink, R. M. de Ridder, K. O. van der Werf, V. Subramaniam, and W. Bogaerts, "Nano-mechanical tuning and imaging of a photonic crystal micro-cavity resonance," *Optics Express*, vol. 14, pp. 8745–8752, Sep 18, 2006.

[59] F. Intonti, S. Vignolini, F. Riboli, *et al.*, "Spectral tuning and near-field imaging of photonic crystal microcavities," *Physical Review B*, vol. 78, pp. 1–4, Jul 2008.

[60] R. Inoue, H. Kitano, and A. Maeda, "Double-sphere model in the enclosed cavity perturbation technique," *Journal of Applied Physics*, vol. 93, pp. 2736–2745, Mar 1, 2003.

[61] X. Y. Chew, G. Y. Zhou, F. S. Chau, and J. Deng, "Enhanced resonance tuning of photonic crystal nanocavities by integration of optimized near-field multitip nanoprobes," *Journal of Nanophotonics*, vol. 5, pp. 1–6, Apr 21, 2011.

[62] R. G. Hunsperger, *Integrated Optics: Theory and Technology*, 6th ed. New York: Springer Science & Business Media, 2009.

[63] X. Y. Chew, G. Y. Zhou, F. S. Chau, J. Deng, X. S. Tang, and Y. C. Loke, "Dynamic tuning of an optical resonator through MEMS-driven coupled photonic crystal nanocavities," *Optics Letters*, vol. 35, pp. 2517–2519, Aug 1, 2010.

[64] P. B. Deotare, M. W. McCutcheon, I. W. Frank, M. Khan, and M. Loncar, "Coupled photonic crystal nanobeam cavities," *Applied Physics Letters*, vol. 95, Jul 20, 2009.

[65] I. W. Frank, P. B. Deotare, M. W. McCutcheon, and M. Loncar, "Programmable photonic crystal nanobeam cavities," *Optics Express*, vol. 18, pp. 8705–8712, Apr 12, 2010.

[66] X. Chew, G. Y. Zhou, F. S. Chau, and J. Deng, "Nanomechanically tunable photonic crystal resonators utilizing triple-beam coupled nanocavities," *IEEE Photonics Technology Letters*, vol. 23, pp. 1310–1312, Sep 15, 2011.

[67] F. Tian, G. Y. Zhou, Y. Du, *et al.*, "Energy-efficient utilization of bipolar optical forces in nano-optomechanical cavities," *Optics Express*, vol. 21, pp. 18398–18407, Jul 29, 2013.

[68] Q. M. Quan, P. B. Deotare, and M. Loncar, "Photonic crystal nanobeam cavity strongly coupled to the feeding waveguide," *Applied Physics Letters*, vol. 96, pp. 1–3, May 17, 2010.

[69] F. Tian, G. Y. Zhou, F. S. Chau, J. Deng, and R. Akkipeddi, "Measurement of coupled cavities' optomechanical coupling coefficient using a nanoelectromechanical actuator," *Applied Physics Letters*, vol. 102, pp. 1–4, Feb 25, 2013.

[70] T. J. Kippenberg and K. J. Vahala, "Cavity opto-mechanics," *Optics Express*, vol. 15, pp. 17172–17205, Dec 10, 2007.

[71] T. J. Kippenberg and K. J. Vahala, "Cavity optomechanics: Back-action at the mesoscale," *Science*, vol. 321, pp. 1172–1176, Aug 29, 2008.

[72] M. Aspelmeyer, T. J. Kippenberg, and F. Marquard, "Cavity optomechanics," *Reviews of Modern Physics*, vol. 86, pp. 1391–1452, Dec 30, 2014.

[73] F. Tian, G. Y. Zhou, F. S. Chau, *et al.*, "Tuning of split-ladder cavity by its intrinsic nano-deformation," *Optics Express*, vol. 20, pp. 27697–27707, Dec 3, 2012.

[74] C. W. Wong, P. T. Rakich, S. G. Johnson, *et al.*, "Strain-tunable silicon photonic band gap microcavities in optical waveguides," *Applied Physics Letters*, vol. 84, pp. 1242–1244, Feb 23, 2004.

Chapter 8

Quantum dot nanophotonics: micropatterned excitation, microarray imaging, and hyperspectral microscopy

Gauri Bhave and John X.J. Zhang

8.1 Introduction

Research in the field of electronics was limited for a long time to three-dimensional systems. In the 1970s, electronic structures confined to two dimensions were first created. These were called *quantum wells* [1, 2]. A quantum well consists of a very thin, flat layer of semiconductor material sandwiched between layers of different semiconductor materials with a higher conduction band energy. The confinement of electrons in one direction at the de Broglie wavelength occurs since the effective mass of an electron in a semiconductor is small and the de Broglie wavelength is relatively large. In a quantum well structure, the motion of electrons bound in a layer of several crystalline monolayer thicknesses is strongly quantized in the perpendicular direction.

The quasi-two-dimensional properties of quantum well structures showed great promise in electronics and optoelectronics. The observed unusual properties of quantum wells gave a boost to research in the field of quantum electronics, leading to the Nobel Prize-winning discovery of the Integer Quantum Hall Effect (IQHE) in 1985 by Klaus von Klitzing [3–6]. Quantum wells have been studied in detail and their effects are well understood. These structures have been used in several devices like laser diodes or microwave receivers.

The 1980s saw great strides in fabrication technology particularly in lithographic processes. This allowed for the creation of a quasi-one-dimensional structure called a *quantum wire*. The next step in the confinement of electronic structures was achieved by Reed *et al.* [7] from Texas Instruments Inc. Complete quantization of the electrons' free motion was achieved by trapping it in a quasi-zero-dimensional structure called a *quantum dot* (QD).

The "dot" referred to in QDs connotes an extremely confined region of space approaching zero dimensions. QDs are nanocrystals made of semiconductor materials whose excitons are confined in all three dimensions [7]. They have

Thayer School of Engineering, Dartmouth College, Hanover, NH, USA properties that are in between bulk semiconductors and discrete molecules due to their nanometer-size regime. QDs are of great interest due to several of their unique properties. By using various techniques, it is possible to control the shape, dimensions, structure of energy levels, and the number of confined electrons in QDs. Typically, these structures have a small number of electrons facilitating *ab initio* calculations. These properties make QDs the ideal mini laboratories of particular interest to theoretical physicists.

Despite the discovery and fabrication of QDs, due to their highly toxic constituents, such as cadmium, QDs were not initially used in biological applications. The inherent inability of QDs to be readily dispersed in aqueous solutions also hindered their usefulness in labeling biological samples. Capitalizing on advancements in QD synthesis techniques aimed to improve QD size monodispersity while maintaining highly luminescent behavior, in 1998 two landmark papers reported the encapsulation of QDs in water soluble coatings in order to allow labeling of both formalin-fixed [7] and live [8] cells.

The decade of 2000–2010 alone saw the publication of almost 100,000 manuscripts relating to the development, characterization, and use of QDs [7]. Current researches focus mainly on their optical and electrical properties. QDs show unique properties which are very attractive for many applications. They emit light in a very narrow, controllable spectral range which can be of a great advantage in construction of displays, semiconductor lasers, and imaging systems [8]. The self-assembling characteristics of QDs in addition to their small dimensions and the possibility of densely packed matrices open up the possibility of their use in memory media and new generations of computers.

One of the simplest applications that QDs have been recently used in is the use of a thin film of QDs fitted on top of a fluorescent or light-emitting diode (LED) lamps to convert light from a blue to a redder color similar to light produced by incandescent lamp while retaining all the advantages of LED lighting technology. QDs are also used in place of pigments and dyes since they are brighter and more controllable than organic dyes.

QDs are being hailed as a breakthrough technology in improving the efficiency of solar cells. Use of QDs in devices such as solar cells could offer up to a 10% increase in efficiency. QDs can also be used in charge-coupled devices to fabricate smaller and more efficient image detecting chips in digital cameras and similar applications.

Recently, a lot of research has focused on the use of QDs in display applications like televisions and computer screens. There are three important advantages in the use of QDs in display applications. First, QDs can be tuned to give pure saturated colors unlike traditional displays such as liquid crystal displays in which the image is made by combinations of red, blue, and green crystals. The second advantage is that QDs can be operated in electroluminescent mode, thus eliminating the need of a backlight and making QD displays much more energy efficient. Third, QDs are much smaller than liquid crystals making it possible to create a very high-resolution image.

There are other promising applications presently pursued such as using colloidal QDs to make semiconductor lasers. QDs have long been proposed as the best semiconductor materials for highly efficient lasers, and molecular beam epitaxy (MBE) has been used to make electrically driven heterostructure QD lasers for nearly a decade. At present, photo pumped lasing has been demonstrated, and electrically pumped lasing is a challenge for the future. Other applications of QDs include their use in chemical sensing, targeted drug delivery, and their use in the study of quantum mechanics.

There are many interesting chemical challenges to further applications of colloidal QDs. Besides size, shape, and surface control, semiconductor composition is an issue that has been explored. Doping n- or p-type impurities in QDs is an important step for exploring their use in advanced electronic applications. Adding magnetic impurities inside a small nanocrystal is being studied to strongly enhance the magnetic response of the nanocrystals. Research is intense in this direction with the goal being to make new magneto-optic and spintronic materials.

8.2 Principles

The different energy states of semiconductor materials can be used to explain the properties of QDs. In semiconductors, there is an energy band gap between the valance band (ground state) and conduction band (excited state). When energy is supplied (e.g., light is introduced), an electron from the valance band absorbs the (photon) energy and jumps up to the conduction band. Once electrons reach the conduction band (10^{-15}–10^{-14} s), they stay there briefly (10^{-8}–10^{-5} s), and relax down to the lowest energy state of the conduction band (10^{-12} s or less) and then to the valance band (10^{-9}–10^{-7} s). If this relaxation is by photon emission, the phenomenon is called fluorescence. When the introduced light is in near or in visible range, the color of light emitted by the semiconductor material is determined by the width of the energy band gap.

However, in QDs, the particle diameter corresponds to the system of quantum confinement, in which the spatial extent of the electronic wave function is related to the dot size. When the size of a semiconductor crystal becomes small enough that it approaches the size of the exciton Bohr radius, then the electron energy levels can no longer be treated as continuous. The energy levels must be treated as discrete, such that a small and finite separation between energy levels is present. This situation of discrete energy levels is called quantum confinement, and under these conditions, the semiconductor material ceases to resemble bulk, and instead can show the properties of a QD. Figure 8.1 shows the difference in the energy band gap structure of QD compared to a bulk semiconductor.

Quantum confinement of both the electrons and holes in all three dimensions leads to a quantized energy band gap and fixed wavelength of light emission. As a result, a decrease in dot size corresponds to an increase in the energy band gap of the semiconductor material. This is called quantum-size effect. As the QD size increases, the gap decreases. Consequently, both the absorption and emission of QDs shift to lower wavelengths (higher energies) as the size of the dots gets smaller. In other words, the absorption and emission wavelengths are size dependent, which can be

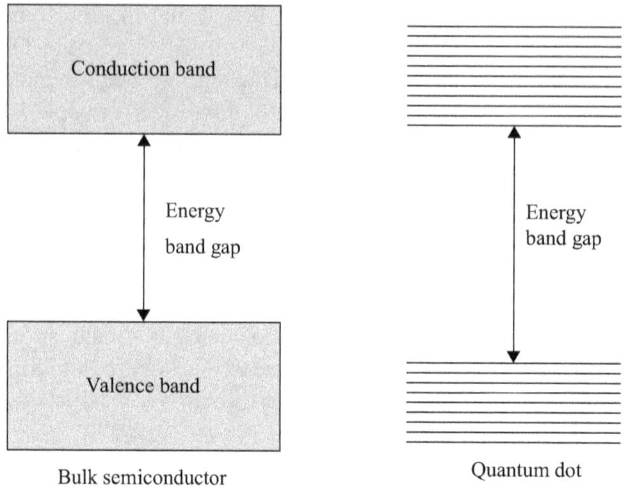

Figure 8.1 Schematic showing differences in energy band gap structure of core: shell QDs and bulk semiconductor material

controlled precisely through synthetic conditions. The emission wavelength of the particle is controlled by the size of the nanoparticles of the same material. QDs can be synthesized to emit anywhere from ultraviolet (UV) to visible all through to infrared regime. These particles have typical photoluminescence (PL) with a full width at half maximum (FWHM) 30 nm or less. This narrow spectral distribution indicates the color purity of the emitted light. These unique characteristics are shown in Figure 8.2.

8.3 Fabrication processes

The earliest method used for the fabrication of QDs was etching. Reed *et al.* [7] etched QDs in a structure containing two-dimensional electron gas. In this process, the surface of a sample containing quantum wells is covered with a polymer mask and a pattern corresponding to the shape of the desired QDs is exposed to create the nanostructure. To obtain the required high resolution, basic optical lithography is replaced by electron or ion beam lithography. The mask is removed from the exposed areas following which the entire area is covered with a thin metal layer. Liftoff technique is used followed by chemical etching to form thin pillars containing the cut out fragments of quantum wells [9].

QDs are currently fabricated using two techniques. These two types of QDs are: (a) epitaxial and (b) colloidal. Epitaxial QDs are formed by depositing a semiconducting material on a substrate by using MBE. The strain causes the top material to buckle and form QDs by the Stranski–Krastanov growth method [10]. Synthesis of colloidal QDs involves the injection of precursors into the organic solvents at high temperatures [11]. The temperature activates nucleation of small crystallites that continue to grow from the unreacted precursors until stopped by

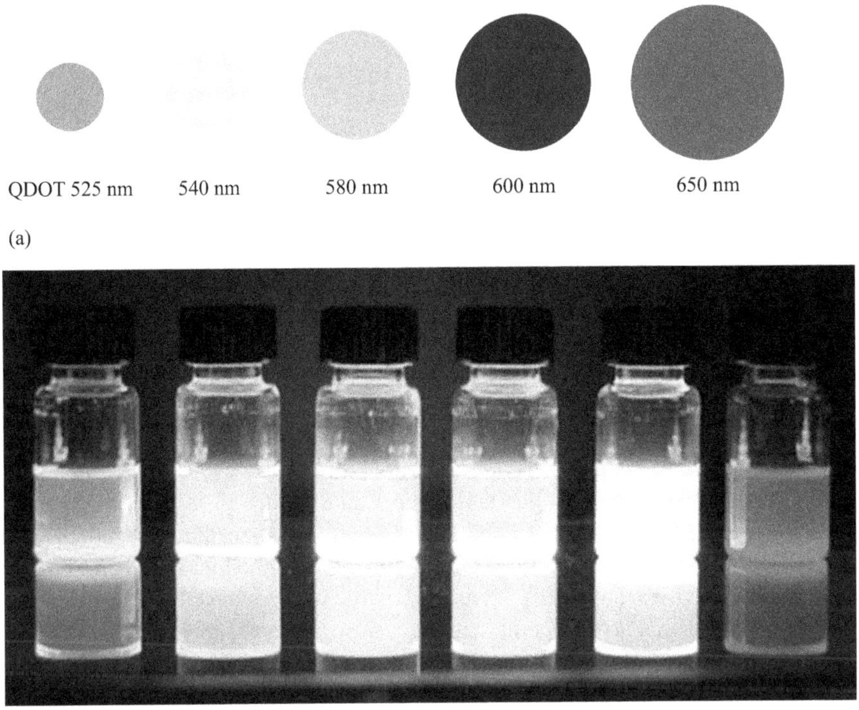

(b)

Figure 8.2 Colloidal QDs made of CdSe/ZnS (core/shell) shows (a) size dependence of emission wavelength from QDs. (b) Photoluminescence from QDs of different wavelengths (sizes)

cooling. The size of the QDs fabricated using this method can thus be finely controlled, by controlling the process time during synthesis. This process results in monodisperse solution. Colloidal QDs are preferred over epitaxial QDs, due to their low cost, ease of fabrication, and integration.

An important feature of the small nanocrystals is the great proportion of surface atoms. Since the goal in fabricating QDs is to achieve a good and stable luminescence, the surface must be inert to the excited electrons. For this, there should not be any surface states to trap electrons. Initially, QDs were fabricated with a simple CdSe core. These particles showed poor PL and moderate photostability. In 1996, this drawback was solved by growing a shell of ZnS, a higher band gap semiconductor, around the CdSe core. The CdSe/ZnS core/shell QDs, and subsequent core/shell systems showed high intensity of emissions.

Typical QDs that are used in the colloidal form have a core:shell structure with an organic capping layer. Core and shell structures are typically composed of types II–VI, IV–VI, and III–V semiconductors. Organic capping is typically provided to prevent the aggregation or precipitation of QDs (Figure 8.3). Organic compounds such as thiols, trioctylphosphine (TOPO) oxide oleic acids have been used to create organic capping.

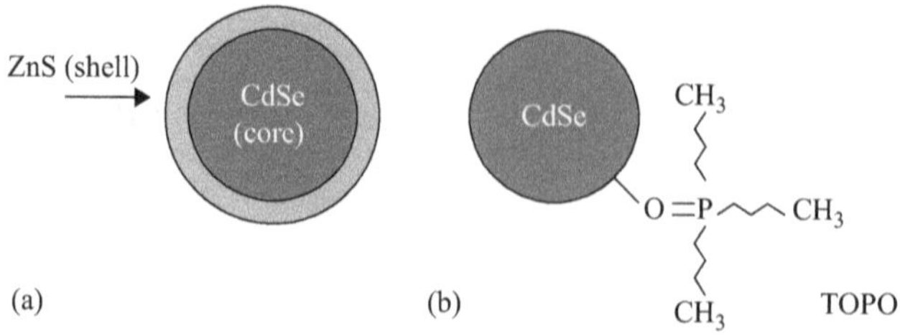

Figure 8.3 (a) Schematic showing structure of core:shell QDs and (b) core:shell CdSe:ZnS configuration with outer TOPO or ligand layer

Core–shell QDs are divided into types based on the relative conduction and valence band edge alignment of the core and shell. The types are: (a) type I and (b) type II. When the energy band gap of the shell material is larger than that of the core material, the electron–hole pair is confined within the core, as shown in Figure 8.4(b). These QDs are called type-I QDs and demonstrate higher PL efficiency as compared to type-II QDs (where the band gap of the shell is smaller than the QD). In type-II QDs, the electrons and holes are confined to the shell and the core, respectively, as shown in Figure 8.4(b). Here, the charge carriers have to cross the core–shell interface for radiative recombination, emitting fluorescence with a wide and tunable wavelength. However, strong interactions also occur between these charge carriers and any surrounding materials, which is highly dependent on the materials themselves and greatly affects the overall electrical properties of QDs. Hence, it is resultantly very difficult to formulate a general many-electron theory of QDs.

8.4 QD excitation

QDs can be excited in two ways to achieve fluorescence from them, namely (a) PL and (b) electroluminescence (EL). In PL, QDs are excited by light of a lower wavelength and red shifted emission of the emission wavelength corresponding to the QD size is obtained. In EL, electric field is applied to a layer of QDs to obtain emission of photons.

8.4.1 Photoluminescence

QDs can be optically excited. The diameter of QDs is smaller than the size of its exciton Bohr radius which leads to quantum confinement. An exciton is a bound state of an electron and a quasi-particle called a hole. After relaxation from the excited state to its lower energy state, the electron and hole recombine (exciton recombination), emitting a photon. The overall process of optical excitation, relaxation of excited state, recombination of electron and hole, and fluorescent emission is called PL. In PL, the initial energy required to move the electron to a higher energy state is provided by a photon of higher energy striking on the QD. The red shifted emission wavelength thus depends on the band gap energy, the confinement energies of the electron–hole pair and bound energy of the exciton. The emission energy thus depends on the size of the QD. The number of photons

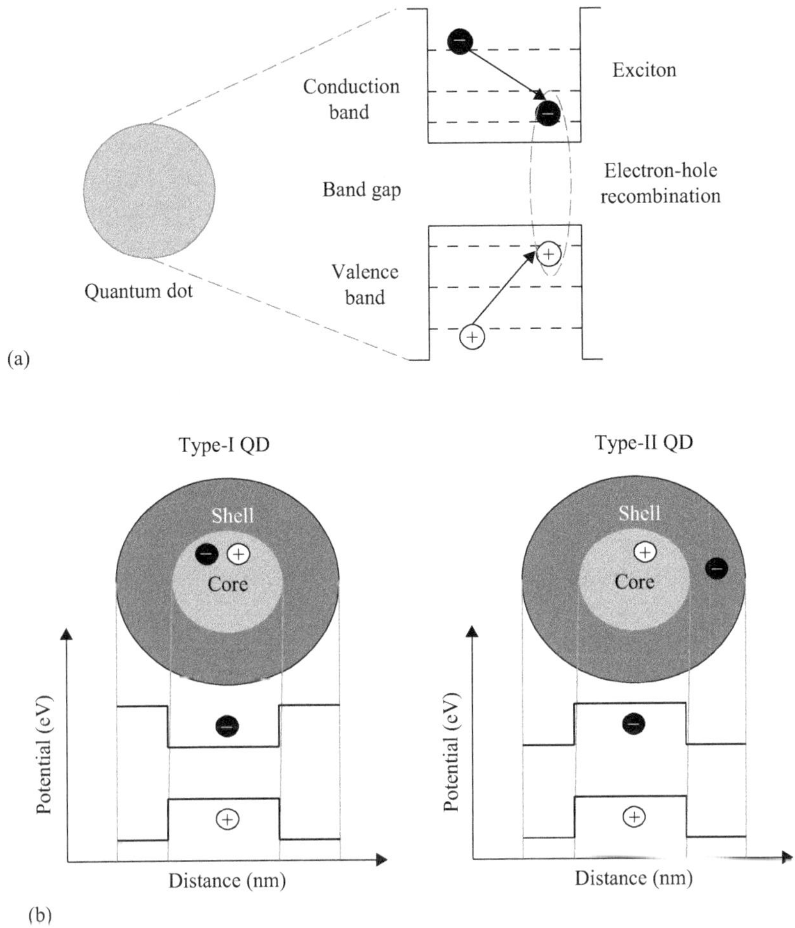

Figure 8.4 Electrical properties of QDs. (a) Schematic of electron–hole recombination and the energy levels. (b) Energy level diagram of type-I and type-II QDs

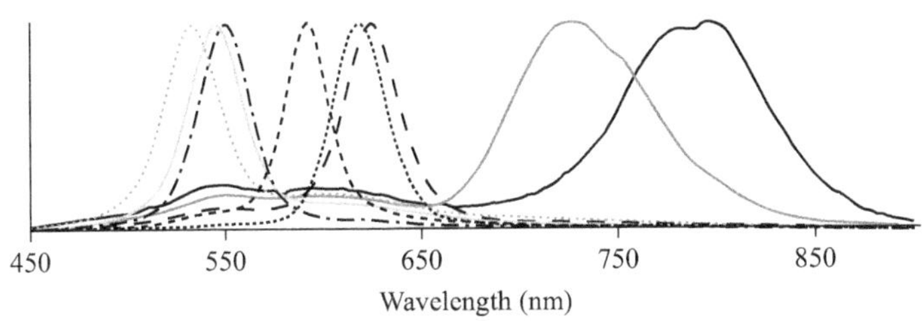

Figure 8.5 Fluorescence of QD with different emission wavelengths. Reprinted with permission from Hoshino [93], copyright the Optical Society

emitted can be measured as a function of energy, which gives the PL spectrum. Different from many organic dyes, QDs can be excited by many light sources within a large wavelength range, since QDs have continuous and broad absorption spectra. Furthermore, many kinds of QDs can be excited by the same light source and their emission can be easily separated. The emission light spectra of QDs are narrower and more symmetric than conventional organic dyes, making the sensitivity of detection higher than for organic dyes. Because of this property, QDs are attractive fluorophores for biological imaging applications. Figure 8.5 shows the emission wavelengths of QDs of different sizes.

8.4.2 Electroluminescence

In EL, the excitement energy for moving an electron to a higher energy level is provided by the recombination of electrons and holes, forming excitons, in the vicinity of the QD, due to the application of an electric field. The excitons dissipate their energy by emitting light characteristics of the optical band gap of the QDs.

8.5 Applications of QDs

PL of QDs can be used for certain types of applications such as staining of biological markers, where PL properties of QDs can solve issues such as requirement of high-intensity emission and in cases where the size of the optical system required for imaging is not of particular importance. A UV source can be used for excitation of QDs and combinations of filters can be used to focus on emissions from a single color of QD emissions. The benefits of QDs for such applications include the ability to tailor emission and absorption wavelengths through proper choice of materials and sizes; they are much more stable and slow to photobleaching compared with commonly used organic fluorescent dyes.

The ease of fabrication and processing of colloidal QDs on silicon, through nanopatterning methods and compatibility with advanced microfabrication techniques, opens up the possibilities for creating nanophotonic microsystems with mass reproducibility and enables robust and tunable imaging, sensing and display applications. An integrated excitation source with a lab-on-a-chip would have impact in many areas including: (1) diagnostics, (2) environmental, and (3) biomedical research by reducing costs for expensive projects such as genomics or proteomics measurements [12].

Site-controlled patterning of LEDs has importance in a number of applications such as micro-electro-mechanical system (MEMS) and multicolor excitation sources in micro-total analysis systems (μ-TASs). Combination of filter cubes or laser sources is typically used for standard fluorescence microscopes, which are not suitable for easy-to-use on-site reduced cost diagnostic tools.

Integration of LEDs for biomedical applications has been demonstrated previously by incorporating organic light-emitting diodes (OLEDs) [13–18]. Integrated OLEDs are placed on the rear side of the glass substrate and a microfluidic channel made out of polydimethylsiloxane (PDMS) on the front side. The emission

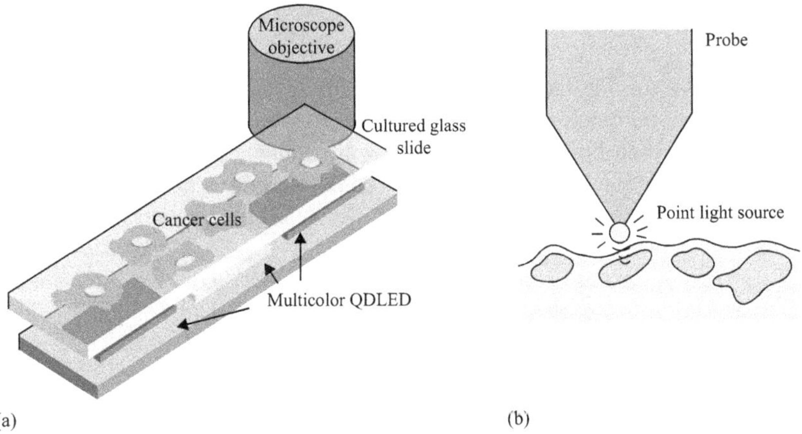

Figure 8.6 Schematics of applications of QDs. (a) Integrated bio-imaging system. (b) Near-field scanning optical microscope scanning a surface. Reprinted with permission from Ashwini Gopal

wavelength of the OLED is 520 nm and has a relatively wide peak of 70 nm. Visual inspection of excitation of fluorescent dye Rhodamine B was demonstrated [19].

Near-field scanning optical microscopy (NSOM) is another application in which QDs can be integrated for additional functionality. NSOM is a microscopic technique for investigating nanostructures that breaks the far field resolution limit [19]. An NSOM can be operated with an aperture and a non-aperture mode. Apertured tips have many issues associated with the tips such as heating, artifacts, contrast, sensitivity, topology, and interference among others. Conventional NSOM has a resolution that is directly related to the size of the light through the probe used. Typical size of the aperture of probe is 50–150 nm in diameter. NSOM have typically been used for a number of applications such as molecular feature characterization of organic interfaces [15,16], polymers [18], and optical recording for data storage devices [19].

Conventional NSOM is a powerful tool for exploring nanoscale features in thin films (Figure 8.6(b)). The major drawback with this system is the manual assembly process of the probe with an illuminating fiber that reduces its capability for large-scale applications. Also, the requirement of an external light source for each probe increases the complexity. A number of techniques for miniaturizing the NSOM probes have been developed using MEMS [20]. Light absorbing gold particles [18] and fluorescent beads [14] have been attached to the scanning tips to improve the resolution of the NSOM tips. The resolutions of these have been determined by the amount of light coupled to the tip. An external laser source is required for excitation that makes the system bulky and complicated. QD EL integrated in this application would provide a great advantage by improving the resolution of the near-field imaging system. Resolution that we could achieve theoretically would be that of a single QD molecule (size: 5–10 nm). Figure 8.6 shows schematics of two applications in which QDs can be used as light sources.

8.6 Micropatterned excitation

A number of approaches have been developed for depositing QDs onto substrates to create confined, multicolor, light-emitting QD patterns on various substrates. QDs typically have a high fluorescence efficiency and tunable light emission depending on the size of the particle. Common QD-based light-emitting devices consist of thin monolayers or multilayers of nanoparticles deposited using controllable methods for site- and size-specific patterning.

Some of the standard methods implemented to create QD thin films are phase separation, spin coating, drop casting, electro deposition, and Langmuir–Blodgett methods.

8.6.1 Phase separation

Coe-Sullivan *et al.* [19] developed a phase separation method for deposition of QDs. During the spin coating process, the QDs separate from the small molecules and form a single monolayer on top of the organic top surface. The phase separation occurs between the QDs and the organic molecule *N,N*-diphenyl-*N,N*-bis(4-*n*-butylphenyl)-(1,1-biphenyl)-4,4-diamine (TPD). This method, however, has a shortcoming because it is necessary for the organic material to be a solvent that is compatible with the QDs. This method does not allow the capability of definition patterns while depositing.

8.6.2 Spin coating

Zhao *et al.* [20] demonstrated the spin-coating the QDs from chloroform solution onto a hydrophobic thermally cross-linked hole transporting layer (HTL) of polystyrene (PS) – TPD – perfluorocyclobutane (PFCB). An electron transport layer (ETL) was deposited on top of the QD layer to complete the device. The external quantum efficiency (EQE) of these devices was 1.6% at 100 cd/m^2.

8.6.3 Langmuir–Blodgett method

A novel fabrication method was introduced by combining colloidal QDs and metal organo-chemical vapor deposition (MOCVD)-based systems. In this system, a Langmuir–Blodgett film of colloidal QDs is sandwiched between a p-type MOCVD GaN layer grown on sapphire and an n-type GaN layer grown using energetic neutral atom beam lithography/epitaxy. These diodes had EQE on the order of 0.001%–0.01% [21].

8.6.4 Micro-contact printing

Micro-contact printing is a method to form patterns using conformal contact of patterns of inked PDMS onto flat substrates [22]. A number of micro-patterning methods have been developed in the past. Some of them are photolithography, inkjet printing, nanoimprinting; and soft lithography has been used for microfabrication of various materials. Optical lithography is a fundamental tool in microfabrication. Nevertheless, these systems are not suitable for non-planar surfaces. Soft lithography has the advantage of high resolution, low cost, and large area of operation.

Micro-contact printing has also been developed as an alternate method for depositing QDs. This method has been used to deposit close packed monolayers of alkane thiol Au nanoparticles [23,24]. Micro-contact printing has also been used for deposition of ($Cd_xZnS_{1-x}Se$) $Cd_yZn_{1-y}S$ core/shell on 4,40-*N*,*N*0-diphenylcarbazole (CBP) for fabrication of green diodes [25]. Typically in micro-contact a PDMS stamp is inked with the polymer and is then transferred onto a solid substrate. PDMS substrates can make conformal contact with surfaces over large areas, enabling patterning in nanometer or micrometer size features.

PDMS stamps have a drawback, which is that they swell easily in contact with non-polar solvents such as toluene and chloroform solutions. In this case, conformal deposition becomes difficult. A number of methods have been developed to overcome this problem; Coe-Sullivan *et al.* have a modified surface of PDMS stamp by coating chemical vapor deposited parylene C layer [24]. The particles are then spin coated onto the PDMS surface, which is micro-contact printed onto the hard substrate. This enables uniform deposition of multilayer of QDs onto the substrates. The process has a drawback such that spin coating consumes a large amount of particles, thereby a significant amount of the material is wasted [26].

An alternate method was demonstrated by Bulovic [27] in which QDs can be inkjet printed onto the PDMS substrate. This is achieved by a thermal inkjet pico fluidic system developed by Hewlett-Packard. Monolayers of particles can be easily achieved in this process. The drawback with the system is that the pattern variation cannot be easily achieved [28].

Kuan *et al.* [28] also prepared a patterned ZnO film by simply printing a pattern using a PDMS mold on a sol–gel derived ZnO film. The ZnO films were spin coated onto the silicon substrate and heated to a certain temperature. Patterned PDMS was brought into contact with the surface and patterns on ZnO were obtained by removing the PDMS stamp. An alternate method to pattern ZnO films was developed by Yacaman *et al.* [29]. In this method, the assembly of nanoparticles was achieved by drying a small amount of colloidal ZnO nanoparticles suspended between a floating template (PDMS mold) and the substrate (silicon wafer).

Gopal *et al.* [30–35] developed a modified micro-contact printing method, where the nanoparticles re-suspended in hexane are used in a modified Langmuir–Schaefer technique to create thin films of particles, which are picked up by PDMS stamps and micro-contact printed on silicon and glass substrates.

8.6.5 Thin film formation

Self-assembly of monolayers and multilayers of nanoparticles have been demonstrated on a variety of substrates, through deposition on surface modified glass and metal substrates [36,37], self-assembly of multilayer polymer superlattices with incorporated nanoparticles [38], induced depositions with an electrical or magnetic field [39,40], direct etching or growth of nanoparticles on substrates [41], and evaporation of nanoparticle dispersions on substrates or on a water surface (Langmuir–Blodgett films) [42,43].

Position control of QDs attachment has yet to be thoroughly investigated to exploit the fascinating nature of colloidal QDs and their integration with existing micro-nanostructures. The applications of the above assembly methods have been limited by the quality of film, the lack of transferability after the thin film self-assembly, or pre-treatments needed for nanoparticle functionalities. Recently, it has been demonstrated that monodisperse, spherical gold nanoparticles can be self-assembled on a convex water–air interface, and subsequently be transferred using the Langmuir–Schaefer technique [23,24].

The morphology of the semiconductor QDs film differs from that of the gold nanoparticles, due to their non-spherical crystalline structure [44]. Here, we report a technique that would enable the controlled deposition of QD nanoparticles. We expand the self-assembly method with the traditional Langmuir–Schaefer technique to rapidly deposit monolayers of cadmium selenide (CdSe):zinc sulfide (ZnS) QDs on flat substrates. A uniform film of QDs self-assembled on water was transferred using hydrophobic PDMS stamps with various micro-scale patterns, and was subsequently stamped on glass or silicon substrates.

The film of nanoparticles forms in a process similar to crystallization [44, 45] as shown in Figure 8.7. First, the hydrophobic solvent holding the nanoparticles is deposited on a convex water surface. The convex water surface can be formed using various techniques including simply pinning the edges of the water surface

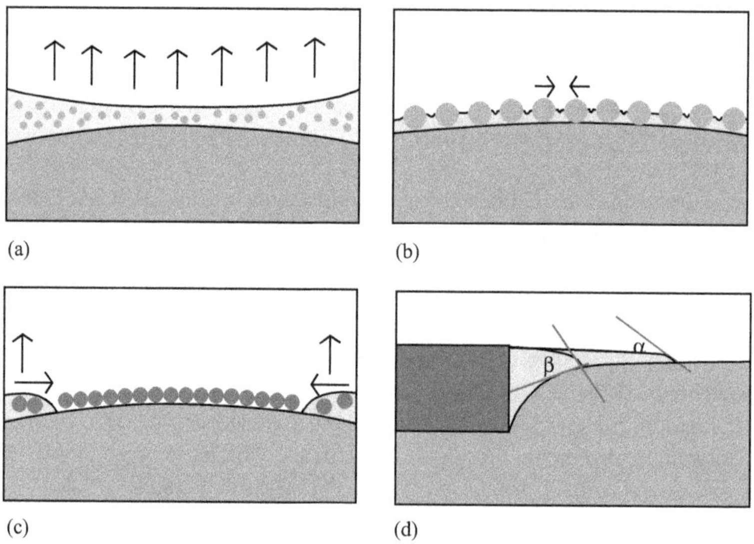

Figure 8.7 Self-assembly mechanism. (a) The solvent evaporates, thinning the layer of colloidal suspension throughout the water surface. (b) Menisci of solvent forms between the exposed nanoparticles and capillary immersion forces pack them together. (c) The contact line grows outward radially due to further evaporation and convective force that draws additional nanoparticles from the suspension to the edge of the growing film. (d) The meniscus slope angle increases, which causes multilayers of nanoparticles to form near the edge ($\beta > \alpha$). Reprinted with permission from Ashwini Gopal

using a solid ring structure. The organic solvent in the deposited colloidal solution begins to evaporate. Because of the convex water surface, the layer in the center of the surface is the thinnest, exposing the nanoparticles and forming solvent menisci between individual exposed nanoparticles (Figure 8.7(a)). As the solvent evaporates in the thinnest region, an attractive capillary immersion force collects the nanoparticles into densely packed hexagonal arrays called the nucleus [46,47] (Figure 8.7(b)). Then, as the solvent evaporates further, the contact line of the colloidal suspension and water grows outward radially and convective flow of the solvent attracts the nanoparticles from the colloidal suspension to the contact line, where they self-assemble onto the outer edge of the growing nucleus [48] (Figure 8.7(c)). As the contact line approaches the Teflon ring, where the water is pinned down, meniscus slope angle at the solvent and water interface increases from slope angle α at the slightly convex water surface that composes most of the surface inside the ring, to slope angle β at the edge where water is pinned down. As a result, multilayers are formed (Figure 8.7(d)).

Various factors contribute to the quality of film formed using this technique. Some of the controllable factors that play a role are the solvent used to form the colloidal solution, evaporation rate, and the curvature of the water surface. The monodisperse colloidal solution can be controlled by the use of various solvents and varying the concentration of the nanoparticles in solution. The evaporation rate can also be controlled by changing surrounding conditions by using fume hoods. For controlling the radius of curvature of the water surface, the convex water surface can be considered to act as a lens and the focal length in air can be calculated using the lens maker's formula.

Lens maker's formula is given by:

$$\frac{1}{f} = (n-1)\left[\frac{1}{R1} - \frac{1}{R2} + \frac{(n-1)}{nR1}\frac{d}{R2}\right] \tag{8.1}$$

where f is the focal length, n is the refractive index of the material, $R1$ is the radius of curvature of the lens surface closest to the light source, $R2$ is the radius of convex water surface curvature of the lens surface farthest from the light source, and d is the thickness of the lens. In this case, $R2 = \infty$, which reduces (8.1) to the following:

$$\frac{1}{f} = (n-1)\left[\frac{1}{R1}\right] \tag{8.2}$$

If we assume the refractive index of water to be $n = 1.31$, the radius of curvature can be calculated and adjusted. A greater radius of curvature causes thinner and more fragile films to be formed. With a higher curvature, the solvent in the center of the water surface is very thin; resulting in less time for convective flow to collect nanoparticles towards the center before the nucleus forms and the contact line begins to grow outward. Also, controlling initial film growth becomes increasingly difficult with greater curvature because the sensitivity of the curvature to the volume of water increases as well. The optimal volume of water can be adjusted depending on the thickness requirements of the nanoparticles film.

8.6.6 Particle transfer

After the formation of the self-assembled QD thin film on the water surface, the QDs can be picked up using PDMS stamps with micro features and deposited onto various substrates like silicon or glass. PDMS stamps can be created with various shapes, sizes, and surface morphologies using standard fabrication techniques, like the use of a silicon master patterned with SU-8 photoresist. The PDMS "stamp" is then removed from the master and cut into the desired shape and size. The QD film that is formed on the convex water is then picked using the PDMS stamp. The inked stamp is then inverted onto a substrate using a setup shown in Figure 8.8. A gentle pressure is applied for 30 s such that the particles transfer completely onto the substrate. The PDMS stamps can be treated with buffered oxide etchant to prevent sticking onto the substrate during micro-contact printing.

The use of PDMS stamps allows the flexibility of integrating QDs on microsystems. The feature size of patterned QDs can be arbitrarily chosen from millimeter scale to single molecular order [48]. The patterned QDs can also electrically excited to compose LEDs that can be used for nanoscale fluorescence excitation [49]. One example of QD arrays is high-definition displays, where red, green, and blue (RGB) QD pixels are patterned in a dense array to allow creation of full color images [50]. Such printing-based techniques can be used to construct the miniature array of QD light sources, chambers, and optical filters. Such integrated microscopic systems enable novel microarrays for cell culture and screening, where cells are investigated while grown in several hundreds of microwells under precisely controlled environments [51]. Another important potential application is microchip-based detection of tumor cells [52], where integrated immunofluorescence imaging is much needed. Imaging of multiple biomarkers enables precise identification and quantification of cells [53].

PL of QDs can be used for certain types of applications such as staining of biological markers, where PL properties of QDs can solve issues such as requirement of high-intensity emission. However, an external excitation source is required for PL of QDs. Hence, some key applications benefit from the electrical excitation

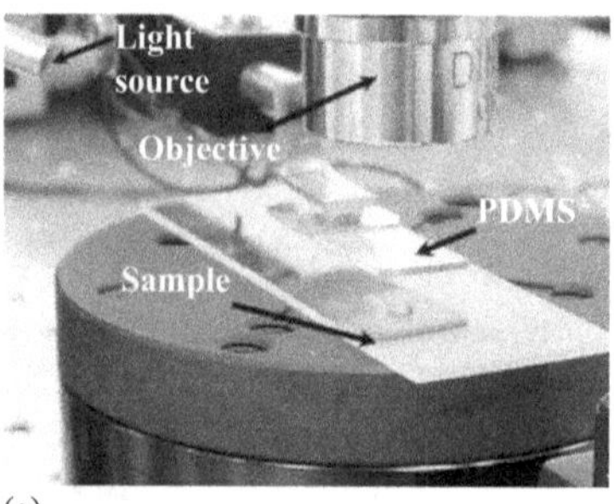

(a)

(b)

Figure 8.8 Micro-contact printing of QDs onto substrate. (a) Procedure of particle transfer. (b) Emission from the micro-contact printed QD array of colloidal QDs. For applications requiring creation of nanophotonic microsystems with mass reproducibility, creation of micron scale light sources with pure color emissions and for high-throughput applications, it is important to fabricate structures to facilitate electrical excitation of colloidal QDs. The individual addressability of these devices eliminates the need for filters and electrical excitation eliminates the use of excitation light source and filters.

8.7 Light-emitting diodes

8.7.1 Inorganic LEDs

Inorganic LEDs are made from crystalline semiconductor materials. The semiconducting material is doped with impurities to create a p–n junction. Upon application of a voltage, current flows from the anode to the cathode. Electrons and holes flow into the junction and when they combine, the electron falls to a lower energy level and releases the excess energy in the form of a photon. This is the radiative nature of electron–hole combination.

The wavelength of the light emitted in this case depends on the energy band gap of the materials forming the p–n junction, since this defines the amount of energy released in the form of a photon. Some materials, like silicon and germanium, are indirect band gap materials. In such materials, non-radiative recombinations are dominant. Radiative recombination lifetimes are on the order of milliseconds. This leads to very low efficiencies, especially at room temperature. All these factors limit the pure wavelengths of light that can be obtained from inorganic LEDs. Materials used to fabricate LEDs have a direct band gap with energies corresponding to near-infrared (NIR), visible, or near-UV wavelengths.

Compound semiconductor materials like gallium arsenide, gallium phosphide, and indium phosphide are used to make the p–n junction. These compound semiconductors are classified according to the valence bands occupied. III–V semiconductor materials are generally used to make LEDs. The wavelength of light obtained from the LEDs is designed by selecting specific materials. Pure gallium arsenide emits in the infrared region. Other materials commonly used are aluminum gallium arsenide (AlGaAs) for visible red region, gallium phosphide for green region, and aluminum indium gallium phosphide for the yellow and orange region.

Silicon is an indirect band gap material. The radiative band-to-band transitions are rare. These materials have radiative recombination lifetime on the order of milliseconds. As a consequence, non-radiative recombination at the defects is dominant. This leads to very low efficiencies, especially at room temperature. A number of different methods have been used to investigate emission from silicon namely, bulk silicon LEDs, rare earth doped nanocrystalline silicon, Silicon On Insulator (SOI)-based LEDs and porous silicon-based LEDs [71].

The advantages of inorganic LEDs are high efficiency, low turn-on voltage, high emission intensity, and long lifetimes. There are, however, some inherent drawbacks due to the nature of operation of inorganic LEDs. LEDs of different emission wavelengths cannot be easily made. Certain wavelengths are produced using either combinations of LEDs emitting primary colors or by the use of a phosphor material in combination with a blue or ultraviolet light-emitting diode (UVLED) to achieve desired wavelength. Using this method, it is not possible to achieve saturated, pure, single wavelengths emissions in all regions of the visible spectrum. The average FWHM for a commercially available inorganic LED is 70–90 nm. Another issue faced by these types of systems is that differences exist in the FWHM and efficiency based on the emission wavelength.

8.7.2 Organic LEDs

LEDs in which the emissive layer is composed of an organic compound are called OLEDs. The basic structure of OLEDs consists of a layer of certain types of organic materials between two electrodes. The organic materials used are electrically conductive with conductivities ranging from highly conductive to highly resistive. These materials are considered to be organic semiconductors. Different strategies are applied to fabricate OLEDs.

The most basic structure of polymer OLED consists of a layer of organic material between two electrodes. Other strategies consist of designs with two or more polymer layers to improve the efficiency of the device. Materials and layers are chosen to improve charge injection from electrodes or to block the charge from reaching the opposite electrode. Since silicon has an indirect band gap nature, it is difficult to get good light emission from silicon. To overcome this drawback, OLEDs have been combined with silicon substrate [72].

OLEDs are cheaper to produce as compared to traditional inorganic LEDs. OLEDs have found wide applications in cell phones, television displays, digital cameras, etc. [72]. OLEDs have the advantage of good color, brightness, contrast, power efficiency, and fast response time. They do not require backlight since each pixel can be controlled individually. This makes it possible to have thinner and more compact OLEDs. OLEDs however face some problems such as instability and lower carrier mobility.

One of the biggest technical issues faced by OLEDs is that of limited lifespan. The lifespan of such devices is limited by the rate of degradation of organic materials. Non-radiative recombination centers and luminescence quenchers accumulate, over time in the emissive layer of the OLED. This results in degradation of the emissive layer. Besides this type of electrochemical degradation, the organic emitters are also susceptible to environmental factors like moisture and thermal conditions.

Due to these drawbacks, organic emitters are gradually being replaced with semiconductor nanocrystals. Nanocrystal emitters have the advantages of well defined and tunable quality of emission along with different emission colors, electron affinities, and ionization potentials.

8.7.3 Quantum dot light-emitting diodes

The structure of quantum dot light-emitting diodes (QDLEDs) is very similar to the structure of OLEDs. The emissive organic layer in the OLED structure is replaced by a layer of QDs. Initial QDLEDs consisted of a basic structure of QDs between the two electrodes. Improvement in the efficiency of QDLEDs has been achieved over time by the addition of electron transporting and hole transporting layers (ETL and HTL) between which the QD layer is sandwiched. When an electric field is applied to this structure, electrons and holes move into the QD layer and recombine, resulting in emission of photons.

Compared to traditional inorganic LEDs and OLEDs, QDLEDs display some inherent advantages like better saturation of colors, smaller FWHM value of about 20–30 nm. QDLEDs can be fabricated into compact systems. Due to the capability of QDLEDs of working without backlight, they have the possibility of fabricating flexible devices and fabricating multicolor QDLEDs on a single substrate. However, QDLEDs face some disadvantages such as low efficiency and lifetime.

QDLEDs are of two basic types: (a) QDLEDs with organic materials for charge transport layers (CTLs) and (b) QDLEDs with inorganic materials for CTLs. For the first type in which organic CTLs are used, QDLEDs show relatively higher efficiency and have a simpler device design and fabrication. However, they face problems like lower maximum current density and device instabilities. To overcome these disadvantages, QDLEDs with inorganic CTLs were designed. These devices show better stability but there is a loss of efficiency due to QD damage during deposition of overlying inorganic layers.

The initial design of the QDLEDs was a simple structure with a QD film between an anode and cathode. When an electric field is applied across the device, the electrons and holes move to the QD layer, which is sandwiched between the anode and cathode, and is captured in the QD, where they recombine to emit a photon. To improve the efficiency of the QDLED and to address the challenge of bringing together the electrons and holes in a small region of the QD layer, hole transporting and electron transporting layers (HTL and ETL) were added to the structure. ETL and HTL effectively bring together the electrons and holes in the small region of the QD layer and suppress the dissipation and escape of photons. A single layer of QDs is used as the emissive layer, to allow electrons and holes to be directly transferred from the surfaces of the ETL and HTL, to provide high recombination efficiency.

The choice of materials for the ETL and HTL depends upon a number of conditions, including the application and desired qualities of the QDLED. Organic, inorganic, or a combination of the two materials can be used. Depending on the use of organic or inorganic materials, injection and transport of either holes or electrons are favored. This results in electron–hole combination, either closer to the cathode or to the anode, which can lead to the quenching of produced excitons. To prevent this, a number of strategies are used. In some cases where organic materials, which favor hole transport are used, an extra hole-blocking layer is introduced. Other strategies include the optimization of HTL and ETL through changing of thin-film thickness and fabrication techniques to balance the charge carrier and transport capacities of the two layers.

The choice of HTL and ETL materials significantly affects the other layers of the QDLED structure. When organic materials are used to form the ETL and HTL, the fabrication of these layers is usually done by spin coating of the CTLs. This imposes certain constraints on the materials that can be chosen and also the fabrication methods used to deposit the underlying and overlying layers. Some solvents can affect the underlying electrode during deposition and if baking is required, this can affect the resistance of the underlying layer. The deposition of QDs also faces some constraints when the CTLs are organic layers since organic solvents have to be deposited before and after the deposition of the QDs which can disturb the QD layer.

When inorganic materials are chosen to be used as the CTLs, the properties of inorganic thin films can affect the performance of the QDLED structure. Different approaches to thin film deposition have to be explored to avoid pinholes and other defects in CTLs to protect QDs from environmental contact and also from direct contact with the electrodes. During deposition of the layer over the QDs, some metal oxide deposition techniques requiring high temperature cannot be used, to avoid degradation of the QD layer. The charge transport capabilities of the ETL and HTL have to be finely tuned to enable the recombination of holes and electrons in the QD layer as recombination in other layers reduces efficiency. Other factors that affect the choice of CTL materials and fabrication methods are the ease of handling and fabrication compatibility, optical and electronic properties of thin films, and physical properties such as uniformity of film obtained, control over thickness, etc.

The choice of the cathode and anode used depends on various desired properties. Usually, indium tin oxide (ITO) is used as an anode material for fabrication of a QDLED on a glass substrate. This is because of the hole injection capabilities of ITO in addition to ITO being optically transparent in the visible wavelength range, allowing the passage of the QD emission through the anode and glass substrate. In some other designs, doped silicon is used as an anode and in this case the cathode is kept transparent to obtain the light emissions. A metal like aluminum or silver is usually used to form the cathode due to its electron injecting capacity. The choice of metal is made depending on the desired fabrication method compatible with the underlying layer, desired lifetime, and thickness. Figure 8.9 shows the basic device of a QDLED.

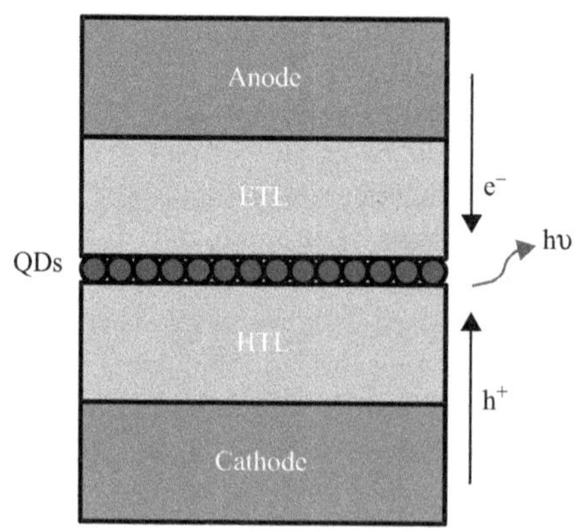

Figure 8.9 Schematic of a typical QDLED structure (HTL: hole transporting layer, ETL: electron transporting layer)

Depending on the application, the sizes, shapes, and emission wavelength of QDLEDs can be tuned. The materials selected to form the HTL and ETL depend to a large extent on the energy band gaps of the QDs. CdSe/ZnS core/shell QDs are commonly used in the fabrication of QDLEDs due to their easy fabrication and excellent optical properties.

The band gap energy of QDs depends on the size of the QDs and thus depends on the emission wavelength of the QDs. The band gap energy of CdSe/ZnS QDs corresponding to various emission wavelengths is given in Table 8.1.

The energy band gap values of the QDs are important for the functioning of the QDLEDs. The choice of the surrounding layers depends on these energy band gap values since the surrounding layers should be able to provide flow of holes and electrons into the QD layer. Figure 8.10 shows an example of the energy band diagram of a QDLED structure.

Combinations of various materials have been explored to optimize the design and performance of QDLEDs. Besides organic materials, inorganic materials have been considered as a transport layer in organic or QDLEDs. Inorganic materials

Table 8.1 Band gap energy of CdSe/ZnS QDs

Wavelength (nm)	Band gap (eV)
560	2.125
580	2.0
600	1.98
620	1.91

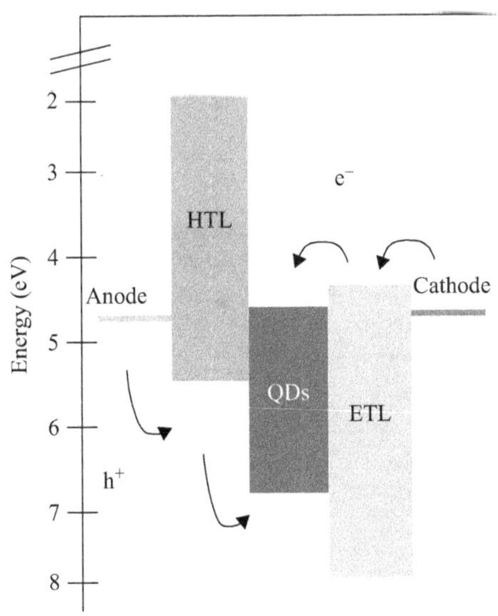

Figure 8.10 Example energy band diagram of a QDLED

enhance the stability of solution-processed LEDs. MoS_2 [88] and NiO [89] have been used as HTLs in OLEDs or QDLEDs, respectively. Some recent publication reports also indicate the use of TiO_2 [73,74] and ZnO [75] as an ETL in LEDs.

One such QDLED fabricated is shown in Figure 8.11. In this case, a sandwich structure was fabricated with a layer of colloidal QDs sandwiched between inorganic CTLs. The QDLED was fabricated on a glass substrate and ITO was used as an anode. Nickel oxide is used as the HTL and is deposited on top of the ITO anode by spin coating of nickel acetate tetrahydrate in methanol. A layer of Al_2O_3 was deposited conformally, using atomic layer deposition (ALD), between the anode and HTL as an isolation layer to protect the QDs from direct contact with the anode in case of defects in the NiO film. On top of this, the QD layer is deposited using two different techniques of micro-contact printing and spin coating. On top of the QD layer, a thin film of zinc oxide nanoparticles is deposited, by spin coating in butanol, which acts as the ETL. Aluminum is used as the top electrode (cathode). Figure 8.11(a) shows the schematic of the cross-section of the device [32].

Another type of QDLED structure explored is based on a silicon substrate. The structure of QD-based LEDs consists of multi-layers of inorganic materials: the combination of gold (thickness: 5 nm) and silver (12 nm) as the cathode, $ZnO:SnO_2$ (ratio 3:1, 40 nm) as the ETL, CdSe:ZnS particles as the light emission layer, 1 nm silicon dioxide, and p-silicon as the bottom electrode. Figure 8.11(b) shows a schematic of a silicon-based QDLED. The emission is observed through the top metal cathode, similar to a top emitting organic light-emitting diode (TEOLED).

The operation and performance of the QDLEDs depend on various factors including the optical and electronic characteristics of the materials used to form the various layers in the QDLED structure. The thicknesses of various films and the materials used can be tuned to obtain the best possible performance. For example, in the QDLED shown in Figure 8.11(a), the following criteria were considered while choosing the material, fabrication method used, and thickness of the various

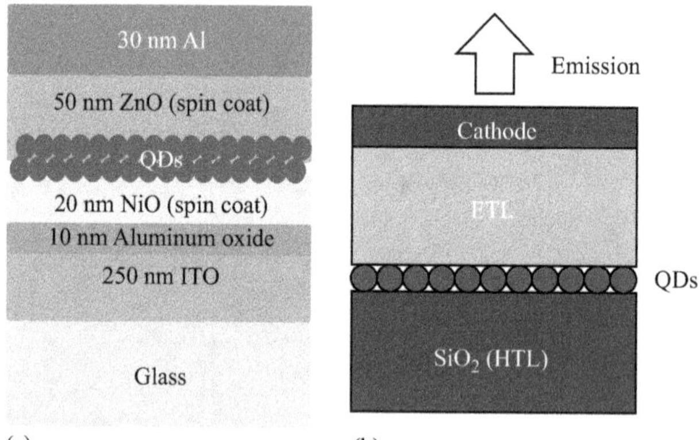

Figure 8.11 Schematics of QDLEDs. (a) QDLED on glass substrate and (b) silicon-based QDLED

layers. ITO was used as the anode due to its transparent optical characteristics and hole injection capabilities. NiO was chosen as the HTL material. E-beam evaporation followed by annealing can be used to deposit NiO. This however limits the maximum possible thickness of the HTL due to the optical properties and annealing and oxidation requirement. Excessive annealing can lead to increase in the resistance of the underlying ITO layer and less annealing can lead to incomplete oxidation of the Ni layer, forming pockets of NiO which disrupts both the optical and electrical characteristics of the device. Deposition of thin films less than 10 nm in thickness using e-beam evaporation causes pinhole defects and patches in the film. This leads to uneven hole transport and patches can cause QDs to come in contact with the environment and also causes direct contact with the anode. At high turn-on voltages, direct contact of QDs with the anode can cause burning and damage of QDs. Contact of QDs with the environment causes degradation of QDs and leads to reduced lifetime of device.

To overcome the constraint of low thickness, and patched layer due to small thickness, the method of NiO layer fabrication was changed. Two choices for replacing the e-beam evaporation technique are sputtering and spin coating. Spin coating nickel acetate tetrahydrate in methanol at 1500 rpm and then annealing at 120°C for 90 min was chosen to form the film. The 20 nm thick film has been shown to have optical properties which do not affect the device in the visible wavelength range and small thicknesses of 20 nm [76]. The decision of using spin coating was taken due to results which show that spin coating followed by annealing gives a smoother surface with fewer holes and other defects as compared to sputtered films of similar thickness.

To protect the QDs from direct contact with the anode, an extra layer of Al_2O_3 was added between the ITO and NiO layers. The Al_2O_3 was deposited using ALD technique in which the surface of the substrate is chemically acted upon by the precursor of the metal oxide to be deposited. Due to this, at each pulse, a single atomic layer of the material is deposited on the substrate. This technique deposits a layer over the entire substrate irrespective of the non-uniformities of the substrate. Thus, Al_2O_3 was conformally deposited using ALD, as a protection layer between the ITO and NiO layers. Due to this, even in areas where defects are formed in the NiO layer, QDs would not come in direct contact with the ITO layer and the environment. In this case, the decision of using Al_2O_3 was taken based on various electronic and optical properties. Al_2O_3 has a band gap structure similar to NiO and can thus act as a hole injection layer. NiO layer is still required, in addition to the Al_2O_3 layer, as the carrier concentration of Al_2O_3, is low and using only Al_2O_3, as the HTL would reduce the efficiency of the device.

ZnO was chosen as an ETL due to its high charge carrier density and electron transport capabilities. ZnO can be deposited using sputtering. The amorphous ETL though effective in electron transport, it does not protect the underlying QD layer from environmental factors such as contact with oxygen. The QD layer also gets affected during the sputtering of the metal oxide film overlying the QD layer. This results in shorter lifetimes of QDLEDs [77]. To overcome these drawbacks, a device structure using a spin-coated ZnO nanoparticles film was introduced as the ETL. The ZnO nanoparticles film offers the advantage of efficient electron transport and is more robust and easier to process as compared to sputtered metal oxide films of comparable thickness [77].

In the device structure shown in Figure 8.11(b), p-silicon was used as the HTL. There are a number of advantages to using p-type silicon as a hole injection layer, as compared to ITO [78]. ITO is inefficient in injecting holes to polymer layers whereby an additional HTL is required before an emitting layer in the system [65,79–81]. Carrier injection from silicon electrode is improved because most of the electric field drop occurs across the highly resistive layer of thin SiO_2. The SiO_2 on silicon acts as a buffer layer for the system. The buffer layer will restrain the injected hole current and then improve the balance between hole- and electron-injection. Resulting high probability of radiative recombination consequently promote luminance and EL efficiency [82]. The Fermi level of the silicon is aligned to that of the electroluminescent material [50]. The excitons produced at QD layers can be easily quenched in close proximity to a metal electrode [83]. The ZnO:SnO_2 layer hence separates the electroluminescent QD layer from the metal layer where by increasing the radiative recombination. ZnO:SnO_2 layer improves the electron injection into the system and simultaneously acts as a hole-blocking layer. A band diagram of this device reveals that the barrier injection from ZnO layer into the emitting QDs is small. The thin SiO layer on p-silicon anode offers a strong hole injection. Characterizations of optical and electronic properties of the ETL films were used to decide the material to be used.

Both the devices were switched on by applying voltage across the devices. Figure 8.12(a) shows image of EL from glass-based QDLEDs and Figure 8.12(b) shows image of EL from silicon-based QDLEDs.

The combination of micro-scale lithographic patterning and nanostamping methods makes it possible to accurately define diffraction-limited multicolor light sources on a single silicon substrate with great accuracy. The ease of fabrication and post-processing of QDLEDs open up the possibilities for creating robust micro imaging systems integrating nanophotonics and Complementary Metal Oxide Semiconductor (CMOS) electronics for advanced displays, spectral analysis, medical imaging, and μ-TASs in biomedicine.

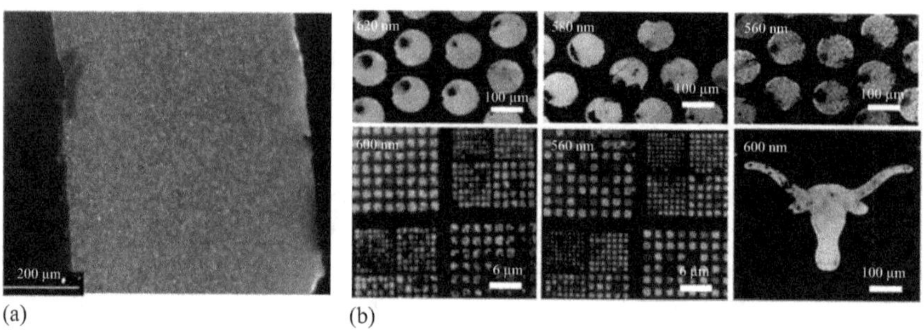

Figure 8.12 Electroluminescence of QDLEDs. (a) QDLED on glass substrate. Reprinted with permission form Bhave [32], copyright IEEE. (b) Silicon-based QDLED. Reprinted with permission from Ashwini Gopal

8.8 Microarray imaging

Microfluidic lab-on-chip systems have gained great importance in the last few years. The advantageous properties of these systems, such as their compactness, usefulness in point-of-care applications, and low cost, among others, have brought research in lab-on-chip systems to the fore [54]. On-chip imaging systems have been recently brought forward but still face some major challenges. Great progress has been made in on-chip bright field imaging with both lensless systems and with image-processing techniques [54–57]. Although many disadvantages of current systems impede wide implementation, a major obstacle in the development of on-chip imaging systems is the availability of miniature, narrow band light sources [58,59]. Current lab-on-chip systems still rely on the use of off-chip, bulky light sources for multicolor imaging, which obviates their advantages of point-of-care applicability, small size, and low cost [60–65].

For such applications, PL from patterned QDs can be used as an on-chip light source. PL properties of QDs can solve issues such as requirement of high-intensity emission. A UV source of excitation can be used for excitation of QDs and combinations of filters can be used to focus on emissions from a single color of QD emissions.

Various modes of imaging can be performed using patterned QDs in combination with a microwell setup, for various applications. Such setups can be used in fluorescence imaging where QDs are used as both, the light source and also the fluorescent marker. Patterned QDs can be used solely as light sources in multi-spectral transmission mode imaging.

In setups where patterned QDs are used as a light source, an external excitation source is required to obtain emissions from the QDs. Since the absorption bandwidths of QDs can be tuned, an external LED in the absorption wavelength of the QDs can be used for this purpose. An advantage of the ability to produce thin, uniform, robust QD films of QDs on glass substrates is that the excitation light source can be introduced into the glass substrate and the evanescent field produced by the total internal reflection can be used to excite the deposited films. Electroluminescence of QDs (QDLEDs) can also be used as light sources in this type of setup for various types of imaging. A schematic of the setup for the use of patterned QDs as light sources for microarray imaging is shown in Figure 8.13.

8.8.1 Immunofluorescence imaging

Immunoassays play a critical role in identification of cancer cells from normal cells as well as profiling of individual cancer cells based on expressions of disease-specific biomarkers. Among such biomarkers, epithelial cell adhesion molecule (EpCAM) is most commonly used to identify tumor cells, since 90% of tumors are epithelial in origin [67]. EpCAM has been used for immunoassay-based separation of circulating tumor cells (CTCs) originating from colon, prostate, breast, and lung carcinomas. Because some tumor cells are reported to be EpCAM negative [68], a combination of multiple biomarkers can provide a more detailed analysis for such EpCAM negative cells. Cytokeratin is another protein of intermediate filaments

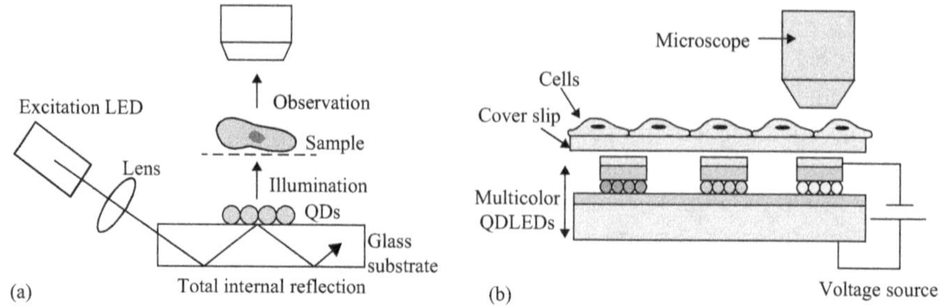

Figure 8.13 Schematic of experimental setup of the arrayed multicolor light source for microarray imaging. (a) Photoluminescence and (b) electroluminescence. Reprinted with permission from Hoshino [34], copyright Elsevier

found in the cytoskeleton of epithelial tissue, and is used for immunofluorescence of cells from epithelial carcinomas. Other disease-specific biomarkers include prostate specific antigen (PSA) for prostate cancer [69] and estrogen receptor (ER), progesterone receptor (PR), and human epidermal growth factor receptor 2 (HER2) for breast cancer [53]. An example of biomarkers for normal cells is CD45 (leukocyte common antigen) used in CTC detection. Anti-CD45 fluorescent marker staining of non-cancerous normal leukocytes is used for negative screening [52].

Imaging of multiple biomarkers enables precise identification and quantification of cells [53]. In such systems, QDs can be useful as excitation sources for such multicolor fluorescence microscopy. QD-based immunofluorescence imaging systems use both, the wide band absorption and narrow band emission properties of QDs where both the excitation source and the fluorescence marker consist of colloidal QDs. Figure 8.14 shows the fluorescence excitation setup which includes a PDMS microwell array, excitation and emission filters, and the QD light source. The samples are placed in an array of five-by-five microwells for cell culture. The PDMS microarray can be easily fabricated by replicating a master mold built by a three-dimensional prototyping printer. The optical filters can also be made by a printing on a transparent film. The photograph in Figure 8.14(a) shows an example of excitation and mission filters printed in arrays which match the pitch of the microwells. In this example, combinations of cyan, magenta, and yellow dyes, which are easily available with most commercial printing systems were used. Other types of dyes with different absorption characteristics may be patterned by inkjet printing.

Figure 8.14 shows setup for a microarray imaging system setup. Figure 8.14(a) shows a picture of a setup with patterned QDs on a glass substrate used as a light source for imaging cells cultured in a microwell array made of PDMS. Excitation and emission filters are used to obtain single wavelength light source. The cells can be stained with traditional fluorescence staining methods or conjugated QDs with emission wavelengths that are red shifted as compared with the emission wavelength of the light source. Multiple films are used to obtain desired transmission characteristics. The emission from the QD light source (blue-green) is first filtered by the excitation filter. A UV-cut plastic thin film is inserted between the light

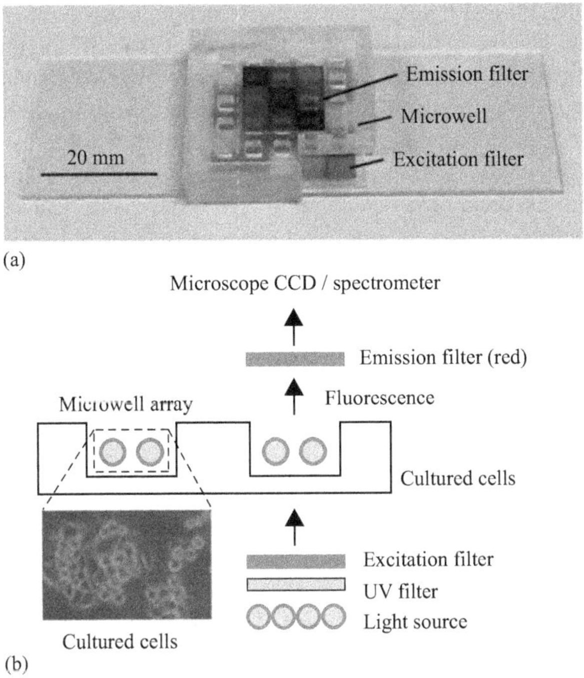

Figure 8.14 Microwell array imaging setup. (a) Microwell array with excitation and emission filter arrays and (b) schematic of the experimental setup. The emission from the light source QD filtered by an excitation filter (UV filter) excites the sample, whose emission is filtered by a red filter to be observed. Reprinted with permission from Hoshino [34], copyright Elsevier

source QDs and the excitation filter. It prevents cross excitation of sample QDs by stray light. The filtered excitation light excites the sample, which is either a fluorescence dye or cells conjugated with another type of QDs with longer emission peak as compared with the excitation source.

8.8.2 Transmission mode imaging

One application in which patterned QDs can be used is continuous monitoring of cultured cells. The effects of various drugs on cultured cells can be tested using standard 96-well plates with integrated QD light sources. In such situations, continuous monitoring of cells is useful in combination with multispectral transmission mode imaging using multicolor light sources for each well [66].

Multispectral imaging of cells has shown great promise. Recent studies have shown that separate data can be obtained using light sources of different wavelengths and the specific spectral signature obtained from biological samples can be used in identifying various biomarkers [70]. The narrow FWHM of the emission wavelength of QDs, which is on the order of 15–20 nm, gives good control on the emission wavelength and light purity for spectroscopic measurement in the visible wavelength region. These light sources due to the capability of micropatterning have the advantage of multicolor, confined emission on as single chip making the integration of QDs ideal for on-chip multicolor transmission mode imaging.

Transmission mode imaging can be used with traditional fluorescence stains to study specific cell characteristics. For example, the nucleus to cytoplasm ratio is a commonly used marker for the detection of most cancer cells. Cancer cells tend to have larger nucleus–cytoplasm ratio than other cells [84]. This criterion is often applied in computer-based cancer cell analysis such as the study of cellular response to drugs. This ratio can be measured by using appropriate combinations of nuclear stains and QD emission wavelengths. Integration of on-chip transmission mode imaging in microfluidic systems can help to achieve high-throughput devices for a variety of applications.

Figure 8.15 shows an example of the use of multispectral imaging of cancer cells. Breast cancer is the most common cause of cancer death in women. It has been shown that early detection can lead to increased survival rates and early screening enables studying the dependence of survival rates on stage of diagnosis. In this application, MDA-MB231 cancer cells (breast cancer cells) are fixed and aligned onto the patterned QD light source. A UV light with the emission wavelength peak at 365 nm from a high-power LED (200 mW) was focused and introduced inside a glass slide. An evanescent field was induced on the glass slide surface due to the total internal reflection of the transmitted UV light. The QD films were excited using this field. Since the UV evanescent field on the glass slide decays quickly in the near-field and does not transmit energy in the far-field, the

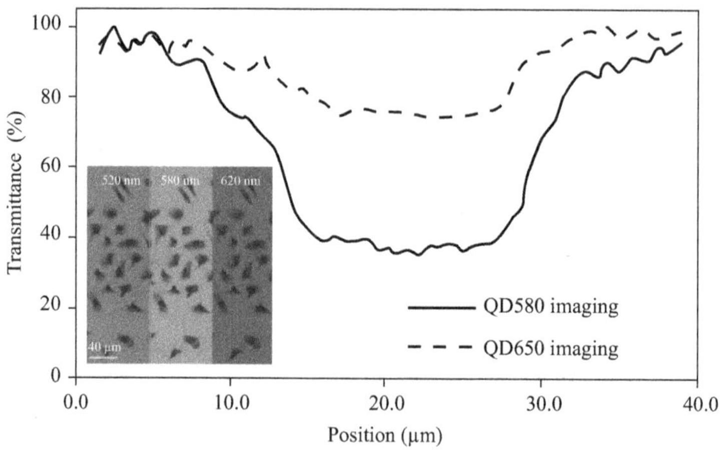

Figure 8.15 Normalized transmittance profiles of a cancer cell with 580 nm (the peak matched with the absorption of the dye) and 650 nm illuminations, the cytoplasm–nucleus ratio is easy to detect at the 580 nm illumination due to higher difference in intensity. Reprinted with permission from Hoshino [34], copyright Elsevier

UV light intensity observed with the microscope was negligible compared to that of the excited QDs. The cells are stained with HEMA3 dye (Fisher Scientific), which is commonly used to visualize the cell nucleus. The dye absorbs visible light and makes the stained area observable under transmission-mode illumination. The contrast between the nucleus and cytoplasm is highest at certain wavelengths, depending on the stain used. Multispectral transmission imaging can be effectively used in such situation, since the wavelength giving the highest contrast can be used to measure the nucleus–cytoplasm ratio and light sources with other emission wavelengths can be used to observe other cell features and bodies within the same cell. Figure 8.15 shows the images of such cells with QDs of various emission wavelengths and the measurement of cytoplasm–nucleus ratio.

Patterning of QDs thus shows great potential in its use as light source in on-chip microarray imaging. The use of inherent properties of QDs like the narrow emission bandwidth and tunable optical properties can be used for high-throughput imaging for identification and observation of cell structures.

QDLEDs were also used to image the same cells. An arrayed pattern of QDLED is used to excite the stained slides placed near the substrate. Patterned substrates give the flexibility to image a single cell under different excitation wavelength. The size of the individual patterned QDLED source can be fabricated to be on the order of a single cell (10 μm). In order to obtain a better understanding of the absorption characteristics of the QDLED via cancer cells, integration of the cells with a larger sized LED is also shown. The sizes of the LEDs were 1 mm × 1 mm.

Emission wavelengths 560 nm, 580 nm, and 600 nm were used as excitation sources for imaging cells. On excitation with different light sources, it was observed that the nucleus absorbs at 580 nm (QDLED excitation) and passes 600 nm (QDLED excitation). The results are very similar to the data obtained for the absorption of light emission from the HEMA3 dye used. Different portions of the cell absorb at different wavelengths. Images of excitation from different light sources (560 nm, 580 nm, and 600 nm) with feature size 100 μm are given in Figure 8.16. Figure 8.16(a) shows the measurement of the nucleus-to-cytoplasm ratio using QDLEDs of different wavelengths, similar to the processing performed for the PL images. Figure 8.16(b) shows imaging of cells using various emission wavelengths of the QDLEDs.

8.9 Hyperspectral microscopy

Hyperspectral imaging is a technique where spectral information is recorded over many wavelengths for each pixel of a high-resolution microscopic image [85,86]. It plays an essential role in multi-biomarker-based imaging such as immunoassay-based disease analysis, where tumor originated cells are isolated [87] and identified [88] through labeling of multiple disease-specific proteins. A swept light source is a key element in hyperspectral microscopy. It usually consists of a wide band light source coupled with mechanically motorized color wheels or gratings coupled with mechanically scanned stages [85,86], which make the setup complicated and

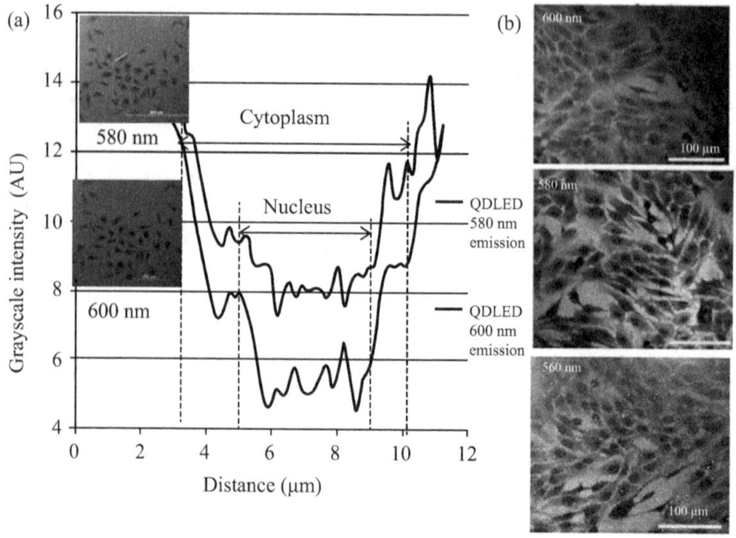

Figure 8.16 Imaging of cells with QDLEDs. (a) Normalized transmittance profiles of a cancer cell with 580 nm (the peak matched with the absorption of the dye) and 600 nm illuminations, the cytoplasm–nucleus ratio is easy to detect at the 580 nm illumination due to higher difference in intensity. (b) MDA 435 cancer cells excited by QDLEDs with emission wavelength 600 nm, 580 nm, and 560 nm (top to bottom). Reprinted with permission form Bhave [32], copyright IEEE

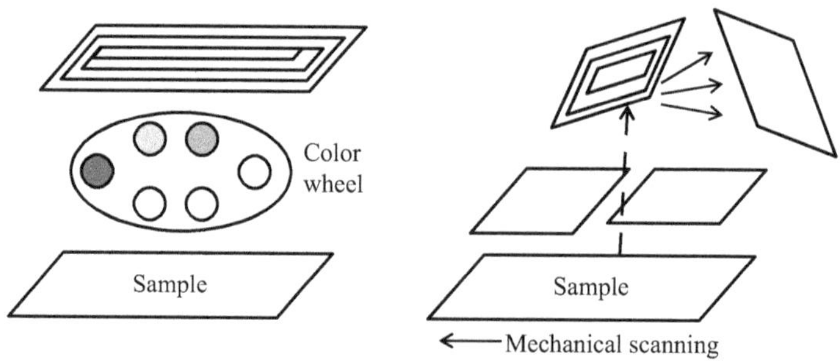

Figure 8.17 Schematic of typical hyperspectral microscopes consisting of a color wheel (left) or a mechanical stage (right). Adapted with permission from Hoshino [93], copyright the Optical Society

expensive. Figure 8.17 shows the schematic of a traditional hyperspectral imaging setup. In addition, the emission spectra of conventional white or wide band light sources, such as white LEDs, mercury vapor lamps, fluorescent tubes, usually contains material-specific non-uniform peaks, which make them difficult to be used for a swept light source. The optical properties of QDs provide a great alternative as a light source for hyperspectral microscopy. The use of QDs allows for easy construction of a compact swept light source that uniformly covers visible-NIR wavelength range. Another unique advantage of the QD-based swept light source is that it does not contain any moving parts. Elimination of mechanically moving parts is a recent trend in developing compact, fast, robust, and low cost devices. Solid-state disks (SSDs) and solid-state relays (SSRs) are examples of commercial success of such schemes. QD light sources can take the advantage of the simple UVLED-based excitation and the wide variety of emission wavelengths of QDs. Emission wavelengths of the light source can be easily changed by digital on–off control of solid-state LEDs.

Different types of setups which use QDs as light source for hyperspectral imaging have been investigated. Figure 8.18 shows one such digitally switchable QD excitation setup. The QDs are packed in a PDMS enclosure that is replicated from a plastic mold fabricated by rapid 3D printing. It includes an array of miniature wells, each of which contains a single color of QDs (see the top view). QDs in a well are excited by a UVLED. The enclosure also serves as a circular light guide plate that transmits emission from the QDs (see the side view of Figure 8.18). The emission is coupled into a circular plate that has a concave surface designed to diffuse the emission from the QDs uniformly onto the top plate. Multiple dimples are created on the concave surface to scatter the light. On top of the concave surface is a top plate consisting of a diffuser and a 400 nm long-pass filter to cut off excitation UV lights.

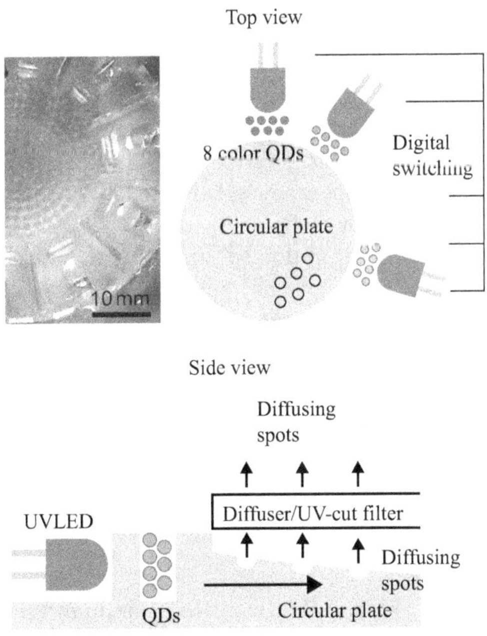

Figure 8.18 Light source using colloidal QDs. Emission of various wavelengths (shown: 520–800 nm) can be chosen by digital on–off control of the excitation UVLEDs. Reprinted with permission from Hoshino [93], copyright the Optical Society

Multiple QDs with the peak emission wavelengths ranging from visible to NIR can be used in the setup. The emission intensity for each color can be controlled by adjusting that of its respective excitation UVLED, which can be driven by a transistor array connected to a digital controller. Since each LED is operated independently, any linear combination of the multiple colors can be created by the light source.

An important application of this QD-based hyperspectral imaging system can be the detection of multiple biomarkers. Gold nanoparticles are widely used for optical labeling. Due to the effect of localized surface plasmon resonance, gold nanoparticles have size- and shape-dependent characteristic absorption spectra. Gold nanoparticles with different absorption peaks can be used for optical labeling of multiple biomarkers used in hyperspectral absorption spectroscopy [93]. The QD light source, which allows for simple spectral measurements, gives a clear edge to such nanoparticle-based imaging schemes.

Cells are labeled with gold nanorods and nanospheres with characteristic absorption and transmission spectra. The QD-based light source is employed in combination with a standard microscope to perform hyperspectral microscopy. Transmission mode imaging of labeled cells is done with the digitally controlled light sources to measure absorption of each wavelength of light by the labeled cells. Transmission intensities of each wavelength can then be plotted to detect the presence of specific nanorods or nanospheres by comparing with the absorption and transmission spectra of the nanoparticles [93].

The use of QDs allows for hyperspectral imaging of biosamples with a simple microscope configuration. In addition, proper choice of different gold nanoparticles, such as nanorods and nanospheres, enables absorption-based analysis of multiple biomarkers. Combination of QDs and gold nanoparticles provides a potential on-chip hyperspectral imaging tool for multi-biomarkers recognition at cellular level. In addition to microscopy, QD light sources can be used for several types of absorption spectroscopy techniques, such as pulse oximetry [90] or tissue spectroscopy [91], where the absorption spectra are correlated with physical characteristics. A possible application area beyond biomedical applications is the use as an illumination source alternative to conventional RGB-LEDs. Conventional RGB color spaces, such as sRGB or Adobe RGB, cover only a portion of visible colors [92], while our light source can simulate any spectrum expressed as linear combinations of multiple QD emission peaks. The QD light source can be a strong illumination tool for medical and industrial imaging.

References

[1] L. L. Chang, L. Esaki, and R., Tsu, "Resonant tunneling in semiconductor double barriers," Applied Physical Letters, vol. 24, p. 593, 1974.

[2] R. Dingle, W. Wiegmann, and C. Henry, "Quantum states of confined carriers in very thin $Al_xGa_{1-x}As$-GaAs-$Al_xGa_{1-x}As$ heterostructures," Physical Reviews Letters, vol. 33, p. 827, 1974.

[3] K. Von Klitzing, G. Dorda, and M. Pepper, "New method for high-accuracy determination of fine-structure constant based on quantized Hall resistance," Physical Reviews Letters, vol. 45, p. 494, 1980.

[4] R. B. Laughlin, "Quantized Hall conductivity in two dimensions," Physical Review B, vol. 23, p. 5632, 1981.

[5] R. B. Laughlin, "Quantized motion of three two-dimensional electrons in a strong magnetic field," Physical Review B, vol. 27, p. 3383, 1983.

[6] R. B. Laughlin, "Anomalous quantum Hall effect: An incompressible quantum fluid with fractionally charged excitations," Physical Reviews Letters, vol. 50, p. 1395, 1983.

[7] M. A. Reed, R. T. Bate, K. Bradshaw, *et al.*, "Spatial quantization in GaAs–AlGaAs multiple quantum dots," Journal of Vacuum Science & Technology B, vol. 4, p. 358, 1986.

[8] A. Franceschetti and A. Zunger, "Direct pseudopotential calculation of exciton Coulomb and exchange energies in semiconductor quantum dots," Physical Reviews Letters, vol. 78, p. 915, 1997.

[9] K. Hoshino, A. Gopal, D. Ostrowski, *et al.*, "Single monolayer nanocrystal led on probe tip for near-field molecular imaging and patterning," Presented at International Conference on Micro-Electro-Mechanical Systems (MEMS'08), Arizona, USA, 2008.

[10] D. Bimberg, M. Grundmann, and N. N. Ledentsov, Quantum Dot Heterostructures, Wiley, New York, NY, 1999.

[11] B. O. Dabbousi, J. Rodriguez-Viejo, F. V. Mikulec, *et al.*, "(CdSe) ZnS core–shell quantum dots: Synthesis and characterization of a size series of highly luminescent nanocrystallites," Journal of Physical Chemistry B, vol. 101, pp. 9463–9475, 1997.

[12] K. S. Shin, Y. H. Kim, K. K. Paek, *et al.*, "Characterization of an integrated fluorescence-detection hybrid device with photodiode and organic light-emitting diode," IEEE Electron Device Letters, vol. 27, pp. 746–748, 2006.

[13] S. Khalfallah, C. Gorecki, J. Podlecki, M. Nishioka, H. Kawakatsu, and Y. Arakawa, "Wet-etching fabrication of multilayer GaAlAs/GaAs microtips for scanning near-field optical microscopy," Applied Physics A: Materials Science & Processing, vol. 71, pp. 223–225, 2000.

[14] O. Hofmann, X. Wang, A. Cornwell, *et al.*, "Monolithically integrated dye-doped PDMS longpass filters for disposable on-chip fluorescence detection," Lab on a Chip, vol. 6, pp. 981–987, 2006.

[15] M. F. Garcia-Parajo, J. A. Veerman, S. J. T. Van Noort, B. G. De Grooth, J. Greve, and N. F. Van Hulst, "Near-field optical microscopy for DNA studies at the single molecular level," Bioimaging, vol. 6, pp. 43–53, 1998.

[16] B. Yao, G. Luo, L. Wang, *et al.*, "A microfluidic device using a green organic light emitting diode as an integrated excitation source," Lab on a Chip, vol. 5, pp. 1041–1047, 2005.

[17] V. Savvate'Ev, Z. Chen-Esterlit, J. W. Aylott, *et al.*, "Integrated organic light emitting device/fluorescence-based chemical sensors," Applied Physics Letters, vol. 81, p. 4652, 2002.

[18] S. Camou, M. Kitamura, G. Jean-Philippe, F. Hiroyuki, A. Yasuhiko, and F. Teruo, "Organic light emitting device as a fluorescence spectroscopy light source: One step towards the lab-on-a-chip device," Applications of Photonic Technology 5, Qubec City, Canada, Proc. SPIE 4833, 2003.

[19] S. Coe-Sullivan, J. S. Steckel, W. K. Woo, M. G. Bawendi, and V. Bulovic, "Large-area ordered quantum-dot monolayers via phase separation during spincasting," Advanced Functional Materials, vol. 15, pp. 1117–1124, 2005.

[20] J. Zhao, J. A. Bardecker, A. M. Munro, *et al.*, "Efficient CdSe/CdS quantum dot light-emitting diodes using a thermally polymerized hole transport layer," Nano Letters, vol. 6, pp. 463–467, 2006.

[21] M. Achermann, M. A. Petruska, D. D. Koleske, M. H. Crawford, and V. I. Klimov, "Nanocrystal-based light-emitting diodes utilizing high-efficiency nonradiative energy transfer for color conversion," Nano Letters, vol. 6, pp. 1396–1400, 2006.

[22] V. Santhanam and R. P. Andres, "Microcontact printing of uniform nanoparticle arrays," Nano Letters, vol. 4, pp. 41–44, 2004.

[23] V. Santhanam, J. Liu, R. Agarwal, and R. P. Andres, "Self-assembly of uniform monolayer arrays of nanoparticles," Langmuir, vol. 19, pp. 7881–7887, 2003.

[24] J. S. Steckel, P. Snee, S. Coe-Sullivan, *et al.*, "Color-saturated green-emitting QDLEDs," Angewandte Chemie, vol. 118, pp. 5928–5931, 2006.

[25] L. A. Kim, P. O. Anikeeva, S. A. Coe-Sullivan, J. S. Steckel, M. G. Bawendi, and V. Bulovic, "Contact printing of quantum dot light-emitting devices," Nano Letters, vol. 8, pp. 4513–4517, 2008.

[26] A. Rizzo, M. Mazzeo, M. Palumbo, *et al.*, "Hybrid light-emitting diodes from microcontact-printing double-transfer of colloidal semiconductor CdSe/ZnS quantum dots onto organic layers," Advanced Materials, vol. 20, pp. 1886–1891, 2008.

[27] J. Yu, J. Chen, H. Huang, M. Bawendi, and V. Bulovic, "Microcontact printing of quantum-dot LEDs using an inkjet-assisted patterning method." MTL Annual Research Report, 2007.

[28] C. Y. Kuan, J. M. Chou, C. Leu, and M. H. Hon, "Sol–gel-derived ZnO coating with nanopatterns fabricated by nanoimprinting," Journal of the American Ceramic Society, vol. 91, pp. 3160–3166, 2008.

[29] S. Sepulveda-Guzman, B. Reeja-Jayan, E. De la Rosa, *et al.*, "Room-temperature deposition of crystalline patterned ZnO films by confined dewetting lithography," Applied Surface Science, vol. 256, pp. 3386–3389, 2009.

[30] A. Gopal, K. Hoshino, S. Kim, and X. Zhang, "Multi-color colloidal quantum dot based light emitting diodes micropatterned on silicon hole transporting layers," Nanotechnology, vol. 20, p. 235201, 2009.

[31] A. Gopal, K. Hoshino, and X. Zhang, "Photolithographic patterning of subwavelength top emitting colloidal quantum dot based inorganic light emitting diodes on silicon," Applied Physics Letters, vol. 96, p. 131109, 2010.

[32] G. Bhave, Y. Lee, K. Hoshino, and X. J. Zhang, "Colloidal quantum dot-based light emitting diodes with solution processed electron transporting

layer for cellular imaging," IEEE Sensors Journal, vol. 15, iss. 1, 234–239, 2015, doi: 10.1109/JSEN.2014.2341675.

[33] G. Bhave, Y. Lee, E. Ng, and X. J. Zhang, "Micro patterned quantum dots excitation for cellular microarray imaging," in Proceedings of the SPIE 9341, Bioinspired, Biointegrated, Bioengineered Photonic Devices III, 93410F, doi:10.1117/12.2083645, March 11, 2015.

[34] K. Hoshino, G. Bhave, E. Ng, and X. J. Zhang, "Micro patterned quantum dots excitation and imaging for cellular microarray screening," Sensors and Actuators A, vol., 216, pp. 301–307, 2014.

[35] G. Chumanov, K. Sokolov, B. W. Gregory, and T. M. Cotton, "Colloidal metal films as a substrate for surface-enhanced spectroscopy," Journal of Physical Chemistry, vol. 99, pp. 9466–9471, 1995.

[36] H. O. Finklea, "Of thiols and related molecules on electrodes," Electroanalytical Chemistry A: Series of Advances, vol. 19, p. 109, 1996.

[37] J. Schmitt, G. Decher, W. J. Dressick, *et al.*, "Metal nanoparticle/polymer superlattice films: Fabrication and control of layer structure," Advanced Materials (Weinheim), vol. 9, pp. 61–65, 1997.

[38] M. Giersig and P. Mulvaney, "Preparation of ordered colloid monolayers by electrophoretic deposition," Langmuir, vol. 9, pp. 3408–3413, 1993.

[39] J. Rybczynski, U. Ebels, and M. Giersig, "Large-scale, 2D arrays of magnetic nanoparticles," Colloids and Surfaces A: Physicochemical and Engineering Aspects, vol. 219, pp. 1–6, 2003.

[40] B. O. Dabbousi, J. Rodriguez-Viejo, F. V. Mikulec, *et al.*, "(CdSe) ZnS core–shell quantum dots: Synthesis and characterization of a size series of highly luminescent nanocrystallites," Journal of Physical Chemistry B, vol. 101, pp. 9463–9475, 1997.

[41] B. O. Dabbousi, M. G. Bawendi, O. Onitsuka, and M. F. Rubner, "Electroluminescence from CdSe quantum dot/polymer composites," Applied Physics Letters, vol. 66, p. 1316, 1995.

[42] T. P. Bigioni, X. M. Lin, T. T. Nguyen, E. I. Corwin, T. A. Witten, and H. M. Jaeger, "Kinetically driven self-assembly of highly ordered nanoparticle monolayers," Nature Materials, vol. 5, pp. 265–270, 2006.

[43] A. L. Efros and A. V. Rodina, "Band-edge absorption and luminescence of nonspherical nanometer-size crystals," Physical Review B, vol. 47, pp. 10005–10007, 1993.

[44] C. D. Dushkin, G. S. Lazarov, S. N. Kotsev, H. Yoshimura, and K. Nagayama, "Effect of growth conditions on the structure of two-dimensional latex crystals: Experiment," Colloid & Polymer Science, vol. 277, pp. 914–930, 1999.

[45] F. Bresme and M. Oettel, "Nanoparticles at fluid interfaces," Journal of Physics: Condensed Matter, vol. 19, p. 413101, 2007.

[46] P. A. Kralchevsky and K. Nagayama, "Capillary interactions between particles bound to interfaces, liquid films and biomembranes," Advances in Colloid and Interface Science, vol. 85, pp. 145–192, 2000.

[47] N. D. Denkov, O. D. Velev, P. A. Kralchevsky, I. B. Ivanov, H. Yoshimura, and K. Nagayama, "Two-dimensional crystallization," Nature 361, 26, doi:10.1038/361026a0, 1993.
[48] K. Hoshino, T. C. Turner, S. Kim, A. Gopal, and X. Zhang, "Single molecular stamping of a sub-10-nm colloidal quantum dot array," Langmuir, vol. 24, pp. 13804–13808, 2008.
[49] K. Hoshino, A. Gopal, M.S. Glaz, D.A. Vanden Bout, and X. Zhang, "Nanoscale fluorescence imaging with quantum dot near-field electroluminescence," Applied Physical Letters, vol. 101, no. 4, p. 043118, 2012.
[50] K. Bourzac, "Quantum dots go on display," Nature, vol. 493, p. 283, 2013.
[51] H. Moeller, M. K. Mian, S. Shrivastava, B. G. Chung, and A. Khademhosseini, "A microwell array system for stem cell culture," Biomaterials, vol. 29, pp. 752–763, 2008.
[52] K. Hoshino, Y. Y. Huang, N. Lane, *et al.*, "Microchip-based immunomagnetic detection of circulating tumor cells," Lab Chip, vol. 11, pp. 3449–3457, 2011.
[53] J. W. Uhr, M. L. Huebschman, E. P. Frenkel, *et al.*, "Molecular profiling of individual tumor cells byhyperspectral microscopic imaging," Translational Research, vol. 159, pp. 366–375, 2012.
[54] X. Cui, X. Heng, and C. Yang, "Optofluidic microscope – A complete on-chip imaging device," Imaging, Manipulation, and Analysis of Biomolecules, Cells, and Tissues VI, vol. 6859, p. 68590Y, 2008.
[55] M. Adams, M. Enzelberger, S. Quake, and A. Scherer, "Microfluidic integration on detector arrays for absorption and fluorescence micro-spectrometers," Sensors and Actuators A, vol. 104, pp. 25–31, 2003.
[56] A. Gurkan, S. Moon, H. Gecki, *et al.*, "Miniaturized lensless imaging systems for cell and microorganism visualization in point-of-care testing," Biotechnology Journal, vol. 6, pp. 138–149, 2011.
[57] C. D. Chin, V. Linder, and S. K. Sia, "Lab-on-a-chip devices for global health: Past studies and future opportunities," Lab Chip, vol. 7, pp. 41–57, 2007.
[58] G. Stybayeva, O. Mudanyali, S. Seo, *et al.*, "Lensfree holographic imaging of antibody microarrays for high-throughput detection of leukocyte numbers and function," Analytical Chemistry, vol. 82, pp. 3736–3744, 2010.
[59] A. Ozcan and U. Demirci, "Ultra wide-field lens-free monitoring of cells on-chip," Lab Chip, vol. 8, pp. 98–106, 2008.
[60] A. Coskun, T.-W. Su, and A. Ozcan, "Wide field-of-view lens-free fluorescent imaging on a chip," Lab Chip, vol. 10, pp. 824–827, 2010.
[61] W. Bishara, T-W. Su, A. F. Coskun, and A. Ozcan, "Lensfree on-chip microscopy over a wide field-of-view using pixel super-resolution," Optics Express, vol. 18, no. 11, pp. 11181–11191, 2010.
[62] G. M. Whitesides, "The origins and the future of microfluidics," Nature, vol. 442, pp. 368–373, 2006.
[63] J. El-Ali, P. K. Sorger, and K. F. Jensen, "Cells on chips," Nature, vol. 442, pp. 403–411, 2006.

[64] X. Heng, D. Erickson, L. R. Baugh, *et al.*, "Optofluidic microscopy – A method for implementing a high resolution optical microscope on a chip," Lab on a Chip, vol. 6, pp. 1274–1276, 2006.

[65] W. Bishara, S. O. Isikman, and A. Ozcan, "Lensfree optofluidic microscopy and tomography," Annals of Biomedical Engineering, vol. 40, no. 2, pp. 251–262, February 2012.

[66] M. Agathocleous and W. A. Harris, "Metabolism in physiological cell proliferation and differentiation," Trends in Cell Biology, vol. 23, no.10, pp. 484–492, 2013.

[67] K. Date, K. Matsumoto, H. Shimura, M. Tanaka, and T. Nakamura, "HGF/NK4 is a specific antagonist for pleiotrophic actions of hepatocyte growth factor," FEBS Letters, vol. 420, pp. 1–6, 1997.

[68] R. Königsberg, E. Obermayr, G. Bises, *et al.*, "Detection of EpCAM positive and negative circulating tumor cells in metastatic breast cancer patients," Acta Oncology, vol. 50, pp. 700–710, 2011.

[69] S. L. Stott, R. J. Lee, S. Nagrath, *et al.*, "Isolation and characterization of circulating tumor cells from patients with localized and metastatic prostate cancer," Science Translational Medicine, vol. 2, pp. 25ra23–25ra23, 2010.

[70] N. Guo, L. Zeng, and L. Wu, "A method based on multispectral imaging technique for white blood cell segmentation," Computers in Biology and Medicine, vol. 37, pp. 70–76, 2006.

[71] A. Gopal, "Multicolor colloidal quantum dot based inorganic light emitting diode on silicon: Design, fabrication and biomedical applications," University of Texas, Austin, TX, Dec. 2010.

[72] H. H. Kim, T. M. Miller, E. H. Westerwick, *et al.*, "Silicon compatible organic light emitting diode," Journal of Lightwave Technology, vol. 12, pp. 2107–2113, 1994.

[73] S. A. Haque, S. Koops, N. Tokmoldin, *et al.*, "A multilayered polymer light-emitting diode using a nanocrystalline metal-oxide film as a charge-injection electrode," Advanced Materials, vol. 19, pp. 683–687, 2007.

[74] S. H. Kang, C. K. Kumar, Z. Lee, K. H. Kim, C. Huh, and E. T. Kim, "Quantumdot light-emitting diodes utilizing CdSe/ZnS nanocrystals embedded in TiO thin film," Applied Physics Letters, vol. 93, p. 191116, 2008.

[75] J. M. Caruge, J. E. Halpert, V. Bulovi, and M. G. Bawendi, "NiO as an inorganic hole-transporting layer in quantum-dot light-emitting devices," Nano Letters, vol. 6, pp. 2991–2994, 2006.

[76] A. A. Al-Ghamdia, W. E. Mahmouda, S. J. Yaghmoura, and F. M. Al-Marzoukia, "Structure and optical properties of nanocrystalline NiO thin film synthesized by sol–gel spin-coating method," Journal of Alloys and Compounds, vol. 486, n. 1–2, pp. 9–13, 2009.

[77] J. Kwak, W. K. Bae, D. Lee, *et al.*, "Bright and efficient full-color colloidal quantum dot light-emitting diodes using an inverted device structure," Nano Letters, vol. 12, pp. 2362–2366, 2012.

[78] X. Zhou, J. He, L. S. Liao, *et al.*, "Enhanced hole injection in a bilayer vacuum deposited organic light-emitting device using a p-type doped silicon anode," Applied Physics Letters, vol. 74, p. 609, 1999.

[79] S. Camou, M. Kitamura, G. Jean-Philippe, F. Hiroyuki, A. Yasuhiko, and F. Teruo, "Organic light emitting device as a fluorescence spectroscopy's light source: One step towards the lab-on-a-chip device," In IC02 (pp. 1–8). International Society for Optics and Photonics, 2003.

[80] M. C. Schlamp, X. Peng, and A. P. Alivisatos, "Improved efficiencies in light emitting diodes made with CdSe (CdS) core/shell type nanocrystals and a semiconducting polymer," Journal of Applied Physics, vol. 82, p. 5837, 1997.

[81] I. D. Parker, "Carrier tunneling and device characteristics in polymer light-emitting diodes," Proceedings of SPIE, vol. 2144, p. 51, 1994.

[82] H. Rong, A. Liu, R. Jones, *et al.*, "An all-silicon Raman laser," Nature, vol. 433, pp. 292–294, 2005.

[83] I. A. Larkin, M. I. Stockman, M. Achermann, and V. I. Klimov, "Dipolar emitters at nanoscale proximity of metal surfaces: Giant enhancement of relaxation in microscopic theory," Physical Review B, vol. 69, p. 121403, 2004.

[84] D. Marrinucci, K. Bethel, M. Luttgen, R. H. Bruce, J. Nieva, and P. Kuhn, "Circulating tumor cells from well-differentiated lung adenocarcinoma retain cytomor-phologic features of primary tumor type," Archives of Pathology & Laboratory Medicine, vol. 133, pp. 1468–1471, 2009.

[85] R. A. Schultz, T. Nielsen, J. R. Zavaleta, R. Ruch, R. Wyatt, and H. R. Garner, "Hyperspectral imaging: A novel approach for microscopic analysis," Cytometry, vol. 43, no. 4, pp. 239–247, 2001.

[86] A. Y. Fong and E. Wachman, "Hyperspectral imaging for the life sciences," Biophotonics International, vol. 15, p. 38, 2008.

[87] K. Hoshino, Y. Y. Huang, N. Lane, *et al.*, "Microchip-based immunomagnetic detection of circulating tumor cells," Lab Chip, vol. 11, no. 20, pp. 3449–3457, 2011.

[88] J. W. Uhr, M. L. Huebschman, E. P. Frenkel, *et al.*, "Molecular profiling of individual tumor cells by hyperspectral microscopic imaging," Translational Research, vol. 159, no. 5, pp. 366–375, 2012.

[89] X. Huang, I. H. El-Sayed, W. Qian, and M. A. El-Sayed, "Cancer cell imaging and photothermal therapy in the near-infrared region by using gold nanorods," Journal of American Chemical Society, vol. 128, no. 6, pp. 2115–2120, 2006.

[90] S. Fantini, B. B. Barbieri, E. Gratton, M. Franceschini, J. S. Maier, and S. A. Walker, "Frequency-domain multichannel optical detector for noninvasive tissue spectroscopy and oximetry," Optical Engineering, vol. 34, no. 1, pp. 32–42, 1995.

[91] F. Bevilacqua, A. J. Berger, A. E. Cerussi, D. Jakubowski, and B. J. Tromberg, "Broadband absorption spectroscopy in turbid media by combined frequency-domain and steady-state methods," Applied Optics, vol. 39, no. 34, pp. 6498–6507, 2000.

[92] S. Susstrunk, R. Buckley, and S. Swen, "Standard RGB color spaces," in Proceedings of Color and Imaging Conference (Society for Imaging Science and Technology, 1999), pp. 127–134.

[93] K. Hoshino, P. Joshi, G. Bhave, K. Sokolov, and X. J. Zhang, "Use of colloidal quantum dots as a digitally switched swept light source for gold nanoparticle based hyperspectral microscopy," Biomedical Optics Express, 5(5), pp. 1610–1615, 2014.

Chapter 9
Photothermal microfluidics

Tuhin S. Santra, T.-H. Wu and Eric P. Y. Chiou

9.1 Introduction

9.1.1 Light is a special form of energy

Light is a special form of energy. Its electromagnetic wave nature allows its energy to be compressed in both temporal and spatial domains (Figure 9.1). This allows for the creation of extremely high energy density in space not possible with other actuation mechanisms commonly used in microfluidics. Through proper engineering designs, this unique feature can bring out new applications in high-speed microfluidics.

Another unique feature of light is that it does not require any propagation media to deliver its energy. Optical systems with a set of lenses and mirrors can be designed to deliver light energy to any arbitrary locations in space. This eliminates the need of wiring (electrical, fluidic, and mechanical) and routing issues for energy delivery. This feature is more appreciated when one needs to address a large array of 2D or 3D elements. Optical actuation can greatly simplify the device and system complexity.

9.1.2 Why photothermal?

Light has been widely used in microfluidics for detection and analysis applications. Examples include fluorescence detection, surface enhanced Raman scattering, and optical imaging. Light for actuation is less common. Optical tweezers are a representative example, in which light moves and traps microscopic objects by transferring optical momentum to the manipulated objects. Through sculpturing optical field landscapes, one can trap, guide, rotate, pattern, and sort small objects in microfluidics [1, 2]. One drawback of optical tweezers is that actuation by directly transferring light momentum to objects is not energy efficient. For example, it requires a 1 mW laser beam to generate ~1 pN force on a microscopic particle with a diameter of 1 μm in fluid. This force is able to move this particle at a speed of ~100 μm/s in fluid, according to the Stoke's Law $F = 6\pi r v \eta$, where F is the drag force, r is the radius of the particle, v is the speed of the moving particle, and η is the viscosity of the water (1.0020 mPa s at 20°C). This corresponds to a power

University of California, Los Angeles, CA, USA

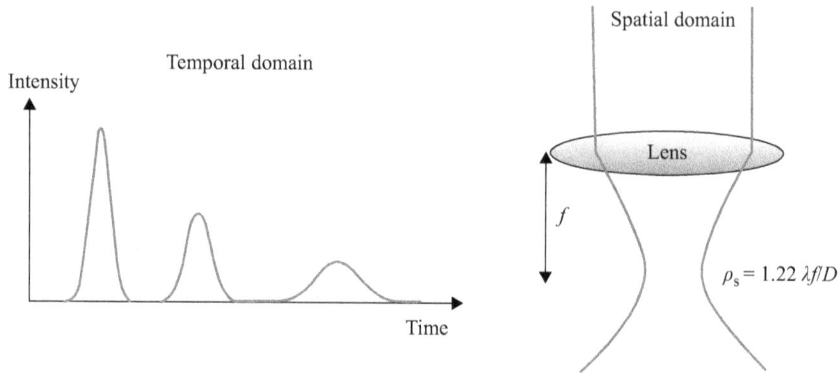

Figure 9.1 Light energy can be compressed in both temporal and spatial domains to generate extremely high instant intensity

dissipation rate of ~10^{-13} mW (Power = Force × Speed). The energy conversion rate from light to mechanical energy is nearly zero.

The reason to emphasize photothermal microfluidics in this chapter is to advocate the idea that converting light energy to thermal energy first before generating mechanical power for actuation is more energy efficient. First, using a light beam, one can dynamically and selectively heat up a local region on a device to control local surface tension. Surface tension is a dominating force in small scales. By changing the temperature of solids or liquids, surface tension is strongly affected and can be utilized to drive the movement of liquids as well as objects in liquids. Second, optical heating can be short and intense. This unique feature allows thermal energy to be quickly deposited to significantly raise the local temperature for effective mechanical actuation before the thermal energy dissipates away due to fast heat transfer in small scales. Third, the ultrahigh energy density produced by focusing a short laser pulse can generate large mechanical forces for unique applications that are not feasible with other mechanisms. Examples include high-speed fluid actuation in highly damped microfluid and nanofluid environments.

This chapter includes three parts. The first part covers the basic physics and principles of photothermal microfluidics. The second part focuses on cell manipulation using photothermal microfluidics and nanofluidics, and the third part covers the applications of photothermal microfluidics and nanofluidics for flow and droplet control.

9.2 Part 1: Basic principles

9.2.1 How much light energy can be compressed in the temporal and spatial domains?

Light is electromagnetic wave with no need of any media to support its propagation. The energy carried by electromagnetic wave can be superpositioned in space without limitation. Rapid advancement in laser technologies in recent decades makes pulse lasers widely available at affordable prices. One can purchase short pulse lasers with pulse duration ranging from ms to fs and pulse energy from pJ to mJ, and wavelengths from UV to IR. To have an idea how much light energy one can create with a focused laser pulse, here is an example. Consider a Q-switched Nd:YAG laser with a wavelength at 1064 nm, pulse duration of 1 ns, and pulse energy of 1 μJ, and focused to a small spot of 1 μm in diameter. The average light intensity at the focal point is 10^{11} W/cm^2 during the pulse illumination period. If the pulse duration is decreased to 100 fs and the pulse energy increased to 1 mJ, the instantaneous illumination intensity can reach 10^{18} W/cm^2. If this light spot is focused on an object which absorbs all incident light energy, the material at the focal spot would be instantly vaporized. This is true even for materials that are transparent to this wavelength of light at low light intensities. Nonlinear optical absorption dominates at such high intensity, at which level electric field strength (E: V/m) of light is strong enough to ionize electrons from atoms, causing breakdown of molecules, and inducing plasma [3].

9.2.2 *How is light converted into heat in microfluidics?*

When fluxes of photons strike an object, they are either scattered or absorbed. The absorbed photons that do not re-radiate eventually turn into heat and raise the temperature of the illuminated object. The easiest way to convert light into heat in microfluidics is by using light-absorbing materials, which can take the form of the device substrates, structures, fluid, or nanoparticles and molecules in fluid. The light absorption is linear, which means the amount of light absorbed is linearly proportional to light intensity. Most photothermal microfluidics applications using continuous wave (CW) lasers belong to this category.

Nonlinear optical absorption is less straightforward but may provide some fascinating features in microfluidics applications, especially with the rapid advancement of laser technologies that make compact, high-power, high-speed, short pulse lasers available at many different wavelengths. In the nonlinear optical absorption domain, an object that is originally transparent to light at low light intensity can become light absorbing. Strong light intensity can produce large electric field to ionize electrons from atoms to form plasma, an ensemble of free electrons and ions. This could dramatically change the local material properties and induce strong and localized absorption at regions where plasma are created. Nonlinear optical absorption is ideal for addressing 3D structures for energy delivery, especially for regions deep under a surface since it minimizes potential energy loss along the pathway before reaching the focal point.

9.2.3 *Pathway 1: Direct water absorption*

The absorption coefficients in water for electromagnetic waves of different frequencies vary a lot as shown in Figure 9.2. Amazingly, it has the lowest absorption spectrum window in the visible light range. In the near infrared (NIR) range (700–1100 nm), the absorption starts to increase. Most optical trapping and manipulation

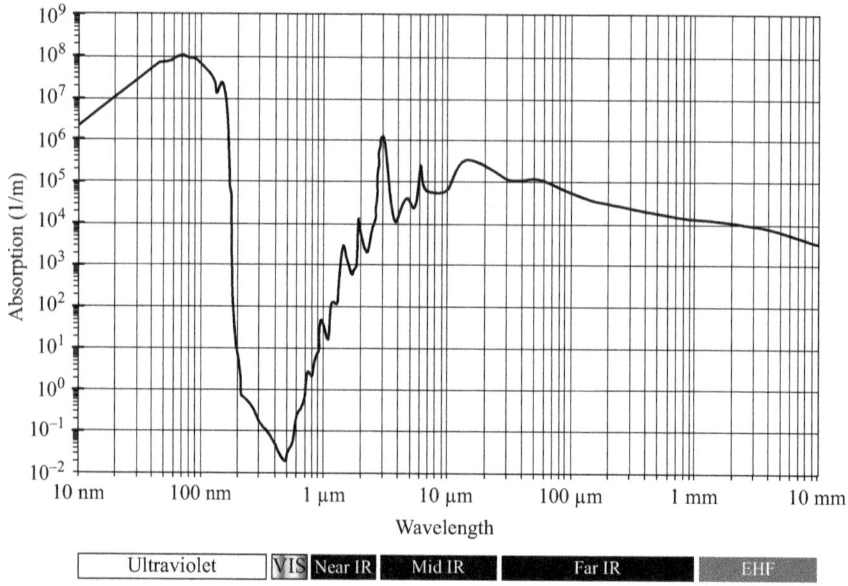

Figure 9.2 Electromagnetic wave absorption spectrum in water (John Bertie. "John Bertie's Download Site – Spectra". Reprinted with permission from Bertie (2012))

are done with lasers with wavelength in the NIR range to avoid photodamage to cells that may contain light-absorbing molecules in the visible spectrum while maintaining low enough absorption in water to prevent cooking cells. Some laser trapping is done by using the 1550 nm wavelength, where the absorption coefficient is already 4 orders of magnitude larger than that of the 500 nm. CO_2 laser is another option. The high-energy conversion feature of CO_2 laser makes it a cost-effective choice for high-power laser manufacturing applications. CO_2 laser has a wavelength of 10.6 μm. The absorption coefficient is 10^5/m, 6 orders of magnitude higher than the absorption at 500 nm. It means a 10-μm-thick water layer is sufficient to absorb light down to 1/e times of its original intensity. It is efficient for photothermal heating application. One drawback of CO_2 laser is that it is also strongly absorbed by glass and plastic, which are commonly used materials in microfluidics. An alternative substrate material to avoid strong absorption is low doping silicon. Heavily doped silicon has many free carriers that could significantly increase the absorption coefficient at the 10.6 μm wavelength.

9.2.4 Pathway 2: Light-absorbing materials

Light absorption can also be achieved by adding light-absorbing materials in water or on device structures. For example, one can use visible light to illuminate a thin amorphous silicon layer coated on the substrate of a microchannel to generate temperature gradient to control microfluidic flows [4]. One can also use NIR laser

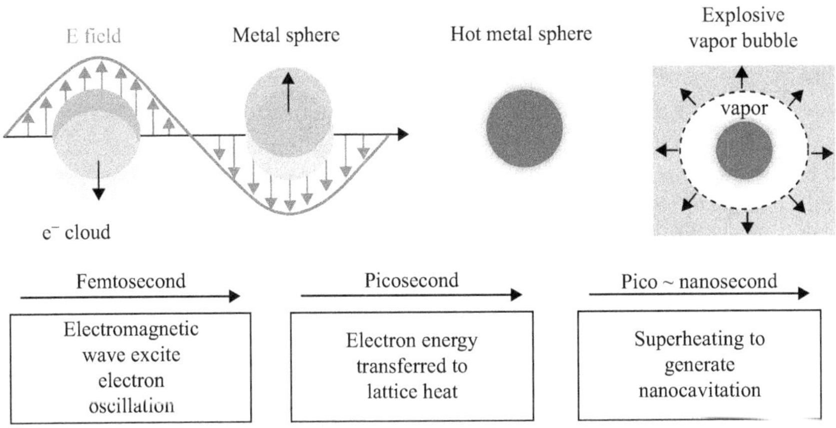

Figure 9.3 Rapid transfer of energy from light to cavitation explosion through metallic nanoparticles

to heat up a transparent indium tin oxide (ITO) electrode to induce electrokinetic flows [5]. The other most commonly used materials are various forms of gold nanoparticles, such as nanospheres, nanoshells, and nanorods. The advantage of using gold nanoparticles is that light absorption spectrum can be easily tuned by changing the shapes of Au nanoparticles. For example, nanosphere has its highest absorption peak at 530 nm in water [6]. Au nanorods can have different light absorption peaks in the NIR range depending on the aspect ratios of the rods [7, 8]. The surface chemistry properties of gold also make them easy to modify with other molecules such as antibodies to selectively target specific cells [9].

Illuminating metallic nanoparticles with short laser pulses has created many interesting applications in the past decade. As illustrated in Figure 9.3, free electrons on metallic nanoparticles oscillate under the illumination of electromagnetic waves. The kinetic energy of these oscillating electrons can convert into lattice heat to rapidly raise the temperature of the particles. If these nanoparticles are immersed in water, heat conduction starts to dissipate thermal energy into the surrounding water layer within few picoseconds [10]. Thermal diffusion length can vary from few nanometers in picoseconds to tens of nanometers in nanoseconds. If the pulse energy is high enough, the temperature in the thin water layer around a metallic nanoparticle can be rapidly raised up to trigger explosive cavitation bubbles [3]. These cavitation bubbles rapidly expand and collapse, inducing strong, highly localized mechanical, thermal, and fluid perturbation to the surrounding fluid and structures. These effects have been widely applied in photothermal therapeutics and gene transfection [8, 11–13].

One thing to be noted here is that the role of these metallic nanoparticles is a transducer that converts light energy into heat and then into mechanical energy through triggering cavitation bubbles in water. If the pulse energy does not surpass the threshold for bubble generation, the heat accumulated in the water layer near the particles simply dissipates away in the form of heat. During the energy conversion process, the metallic nanoparticle does not store much energy even though its temperature is high. This is because metal usually has a much smaller heat capacity than water (e.g. gold has specific heat capacity of 0.129 J/g°C at 25°C and water is 4.184 J/g°C). This implies that energy used to support an explosive bubble mainly comes from the thermal energy stored in the thin water layer before the bubble is initiated. This also means that one cannot use a very short laser pulse to excite a large bubble using metallic nanoparticles even when high pulse energy is applied, in which case the pulse will melt and vaporize the metallic structures to cause fragmentation, instead of triggering a larger bubble. On the other hand, this also implies that the bubble size triggered by a long laser pulse would be larger than by a short laser pulse on the same metallic nanoparticles. As the long laser pulse heats up the metallic nanoparticles and the surrounding water layer, much of the energy is already stored in the relatively thick water layer. The thicker water layer gives rise to a larger bubble as it reaches the critical temperature for bubble cavitation. For example, the size of cavitation bubbles triggered by nanosecond laser pulses is usually ~1 μm [6, 10, 14]. When a femtosecond laser pulse is used, the size of cavitation bubble can be sub-micron.

9.2.5 *Pathway 3: Nonlinear optical absorption*

Nonlinear optical absorption is typically achieved by using pulse laser due to its requirement of strong electric field to induce nonlinear effects. A dielectric material such as water can become highly light absorbing under the illumination of a high-intensity light beam. Strong light intensity ($I \sim E^2$) translates to large electric field oscillations in the EM wave. When the light energy is highly compressed in the temporal domain and tightly focused in the spatial domain, the instantaneous electric field can easily ionize electrons from atoms. The ionized free electrons are charged particles and can be further accelerated by the light pulses to accumulate sufficient energy to knock out more electrons from other atoms as they collide. This creates a plasma, an ensemble of charged particles containing both negatively charged electrons and positively charged ionized atoms at the laser focal point. A plasma is strongly light absorbing and its temperature can go up to higher than 10,000 K, a temperature higher than the surface of Sun [15]. Such a hot plasma creates a cavitation bubble expanding and collapsing rapidly and strongly perturbing the surrounding fluid flows (Figure 9.4). The dynamics of bubble expansion and collapse induced by this direct laser-induced breakdown (LIB) effect and that generated on metal nanoparticle heated by short pulse lasers is similar. However, the size and the energy of cavitation bubbles induced by LIB are orders of magnitude larger than those induced by photothermal absorption on metal nanoparticles. For example, a focused nanosecond laser pulse can trigger a cavitation bubble spanning hundreds of micrometers. While the bubble size triggered by illuminating a metal nanoparticle is in the order of 1 μm. The energy of a cavitation bubble is proportional to its volume, $E_{\mathrm{B}} = \frac{4}{123}\pi R_{\max}^3 (p_\infty - p_{\mathrm{v}})$, where E_{B} is the bubble energy, $R_{\max}$ is the maximum radius of the bubble, p_∞ is the atmospheric pressure, and p_{v} is the pressure inside the maximally expanded bubble (~2330 Pa at 20°C, saturated vapor pressure of water) [16]. Before a nucleated bubble expands to its maximum size and starts to collapse, the pressure inside the bubble can be high. For a cavitation bubble with a

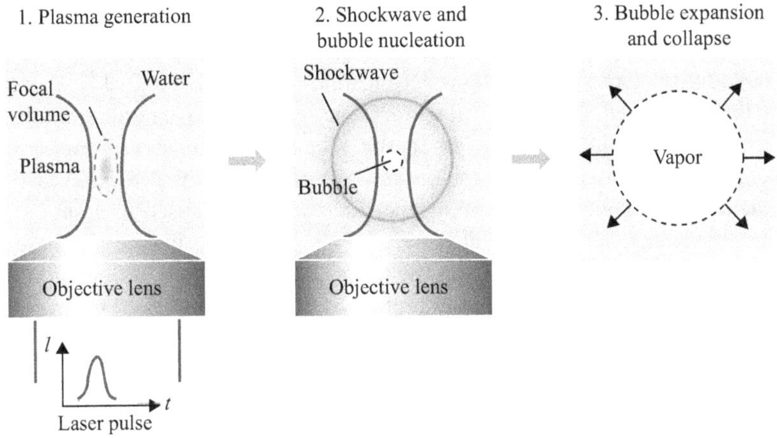

Figure 9.4 *LIB in water. An intense and tightly focused laser pulse can provide strong enough electric field to ionize water molecules and generate hot plasma and followed by high-speed cavitation bubble expansion to strongly perturb neighboring fluid flows*

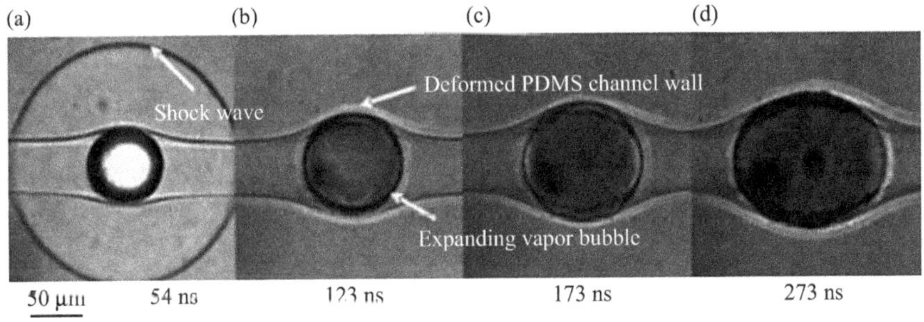

Figure 9.5 *An example of nanosecond pulse laser triggered cavitation bubble in a PDMS microfluidic channel. The large bubble pressure is strong enough to significantly deform the local PDMS channel in less than a μs. Reprinted with permission from Chen (2012) [18], © The Royal Society of Chemistry*

maximum size of 100 μm, the pressure inside this bubble can be in the MPa level during the initial expansion period. Such a large pressure can push the neighboring fluid to flow at a speed of hundreds of m/s or higher [17].

Figure 9.5 shows the time-resolved images of a cavitation bubble excited in the center of a microfluidic channel. The vapor pressure inside an expanding bubble is large enough to deform a solid polydimethylsilxane (PDMS) channel wall from its original width of 50–100 μm in less than a μs. Such a fast, large-force, localized, and on-demand actuation mechanism provides new opportunities in high-speed microfluidics and nanofluidics applications.

The size of bubbles triggered by LIB is both pulse duration and pulse energy dependent. There exist minimum threshold light intensity and energy for pulse of different durations for inducing the water breakdown phenomena [3, 17]. Figure 9.6 shows the calculated threshold value for laser pulse of different durations and wavelengths. Surprisingly, the pulse energy required to trigger this event is less wavelength dependent. For a laser pulse with duration of 1 ns, the required fluence is ~500 J/cm^2 (Figure 9.6(b)). Assuming the laser beam is focused to a spot size of 1 μm in diameter, one needs a pulse with energy of 5 μJ to induce a bubble. The shorter the pulse duration, the smaller the pulse energy required. The relationship is inversely proportional until the pulse duration is shorter than 10 ps. For pulse duration below 10 ps, the avalanching ionization process is less efficient since the ionized

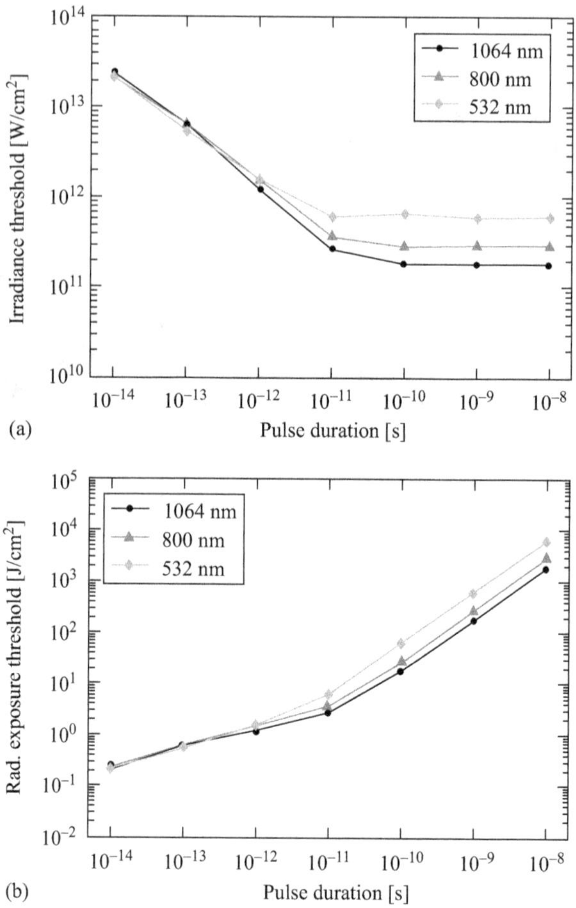

Figure 9.6 Threshold light intensity and pulse energy required to induce water breakdown and trigger cavitation bubbles using light pulses with different durations and wavelengths. Reprinted with permission from Vogel (2005) [3], © Springer

	E_p [μJ]	R_{max} [μm]	T_{osc} [μs]	E_B [μJ]	E_B/E_p [%]
0.7 × Threshold	5.6	95	17	0.18	3.2
1 × Threshold	8	118	21	0.34	4.3
2 × Threshold	16	140	25	0.57	3.6
3 × Threshold	24	200	37	1.66	6.9

Figure 9.7 Energy conversion efficiency of 6.9% from a light pulse to a cavitation bubble has been achieved. The table shows the results of using 6-ns pulses from a Q-Switch Nd:YAG laser focused by a 0.9 NA objective lens to trigger cavitation bubbles. Reprinted with permission from Rau (2006) [17], © Elsevier

electrons were not given sufficient time to collide with many atoms and thus requires higher light intensity to induce the same breakdown effect (Figure 9.6(a)).

Energy conversion efficiency from light to mechanical actuation is a major advantage of photothermal microfluidics compared to conventional direct optical momentum transfer. Figure 9.7 presents a table showing the conversion efficiency from light pulse (E_p) to bubble energy (E_B) for driving fluid flow. Conversion efficiency of 6.9% can be experimentally achieved [17], while in optical tweezers' case, this energy utilization efficiency is nearly zero percent.

9.3 Part 2: Photothermal microfluidics and nanofluidics for cell manipulation

9.3.1 Femtosecond laser transfection and subcellular surgery

Femtosecond laser surgery relies on nonlinear light absorption by the ordinarily non-light-absorbing water or tissue molecules [3]. A tightly focused laser beam is required in order to elicit the nonlinear effects and the resultant instantaneous peak optical field at the laser focus to generate plasma and subsequently a cavitation bubble through optical breakdown of the biological medium. Depending on the laser pulse repetition rate, different dissecting mechanisms are engaged. For MHz repetition rate, low pulse energies slightly below the optical breakdown threshold are typically used. Photochemical and accumulated thermal effects from the created low-density plasma ablate the biological material at the laser focus. For kHz laser operation, disruption of the cellular structure is largely attributed to the mechanical effects from the induced thermo-elastic stress and cavitation bubbles at pulse energies above the breakdown threshold [3]. Compared to nanosecond or picosecond laser surgeries, femtosecond lasers require significantly lower optical breakdown threshold energy – few nJ per laser pulse whereas breakdown threshold is hundreds of nJ for picosecond lasers and few μJ for nanosecond lasers. As a result, the size of the induced cavitation bubble can be as small as few hundred nanometers by femtosecond pulses [19]. This capability allows for ablating subcellular structures with a spatial resolution below the optical diffraction limit. König *et al.* demonstrated dissection of isolated human chromosomes using a 800-nm wavelength Ti:sapphire laser operating at a repetition rate of 80 MHz with 170-fs pulse width. The full width half maximum (FWHM) of the cut line on a chromosome was measured below 200 nm [20]. For intracellular surgeries in live cells, ablation of mitochondrion [21, 22] as well as single microtubule fiber [23] can be achieved using 1-kHz repetition rate femtosecond laser without compromising cell viability (Figure 9.8(a)). Yanik *et al.* used femtosecond laser to perform axontomy in *Caenorhabditis elegans* and showed the severed neuron was capable of regeneration at certain frequency as evaluated by its fluorescence image (tagged with a green fluorescent protein or GFP) as well as by functional recovery in the animal's locomotion (Figure 9.8(b)) [24].

By positioning the laser focus on the cell membrane, femtosecond laser surgery can be used to create transient membrane pores for cargo uptake from surrounding medium. Tirlapur and König first demonstrated femtosecond laser transfection of Chinese hamster ovarian (CHO) and rat-kangaroo kidney epithelial (PtK2) cells [25]. After cell membrane poration, plasmid DNA molecules in the medium were able to enter the cell cytoplasm by diffusion and the encoded eGFP gene expression could be observed after delivery. Stevenson *et al.* performed systematic evaluation of transfection efficiency and cell viability using a 80 MHz, 120 fs laser operating at a wavelength of 800 nm [26]. Laser exposure duration was evaluated at 10–250 ms, and the average laser power was tested between 50 mW and 225 mW. Transfection efficiency depended strongly on the laser fluence which

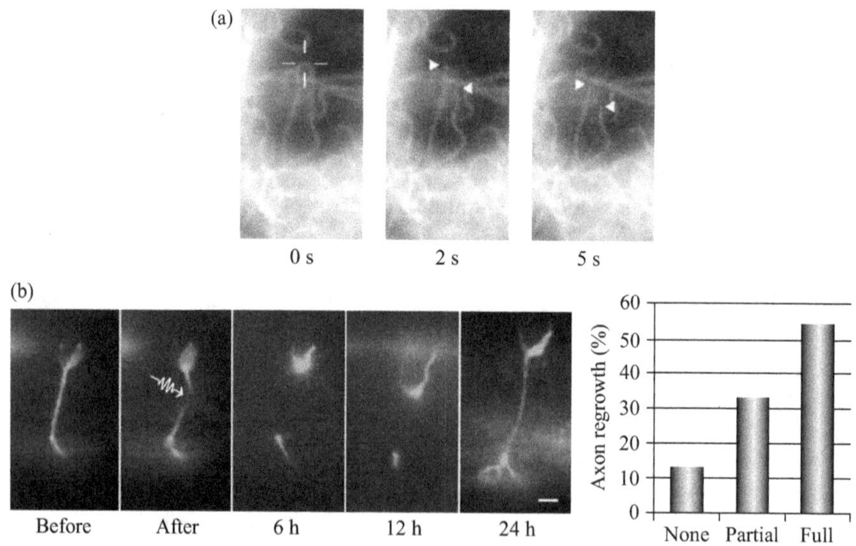

Figure 9.8 Examples of femtosecond subcellular surgery. (a) Single microtubule fiber was ablated at 0 s indicated by the cross hair. Seconds after ablation, the broken ends recoiled due to existing tension. Reprinted with permission from Heisterkamp (2005) [23], © Optical Society (b) Neuron growth and recovery after femtosecond laser axotomy in C. elegans. Right panel shows statistics of axon growth 24 h after femtosecond laser surgery. Reprinted with permission from Yanik (2004) [24], © Nature Publishing Group

is the cumulative laser energy per unit area at the laser focus. Under optimized laser condition, a 50% transfection efficiency and corresponding cell viability at 65% was obtained in CHO cells. By incorporating a patch clamp setup into the femtosecond laser transfection system, Baumgart *et al.* were able to monitor simultaneous cell membrane potential changes resulting from laser poration [27]. They measured cell membrane depolarization time of 40–50 ms, indicative of the transient membrane pore lifetime, which is some milliseconds longer than the laser exposure time. From these measurements, they estimated volume exchange between inside and outside the cell was ~0.4 times the cell volume.

Another example of femtosecond laser cell manipulation is cell fusion [28, 29]. By laser poration of the interface of two adjacent cells, the disrupted plasma membrane can reseal with neighbor cell membrane and merge two cells into one fused cell (Figure 9.9(a)). Fusion of yeast protoplasts, human leukemia cells and two-cell porcine embryos were demonstrated. Multifunctional femtosecond laser surgeries have been demonstrated. By switching between mode-lock and CW operation of the Ti:sapphire laser, optical trapping function can be incorporated into the femtosecond laser setup. Brown *et al.* demonstrated trapping of a human leukemia HL60 cell using CW mode of the laser through a high NA objective lens [30]. The laser was briefly switched to mode-lock mode outputting femtosecond pulse trains for a duration <100 ms. Membrane poration was verified by Trypan Blue dye uptake, and the cell could be continuously trapped and tweezed following laser poration. In order to bypass restriction of short working distance as a result of using high NA objective lens for

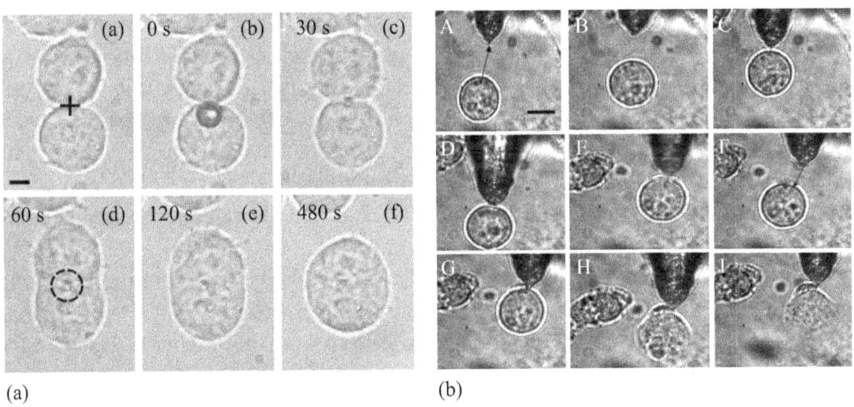

Figure 9.9 *(a) Femtosecond laser cell fusion. Laser focus is indicated by cross. Two-cell fusion began around 30 s after laser ablation of cell membranes and completed in 8 min. Reprinted with permission from Kuetemeyer (2011) [29], © SPIE – the international society for optics and photonics (b) Single cell trapping and lysis using an axicon-tip optical fiber. Target cell was drawn to the fiber tip when the laser was operating in CW mode. After switching to mode-lock mode, trains of femtosecond pulses lysed the target cell. Reprinted with permission from Mohanty (2008) [31], © SPIE – the international society for optics and photonics*

laser focusing, Mohanty *et al.* used a tapered, axicon-tip single mode optical fiber for laser tweezing and dissection [31]. Trapping the target cell at the tip of the optical fiber under CW mode followed by femtosecond laser lysing of the trapped cell was demonstrated in Figure 9.2(b). Using this setup, guiding and stretching of neurons as well as laser dissection of neuronal processes could be accomplished. Some potential cytotoxic side effects of femtosecond laser surgery include DNA damage [32], reactive oxygen species generation [33, 34], mitochondria swelling, and depolarization [35].

9.3.2 Nanosecond pulsed laser-induced cavitation bubbles for transfection

Nanosecond lasers have been used in biomedical applications such as intraocular surgeries and laser-assisted in situ keratomileusis since 1980s [36, 37]. Mechanisms of such pulsed laser ablation of tissues have been reviewed in details by Vogel and Venugopalan [16]. Due to the higher optical breakdown threshold energy of nanosecond lasers compared to femtosecond lasers, minimum plasma size is on the order of few microns to tens of microns [38]. The rapid thermo-elastic stress built-up within the plasma results in a fast expanding shockwave and cavitation bubble [39]. The maximally expanded bubble size ranges from tens of microns to millimeters, depending on the deposited laser pulse energy. By controlling the nanosecond laser energy and the distance of laser focus to the target cell layer, cell membrane permeabilization or lysis can be accomplished in multiple cells by a single laser pulse. As the cavitation bubble expands from the laser focus, surrounding fluid is accelerated outward and exerts shear stress on the nearby cells. When the bubble collapses, formation of a high-speed fluid jet can exert further shear stress on the cells that lead to cell detachment, lysis or membrane permeabilization [17, 40–42]. Dijkink *et al.* investigated the effect of varying the vertical distance of laser focus to the cell monolayer [40]. At a stand-off distance of ~0.65 times the maximum bubble radius, the cavitation bubble exerted the highest shear stress on the cells below and resulted in the largest cell detachment zone. However, in order to achieve the largest area of cell permeabilization as indicated by the calcein dye uptake zone (Figure 9.10(a)), the stand-off distance should be close to 0 where the laser focus is right above the cell layer and solid boundary. Under this configuration, Hellman *et al.* estimated the maximum shear stress on the cell monolayer due to laser-induced bubble expansion [42]. As depicted in Figure 9.11, maximum shear stress decays radially away from the laser focus and increases with laser energy. Different zones of cellular injury: lysis, necrosis, permeabilization, or molecular delivery, can be characterized by their corresponding critical shear stress levels.

Clark *et al.* increased laser cell transfection throughput by incorporating a pair of galvo scanning mirrors and a large F-theta scanning lens to steer the focused laser beam in a grid pattern over large areas, yielding a throughput of 2000–5000 cells/s [43]. Optoinjection efficiency of 85–100% was obtained with up to 80% cell viability. Targeted cell injection can be achieved by controlling the scanning pattern (Figure 9.12(a,b)). By tuning the laser energy, selective cell killing using the laser system to purify a heterogeneous cell population was also demonstrated [44]. He *et al.* demonstrated nanosecond laser lysis of single cell encapsulated in a picoliter

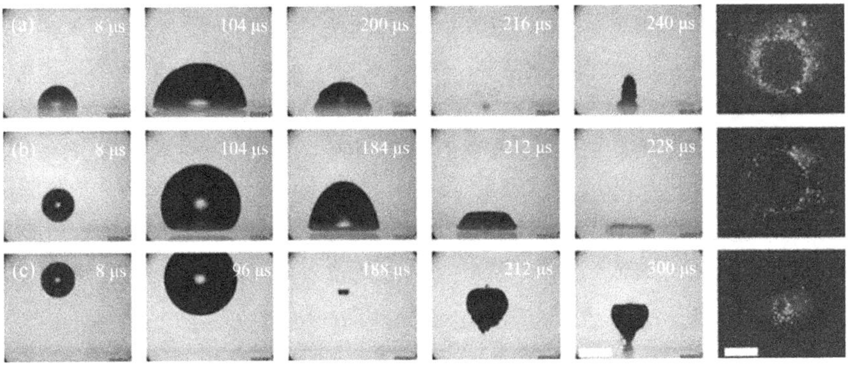

Figure 9.10 *Nanosecond laser-induced cavitation bubble dynamics near a cell monolayer on a solid boundary [40]. Back and white panels are side view images of bubble evolution. Cell monolayer is at the bottom of each frame. Fluorescent images show calcein uptake pattern in the cell monolayer after laser optoporation. Laser focus positioned at a vertical distance of (a) 0.1×, (b) 0.9×, and (c) 2× of maximum bubble radius. Reprinted with permission from Dijkink (2008) [40], © IOP*

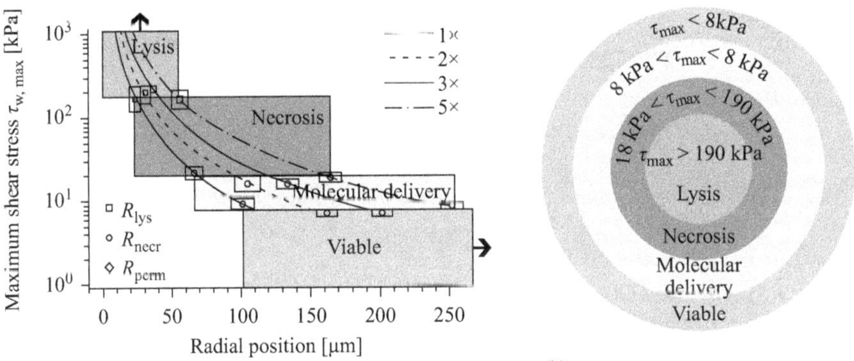

(a) (b)

Figure 9.11 *(a) Estimated maximum shear stress on cell monolayer due to laser-induced cavitation bubble expansion. (b) Different zones of cellular injury characterized by critical shear stress levels. Reprinted with permission from Hellman (2008) [42], © John Wiley and Sons*

droplet [45]. Rapid laser-induced cell lysing (Figure 9.12(c)) allowed for preserving cellular materials at the moment of lysis compared to chemical lysis where the slow reaction might alter the cell state. Besides cavitation bubble, laser-induced stress waves can also cause cell membrane permeability. By irradiating nanosecond laser pulses on a light-absorbing black rubber disk positioned on top of the cell layer, Terakawa *et al.* showed stress wave mediated transfection can achieve ~3% gene expression in NIH 3T3 cells [46]. Nanosecond LIB has also been widely used in tissue microdissection [47, 48]. Figure 9.12(d) depicts the dissection setup. The thin

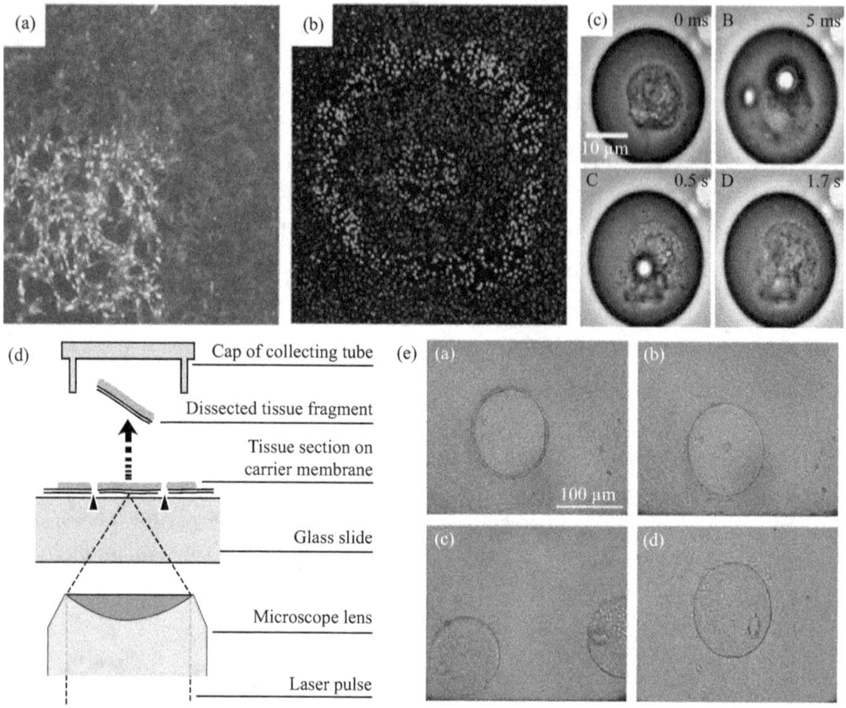

Figure 9.12 (a,b) Scanning nanosecond optoinjection. Targeted delivery was performed by scanning nanosecond laser beam over desired cell culture regions. Multiple dye deliveries were achieved by sequentially swapping out the dye containing medium during optoinjection. Reprinted with permission from Clark (2006) [43], © SPIE – the international society for optics and photonics (c) Laser lysing of a single cell encapsulated in a picoliter droplet. Reprinted with permission from He (2005) [45], © American Chemical Society (d) Schematic of laser microdissection and sample capture setup. Reprinted with permission from Vogel (2007) [47], © Elsevier (e) Live cell culture dissected by scanning nanosecond laser pulses (circular region). After catapulting into the collection tube and recultivating, growth, and expansion of the dissected cell population was observed. Reprinted with permission from Horneffer (2007) [48], © SPIE – the international society for optics and photonics

tissue sample is mounted on a UV-light-absorbing carrier membrane. Dissection is performed by scanning a focused nanosecond UV laser beam along the outline of the desired tissue region. Lastly a higher energy laser pulse is focused at the periphery of the sample and the strong bubble cavitation catapulted the dissected section upward into a collection vessel [47]. Besides fixed cells, live cell culture dissection and capture was also demonstrated using the technique [48].

9.3.3 Nanoparticle-assisted photothermal therapy and cell manipulation

Light-absorbing nanoparticles such as metallic plasmonic nanoparticles act as efficient transducers converting absorbed light energy into heat. When excited with a CW laser, heat generation within the nanoparticle and heat conduction into the surrounding medium reach equilibrium, producing a steady temperature increase in the vicinity of the nanoparticle [11]. When the nanoparticle is excited with a short laser pulse, most of the generated heat is accumulated in the nanoparticle due to the slower heat diffusion speed of the surrounding environment. As a result, large transient temperature increase is confined within the nanoparticle and a thin layer of surrounding medium characterized by $\sqrt{D\tau}$, where D is the thermal diffusivity of the medium and τ is the laser pulse duration. For example, heat diffusion length in water during 1 ns laser pulse irradiation is approximately 10 nm. Superheating of this thin liquid medium layer results in a cavitation bubble whose size depends on the nanoparticle size and absorption coefficient at the laser wavelength, laser pulsewidth, and laser fluence [10]. By utilizing these photothermal effects of nanoparticles on a cell, light-induced cell killing, transient cell membrane poration for cargo delivery or gene expression regulation have been demonstrated.

9.3.4 Photothermal therapy

Nanoparticle-assisted photothermal therapy uses plasmonic nanoparticles such as gold nanoshells [12, 49] or nanorods [8] with a peak absorption around NIR to maximize light penetration into the tissues. For in vivo delivery of nanoparticles to the tumor site, both passive and active targeting have been used [50]. In passive targeting, nanoparticles accumulate in the tumor tissues by the leaky vasculature in the tumor microenvironment. In active targeting, nanoparticles are functionalized with tumor cell targeting antibodies or peptides for a more efficient and rapid uptake by the cancer cells. Besides surface properties of the nanoparticles, size and shape of the nanoparticles also have an effect on cell uptake. Chithrani *et al.* found 50 nm gold spheres are most readily taken up by HeLa cells compared to other gold nanospheres with diameters ranging from 10 to 100 nm and gold nanorods of similar sizes [51]. Similar result was also reported in nanoparticle uptake by tumor tissues in vivo [50]. After the nanoparticles are bound to the target cell, excitation laser irradiates the target tissues to generate heating or bubble cavitation to kill the cancer cells. For CW laser phototherapy, an estimated temperature increase to 70–80°C by light-induced nanoparticle heating is required to effectively kill the target cells [11]. Huang *et al.* showed targeting of cancer cells using antibody-conjugated gold nanospheres and nanorods. As shown in Figure 9.3(a), gold nanospheres bound cancer cells exhibited strong optical absorption at 550 nm while gold nanorods bound cells exhibited peak absorption around 900 nm. After illumination by 800-nm NIR laser, gold nanorod bound cancer cells were killed while non-targeted normal cells remained viable. By using a pulsed laser irradiation, Lapotko *et al.* demonstrated selective laser lysis of tumor cells by induced cavitation bubbles [14]. They found that subsequent to nanoparticle binding to the cell membrane, particles were internalized by the cell and formed larger aggregates. These particle aggregates were able to induce microscale cavitation bubbles upon laser excitation and effectively lyse and kill the cancer cell (Figure 9.6(b)). Zharov *et al.* also described similar effects of laser-induced bubbles overlapping for efficient cell lysis [6]. They showed in MDA-MB-231 breast cancer cells that optimal gold nanoparticle size was 30–40 nm, and minimum particle density was ~2.44×10^{11} cm^{-3}, corresponding to ~430 particles per cell. Similar nanoparticle generated bubble has also been used in selective laser lysis of single bacterium as shown in Figure 9.13(c) [13]. Wang *et al.* demonstrated synthesis of supramolecular gold nanoparticles of controlled sizes through multivalent molecular recognition based on β-cyclodextrin and adamantane motifs [9]. Using this approach, 2-nm gold colloids self-assembled into 118 nm supramolecular nanoparticles. Enhanced photothermal effect and microbubble induction were observed in 118 nm supramolecular nanoparticles but not in 2 nm gold colloids. To enhance photothermal therapy efficacy, gold nanoparticles can be conjugated with antitumor drug molecules. Light stimulated physical effects such as heating and bubble cavitation combined with biological effects from the released drug molecules have shown synergistic results for cancer treatment [52].

Spatially targeting specific cells can be achieved by controlling the laser exposure pattern. Wu *et al.* demonstrated light image patterned dye delivery into cells cultured on a gold nanoparticles coated substrate [53]. Riahi *et al.* used photothermal ablation to specifically disrupt leader cells on the leading edge of epithelial cell cultures, which helped elucidate initiation and regulation of leader cells during collective cell migration [54]. Besides photothermal effects, additional functionalities can be incorporated by synthesis of hybrid nanoparticles. Kim *et al.* demonstrated synthesis of magnetic gold nanoshells [55]. As illustrated in Figure 9.14(a), 7-nm superparamagnetic magnetite nanoparticles were embedded in a 15-nm-thick gold nanoshell on a 100-nm silica core. The resultant hybrid nanoparticle has dual functionality in photothermal therapy as well as magnetic resonance imaging (MRI) contrast agent. Cheng *et al.* [56] demonstrated image-guided photothermal cancer therapy (Figure 9.14(b)). First hybrid magnetic gold nanoparticles were injected into the animal. An external magnet was placed near the tumor site to facilitate nanoparticle accumulation in the tumor tissue. As high as eightfold increase in nanoparticle accumulation was observed under the external magnetic field. Finally photothermal therapy progression can be monitored by MRI or by nanoparticles' upconversion luminescence. Quantum dots have also been investigated as multifunctional agents for fluorescence imaging and photothermal therapy [57].

9.3.5 Photothermal cargo delivery

By carefully controlling the laser-induced bubble size and density on the cell membrane, transient pores can be generated in the cell membrane for cargo delivery without killing the cells [58]. Delivery of plasmid DNA [7] and small interfering RNA (siRNA) [59] has been demonstrated using this approach. Cargo delivery mainly relies on passive diffusion across the transient membrane pores, which becomes inefficient as the cargo size increases. Wu *et al.* demonstrated cargo

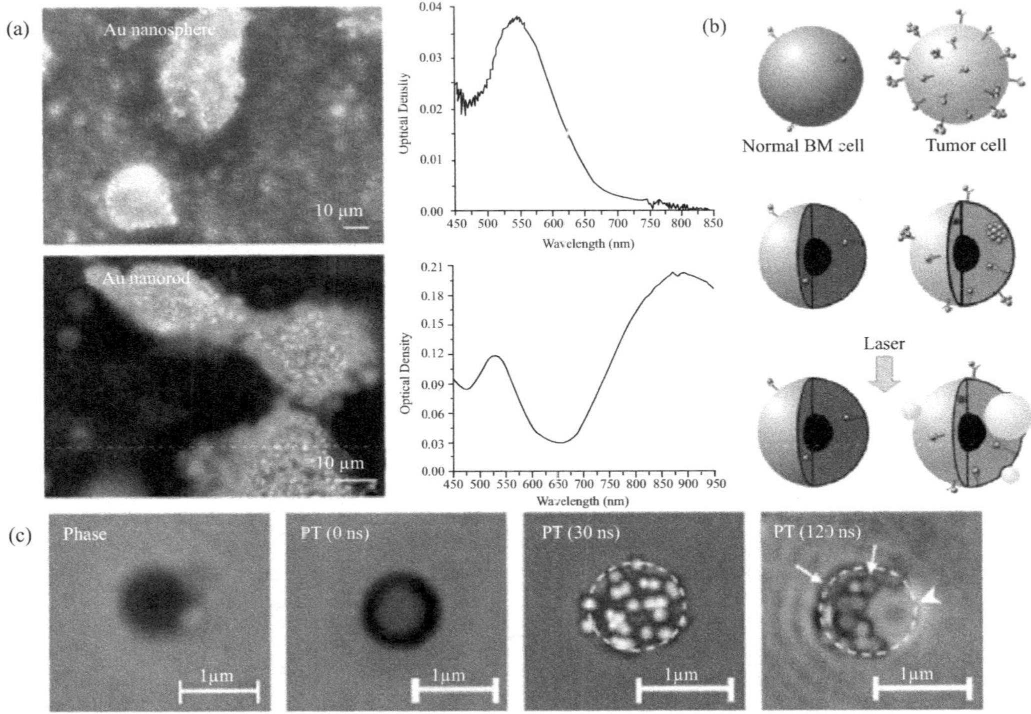

Figure 9.13 *(a) Cancer cells targeted with gold nanospheres (upper panel) exhibited light scattering and absorption in the visible spectrum. Cancer cells targeted with gold nanorods (lower panel) exhibited scattering and absorption in the NIR. Reprinted with permission from Huang (2006) [8], © American Chemical Society (b) Two-stage photothermal lysis of cancer cells. First, gold nanoparticles were bound to surface of cancer cells. After internalized by cells, gold nanoparticles formed aggregates, creating large cavitation bubbles for efficient cell killing upon laser excitation. Reprinted with permission from Lapotko (2006) [14], © John Wiley and Sons (c) Single bacterium laser lysis using gold nanoparticles. Left panel shows phase contrast image of* Staphylococcus aureus *bacterium coated with gold nanoparticles. Photothermal (PT) images show induced cavitation bubbles at 0 ns, 30 ns, and 120 ns after laser pulsing. Arrows point to bubbles around single nanoparticles. The arrowhead indicates bubble around nanoparticle a cluster. Reprinted with permission from Zharov (2006) [13], © Elsevier*

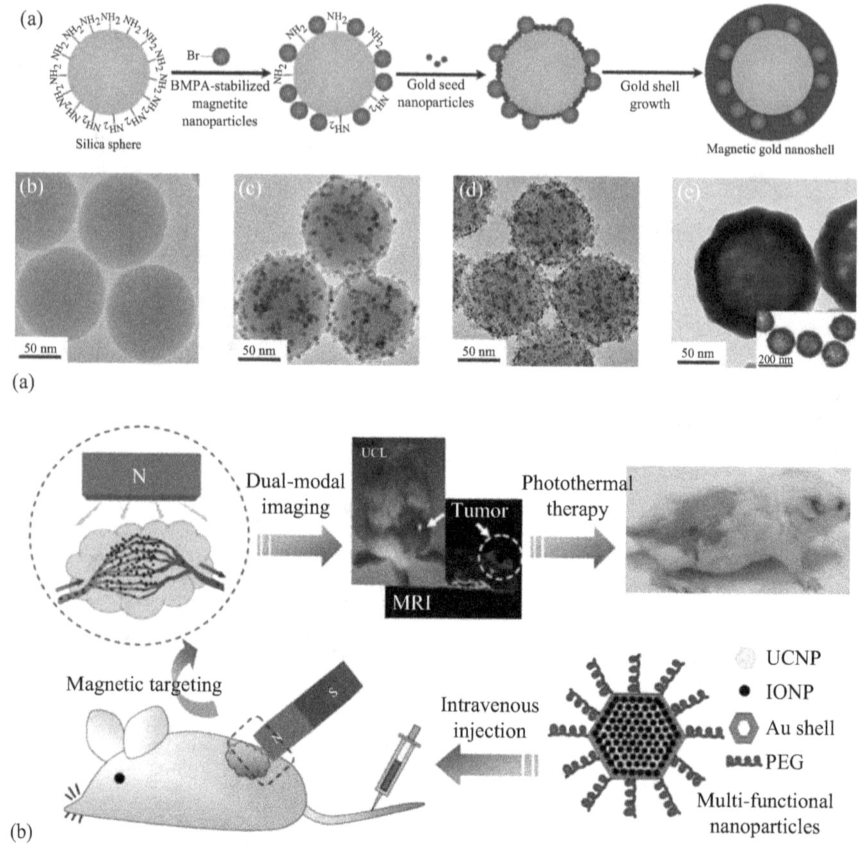

Figure 9.14 *(a) Schematic and corresponding SEM images showing fabrication of multifunctional magnetic gold nanoshells. First, 100-nm silica nanospheres were amino modified on the surface. Second, 7-nm superparamagnetic iron oxide nanoparticles were immobilized on the silica core. Lastly, gold nanoparticles were seeded onto the silica core until a 15-nm-thick gold nanoshell formed and completely embedded the magnetic nanoparticles. Reprinted with permission from Kim (2006) [55], © Wiley and Sons (b) Schematic of multifunctional composite nanoparticles for image-guided photothermal therapy. Particles were guided to the tumor site by an external magnet followed by MRI tumor imaging and NIR laser killing tumor cells. Reprinted with permission from Cheng (2012) [56], © Elsevier*

delivery using a photothermal nanoblade [60]. By incorporating a titanium thin film on the tip of a glass microcapillary pipette, cell membrane can be efficiently cut open using the laser-induced cavitation bubble on the pipette tip. Active cargo delivery into the cell cytosol was done by pressure-controlled fluid pumping

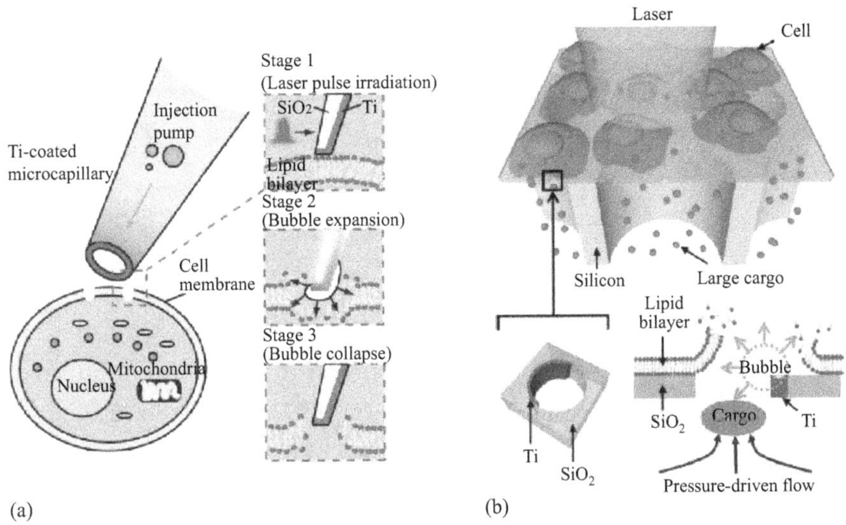

Figure 9.15 *(a) Cargo delivery into cells by the photothermal nanoblade. Upon laser excitation, the titanium thin film on the microcapillary tip generated a vapor bubble which disrupted the cell membrane for delivery. A synchronized pressure-driven flow delivered the cargo from the capillary into the cell cytosol. Reprinted with permission from Wu (2011) [60], © American Chemical Society (b) Schematic of the biophotonic laser-assisted surgery tool. The microfabricated silicon platform consists of an array of holes patterned on a 1.5-μm-thick SiO_2 film. Crescent-shaped titanium films are coated on the sidewalls of these holes for bubble induction followed by active pressure-driven cargo delivery. Reprinted with permission from Wu (2015) [61], © Nature Publishing Group*

through the capillary pipette (Figure 9.15(a)). Cargo as large as micron-sized bacteria can be delivered into cells while maintaining cell viability. Using a microfabricated silicon platform with precisely engineered titanium-coated micron pores, Wu *et al.* demonstrated delivery of large cargo into 100,000 cells in 1 min (Figure 9.15(b)) [61]. Wild-type and mutant *Francisella novicida* bacteria were delivered into HeLa cells for investigating genes required for bacteria phagosomal escape and intracellular replication.

9.3.6 Photothermal gene regulation in cells

Light-controlled DNA or RNA release from the surface of the nanoparticles has been utilized in regulating gene expression in cells [62]. As illustrated in Figure 9.16(a) [63], siRNA molecules are immobilized onto the plasmonic nanoparticles and delivered into cells. Upon laser-induced heating, the siRNA molecule dissociates

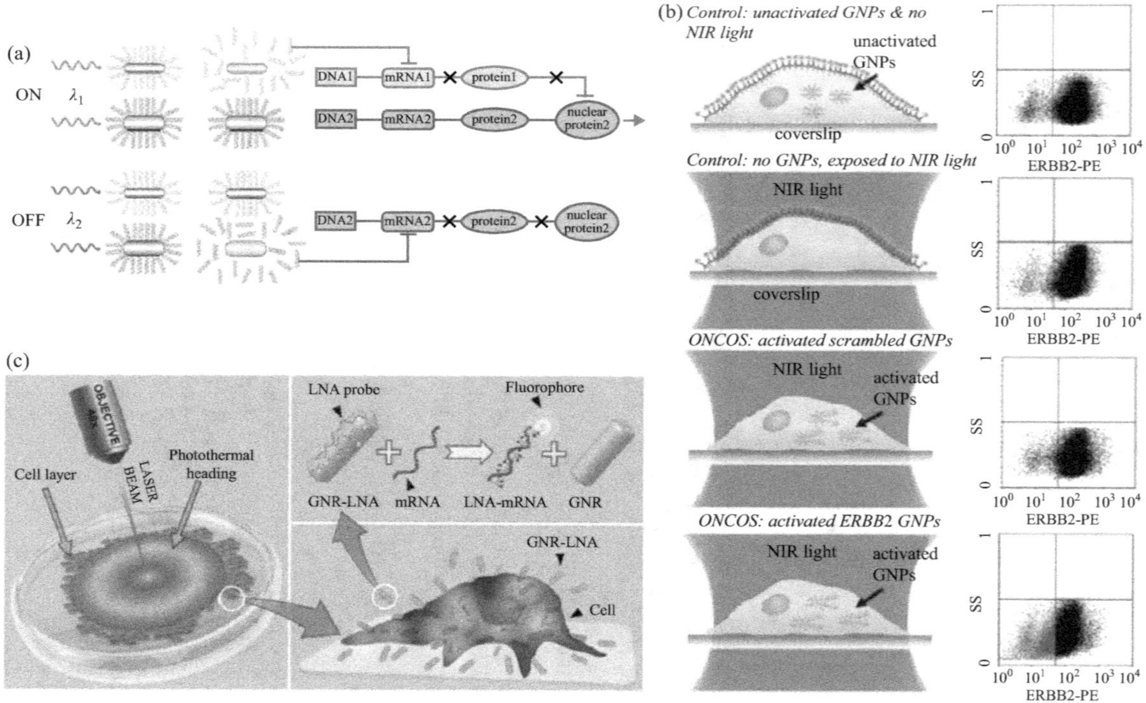

Figure 9.16 *(a) Schematic of photothermal gene regulation by wavelength dependent heating of plasmonic nanoparticles and subsequent releasing of siRNA molecules for target gene interference. Reprinted with permission from Anikeeva (2012) [63], © American Chemical Society (b) Photothermal down regulation of ERBB2 gene in BT474 breast cancer cells. Flow cytometry results showed decrease in ERBB2 expression in the cells treated by NIR light with gold nanorods coated with ERBB2 targeting oligonucleotides. Reprinted with permission from Lee (2009) [64], © American Chemical Society (c) The laser energy is absorbed by the plasmonic GNRs to release the LNA probes in the cells to quench the fluorophore conjugated to the probe. With a target mRNA, the probe displaces from the GNR and hybridizes to the target, allowing intracellular gene expression analysis in living cells and tissues. Reprinted with permission from Riahi (2014) [65], © American Chemical Society*

from the nanoparticle surface and interacts with the target mRNA in the cytosol which stops the translation of mRNA-encoded protein and inhibits target gene expression. By using different shaped nanoparticles functionalized with different siRNA sequences, multiplexed gene regulation can be achieved by switching excitation laser wavelengths that target the desired nanoparticles. Lee *et al.* demonstrated gene interference in breast cancer cells using gold nanorods conjugated to double stranded oligonucleotides [64]. After heating with NIR CW laser, the antisense oligonucleotide was released and bound to the target ERBB2 mRNA, inhibiting its protein translation. As shown in Figure 9.16(b), decrease in ERBB2 expression as quantified by flow cytometry was only observed in light-activated gold nanorods immobilized with ERBB2 targeting oligonucleotides but not in the other control settings. Besides siRNA, molecular beacon probes can also be activated in similar fashion for studying spatiotemporal gene expression in cells (Figure 9.16(c)). Riahi *et al.* [65] used locked nucleic acid probes immobilized on gold nanorods for detecting mRNA molecules in cells. This technique allowed for real-time monitoring cellular heat shock responses from photothermal operation.

9.4 Part 3: Photothermal microfluidics for fluid control

9.4.1 Surface tension-based photothermal microfluidics

Surface tension is a contractive tendency of surface of a liquid. It is a dominating force in small scales. In microfluidics and nanofluidics devices, surface tension can be controlled through electrostatic forces such as electrowetting [66, 67] or thermal mechanisms such as the Marangoni effect [68]. Since surface tension is temperature dependent. By inducing temperature gradient, one can induce surface tension difference at interfaces to drive the motion of fluids, bubbles, and droplets [69–71]. Ohta *et al.* [4, 70] demonstrated trapping and transport of air bubbles by laser heating a light-absorbing substrate (hydrogenated amorphous silicon a-Si:H) to induce thermocapillary forces. More details can be found in Chapter 10.

Liu *et al.* demonstrated optofluidic control of fluid in microchannels by laser heating suspended gold nanoparticles near liquid–air interface [72]. This technique can easily integrate microfluidic circuits for biomolecular and cellular processing without using any complex pump and valve devices [73, 74] or electrode patterning [75, 76]. This optofluidic control technique is different from Marangoni flow, where liquid is driven by surface tension gradient. Figure 9.17 illustrates such nanoparticles mediated optofluidic control. Laser heating the liquid–air interface creates water vapor that condenses and forms tiny water droplets near the interface. The sizes of these droplets grow and eventually coalesce with the original fluid body to drive the flow.

Zheng *et al.* have shown laser heating of a silver film to generate a vapor bubble and induce convective flows around it. This creates a circulation flow to concentrate microparticles to form a ring pattern on the silver film (Figure 9.18) [77]. By moving the laser beam, one can draw patterns formed by microparticles on the surface.

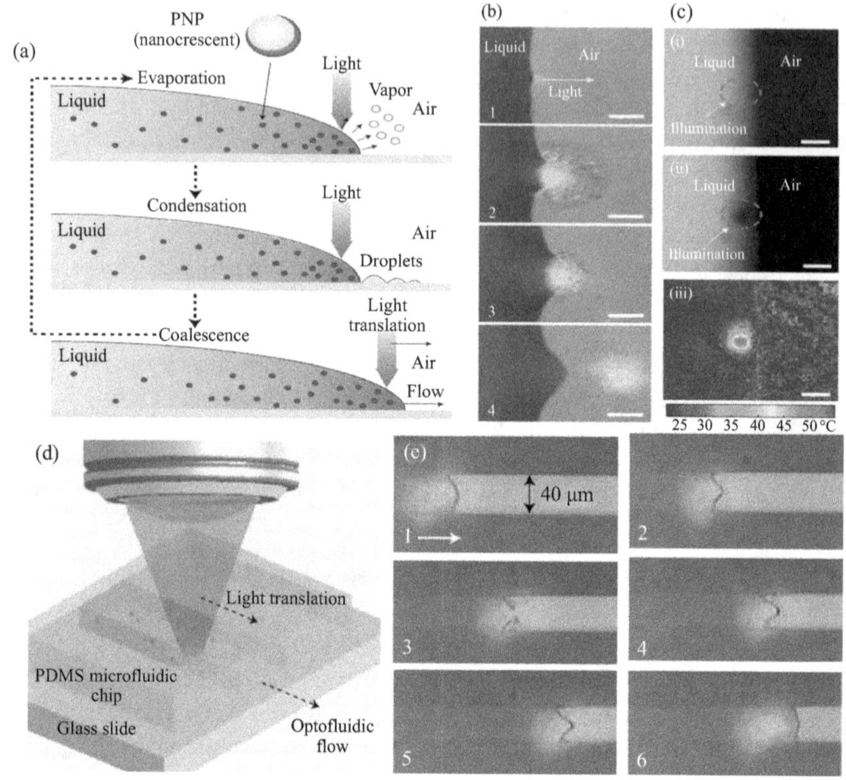

Figure 9.17 Photothermal nanoparticles (PNPs) activated optofluidic flow control. (a) Optically controlled fluid flow at liquid–air interface (b) Light-actuated fluid advance of the liquid–air interface of a 1 nM PNP water solution on a glass surface. (c) Temperature distribution in PNP-suspended liquid with a focused light beam illumination. (d) Illustration of optofluidic control inside a microfluidic channel. (e) Optofluidic control in a 40-μm-wide channel using 0.5 nM PNP-suspended 1× PBS buffer solution, and a 10 mW, 785 nm laser spot moving at a speed of ~50 μm/s. All scale bars are 10 μm. Reprinted with permission from Liu (2006) [72], © Nature Publishing Group

9.4.2 Photothermal-induced material change

Shirasaki *et al.* [78] and Ozaki *et al.* [79] reported microfluidic cell sorters with flow switching triggered by using themoreversible gelation polymer (TGP). TGP solution is in liquid form at low temperature but turns into gel upon heating. As shown in Figure 9.19, upon laser heating of the TGP solution filled channel to induce sol–gel transition, local flow resistance can be changed. This allows one to control the pathway of cells or particles flowing in the upstream. The response time of TGP is on the scale of milliseconds. A sorting throughput of 5 cells/s has been demonstrated.

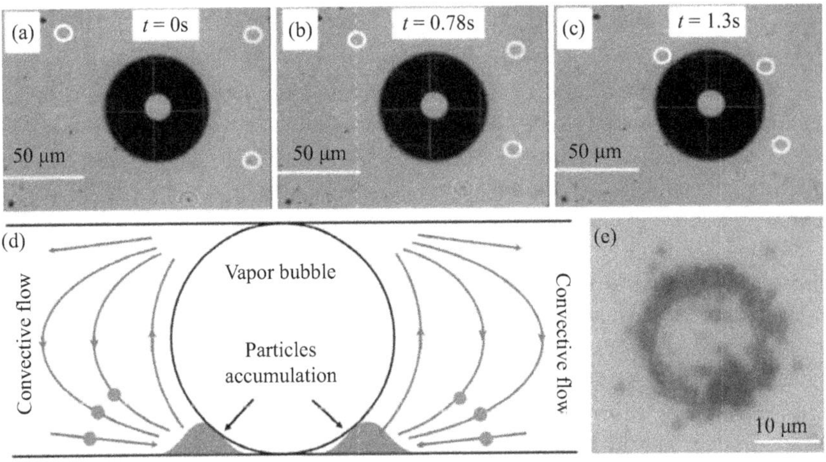

Figure 9.18 *Laser heating a silver film creates a microbubble and induces circulating convective flows to concentrate microparticles on the silver surface. Reprinted with permission from Zheng (2011) [77], © The Royal Society of Chemistry*

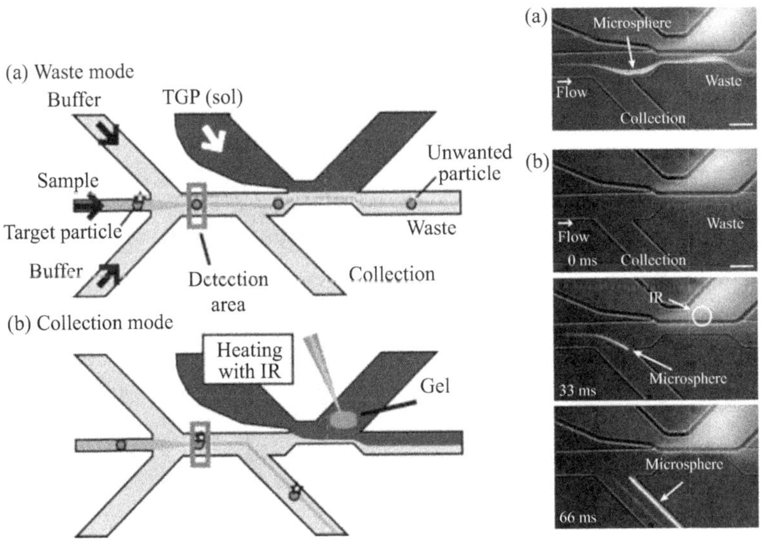

Figure 9.19 *(Left panel) Schematic of a fluorescence-activated cell sorter (FACS) realized by applying TGP. By using an IR (wavelength 1480 nm) laser to heat up the TGP solution to induce sol–gel transition and create a valve function, the downstream flow resistance is changed. The pathway of a target cell or a particle in the upstream can be controlled. (Right panel) The response time of TGP is in the ms scale. Reprinted with permission from Ozaki (2010) [79], © Elsevier*

9.4.3 Photothermal-driven electrokinetics

Increasing the temperature of fluid not only changes the surface tension at interfaces but also the physical properties of the bulk fluid such as the electrical conductivity and permittivity of fluid. If a uniform electric field is applied across a fluid with temperature gradient, net electrostatic forces can be generated to induce fluid flows through electrothermal mechanism. If a light pattern can be used to create temperature gradient profiles, one can control fluid flows using light beams.

William *et al.* demonstrated a rapid electrokinetic patterning (REP) mechanism that allows light beams to manipulate and patterning micro/nanoparticles in a fluid chamber sandwiched by two parallel flat electrodes biased with alternating current (AC) [5]. Figure 9.20(a) shows the schematic of an REP device where two conductive ITO glass substrates sandwiching microparticles suspended in deionized (DI) water. These two substrates are separated by a 50-μm spacer. When the device is under the illumination of an infrared (1064 nm) laser beam, the ITO substrate absorbs the laser beam and heats up at the illuminated area. The heat propagates into the neighboring fluid layer creating a temperature gradient. Under the application of AC electric field, microfluid vortices are created to circulate micro/nanoparticles to the center of the heated spot. The microparticles are trapped by dielectrophoretic forces near the electrode surface. Particles such as polystyrene beads and silica particles (300 nm to 3 μm) have been trapped. The applied AC signals are in the range of 1–100 kHz with 0–20 Vpp. The accumulated particles can be easily assembled, translated, and patterned on the ITO surface by changing the illumination light patterns.

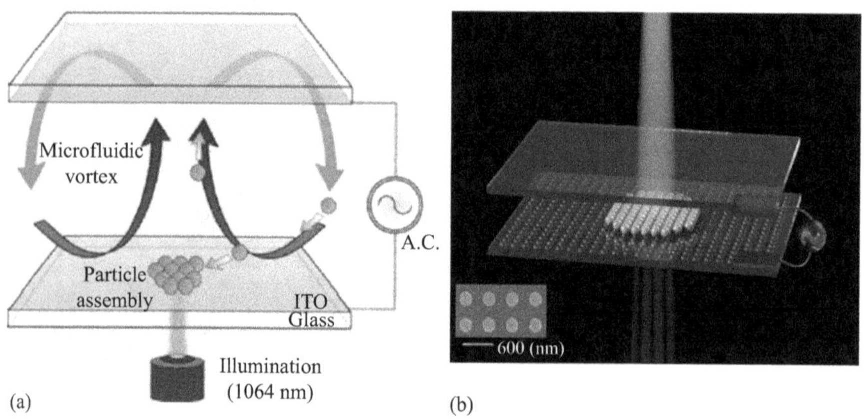

Figure 9.20 (a) A schematic illustration of electrokinetic patterning of colloidal particles with optical landscape. Reprinted with permission from Williams (2008) [5], © The Royal Society of Chemistry (b) Dynamic particle manipulation on a plasmonic nanostructure substrate (inset shows SEM image of the fabricated gold nanodisk array). Reprinted with permission from Ndukaife (2014) [80], © American Chemical Society

Ndukaife *et al.* [80] demonstrated an REP by utilizing gold nanostructures to more efficiently harvest light energy. Figure 9.20(b) shows the schematic of such a plasmonic nanostructure substrate for dynamic particle manipulation. A gold nanodisk array with 240 nm in diameter and 450 nm of lattice constant was fabricated on an ITO-coated glass substrate. Another ITO-coated glass substrate was placed on top and separated by a 90-μm spacer. An AC electric field is applied between these two ITO-coated substrates. Fluorescent polystyrene particles with 1 μm in diameter were suspended in DI water. Under light illumination, these gold nanodisks heat up, as well as the surrounding fluid. This creates local temperature gradient at the illuminated spot and induces microvortex flows to circulate microparticles onto the surface for trapping. When the applied AC signal frequency was reduced to 25 kHz, particles can be concentrated at the laser-illuminated spot to form particle cluster structures.

9.4.4 Cavitation bubble-driven ultrahigh-speed microfluidics

LIB is an efficient mechanism to convert light energy into mechanical energy. The induced cavitation bubbles can produce large mechanical forces in a short time. Through proper engineering design to fully utilize this cavitation mechanism, ultrafast microfluidic devices can be achieved. One application that needs ultrahigh-speed microfluidics is FACS.

Several microfluidic FACS (μFACS) devices and mechanisms have been proposed for sorting cells on a microfluidic chip in the past decade. Most of them aim to develop inexpensive, self-contained, and disposable devices [78, 81–88] for sorting cells and particles. At present, cell viability, purity, and throughput remain major technical barriers for μFACS. Prior μFACS systems have demonstrated sorting speed ranging from 25 cells/s to 1,500 cells/s using different mechanisms [88]. Such throughputs are orders of magnitude lower than conventional droplet-based FACS systems (20,000 cells/s). Despite the great advantage of sterile sorting provided by fully enclosed, disposable, microfluidic chips, their low throughputs greatly limit their applications. The major technical barrier is the lack of a high-speed microfluidic switch that can provide fast and effective switching in a highly viscous microfluidic environment. Small perturbation volume is required in order not to affect the detection and switching of neighboring cells, which is an important factor for achieving high purity.

Wu *et al.* demonstrated a high-speed microfluidic cell sorter that utilizes laser-induced cavitation bubble for cell sorting. This sorter is called pulse laser-activated cell sorter (PLACS) [89]. Figure 9.21 shows the PLACS design and sorting mechanism. The microfluidic chip consists of a main channel with two outlets, collection and waste. The pulsed channel runs in parallel with the main channel and is connected to the main channel through a straight narrow nozzle at the tip of the Y junction. Using hydrodynamic focusing, the input samples are focused to a narrow stream and flow into the left waste outlet. As a desired object flows through the fluorescence detection region upstream of the Y junction, a laser pulse is triggered and focused through a high NA objective lens into the pulsed channel after an

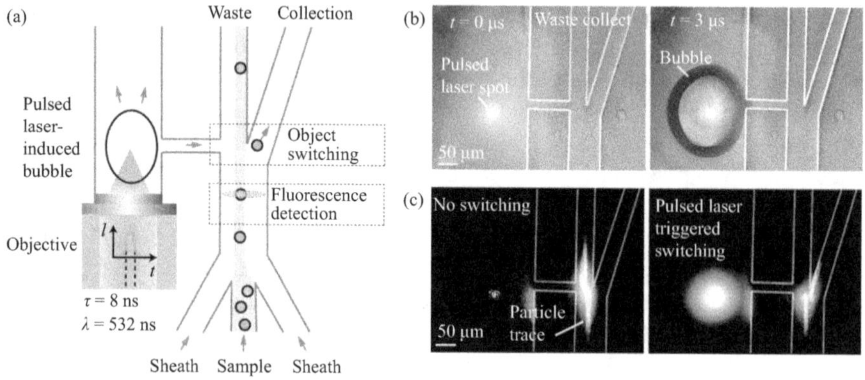

Figure 9.21 PLACS operation. (a) Schematic of the cell sorter. The sample flow is hydrodynamically focused into the waste channel. As the desired object flows through the fluorescence detection region upstream from the Y junction, a laser pulse is triggered and induces a cavitation bubble in the adjacent pulsed channel. The bubble expansion produces a high-speed liquid jet that deflects the desired sample toward the collection channel for sorting. (b) Images of the cavitation bubble generated by the focused pulsed laser beam in PLACS. (c) Fluorescent particle switching using PLACS. Reprinted with permission from Wu (2012) [89], © The Royal Society of Chemistry

optimal delay time. This induces an explosive cavitation bubble which expands and displaces the surrounding fluid creating a high-speed liquid jet through the nozzle into the main channel. This liquid jet deflects the desired object into the collection channel. A rapid switching cycle of 30 μs is demonstrated based on the expansion and collapse rate of a cavitation bubble. Using a nozzle to focus the deflecting liquid jet narrows the switching region in the sample stream and minimizes the disturbance range on neighboring objects for high throughput and high sorting purity.

To demonstrate mammalian cell sorting, human pre-B Nalm-6 or Ramos B lymphoma cells were stained and mixed with untreated cells at various ratios. Results showed at sorting speeds of 1500 cells/s, 98% sort purity, and ~90% cell viability were obtained for sorting fluorescent Nalm-6 cells, a performance comparable to similar experiments performed on a high-end commercial droplet-based FACS. For high-speed cell enrichment, Ramos cells were sorted at 20,000 cells/s. The obtained sort purity is 37.4%, corresponding to enrichment factors of 298.5. Under these sorting conditions, cells sorted into the collection or waste channels exhibited the same stress levels seen in unsorted cells, as measured by *HSPA6* and *FOS* stress–response gene expression [89]. These results showed that cavitation bubble triggered liquid jet switching did not exert additional stress on the cells being sorted. The reproducibility of PLACS sorting under repeated laser pulsing and bubble cycles was tested. It was verified that the microchannels remained intact and exhibited no burning or leakage after 100 million (10^8) actuations.

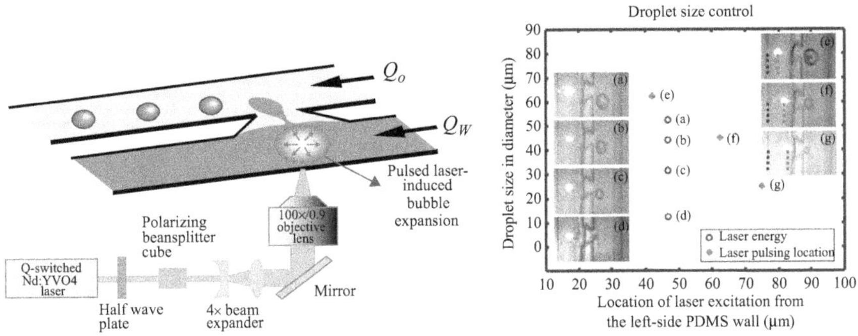

Figure 9.22 *Schematic of a pulsed laser-driven droplet generator. It consists of two microfluidic channels connected by a nozzle-like opening. Pulsed laser induces a rapidly expanding cavitation bubble to push water into the oil channel for droplet formation. Droplet volume is dependent upon the laser pulse energy and pulsing location. Reprinted with permission from Park (2011) [91], © The Royal Society of Chemistry*

Chen *et al.* further improved the sorting performance and sorting purity at high throughputs by introducing 3D hydrodynamic focusing in the upstream [90]. This allows a better synchronization of the cell arrival time and reduces the amount of liquid required to inject to steer target cells into the collection channel. An ultrafast switching with 20 μs complete on–off cycle was demonstrated. This 3D PLACS system has achieved a throughput of 23,000 cells/s with 90% purity in high-purity mode and 45,000 cells/s with 45% purity in the enrichment mode in one stage and in a single channel. PLACS has also been integrated with inertial microfluidic flows for 3D focusing which reduces the sample dilution effects from 3D hydrodynamic focusing [90].

Park *et al.* further demonstrated a high-speed droplet generation by laser-induced cavitation [91]. Figure 9.22 shows the schematic of a pulsed laser-driven droplet generation, which consists of two microfluidic channels (one for water and another for oil) connected by a nozzle-like opening. The stable water–oil interface is formed at the junction by adjusting the flow rates of water and oil. When a pulse laser focused in the water channel near the nozzle, it triggers a cavitation bubble expanding rapidly to perturb the oil–water interface and push the neighboring water into the oil channel to form an aqueous droplet. By using a pulse laser with a repetition rate of 10,000 pulses/s, high-speed droplet generation at a speed of 10,000 droplets on demand has been accomplished. Injected droplet volumes can be continuously tuned between 1 pL and 150 pL with less than 1% volume variation.

9.4.5 *Cavitation bubble-based fluid pumping*

Laser-induced cavitation bubbles expand and collapse symmetrically if there is no boundary near these bubbles to alter the hydrodynamic flows. However, if there is a

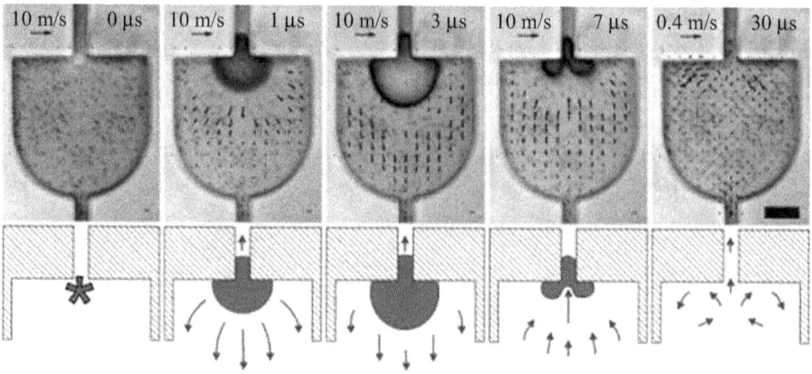

Figure 9.23 High-speed movie frames showing different stages of the bubble growth and collapse. Reprinted with permission from Dijkink (2008) [92], © The Royal Society of Chemistry

structure nearby a bubble, net fluidic flows can be generated by properly designing the boundary structure. Dijkink *et al.* demonstrated a laser-induced cavitation-based micropump [92] using this concept. Shown in Figure 9.23 is a single-layer PDMS microfluidic structure with a narrow channel connecting with a wide reservoir. When a laser pulse is focused near the entrance of the channel, a cavitation bubble is triggered. Due to the non-symmetrical structure around the bubble, the bubble expansion and collapse pattern are significantly affected. During the bubble collapse process, more fluid is attracted from the wide chamber side and results in a net fluid flow pumping into the channel. It has been shown that such mechanism can pump 4000 μm^3 volume within 75 μs against 3 bar pressure head. The optimum pumping was achieved when the distance between the channel inlet and the laser focus is about 80% of the maximum bubble radius. For continuous pumping, a pulse laser repetition rate of 5 Hz is applied to obtain repeatable operation without residual gas bubble formation.

Chen *et al.* also demonstrated another microfluidic pump using scanning laser pulses to sequentially drive the deforming PDMS membrane-based peristaltic pump [18]. In this device, cavitation bubbles were triggered in a separate channel from the sample channel to avoid contamination such as heat, dye, reactive chemicals induced during laser cavitation process. Figure 9.24 shows the schematic of this pulse laser-driven peristaltic membrane pump where bottom channel is about 75 μm in height, 500 μm in width and the top channel is about 16 μm in height and 100 μm in width. The sample channel was separated by a 50-μm-thick membrane to avoid contamination. The pulse laser was focused at the middle of the control channel to induce cavitation bubble under membrane valves. Laser pulses sequentially scan and deform linearly arranged membrane valves. Pulsed laser with 100 μJ energy can generate 370-μm bubble. This cavitation bubble can generate sufficient force to deform membrane for fluid pumping. The maximum flow rate obtained was 460 pL/s at a 557 Hz scanning frequency.

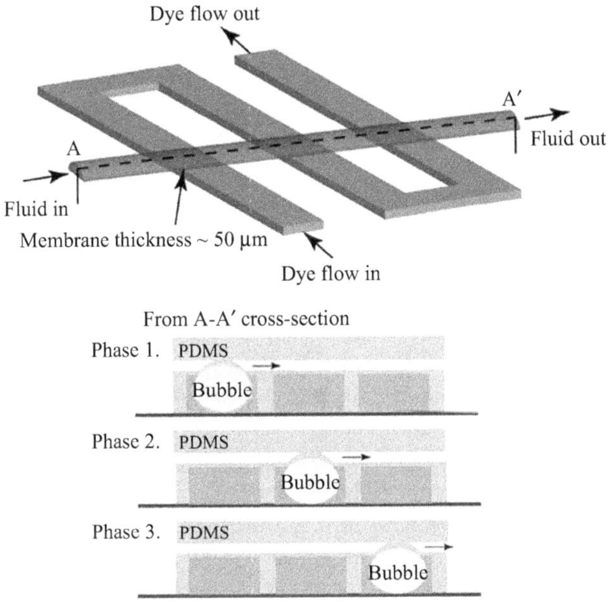

Figure 9.24 Scanning laser pulses-driven peristaltic pump. Cavitation bubbles induced by LIB can produce large enough force to deform the PDMS membrane to seal the top channel. By sequentially deforming the membranes, fluid can be squeezed forward. Reprinted with permission from Chen (2012) [18], © The Royal Society of Chemistry

References

[1] O. M. Marago, P. H. Jones, P. G. Gucciardi, G. Volpe, and A. C. Ferrari, "Optical trapping and manipulation of nanostructures," *Nature Nanotechnology*, vol. 8, pp. 807–819, Nov 2013.

[2] D. G. Grier, "A revolution in optical manipulation," *Nature*, vol. 424, pp. 810–816, Aug 14, 2003.

[3] A. Vogel, J. Noack, G. Huttman, and G. Paltauf, "Mechanisms of femtosecond laser nanosurgery of cells and tissues," *Applied Physics B – Lasers and Optics*, vol. 81, pp. 1015–1047, Dec 2005.

[4] A. T. Ohta, A. Jamshidi, J. K. Valley, H. Y. Hsu, and M. C. Wu, "Optically actuated thermocapillary movement of gas bubbles on an absorbing substrate," *Applied Physics Letters*, vol. 91, Aug 13, 2007.

[5] S. J. Williams, A. Kumar, and S. T. Wereley, "Electrokinetic patterning of colloidal particles with optical landscapes," *Lab on a Chip*, vol. 8, pp. 1879–1882, 2008.

[6] V. P. Zharov, R. R. Letfullin, and E. N. Galitovskaya, "Microbubbles-overlapping mode for laser killing of cancer cells with absorbing

nanoparticle clusters," *Journal of Physics D – Applied Physics*, vol. 38, pp. 2571–2581, Aug 7, 2005.

[7] C.-C. Chen, Y.-P. Lin, C.-W. Wang, *et al.*, "DNA-gold nanorod conjugates for remote control of localized gene expression by near infrared irradiation," *Journal of the American Chemical Society*, vol. 128, pp. 3709–3715, 2006.

[8] X. H. Huang, I. H. El-Sayed, W. Qian, and M. A. El-Sayed, "Cancer cell imaging and photothermal therapy in the near-infrared region by using gold nanorods," *Journal of the American Chemical Society*, vol. 128, pp. 2115–2120, Feb 15, 2006.

[9] S. T. Wang, K. J. Chen, T. H. Wu, *et al.*, "Photothermal effects of supramolecularly assembled gold nanoparticles for the targeted treatment of cancer cells," *Angewandte Chemie – International Edition*, vol. 49, pp. 3777–3781, 2010.

[10] V. Kotaidis, C. Dahmen, G. von Plessen, F. Springer, and A. Plech, "Excitation of nanoscale vapor bubbles at the surface of gold nanoparticles in water," *Journal of Chemical Physics*, vol. 124, May 14, 2006.

[11] X. H. Huang, P. K. Jain, I. H. El-Sayed, and M. A. El-Sayed, "Determination of the minimum temperature required for selective photothermal destruction of cancer cells with the use of immunotargeted gold nanoparticles," *Photochemistry and Photobiology*, vol. 82, pp. 412–417, Mar–Apr 2006.

[12] D. P. O'neal, L. R. Hirsch, N. J. Halas, J. D. Payne, and J. L. West, "Photothermal tumor ablation in mice using near infrared-absorbing nanoparticles," *Cancer Letters*, vol. 209, pp. 171–176, Jun 25, 2004.

[13] V. P. Zharov, K. E. Mercer, E. N. Galitovskaya, and M. S. Smeltzer, "Photothermal nanotherapeutics and nanodiagnostics for selective killing of bacteria targeted with gold nanoparticles," *Biophysical Journal*, vol. 90, pp. 619–627, Jan 2006.

[14] D. O. Lapotko, E. Lukianova, and A. A. Oraevsky, "Selective laser nanothermolysis of human leukemia cells with microbubbles generated around clusters of gold nanoparticles," *Lasers in Surgery and Medicine*, vol. 38, pp. 631–642, Jul 2006.

[15] R. R. Gattass and E. Mazur, "Femtosecond laser micromachining in transparent materials," *Nature Photonics*, vol. 2, pp. 219–225, Apr 2008.

[16] A. Vogel and V. Venugopalan, "Mechanisms of pulsed laser ablation of biological tissues," *Chemical Reviews*, vol. 103, pp. 577–644, Feb 2003.

[17] K. R. Rau, P. A. Quinto-Su, A. N. Hellman, and V. Venugopalan, "Pulsed laser microbeam-induced cell lysis: Time-resolved imaging and analysis of hydrodynamic effects," *Biophysical Journal*, vol. 91, pp. 317–329, Jul 2006.

[18] Y. Chen, T.-H. Wu, and P.-Y. Chiou, "Scanning laser pulses driven microfluidic peristaltic membrane pump," *Lab on a Chip*, vol. 12, pp. 1771–1774, 2012.

[19] A. Vogel, N. Linz, S. Freidank, and G. Paltauf, "Femtosecond-laser-induced nanocavitation in water: Implications for optical breakdown threshold and cell surgery," *Physical Review Letters*, vol. 100, Jan 25, 2008.

[20] K. Konig, I. Riemann, and W. Fritzsche, "Nanodissection of human chromosomes with near-infrared femtosecond laser pulses," *Optics Letters*, vol. 26, pp. 819–821, Jun 1, 2001.

[21] N. Shen, D. Datta, C. B. Schaffer, P. LeDuc, D. E. Ingber, and E. Mazur, "Ablation of cytoskeletal filaments and mitochondria in live cells using a femtosecond laser nanoscissor," *Mechanics & Chemistry of Biosystems*, vol. 2, pp. 17–25, 2005.

[22] T. Shimada, W. Watanabe, S. Matsunaga, *et al.*, "Intracellular disruption of mitochondria in a living HeLa cell with a 76-MHz femtosecond laser oscillator," *Optics Express*, vol. 13, pp. 9869–9880, Nov 28, 2005.

[23] A. Heisterkamp, I. Z. Maxwell, E. Mazur, *et al.*, "Pulse energy dependence of subcellular dissection by femtosecond laser pulses," *Optics Express*, vol. 13, pp. 3690–3696, May 16, 2005.

[24] M. F. Yanik, H. Cinar, H. N. Cinar, A. D. Chisholm, Y. S. Jin, and A. Ben-Yakar, "Neurosurgery – Functional regeneration after laser axotomy," *Nature*, vol. 432, pp. 822–822, Dec 16, 2004.

[25] U. K. Tirlapur and K. Konig, "Cell biology – Targeted transfection by femtosecond laser," *Nature*, vol. 418, pp. 290–291, Jul 18, 2002.

[26] D. Stevenson, B. Agate, X. Tsampoula, *et al.*, "Femtosecond optical transfection of cells: Viability and efficiency," *Optics Express*, vol. 14, pp. 7125–7133, Aug 7, 2006.

[27] J. Baumgart, W. Bintig, A. Ngezahayo, *et al.*, "Quantified femtosecond laser based opto-perforation of living GFSHR-17 and MTH53a cells," *Optics Express*, vol. 16, pp. 3021–3031, Mar 3, 2008.

[28] J. X. Gong, X. M. Zhao, Q. R. Xing, *et al.*, "Femtosecond laser-induced cell fusion," *Applied Physics Letters*, vol. 92, Mar 3, 2008.

[29] K. Kuetemeyer, A. Lucas-Hahn, B. Petersen, H. Niemann, and A. Heisterkamp, "Femtosecond laser-induced fusion of nonadherent cells and two-cell porcine embryos," *Journal of Biomedical Optics*, vol. 16, Aug 2011.

[30] C. T. A. Brown, D. J. Stevenson, X. Tsampoula, *et al.*, "Enhanced operation of femtosecond lasers and applications in cell transfection," *Journal of Biophotonics*, vol. 1, pp. 183–199, Aug 2008.

[31] S. K. Mohanty, K. S. Mohanty, and M. W. Berns, "Manipulation of mammalian cells using a single-fiber optical microbeam," *Journal of Biomedical Optics*, vol. 13, Sep–Oct 2008.

[32] X. Kong, S. K. Mohanty, J. Stephens, *et al.*, "Comparative analysis of different laser systems to study cellular responses to DNA damage in mammalian cells," *Nucleic Acids Research*, vol. 37, May 2009.

[33] U. K. Tirlapur, K. Konig, C. Peuckert, R. Krieg, and K. J. Halbhuber, "Femtosecond near-infrared laser pulses elicit generation of reactive oxygen species in mammalian cells leading to apoptosis-like death," *Experimental Cell Research*, vol. 263, pp. 88–97, Feb 1, 2001.

[34] H. He, K. T. Chan, S. K. Kong, and R. K. Y. Lee, "Mechanism of oxidative stress generation in cells by localized near-infrared femtosecond laser excitation," *Applied Physics Letters*, vol. 95, Dec 7, 2009.

[35] J. Yoon, S. W. Ryu, S. Lee, and C. Choi, "Cytosolic irradiation of femtosecond laser induces mitochondria-dependent apoptosis-like cell death via intrinsic reactive oxygen cascades," *Scientific Reports*, vol. 5, Feb 4, 2015.

[36] A. Vogel, S. Busch, K. Jungnickel, and R. Birngruber, "Mechanisms of intraocular photodisruption with picosecond and nanosecond laser-pulses," *Lasers in Surgery and Medicine*, vol. 15, pp. 32–43, 1994.

[37] A. M. Cotliar, H. D. Schubert, E. R. Mandel, and S. L. Trokel, "Excimer laser radial keratotomy," *Ophthalmology*, vol. 92, pp. 206–208, 1985.

[38] V. Venugopalan, A. Guerra, K. Nahen, and A. Vogel, "Role of laser-induced plasma formation in pulsed cellular microsurgery and micromanipulation," *Physical Review Letters*, vol. 88, Feb 18, 2002.

[39] A. Vogel, S. Busch, and U. Parlitz, "Shock wave emission and cavitation bubble generation by picosecond and nanosecond optical breakdown in water," *Journal of the Acoustical Society of America*, vol. 100, pp. 148–165, Jul 1996.

[40] R. Dijkink, S. Le Gac, E. Nijhuis, *et al.*, "Controlled cavitation-cell interaction: Trans-membrane transport and viability studies," *Physics in Medicine and Biology*, vol. 53, pp. 375–390, Jan 21, 2008.

[41] J. S. Soughayer, T. Krasieva, S. C. Jacobson, J. M. Ramsey, B. J. Tromberg, and N. L. Allbritton," Characterization of cellular optoporation with distance," *Analytical Chemistry*, vol. 72, pp. 1342–1347, Mar 15, 2000.

[42] A. N. Hellman, K. R. Rau, H. H. Yoon, and V. Venugopalan, "Biophysical response to pulsed laser microbeam-induced cell lysis and molecular delivery," *Journal of Biophotonics*, vol. 1, pp. 24–35, Feb 2008.

[43] I. B. Clark, E. G. Hanania, J. Stevens, *et al.*, "Optoinjection for efficient targeted delivery of a broad range of compounds and macromolecules into diverse cell types," *Journal of Biomedical Optics*, vol. 11, Jan–Feb 2006.

[44] M. R. Koller, E. G. Hanania, J. Stevens, *et al.*, "High-throughput laser-mediated in situ cell purification with high purity and yield," *Cytometry Part A*, vol. 61A, pp. 153–161, Oct 2004.

[45] M. Y. He, J. S. Edgar, G. D. M. Jeffries, R. M. Lorenz, J. P. Shelby, and D. T. Chiu, "Selective encapsulation of single cells and subcellular organelles into picoliter- and femtoliter-volume droplets," *Analytical Chemistry*, vol. 77, pp. 1539–1544, Mar 15, 2005.

[46] M. Terakawa, S. Sato, H. Ashida, *et al.*, "In vitro gene transfer to mammalian cells by the use of laser-induced stress waves: Effects of stress wave parameters, ambient temperature, and cell type," *Journal of Biomedical Optics*, vol. 11, Jan–Feb 2006.

[47] A. Vogel, K. Lorenz, V. Horneffer, G. Huttmann, D. von Smolinski, and A. Geberty, "Mechanisms of laser-induced dissection and transport of histologic specimens," *Biophysical Journal*, vol. 93, pp. 4481–4500, Dec 15, 2007.

[48] V. Horneffer, N. Linz, and A. Vogel, "Principles of laser-induced separation and transport of living cells," *Journal of Biomedical Optics*, vol. 12, Sep–Oct 2007.

[49] H. Wang, D. W. Brandl, P. Nordlander, and N. J. Halas, "Plasmonic nanostructures: Artificial molecules," *Accounts of Chemical Research*, vol. 40, pp. 53–62, Jan 2007.

[50] E. A. Sykes, J. Chen, G. Zheng, and W. C. W. Chan, "Investigating the impact of nanoparticle size on active and passive tumor targeting efficiency," *ACS Nano*, vol. 8, pp. 5696–5706, Jun 2014.

[51] B. D. Chithrani, A. A. Ghazani, and W. C. W. Chan, "Determining the size and shape dependence of gold nanoparticle uptake into mammalian cells," *Nano Letters*, vol. 6, pp. 662–668, Apr 2006.

[52] J. W. Shao, R. J. Griffin, E. I. Galanzha, *et al.*, "Photothermal nanodrugs: Potential of TNF-gold nanospheres for cancer theranostics," *Scientific Reports*, vol. 3, Feb 27, 2013.

[53] T. H. Wu, S. Kalim, C. Callahan, M. A. Teitell, and P. Y. Chiou, "Image patterned molecular delivery into live cells using gold particle coated substrates," *Optics Express*, vol. 18, pp. 938–946, Jan 18, 2010.

[54] R. Riahi, J. Sun, S. Wang, M. Long, D. D. Zhang, and P. K. Wong, "Notch1-Dll4 signalling and mechanical force regulate leader cell formation during collective cell migration," *Nature Communications*, vol. 6, p. 6556, 2015.

[55] J. Kim, S. Park, J. E. Lee, *et al.*, "Designed fabrication of multifunctional magnetic gold nanoshells and their application to magnetic resonance imaging and photothermal therapy," *Angewandte Chemie – International Edition*, vol. 45, pp. 7754–7758, 2006.

[56] L. Cheng, K. Yang, Y. Li, *et al.*, "Multifunctional nanoparticles for upconversion luminescence/MR multimodal imaging and magnetically targeted photothermal therapy," *Biomaterials*, vol. 33, pp. 2215–2222, Mar 2012.

[57] M. Q. Chu, X. J. Pan, D. Zhang, Q. Wu, J. L. Peng, and W. X. Hai, "The therapeutic efficacy of CdTe and CdSe quantum dots for photothermal cancer therapy," *Biomaterials*, vol. 33, pp. 7071–7083, Oct 2012.

[58] C. P. Yao, R. Rahmanzadeh, E. Endl, Z. X. Zhang, J. Gerdes, and G. Huttmann, "Elevation of plasma membrane permeability by laser irradiation of selectively bound nanoparticles," *Journal of Biomedical Optics*, vol. 10, Nov–Dec 2005.

[59] D. Heinemann, M. Schomaker, S. Kalies, *et al.*, "Gold nanoparticle mediated laser transfection for efficient siRNA mediated gene knock down," *PLoS ONE*, vol. 8, Mar 11, 2013.

[60] T. H. Wu, T. Teslaa, S. Kalim, *et al.*, "Photothermal nanoblade for large cargo delivery into mammalian cells," *Analytical Chemistry*, vol. 83, pp. 1321–1327, Feb 15, 2011.

[61] Y. C. Wu, T. H. Wu, D. L. Clemens, *et al.*, "Massively parallel delivery of large cargo into mammalian cells with light pulses," *Nature Methods*, vol. 12, pp. 439–444, Apr 6, 2015.

[62] S. E. Lee and L. P. Lee, "Nanoplasmonic gene regulation," *Current Opinion in Chemical Biology*, vol. 14, pp. 623–633, Oct 2010.

[63] P. Anikeeva, and K. Deisseroth, "Photothermal genetic engineering," *ACS Nano*, vol. 6, pp. 7548–7552, 2012.

[64] S. E. Lee, G. L. Liu, F. Kim, and L. P. Lee, "Remote optical switch for localized and selective control of gene interference," *Nano Letters*, vol. 9, pp. 562–570, 2009.

[65] R. Riahi, S. Wang, M. Long, N. Li, P.-Y. Chiou, D. D. Zhang, and P. K. Wong, "Mapping photothermally induced gene expression in living cells and tissues by nanorod-locked nucleic Acid complexes," *ACS Nano*, vol. 8, pp. 3597–3605, 2014.

[66] J. Lee, H. Moon, J. Fowler, T. Schoellhammer, and C. J. Kim, "Electrowetting and electrowetting-on-dielectric for microscale liquid handling," *Sensors and Actuators A – Physical*, vol. 95, pp. 259–268, Jan 1, 2002.

[67] S. K. Cho, S. K. Fan, H. J. Moon, and C. J. Kim, "Towards digital microfluidic circuits: Creating, transporting, cutting and merging liquid droplets by electrowetting-based actuation," *Fifteenth IEEE International Conference on Micro Electro Mechanical Systems, Technical Digest*, pp. 32–35, 2002.

[68] N. O. Young, J. S. Goldstein, and M. J. Block, "The motion of bubbles in a vertical temperature gradient," *Journal of Fluid Mechanics*, vol. 6, pp. 350–356, 1959.

[69] D. E. Kataoka and S. M. Troian, "Patterning liquid flow on the microscopic scale," *Nature*, vol. 402, pp. 794–797, Dec 16, 1999.

[70] T. Sakata, H. Togo, M. Makihara, F. Shimokawa, and K. Kaneko, "Improvement of switching time in a thermocapillarity optical switch," *Journal of Lightwave Technology*, vol. 19, pp. 1023–1027, Jul 2001.

[71] K. T. Kotz, K. A. Noble, and G. W. Faris, "Optical microfluidics," *Applied Physics Letters*, vol. 85, pp. 2658–2660, Sep 27, 2004.

[72] G. L. Liu, J. Kim, Y. Lu, and L. P. Lee, "Optofluidic control using photothermal nanoparticles," *Nature Materials*, vol. 5, pp. 27–32, 2006.

[73] C. M. Ho and Y. C. Tai, "Micro-electro-mechanical-systems (MEMS) and fluid flows," *Annual Review of Fluid Mechanics*, vol. 30, pp. 579–612, 1998.

[74] S. L. Neale, M. P. Macdonald, K. Dholakia, and T. F. Krauss, "All-optical control of microfluidic components using form birefringence," *Nature Materials*, vol. 4, pp. 530–533, Jul 2005.

[75] D. J. Harrison, K. Fluri, K. Seiler, Z. Fan, C. S. Effenhauser, and A. Manz, "Micromachining a miniaturized capillary electrophoresis-based chemical analysis system on a chip," *Science*, vol. 261, pp. 895–897, Aug 13, 1993.

[76] M. A. Burns, C. H. Mastrangelo, T. S. Sammarco, *et al.*, "Microfabricated structures for integrated DNA analysis," *Proceedings of the National Academy of Sciences of the United States of America*, vol. 93, pp. 5556–5561, May 28, 1996.

[77] Y. J. Zheng, H. Liu, Y. Wang, *et al.*, "Accumulating microparticles and direct-writing micropatterns using a continuous-wave laser-induced vapor bubble," *Lab on a Chip*, vol. 11, pp. 3816–3820, 2011.

[78] Y. Shirasaki, J. Tanaka, H. Makazu, *et al.*, "On-chip cell sorting system using laser-induced heating of a thermoreversible gelation polymer to control flow," *Analytical Chemistry*, vol. 78, pp. 695–701, Feb 2006.

[79] K. Ozaki, H. Sugino, Y. Shirasaki, *et al.*, "Microfluidic cell sorter with flow switching triggered by a sol–gel transition of a thermo-reversible gelation polymer," *Sensors and Actuators B: Chemical*, vol. 150, pp. 449–455, Sep 21, 2010.

[80] J. C. Ndukaife, A. Mishra, U. Guler, A. G. A. Nnanna, S. T. Wereley, and A. Boltasseva, "Photothermal heating enabled by plasmonic nanostructures for electrokinetic manipulation and sorting of particles," *ACS Nano*, vol. 8, pp. 9035–9043, Sep 23, 2014.

[81] R. W. Applegate, J. Squier, T. Vestad, *et al.*, "Microfluidic sorting system based on optical waveguide integration and diode laser bar trapping," *Lab on a Chip*, vol. 6, pp. 422–426, Mar 2006.

[82] A. Y. Fu, H. P. Chou, C. Spence, F. H. Arnold, and S. R. Quake, "An integrated microfabricated cell sorter," *Analytical Chemistry*, vol. 74, pp. 2451–2457, Jun 2002.

[83] A. Y. Fu, C. Spence, A. Scherer, F. H. Arnold, and S. R. Quake, "A microfabricated fluorescence-activated cell sorter," *Nature Biotechnology*, vol. 17, pp. 1109–1111, Nov 1999.

[84] C. T. Ho, R. Z. Lin, H. Y. Chang, and C. H. Liu, "Micromachined electrochemical T-switches for cell sorting applications," *Lab on a Chip*, vol. 5, pp. 1248–1258, 2005.

[85] D. Holmes, J. K. She, P. L. Roach, and H. Morgan, "Bead-based immunoassays using a micro-chip flow cytometer," *Lab on a Chip*, vol. 7, pp. 1048–1056, 2007.

[86] N. Idota, A. Kikuchi, J. Kobayashi, K. Sakai, and T. Okano, "Microfluidic valves comprising nanolayered thermoresponsive polymer-grafted capillaries," *Advanced Materials*, vol. 17, pp. 2723–2727, Nov 2005.

[87] Y. Kim, S. Hong, S. H. Lee, *et al.*, "Novel platform for minimizing cell loss on separation process: Droplet-based magnetically activated cell separator," *Review of Scientific Instruments*, vol. 78, Jul 2007.

[88] M. M. Wang, E. Tu, D. E. Raymond, *et al.*, "Microfluidic sorting of mammalian cells by optical force switching," *Nature Biotechnology*, vol. 23, pp. 83–87, Jan 2005.

[89] T.-H. Wu, Y. Chen, S.-Y. Park, *et al.*, "Pulsed laser triggered high speed microfluidic fluorescence activated cell sorter," *Lab on a Chip*, vol. 12, pp. 1378–1383, 2012.

[90] Y. Chen, A. J. Chung, T.-H. Wu, M. A. Teitell, D. Di Carlo, and P.-Y. Chiou, "Pulsed laser activated cell sorting with three dimensional sheathless inertial focusing," *Small*, vol. 10, pp. 1746–1751, 2014.

[91] S.-Y. Park, T.-H. Wu, Y. Chen, M. A. Teitell, and P.-Y. Chiou, "High-speed droplet generation on demand driven by pulse laser-induced cavitation," *Lab on a Chip*, vol. 11, pp. 1010–1012, 2011.

[92] R. Dijkink and C.-D. Ohl, "Laser-induced cavitation based micropump," *Lab on a Chip*, vol. 8, pp. 1676–1681, 2008.

Chapter 10

Optical manipulation for biomedical applications

Wenqi Hu, Qihui Fan and Aaron T. Ohta

10.1 Introduction

The ability to manipulate micro- and nanoscale biological objects is an asset for microbiology and many aspects of biomedical research. While there are a variety of ways to address particles in these size scales, one of the most attractive methods is optical manipulation. The manipulation of biological objects, especially live cells, primarily occurs in aqueous media. Thus, optical manipulation for biomedical applications is intimately integrated with fluidics. This has given rise to the term "optofluidics" [1–5], which covers light interacting or integrated with fluidics, for both biological and nonbiological applications. Here, optical manipulation and optofluidics is reviewed in the context of biomedical applications, so topics that have primarily engineering applications are omitted. The reader is instead referred to References [3] and [5] for an overview of these types of optofluidic applications. Optical manipulation in the context of this chapter includes the use of optically controlled phenomena, such as optically induced electrokinctic forces, as well as purely optical forces, such as optical gradient and optical scattering forces.

One advantage of optical manipulation is its flexibility. Unlike mechanical micromanipulators or other microfabricated structures, the manipulators in an optical system are made of photons and can thus be reconfigured on the fly. Optical manipulation patterns can be dynamic, to enable transport or sorting functions, and can be easily adapted and customized to a specific application or experiment. It is straightforward to use optical patterns to address specific spatial locations with submicron resolution. Beam-shaping technology can also be used to generate complex optical patterns, making it possible to address many spatial locations in parallel. It is also possible to integrate optical manipulation with microfabricated structures such as microfluidic channels, either through the use of transparent materials or the addition of optical waveguides. Furthermore, dynamic optical patterns can be programmed, making it possible to automate complex operations.

University of Hawaii, Honolulu, Hawaii, USA

Since the use of optical patterns can enable much functionality, the accompanying microfluidic devices can often be made relatively simple, reducing fabrication costs. For biomedical applications, microfluidic devices are often designed to be consumable to reduce the possibility of sample cross-contamination, so reducing device production costs results in lower operating costs.

There are some limitations to the use of optical manipulation. An obstacle to deploying any optical system is the setup associated with it; optical systems typically require skilled technicians to set them up, and are relatively expensive. Commercial optical systems have preassembled optical components, but have increased cost and space requirements. This chapter will discuss some optical manipulation systems designed specifically to address these limitations, using lower cost, compact components. Another concern for biological applications is the effect of light upon the objects under manipulation; some types of optical manipulation require highly intense monochromatic light, which can affect cells [6–9]. However, not all methods of optical manipulation require such intense light. Finally, the area over which optical manipulators can operate is constrained by the optical setup. Although some types of optical manipulators can only operate over an area of a fraction of a square millimeter, other optical systems can operate over hundreds of square millimeters.

Despite these limitations, the advantages of optical manipulation are compelling enough to make this an attractive technology for single-cell analysis. There are several functions required in a complete single-cell analysis process, all of which can be realized in an optical system (Figure 10.1). First, cells of interest must be

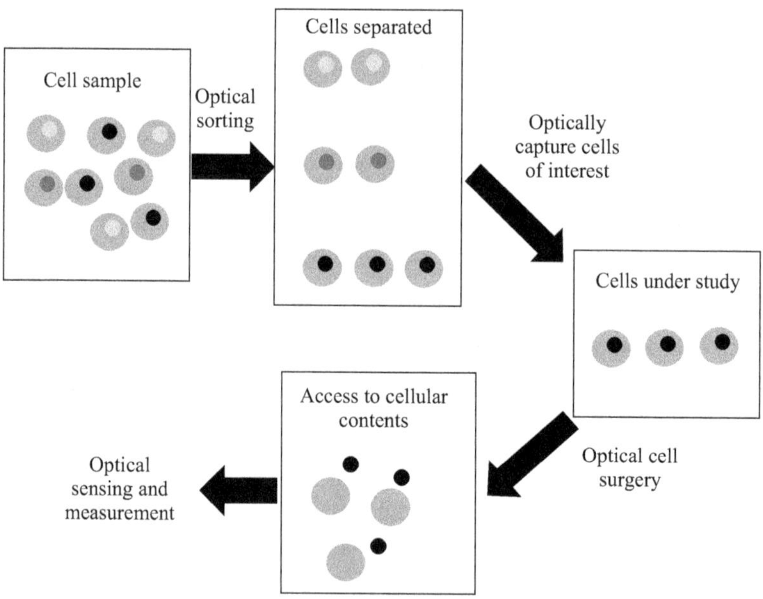

Figure 10.1 Optical manipulation for single-cell analysis. All of the actions necessary for this process can be accomplished by optical means

sorted from a larger population. Then, specific cells of interest are trapped, transported, and isolated. The further study of these cells may then require cellular surgery or cellular lysis. Finally, the cells or the contents of the cells can be measured and quantified. This chapter will focus on optical manipulation in the context of cell trapping, transportation, and sorting. It will also touch upon some functions in cell surgery and lysis. Optical cell sensing and measurement are covered in Chapters 11 and 12.

This chapter will first cover the use of optical gradient forces for the trapping of cells and other biological objects. This method is also widely known as optical tweezers (OT). Other types of direct optical manipulation will be discussed, such as optical cell sorters and optically driven fluids, droplets, and bubbles. Indirect optical manipulation via optically induced electrokinetic forces will be reviewed, in particular, optically induced dielectrophoresis (ODEP). The chapter will also discuss the use of optical manipulation for cell poration, lysis, and surgery. The chapter will finish with a brief discussion of future directions in this field.

10.2 Optical tweezers (OT)

OT uses the force created from photons impinging on a dielectric particle to create traps with strengths ranging from femtonewtons to hundreds of piconewtons. An OT trap can achieve a sub-nanometer spatial resolution, and can trap objects from few nanometers up to tens of micrometers, including many biomolecules, cell organelles, mammalian cells, and microorganism. Compared with other microscale manipulation and measurement tools, such as atomic force microscopy and magnetic tweezers [10], OT can achieve full 3D trapping and control of particles under manipulation.

A typical single-beam OT trap is formed by focusing a laser beam with a Gaussian intensity profile through an objective lens (Figure 10.2). The conservation of photon momentum in this situation gives rise to two major forces: radiation pressure and gradient force [11–13]. The radiation pressure is due to the reflection and absorption of the incoming beam by the particle, and it pushes the particle along the propagation of the beam. The gradient force is due to the refraction of the light as it passes through the particle, and this force pulls the particle toward the highest light intensity region, the focal point [14]. If the laser beam is not strongly focused, the gradient force is significant only in the transverse plane, and it serves to confine the particle to a position in or near the laser beam. However, the particle can still be pushed by the radiation pressure along the optical axis, so true 3D trapping is not achieved. The axial gradient force does become significant in tightly focused laser beam, and is now strong enough to pull the particle toward the focal point, against the radiation pressure. Thus, a particle in a single-beam OT trap is stably trapped at a point slightly behind the focal point, along the direction of light propagation.

The strength of the radiation pressure and the gradient force can be described by Rayleigh scattering theory, where the particle is treated as a dipole far smaller than the light wavelength [12], or Mie scattering theory, which covers micron-sized

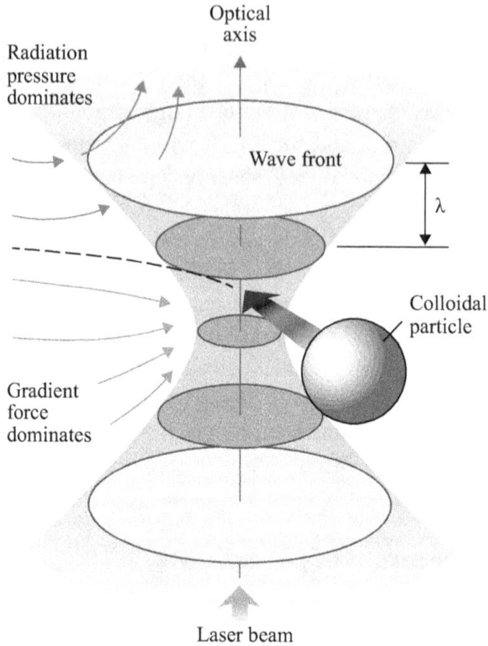

Figure 10.2 A single-beam OT trap. Reprinted with permission from Grier (2003) [15], © Nature Publishing Group

dielectric particles such as living cells and organelles [13]. Regardless, higher optical power results in higher trapping forces, while the index of refraction of the particle to relative to the media dictates the direction of the force. If the particle has a larger index of refraction, it will tend to move toward the region of highest light intensity and be stably trapped in a single-beam OT trap. OT is usually used with particles in a liquid medium, as the liquid provides both buoyancy and damping force for stable particle trapping. The most popular model to describe a particle in an OT trap is a spring harmonic oscillator, described by Hooke's Law [12]. The force of the trap can be determined by multiplying the displacement of bead from the stable trapping point by a spring constant, which is determined by the properties of the trapping laser. Typically, a few milliwatts of power are enough to trap microscale particles, while higher power levels provide the stiffer traps necessary for position and force measurement.

In practice, OT is created by focusing a laser using a focusing lens or fiber optics. A focusing lens with a numerical aperture of at least 1.2, with the rear aperture slightly over-filled to ensure tight focusing, is typically used. Fiber optics is also employed to create optical traps, and is preferred for use with microfluidic devices due to simpler integration [16]. As mentioned earlier, an OT trap requires intense light to create sufficient trapping forces. Low-power OT still utilizes intensities on the order of megawatts per square centimeter at the laser focal point, which raises concerns about photodamage to biological particles [6–9, 17].

Photodamage on cells is related to the laser wavelength and power, and the environmental oxygen concentration [9]. The photodamage is primarily from photochemical effects for wavelengths less than 760 nm, while wavelengths above 800 nm can create damage due to photothermal effects. There are two wavelength ranges, 830 to 970 nm and 790 to 1064 nm, in which minimal photodamage was measured on *Escherichia coli* and Chinese hamster ovary cells [18]. These wavelength ranges contain commercially available laser wavelengths (820 and 1064 nm). Minimal effects on cells trapped using these wavelengths was measured through cloning [6, 19]. An additional advantage is that these near-infrared laser wavelengths do not interfere with the excitation of fluorescent dyes.

10.2.1 Multiple OT

Multiple OT traps can also be created to enable parallel and cooperative manipulation, and to increase throughput of manipulation operations. Cooperative manipulation can be useful in reducing fluctuations in position when trapping heterogeneous particles like cells. A single OT trap will pin the region of the cell with the highest refraction index, which can causes rolling of the cell during transportation. This could be addressed by using multiple OTs to anchor the cell [20].

Dual OTs can be generated by splitting the incoming beam into two polarized components [21, 22]. To create more OTs, the incoming light beam can be time-shared: a single beam is redirected to multiple locations over a short period of time. Stable trapping of particles using this technique requires the laser to revisit a trap location before a particle escapes by Brownian motion. Thus, the speed of the laser deflection technology limits the overall trap number. Scanning mirrors actuated with electrical motors can achieve rates of 1 to 2 kHz, while acousto-optic deflectors (AOD) and electro-optic deflectors (EOD) can beyond 10 kHz [23]. Up to 100 particles have been successfully trapped using an AOD [21]. However, AODs and EODs reduce the available optical power for trapping, and the deflection range of the EOD is also limited to small angles.

A popular method of generating multiple OTs is using a diffraction grating to modulate the wavefront of the incoming light and split it into multiple beams at the image plane (Figure 10.3). To dynamically split the beam, a spatial light modulator (SLM) can be used to create the diffraction grating. The grating hologram can then be easily reconfigured by a computer. This technique is termed as holographic optic tweezers (HOT) [24]. The holograms displayed by the SLM can be generated by mimicking static optical gratings [25]. Multiple OTs can be generated by summing the holograms calculated for each single OT. However, this does not guarantee trap homogeneity, and can also result in traps in undesired locations [26]. To correct this, iterative algorithms can be used [27]. While these algorithms add to the computing time, consumer graphics cards and GPU-based algorithms are sufficient to enable the generation of these more advanced holographs rapidly enough to beat the timescale of the Brownian motion of the particle [14, 26, 28].

In addition to steering and splitting the laser beam, the SLM can enable defocusing of optical traps, and the actuation of the particle in the axial direction, enabling

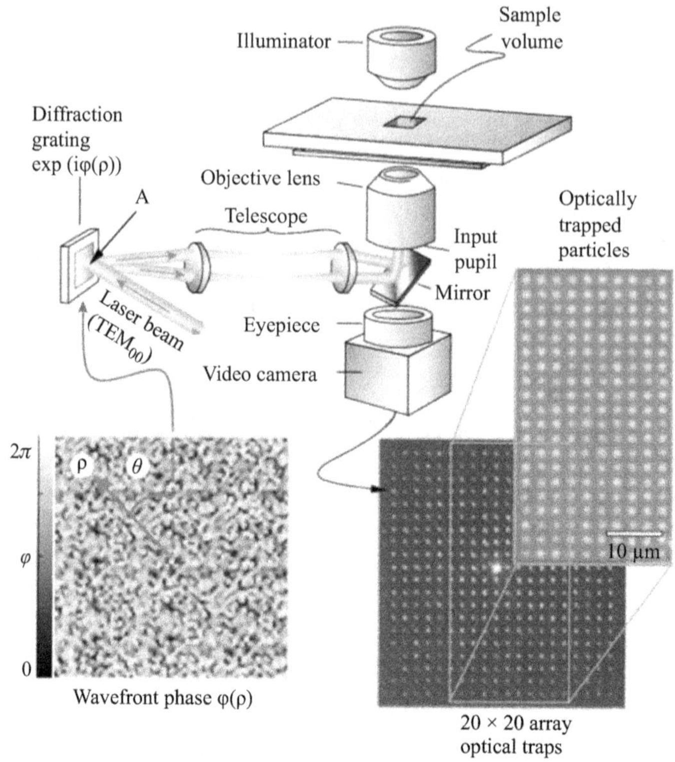

Figure 10.3 HOT with a reflective SLM in the Fourier plane of the light path for the beam shaping of the incoming light. Reprinted with permission from Grier (2003) [15], © copyright Nature Publishing Group

3D positioning [29–31]. The SLM can also correct for the intrinsic or induced optical aberration in optical path [32]. This has been used to reconstruct the beam after passing through turbid material [33]. To realize this, the optical field propagating through the optical system is expressed as a composition of modes. The phase deviation of each mode away from a reference mode is independently investigated and compensated. At last, all modes are turned on to create the compensated pattern. This technique could be useful for investigating biological particles in situ.

The SLM can also sculpt laser beams with special transverse modes [18]. For example, the SLM can transform a Gaussian laser beam into a corkscrew Laguerre-Gaussian beam that carries orbital angular momentum. This "optical vortex" beam can rotate trapped particles [15], trap low-refractive index objects [34], and drive microfluidic pumps [35]. Another interesting trapping beam is the Bessel beam, which features concentric light rings around the main light lobe. This beam profile matches the disk-like shape of erythrocytes, enabling separation from lymphocytes [36]. Another unique feature of the Bessel beam is that it continuously reconstructs its central maximum during propagation. This makes the beam resistant to spreading after passing obstructions. Trapped particles are transversely restricted within the beam, but are free to travel long distances in the axial direction. Bessel beams have been used to successfully trap and axially transfer particles for several millimeters [37].

10.2.2 Biological applications of OT

One application of OT is the measurement of the mechanical properties of biomolecules. Various biomolecules and their motions have been characterized, including the step motion of kinesin, microtubule stiffness, the motion of myosin on actin, the motion of processive nucleic acid enzymes along the DNA, and the bonding strength of complementary DNA strands [38]. Since direct trapping of irregularly shaped molecules is challenging, a "handle" particle such as a microbead is usually attached to the molecule of interest. Figure 10.4 shows several common measurement schemes for investigating the mechanical properties of a biomolecule. Using these schemes, the molecule can be stretched and rotated, measuring both force and torque.

The simplest model for an OT-trapped particle is a spring-mass system. Thus, the force can be quantified by multiplying the displacement of a bead by a spring constant. Many methods are available for calibrating the spring constant, and typically involve measuring the response of the trapped bead to a known force [21]. For example, a constant fluid flow can apply a known viscous drag force onto the bead, and the stiffness can be determined by the displacement of the trapped bead. The stiffness of most OT traps is usually on the order of $1\ \mu N \cdot m^{-1}$.

The measurement of bead displacement can be done with a resolution of a few nanometers, or even to the angstrom level [39]. Two-dimensional detection particles in an OT trap can be achieved with a position-sensitive detector and a quadrant photodiode aligned with the optical axis. This system can detect the divergence of the light and correlate it to the bead displacement [21], achieving a sub angstrom spatial resolution. However, in order to track multiple particles, standard optical imaging is necessary. Particle-tracking algorithms can achieve nearly 1-nm resolution, which is close to the thermal limit [40]. To measure axial displacement of

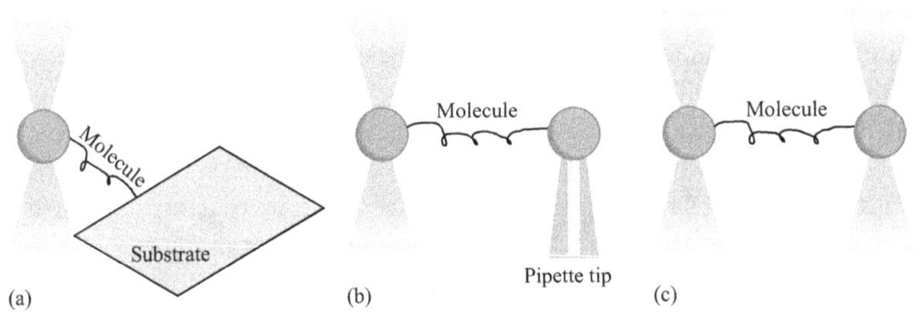

Figure 10.4 Three typical schemes for measuring the mechanical properties of biomolecules using optically trapped beads. The molecule can be anchored between: (a) an OT-trapped microbead and the substrate; (b) an OT-trapped microbead and a microbead trapped by a micropipette; (c) two OT-trapped microbeads

the particle, the visual appearance of the particle can be compared to a prerecorded image at a known depth. Alternatively, advanced imaging techniques such as digital holographic microscopy [31] or stereoscopic imaging [41] could be used to recover depth information.

In some cases, the alternate handles can be used. For example, when studying bacterial flagella, the bacterium itself can act as the handle. After carefully calibrating the trapping stiffness of the bacterium in the optical trap, rotating the bacterium can lead to information about the torsional compliance of its flagella [42, 43].

OT can also be used to measure the mechanical properties of cells. Properties such as elasticity and viscoelasticity usually correlate to the progression of a disease. For example, the fibroblast cell cytoskeleton softens when it deteriorates into cancerous and metastatic state [44], and red blood cells become more rigid after being infected by malaria [45].

Various measurement schemes can analyze these mechanical cell properties. The simplest is trapping a cell by a single-beam OT, and analyzing its motion. For example, a single Gaussian-beam OT can rotate particles with specific shapes, such as healthy red blood cells swollen in a hypertonic solution. In contrast, malaria-infected red blood cells do not display this behavior [46]. Alternatively, an optical trap formed by two counter-propagating beams can also stretch the red blood cells and effectively discriminate malignant cells by the degree of their change in shape [47].

Optically trapped handle particles can also be used to interact with the cell, in a similar fashion as the handle particles for biomolecules [48]. These handles are important for probing cells that are not easily trapped, such as cells adhered to a substrate. The handles also prevent the most intense laser light from being directly incident on the cell, helping to protect sensitive samples. Specially designed microstructures can also function as handles and can actuate biological samples with improved accuracy [49, 50].

OT can spatially transport microscale or nanoscale particles to study particle–particle and particle–environment interactions. For example, OT was used to transport floating neural cells into contact with adherent neural cells to investigate their interaction [51]. In another example, OT was used to trap canine kidney cells (MDCK) against fluidic flow salt with a salt concentration variation, for long-term monitoring of the cells reaction to osmotic shock [52].

Although most cell manipulation occurs in vitro, infrared OT has been used to trap and manipulate red blood cells within subdermal capillaries of living mice [53]. The demonstrated in vivo optical trap stiffness was approximately 10 $\mathrm{pN{\cdot}\mu m^{-1}}$. Optical manipulation can be useful for in vivo manipulation, as there is no mechanical connection to the trap sites, unlike atomic force microscopes or other mechanical micromanipulators although the current trapping depth was limited to approximately 40 μm. Integration of this type of OT with multiphoton imaging [54] and SLM optical aberration correction [33] could enable the in vivo measurement of cell properties in a physiological environment.

In addition to simple transportation, OT can artificially assemble cells into more complex geometries. As a top-down method, OT can be used to create many cellular patterns with biological significance. Mouse embryonic stem cells have been patterned

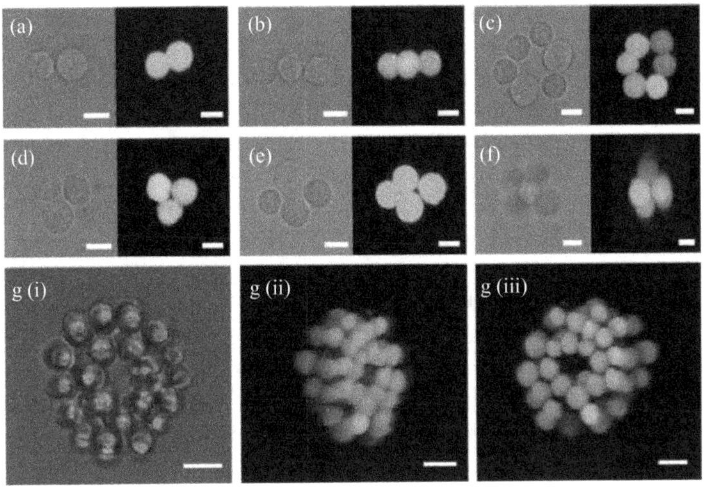

Figure 10.5 *Images of mouse embryonic stem cells patterned into precise 3D structures using holographic OT. The structures are stabilized using an avidin-biotin method of cell adhesion (a–f) or a modular hydrogel matrix (g). Reprinted with permission from Kirkham (2015) [29], © Kirkham et al.*

into a 3D structure by holographic OT (Figure 10.5) [29]. In another example, holographic OT was used to surround fibroblasts by a ring of bacteria, for investigating disease progression [55]. Holographic OT is very useful in microassembly as it increases assembly throughput through parallel and independent 3D manipulation [29].

Onc feature of OT is ability to perform label-free particle sorting. OTs can discriminate particles based upon physical properties such as size and refraction index [56–59]. For example, particle sorting can be achieved in an optical lattice, which is an interference pattern of with periodic intensity maxima [60]. The optical lattice can continuously deflect smaller particles flowing through the interference pattern from larger particles that travel straight through, with a sorting efficiency close to 100%. Optical lattices have successfully sorted erythrocytes from lymphocytes [61]. However, the sorting throughput is low, especially when compared to commercial technologies such as fluorescence activated cell sorters, which have a throughput of 10^6 cells s^{-1} [62]. Thus, this type of sorting may be more suitable for samples of rare or precious cells.

10.2.3 Integration with other technologies

OT control systems borrow from robotic control systems to improve the system stiffness and to add the ability of automatic processing. An intuitive adaption is the interactive control of OT by multitouch technology [63, 64]. In these systems, the real-time microscope images are displayed on a touch screen, with the position of the OTs indicated. Moving the trap locations on the touch screen physically moves the OT traps, making manipulation straightforward.

Additionally, implementation of haptic feedback is useful when the optical force has to balance the fluidic drag to keep particles from escaping the trap. Haptic feedback has been proved to increase the efficiency and safety of interactive microobject manipulation [65, 66]. The manipulation force acting on the trapped particle is scaled up and conveyed to the user through the same haptic device used to control the OT manipulation. This is especially beneficial for avoiding damage when manipulating delicate biosamples. In the future, further automation of OT can be envisioned. Fully automated OT manipulation could serve as a factory for producing microscale biomedical assemblies.

Other optical sensing and actuation mechanisms can also be integrated with OT. For example, laser tweezers Raman spectroscopy uses a tightly focused laser beam to simultaneously optically trap an individual cell and acquire its Raman molecular fingerprint [67]. This technique has been used to monitor real-time cell dynamics such as oxygenation cycle of a single red blood cell.

10.3 Other types of optical manipulation

10.3.1 Near-field particle trapping

Another optical trapping mechanism, especially suited to the manipulation of nanoscale particles, is known as near-field optical trapping. In this type of optical manipulation, light is confined to a waveguide structure. An exponentially decaying evanescent optical field extends out of the waveguide, and can generate sufficient optical gradient force to be used to trap particles. Various types of waveguides have been used to create near-field optical traps, including slot waveguides [68], resonators [69–71], and liquid-core waveguides [72]. Particles can be trapped and transported [73], including live cells [74]. Arrays of traps have also been created using waveguides [75].

10.3.2 Optical cell sorters

It is possible to create cell sorters that use mechanisms distinct from OT (Figure 10.6). Instead of using optical gradient force, selected cells are pushed out of microwells using optical scattering force and transported by a fluid flow for disposal or further study [76]. This setup has the advantage of being able to operate over an extremely large area.

The idea of sorting particles based on optical scattering force can also be integrated into a continuous-flow type microfluidic sorter [77]. It was also found that the amount of scattering force varies for different types of blood cells [78], which gives an additional parameter for controlling cell sorting.

Another interesting method of achieving cell sorting is by creating a moving interference pattern. The moving fringes of the interference pattern create potential barriers. Objects with different sizes will interact with the potential barriers in different ways, providing a means of sorting objects such as algal cells [79].

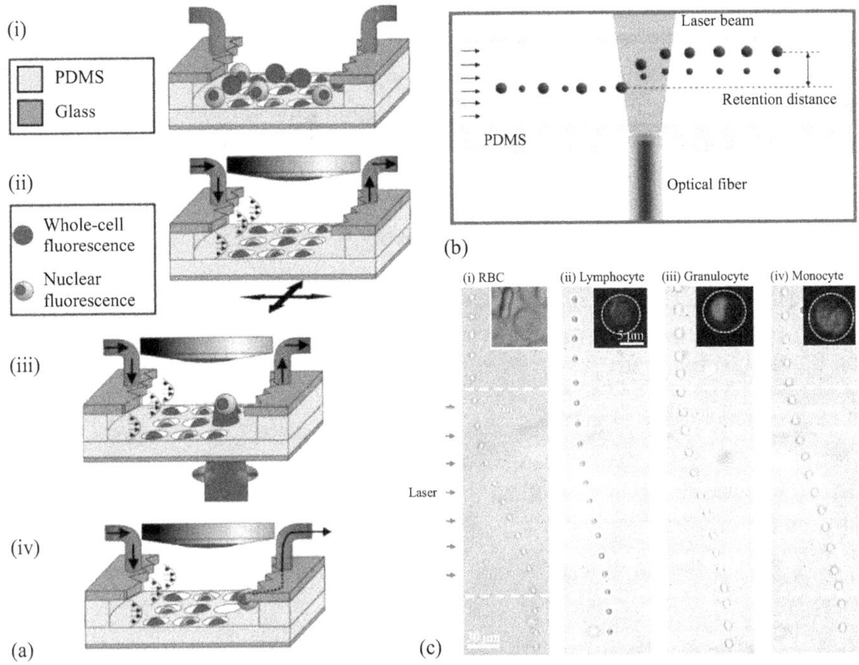

Figure 10.6 *Various types of optical sorters. (a) Radiation pressure is used to eject cells from microwells into a flow channel for downstream collection. Desired cells can be identified from direct imaging. Reprinted with permission from Reference 76. (b) A continuous-flow cell sorter that uses radiation pressure to redirect cells into a different streamline within a flow channel. Reprinted with permission from Reference 77. (c) The variation of optical scattering force by cell type enables sorting. Reprinted with permission from Lee (2013) [78], © AIP Publishing LLC*

10.3.3 Optical actuation of fluids, droplets, and bubbles

It is quite challenging to achieve aqueous fluid actuation using direct optical force. The difference in refractive index between air and water will result in a deformation of the water surface when a laser is incident on the interface; however, this deformation is suppressed by the surface tension of the liquid [80]. Reducing this interfacial tension can result in greater deformations of tens of micrometers along a liquid–liquid interface, although this liquid mixture of surfactant, water, oil, and alcohol is no longer suitable for many biological experiments [81]. Thus, most attempts at driving fluid flows using direct optical force typically use some type of microstructure to convert an OT trap into a physical pump [82–86], or to achieve pumping motion using colloidal particles [87–89].

It is also challenging to achieve aqueous droplet actuation using direct optical force. The refractive index of water is usually less than that of other immiscible fluids, meaning that a standard OT trap will repel water droplets, rather than stably trapping them. The same situation is true for bubbles in a liquid. While stable trapping can be achieved using an optical vortex trap [90], usually other methods are employed to move droplets and bubbles under optical control.

Light can also be used to alter the interfacial tension of the liquid, causing its motion. This can be done by setting up temperature gradients through optical absorption, which then create surface tension gradients that drive fluid motion via the thermocapillary effect [91–97]. Alternatively, the surface tension can be altered using photosensitive surfactants [98, 99].

Finally, optically controlled electrothermal flows can be created if a laser is used to heat the liquid in the presence of an electric field [100, 101]. These vortices can be used to trap particles, or to sort them based upon particle size or material properties [102].

Optically controlled electrical fields can also be used to move droplets using electrowetting, or the control of the contact angle of on a surface [103, 104]. This method requires much lower optical intensities, as the light is not used to generate force on the droplets directly and is instead used to control the electric field within the device.

10.4 Optically induced dielectrophoresis (ODEP)

Optical manipulation can also include indirect methods, where the optical source controls another mechanism as in the optically controlled electrowetting devices. A common force used for biological particles is dielectrophoresis (DEP), which is analogous to OT, as it relies upon the gradient force of an electric field [105]. The induced dipole of a particle experiences a force due to the electric field gradient, resulting in dielectrophoretic force that acts on micro- and nanoscale particles. Traditionally, the electric fields necessary for DEP have been created by microfabricated electrodes [106, 107], but a hybrid optoelectrical approach exists to control electric fields using light: ODEP, otherwise known as optoelectronic tweezers [108, 109].

ODEP integrates the flexibility and control of direct optical manipulation with the properties of DEP. In ODEP, a photosensitive material is used to enable control of the electric field landscape with optical patterns. The resulting nonuniform electric field generates a DEP force. ODEP does not use optical force, so much lower optical intensities can be used as compared to optical gradient traps. The typical optical intensities used for ODEP, 10 to 1000 mW/cm^2, can be generated by noncoherent sources such as computer projectors, LEDs, and liquid–crystal displays. In addition, the light patterns used for ODEP do not need to be tightly focused, as the optical gradient force is not utilized. The combination of low-intensity and loosely focused light means that ODEP manipulation can occur over a much larger area than optical gradient trapping performed with a single focusing lens. ODEP manipulation shares the flexibility and dynamic control of OT and properties of DEP, which allows discrimination of cells based upon their electrical properties.

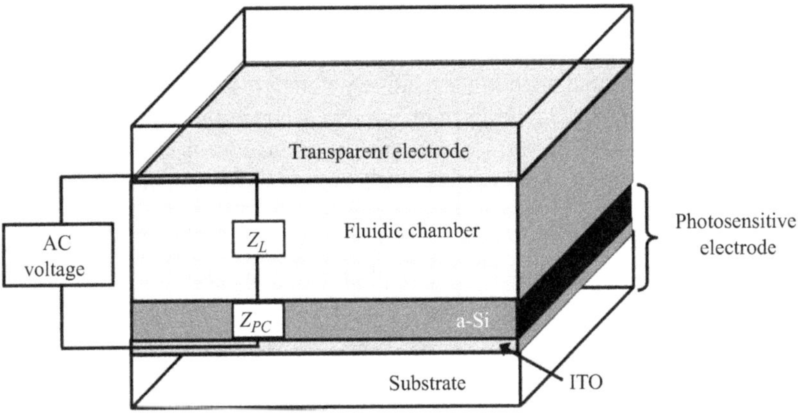

Figure 10.7 The standard device for ODEP. It consists of a fluidic chamber bounded by a transparent electrode and a photosensitive electrode. The photosensitive electrode is usually a-Si on ITO

The standard OET device consists of two electrodes, a transparent planar electrode and a photosensitive electrode (Figure 10.7). The two electrodes are separated by spacers, creating a fluidic chamber between the electrodes that contains the cells under manipulation. An electric field is created in the fluidic chamber by applying an AC voltage across the electrodes. Optical patterns are focused onto the photosensitive electrode, modulating the electrical conductivity. The most widely used photosensitive electrode consists of a layer of intrinsic amorphous silicon (a-Si) deposited on a layer of indium–tin oxide (ITO). A-Si has an electrical conductivity proportional to the intensity of the light that it absorbs. Under ambient lighting conditions, a-Si has a low conductivity. In this case, the impedance of the a-Si, Z_{PC} is larger than the impedance of the liquid layer, Z_L, so most of the applied AC voltage is across the a-Si layer (Figure 10.7). However, when illuminated with light, photogenerated charge carriers increase the conductivity of a-Si by a few orders of magnitude, lowering Z_{PC} and creating a significant voltage across the liquid layer. This switching between a normally low-conductivity state and high-conductivity illuminated state allows the a-Si–coated ITO to act as an optically defined electrode that is a function of the size and intensity of the optical patterns [110], and creates an electric field gradient in the liquid layer between the illuminated and dark areas of the photosensitive electrode. The voltage applied to create the electric field is typically a sinusoidal function with a frequency in the range of 10s to 100s of kilohertz, although square waves can result in more induced DEP force [111].

The light used to alter the conductivity of the photosensitive electrode can include lasers [112], but incoherent light sources can also be used. This includes the mercury lamp present in most computer projectors [110], LEDs [108] and LED arrays [113], and liquid–crystal displays [114, 115]. These incoherent sources can be made to directly form patterns using micromirror arrays or other types of SLMs.

If the output of a computer projector or LCD display is used, the light source and the mechanism for patterning the light are already integrated, facilitating the setup of the ODEP system, and enabling a relatively compact footprint.

Several other electrokinetic phenomena can occur in the ODEP device [116]. In low-frequency electric fields, electrolysis can occur. This is usually undesirable, as the bubble formation can interfere with normal ODEP operation, harm cell membranes, and damage photosensitive electrode layer. Electrothermal heating can also be observed under high optical intensities or large electric fields. This can result in controllable fluid flow [100–102], but the temperature must be monitored to ensure that there are no adverse effects on cells.

Light-induced AC electroosmosis (LACE) is another electrokinetic effect that can be produced in ODEP devices, typically occurring in low-frequency electric fields, on the order of a few kilohertz. In LACE, optical patterns control the electric field and the zeta potential at the liquid–photoconductor interface, resulting in electroosmotic flow [117, 118].

Thus, using the standard ODEP device, it is possible to switch between several types of electrokinetic phenomena, simply by adjusting the optical source and applied AC voltage [116]. If pure ODEP force is desired, electric field frequencies of greater than 50 kHz and optical intensities of less than 100 W/cm^2 ensures that ODEP is the dominant force (Figure 10.8).

Other materials can also be used for the photosensitive electrode. An example is cadmium sulfide (CdS), another photoconductive semiconductor [119].

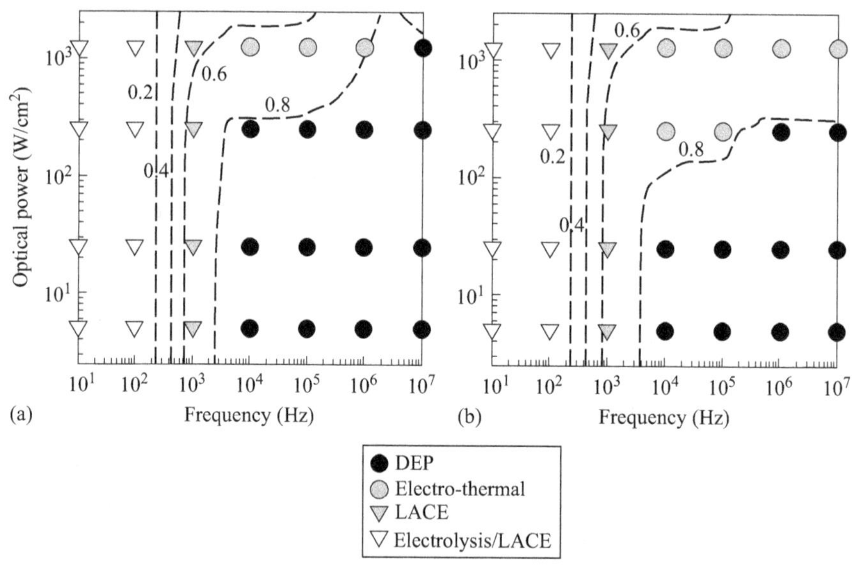

Figure 10.8 Electrokinetic phenomena occurring in the ODEP device with a liquid conductivity of 1 mS/m and an applied voltage of (a) 20 Vpp and (b) 10 Vpp. Operating parameters are usually chosen so that DEP is dominant. Reprinted with permission from Valley (2008) [116], © IEEE

The deposition of CdS can be achieved with a chemical process, as opposed to the plasma-enhanced chemical vapor deposition normally used for a-Si. Lithium niobate can also be used as the photosensitive material. The switching between light and dark states is longer (60 seconds to a few minutes), but the space–charge fields generated in the lithium niobate can remain even if the illumination is turned off [120].

Organic photoconductors such as titanium oxide phthalocyanine can also serve as the photosensitive material and can be spin-coated, simplifying the fabrication process [121, 122]. Photosensitive polymers, such as a mixture of regioregular poly (3-hexylthiophene) (rr-P3HT) and phenyl-C61-butyric acid methyl ester (PCBM) that forms a P3HT/PCBM blend [123] can also be spin-coated. These materials open up other device architectures, including the use of flexible substrates. This enables the photosensitive electrode to be bent, which allows gravitational forces to work with ODEP forces for the size-based separation of microparticles [124].

Additional architectures of the basic ODEP device have been created to add more functionality. The transparent planar electrode, which usually consists of an ITO layer on glass, was replaced with carbon nanotubes embedded in PDMS [125]. This resulted in a transparent electrode that can be integrated with microfluidic channels.

Confining the ODEP electrodes to a single surface is another way to facilitate the integration of ODEP with other microstructures. The two discrete electrodes of the basic ODEP device can be combined onto a single substrate by using an interdigitated array of a-Si-coated electrodes (Figure 10.9(a)) [126]. In this architecture, the electric field lines are primarily parallel to the substrate, in contrast to the electric field lines normal to the electrodes of the standard ODEP device. Thus, anisotropic particles such as microdisks [127] and nanowires [128–130] align with their long axis parallel to the surface of the substrate, unlike in the standard ODEP device [131], enabling the microassembly of anisotropic objects with their major axis parallel to the plane of the substrate. The single-substrate ODEP device also facilitates integration with other technologies. For example, cell transportation using ODEP and droplet manipulation using electrowetting-on-dielectric (EWOD) was achieved by sandwiching an LOET device and an EWOD device together [132].

Droplet manipulation can also be achieved using other variants of the ODEP architecture. A floating-electrode device was designed to manipulate aqueous droplets in oil for droplet-based assays [133, 134]. In this device, the a-Si photosensitive layer is coated with a 15-μm thick layer of polydimethylsiloxane (PDMS) polymer (Figure 10.9(b)). The voltage is applied laterally, along the a-Si layer. The response of the a-Si to light alters the applied electric field and results in DEP force on droplets suspended in oil above the PDMS layer.

Droplet manipulation can also be performed in a device more similar to the standard ODEP device (Figure 10.9(c). In this case, a dielectric layer and hydrophobic coating are added to the photosensitive substrate [135]. At low applied AC voltages (~10 kHz), this device functions as an optoelectrowetting device, allowing the light patterns to control the movement of droplets (Figure 10.9(c-i)). As the frequency of the applied voltage increases (~200 kHz), the device can achieve operation as a standard ODEP device and is capable of moving particles within the droplets using DEP force (Figure 10.9(c-ii)).

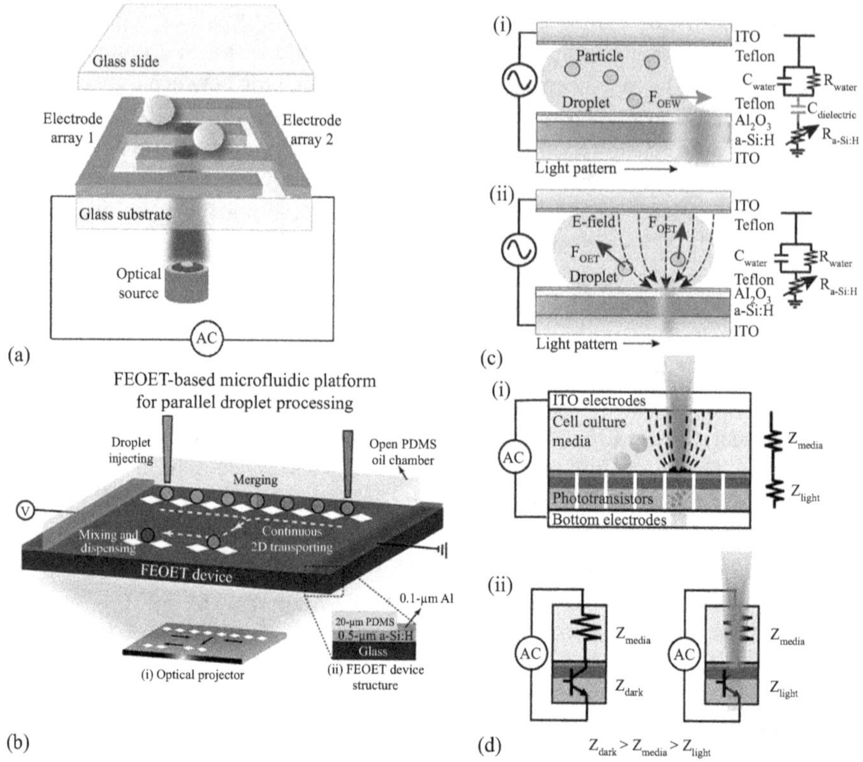

Figure 10.9 Various configurations of ODEP devices for increased functionality. (a) Lateral-field optoelectronic tweezers creates electric fields that are primarily parallel to the surface, and consolidates the ODEP electrodes to one surface. Reprinted with permission from Reference 126. (b) Floating-electrode optoelectronic tweezers enables droplet manipulation. Reprinted with permission from Reference 134. (c) Optoelectrowetting and ODEP can be realized on the same device by changing the frequency of the electric field. Reprinted with permission from Reference 135. (d) Phototransistor-based optoelectronic tweezers have increased conductivity under illumination, enabling operation in cell-culture media and other physiological solutions. Reprinted with permission from Hsu (2010) [136], © The Royal Society of Chemistry

One of the major limitations faced by all the ODEP devices mentioned thus far is the conductivity of the photosensitive electrode under illumination. The photoconductivity of amorphous silicon is not high enough to enable operation in many types of physiological liquids, including phosphate-buffered saline cell-culture media, as the liquid conductivity is too high. This is currently addressed by adding cells to osmotically balanced low-conductivity buffers. Although cells can typically survive for several hours in this media, it is not an ideal environment for cell health.

An ODEP device architecture that addresses this issue must then have a lower illuminated (on-state) conductivity that physiological media (over 1 S/m). To achieve a high on-state conductivity while maintaining a low off-state conductivity, a BJT phototransistor structure was used [136]. In this structure, photogenerated charge carriers create a base current, switching on the transistor, greatly increasing the conductivity of the structure (Figure 10.9(d)). Successful ODEP operation in PBS and one type of cell culture media was demonstrated.

A disadvantage to ODEP operation in physiological media is increased joule heating. Physiological media is more electrically conductive, and thus conducts more current than the lower-conductivity buffers. Joule heating in physiological media is the dominant cause of temperature increase in the ODEP platform (optical absorption heating is negligible at the low optical intensities used for ODEP). It was found that the temperature increase in the PhOET device during operation was approximately 1°C to 2°C [137]. This temperature increase keeps the cells well within the physiological temperature, although if further temperature regulation is desired, then the use of a heat sink can keep the temperature increase below 1°C.

A limitation of optical manipulation is that throughput is limited by the area over which the optical patterns can be operated. This is less of a limitation with ODEP, as the optical intensity and focusing requirements are relaxed compared to trapping with optical gradient forces. Thus, the optical patterns that control ODEP manipulation can be created over relatively large areas. However, the bottleneck in a conventional ODEP system is the sensing system, which usually consists of a microscope capable of imaging an area of a few square millimeters. Thus, combining an ODEP setup with lens-free holographic imaging can greatly increase area over which cells and other particles can be imaged [138], improving the throughput and the parallel manipulation capabilities of ODEP (Figure 10.10(a)). The reported working area of 240 mm^2 is the largest effective area of any optical manipulator.

One application of ODEP is the trapping and transport of cells. Many cell types have been manipulated by ODEP, including *E. coli* bacteria [139], red and white blood cells [110, 115], HeLa cells [126, 140], Jurkat cells [126], and protozoa [141]. Other bioparticles have also been manipulated, including DNA [142] and hydrogel beads with encapsulated cells [143].

One issue that arises when trapping mammalian cells is that of surface fouling, where cells nonspecifically adhere to the surface of ODEP devices. Cellular adhesion forces extend into the nanonewton range, whereas ODEP forces are typically tens to hundreds of piconewtons. Thus, once cells are fully adhered, it becomes difficult or impossible to move them.

It was shown that nonspecific cell adhesion in ODEP devices can be reduced by coating the electrodes with a layer of poly(ethylene glycol) (PEG), a common antifouling polymer [144]. The PEG coating reduced nonspecific cell adhesion of HeLa cells by a factor of 30, as compared to uncoated OET devices. An alternative is to use two photosensitive electrodes and negative DEP forces, so the cells under manipulation do not contact the surfaces of the ODEP electrodes [145]. However, imaging through this second photosensitive layer is more challenging, and cells not actively under manipulation could contact and adhere to the electrodes.

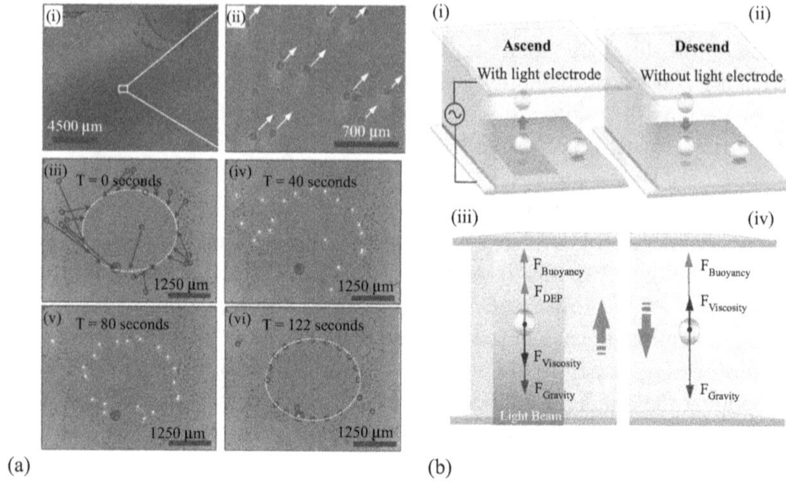

Figure 10.10 Biomedical applications of ODEP. (a) ODEP manipulation of micro-objects over a large area, enabling by lens-free microscopy. (i) The full effective manipulation area. The boxed area is enlarged in (ii). (iii–vi) Manipulation of particles within the field of view. Reprinted with permission from Reference 138. (b) Measuring the mass and density of single cells by levitating the cells using ODEP, then observing the descent of the cells. Reprinted with permission from Zhao (2014) [161], © The Royal Society of Chemistry

Cell sorting using the ODEP system can be based upon visual characteristics. In one experiment, T cells and dendritic cells had similar responses to the ODEP force, but could be distinguished via fluorescent tags. Appropriate ODEP patterns could then be created to isolate the two cell types from one another [146]. Thus, ODEP inherits the sorting capabilities of optical manipulation and also adds features of DEP force. Dielectrophoretic forces are very useful for the discrimination and sorting of cells. DEP force is size-dependent, but more uniquely is also a function of the electrical properties of the cells and the liquid media. Thus, DEP can be used to sort different types of cells or distinguish variations among cells of the same type [147–151].

ODEP has been used in several situations to separate healthy cells from unhealthy or dead cells of the same cell type. A live cell maintains an ion differential between its interior and the surrounding liquid medium. If the cell is suspended in a low-conductivity isotonic buffer, the interior of the cell will have an electrical conductivity greater than the liquid. However, once the cell dies, the ion concentration difference is not maintained, so the conductivity of the cell interior becomes similar to that of the surrounding liquid. This makes the DEP force exerted on live and dead cells different and can even manifest itself as change in the direction of the force.

Live and dead B cells (white blood cells) were selectively concentrated at an applied frequency of 120 kHz [108]. This was performed with an optical pattern consisting of broken concentric rings that constantly decreased in diameter. This caused live cells to be focused at the center of the pattern as those cells experienced a positive DEP force. The dead cells had a negative DEP force and slipped through the gaps in the rings.

DEP force can also have varied effects on normal and abnormal live cells. This was used to discriminate between oocytes cultured under normal conditions and oocytes that had been kept in a nutrient-free solution for 3 days [152]. The normal oocytes could be moved using ODEP at approximately 12 μm/s, while the abnormal oocytes were unresponsive. Valley *et al.* took this a step further and was differentiating between healthy and unhealthy embryos by observing their response to ODEP [153].

A complement to this was the discrimination of live and dead sperm cells. Similarly to the live and dead white blood cells, it was found that live sperm cells experience positive DEP and dead sperm cells experience negative DEP. This can be used to detect live sperm, even if the cells are not motile [154].

DEP force can also be used to distinguish different types of cells, due to variations in their electrical properties. For example, B lymphocytes and red blood cells experience a difference in DEP force due to physiological differences between those cell types [155]. Another example is Jurkat cells and HeLa cells in a solution with a conductivity of 2 mS/m: both cell types experience positive DEP force and are attracted toward ODEP manipulation patterns. However, the strength of the ODEP force depends upon the cell type and the frequency of the electric field [126]. The difference in ODEP force was used to discriminate the two cell types using optical patterns consisting of two scanning lines. A thinner leading line and a wider trailing line were simultaneously scanned across the manipulation area. The leading line produces a weaker ODEP force than the wider trailing line as it does not trigger as much of a change in the a-Si impedance. The Jurkat cells, which experience stronger DEP forces, can be transported by the leading line. However, the leading line lacks the force to transport the HeLa cells. The trailing line does have sufficient force to transport the HeLa cells, so after the scan is completed, the cells retain a spatial separation equal to the spacing of the two scanning lines.

A similar method was used to screen prostate cancer cells from leukocytes [156]. In this case, the prostate cancer cells experienced a stronger ODEP force compared to the leukocytes, although the both cell types experienced an attractive force toward the ODEP patterns. To separate these cells, a scanning line optical pattern was moved across the manipulation area at a velocity sufficient to transport all the cells. After this operation, all the cells of interest are at one side of the manipulation area. Another scanning line was then moved across the manipulation area, in the opposite direction, at a faster velocity. In this step, only the prostate cancer cells experience enough force to be transported by the scanning line, achieving spatial separation of the two cell types.

The same group extended upon this research by differentiating between prostate cancer cells and leukocytes, and oral cancer cells and leukocytes, in a constant-flow sorting platform [157]. Scanning line patterns were again used to separate the faster moving cells (prostate cancer, oral cancer) from the slower moving cells (leukocytes). However, the scanning lines here move orthogonally to the flow of cell solution in a central channel with two outlets. Thus, when the scanning lines move the faster moving cells toward a different location in the channel, these cells are collected in a separate outlet from the slower moving leukocytes. Up to 80% cell purity and 80% cell recovery was achieved using this device. This ODEP device has applications toward enriching circulating tumor cells from samples, and thus it is also significant that approximately 95% of the sorted cells were still viable, enabling further study.

It can be envisioned that these cell sorting capabilities can eventually be integrated in more complex sorting platforms, such as the gravity-driven ODEP-based flow-cytometry platform with integrated particle counting via a fiber-coupled optical detection system [158]. Although this system was optimized for microparticles, not cells, further work could result in an efficient sorting platform for samples of rare or valuable cells. It has also been shown that sorting platforms can be configured so that the exchange between cell-culture media and low-conductivity media for ODEP operations can be automated on chip, making sample preparation easier, and reducing adverse effects of the low-conductivity media on the cells under manipulation [159].

Other important biomedical functions can be performed using ODEP platforms. For example, cell stiffness can be quantified. Cell stiffness is an indicator of cellular behavior and cellular health. The relative stiffness of red blood cells could be measured by observing the change in the diameter of the cells in the presence of an ODEP electric field as compared to the case with no electric field [160].

The ODEP platform was also recently demonstrated as a way to measure the mass of single cells [161]. ODEP force was used to levitate cells above the photosensitive surface, and then the optical patterns and applied voltage was switched off. By measuring the rate of the descent of the cells, the density of the cells could be calculated (Figure 10.10(b)).

10.5 Optical cell surgery

Optical manipulation for cell sorting and trapping has been discussed. Further steps in the single-cell analysis process include operations on the cell itself, such as cell membrane poration for molecular delivery, extraction of cellular contents, or the implantation of objects into the cell. All of these operations can be achieved by optical means.

10.5.1 Optical cell poration

The ODEP platform can use pulses of the electric field to open pores in the membrane of cells, enabling molecular delivery in a process known as electroporation [162]. The electroporation can be reversible, where the induced pores reseal or irreversible. To perform reversible electroporation of specific cells, an AC electric field of 1.5 kV/cm at 100 kHz is applied for 5 seconds [163]. However, most of the electric field is only transferred to cells in areas illuminated by optical patterns on the ODEP device. Thus, electroporation of specific cells is performed

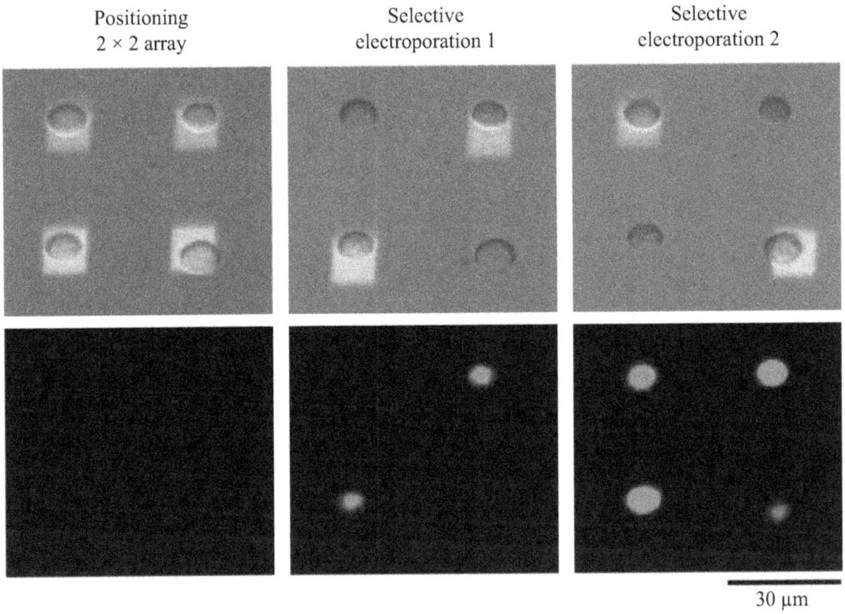

Figure 10.11 *Parallel single-cell electroporation using optically induced virtual microelectrodes. The top row shows the bright field images of cells and the optical patterns. The bottom row shows fluorescent images; cells are only visible in these images after uptaking propidium iodide (PI) dye. Cells are first patterned using ODEP, which does not cause electroporation. Two cells are then subjected to the optically controlled electroporation, and uptake the fluorescent PI dye. Finally, the remaining two cells are porated, resulting in the fluorescence of all cells. Reprinted with permission from Valley (2009) [163], © The Royal Society of Chemistry*

and was demonstrated by the uptake of fluorescent dyes (Figure 10.11). A similar system was also demonstrated for transfection of 4 to 5 kbp plasmids [164].

Pushing irreversible electroporation to an extreme results in cell lysis, where the cellular membrane is degraded to the point that contents of the cell interior may be accessed. This is an important step in a single-cell analysis process and can also be achieved using the ODEP platform. Lin and Lee were the first to show that this lysis can be achieved using ODEP, and that specific cells could be lysed by illuminating them with ODEP optical patterns [165]. Witte *et al.* were able to reduce the size of the optical patterns needed for lysis, achieving single-cell resolution using ODEP for cell lysis [166]. Lysing of a single cell within a group of adjacent cells was demonstrated. It was also shown that the shape of the cell can have an effect on cell lysis, as this affects the distribution of the electric field. Cells that cause localized current crowding can thus effectively be lysis before cells with less of an effect on the electric field [167]. This was used to show that red blood cells could be selectively lysed before white blood cells.

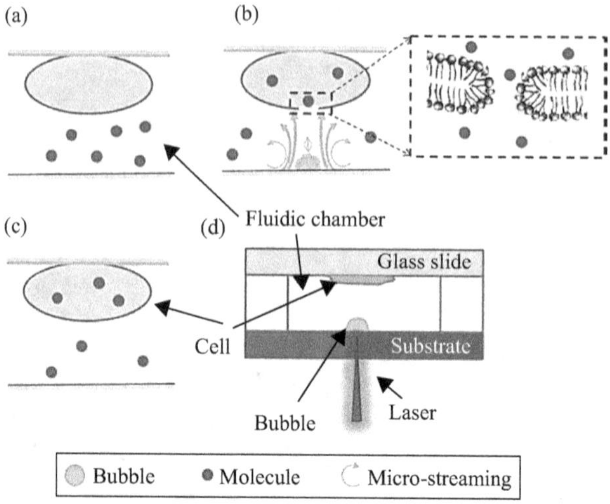

Figure 10.12 *Cell poration by microsecond laser–induced microbubbles. (a) Membrane-impermeable molecules are in the working solution surrounding the cell. (b) A size-oscillating microbubble is induced below the targeted cell. Shear stress from microstreaming flow surrounding the bubble opens transient pores in the cell membrane, as shown in the inset, allowing molecules to enter the cell. (c) The microbubble dissolves rapidly and the cell membrane reseals when the laser pulses cease. (d) Cross-section view of the bubble–cell interaction. Reprinted with permission from Fan (2014) [173], © The Royal Society of Chemistry*

Other optical cell poration and cell lysis techniques can use continuous-wave (CW) lasers, or lasers pulses with microsecond to femtosecond pulse widths. In optical cell poration with CW lasers, usually a 488-nm argon-ion laser or a 405-nm diode laser is focused on the cell membrane. Localized heating of the cell membrane, facilitated by dyes in the liquid media, causes a temporary phase transition of the cell membrane, increasing its permeability [168–171]. This reversible change of the cell membrane permeability allows foreign molecules to enter cells, while allowing cells to survive the poration process [168–170]. This method has single-cell resolution, meaning that a cell can be porated without affecting neighboring cells [170]. Although CW laser poration is relatively harmless to cells, the transfection efficiency is less than 30% for younger cell subcultures and less than 10% for aging cells, which is low compared to other optical cell poration methods [169, 170, 172].

Microsecond-laser pulses cause cell poration by inducing size-oscillating microbubbles [173, 174]. The laser energy is converted into heat, which creates vapor microbubbles in the working solution. The induced microbubbles expand and collapse repeatedly, corresponding to the timing of the laser pulses. This produces a shear stress around the microbubble [175] and can induce nanoscale transient pores in nearby cell membranes (Figure 10.12) [174, 176]. This method provides both

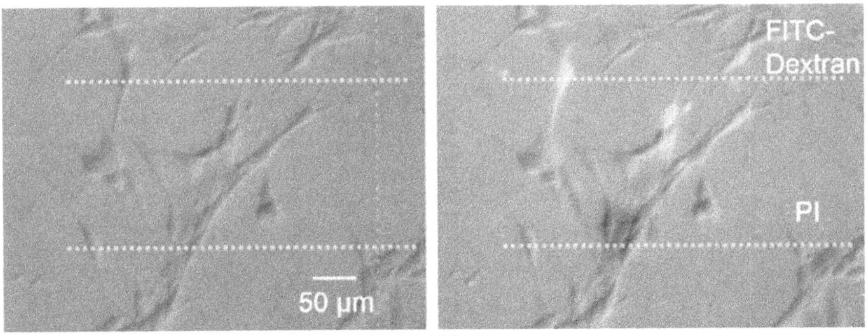

Figure 10.13 Patterned delivery of two types of molecules. 3-kDa FITC-Dextran dye was delivered into the cells along the pattern marked by the upper dashed line, while propidium iodide (PI) dye was delivered into cells along the pattern marked by the lowered dashed line. (a) Bright field image, (b) composite fluorescent image shows the successful delivery of both FITC-Dextran (green fluorescence) and PI (red fluorescence). Reprinted with permission from Fan (2015) [174], © The Royal Society of Chemistry

high poration efficiency and cell viability after the poration; both parameters can reach over 95% [174]. Microsecond-laser poration can also porate single cells while leaving all neighboring cells unaffected. Multiple payloads can also be delivered to the same cell sample in arbitrary patterns (Figure 10.13).

Nanosecond-laser pulses porate cells by a shockwave generated by laser-induced cavitation bubble expansion [176–180]. The plasma formation and shear stress from rapid expansion of the microbubble induced by the ns laser pulses causes cell poration [177, 180]. A hydrodynamic model was created to study the relationship of the generated shear stress and cell poration [176, 181]. The model revealed that the range of maximum bubble wall shear stress (τ_{max}) for cell poration is 8 to 18 kPa. The cells remain unaffected if τ_{max} is lower than 8 ± 1 kPa, while cells experience necrosis if τ_{max} is higher than 18 ± 2 kPa. Cell lysis occurs if the shear stress exceeds 190 kPa [176]. Thus, the range of cell poration or cell lysis can be controlled by modifying the laser energy. In a typical setup, molecular delivery occurs in the range of 30–60 µm from the laser focus point using 10 µJ laser pulse, without loss of cell viability [180]. If the distance between cells and laser spot was less than 30 µm, cells were lysed or nonviable, and if the distance is larger than 60 µm, the cells were unaffected. Poration efficiency can reach to 80% when combined with a laser-enabled analysis and processing system that focuses the laser on targeted cells identified by fluorescent staining [182].

Suspension cells confined by microfluidic structures can also be porated by single ns-laser–induced cavitation bubbles located near the targeted cell [183]. However, cells will experience irreversible poration (cell death) if the cell position is too close to the cavitation bubble. This poration method is also able to focus on

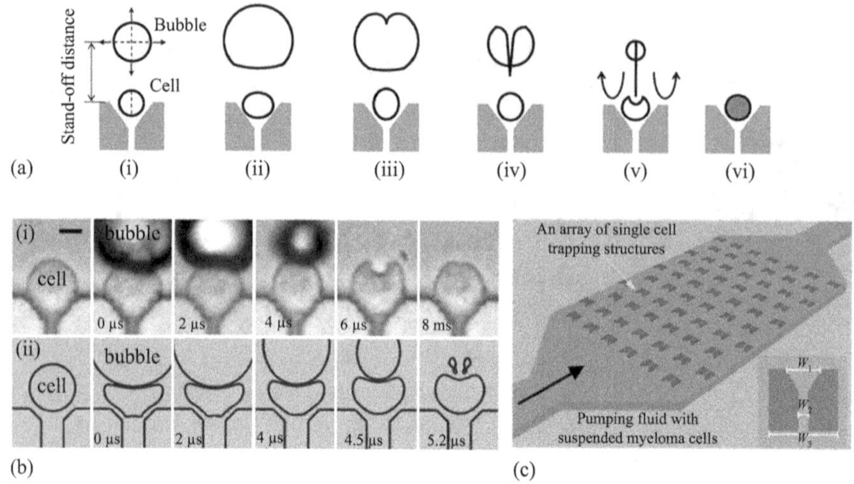

Figure 10.14 Single suspension-cell poration by nanosecond laser. (a) The suspension cell poration uses cavitation bubble-induced microjetting flow. (i) A cavitation bubble is created by a nanosecond laser nearby the trapped cell. (ii) The bubble expands to its maximum size. (iii) The bubble collapses asymmetrically. (iv and v) A microjetting flow is induced by the asymmetric collapse. The trapped cell is deformed and porated by the microjetting flow. (vi) Dye molecules can diffuse into the porated cell. (b) Experimental (i) and simulated (ii) results of single-cell deformation and poration by a cavitation bubble. The scale bar is 10 μm. (c) Schematic of the microfluidic chip with microfluidic structures for trapping single suspension cells. Reprinted with permission from Li (2013) [185], © The Royal Society of Chemistry

localized cells; pulsating tandem microbubbles generated by two Q-switched Nd:YAG lasers with 5-ns pulses were used to porate nearby localized single cells [184]. Single suspension cells were also porated by microjets during collapse of a ns-laser–induced microbubble (Figure 10.14) [185]. The cells were trapped in a microfluidic structure. When the induced cavitation bubbles collapsed, microjets were created that deformed and porated the cell membrane. The bubble distance from the target cell was strongly related to the induced membrane pore size and the cell viability.

Another method for single-cell poration uses femotosecond laser pulses [172, 186–190]. This method utilizes ultrashort laser pulses combined with high-NA objective lenses to focus the laser light onto the targeted cell membrane. Laser-induced multiphoton ionization and plasma formation can open transient pores in cell membranes [190]. Femtosecond-laser–induced single-cell poration was first demonstrated by Tirlapur *et al.* in 2002, and reached 100% transfection efficiency [186]. Reported femtosecond-laser transfection efficiencies are greater than 90% for various cell lines [186–189, 191]. The transfection efficiency on stem cells and primary cells is also better than other techniques such as chemical poration or electroporation [188, 189, 192]. However, submicron accuracy is required to focus the laser on the cell membrane for two photon absorption [193]. A focusing error of 3 μm may result in a decrease of greater than 50% in transfection efficiency [194]. To alleviate this focusing issue, microparticles as large as 1 μm can be trapped by OT to focus the femtosecond-laser precisely on the cell membrane [187]. A Bessel beam is another way to make femtosecond-laser poration less sensitive to focusing as opposed to a Gaussian beam. The effective focal depth was increased to 100 μm in the vertical axis with a Bessel laser beam [172, 194]. Multiple Bessel beams generated for parallel poration by SLM were also demonstrated, which provide the possibility for automated cell transfection with high throughput [195].

The heating from the laser exposure in femtosecond-laser poration is negligible, since the pulse duration is shorter than the thermal conduction time [188]. The formation of a visible microbubble induced by the femtosecond-laser can be an indicator for successful cell poration [188], but if the bubble is too large, severe cell damage may be caused within the targeted region [193].

In vivo transfection with femtosecond lasers has also been demonstrated, such as with tibial mouse muscles [196, 197]. GFP, luciferase, or other molecules were injected into mouse tibial muscle, which was exposed to the femtosecond laser for 5 seconds, while the laser raster scanned over an area of 95 μm by 95 μm at a tissue depth of 2 mm. This in vivo femtosecond-laser transfection result presents better cell morphology, significantly higher cell viability, and less cell apoptosis compared to the electroporation. Meanwhile, since this kind of optoporation does not require high voltage applied to animal or human bodies, there is much less possibilities for tissue damage. Besides the enhancement of intradermal transfection described previously, femtosecond lasers can also improve the intratumoral DNA delivery, which has the potential for DNA vaccination as well as for plasmid DNA gene therapy in the clinical setting [198].

The main limitation of femtosecond-laser poration is the high cost of the laser and other equipment, which is on the order of $100,000. This cost can be a barrier for certain applications.

10.5.2 Optical cell surgery

Lasers can be used as a tool for the precision cutting of cells or organelles [199]. This optical cell surgery includes cell lysis or the controlled breakdown of the cell membrane. Optical cell lysis can be confined to precise and localized regions in a rapid (microseconds) process [200]. Laser lysis is usually conducted with a Q-switched Nd:YAG laser using picosecond to nanosecond laser pulses. The laser pulses create cavitation bubbles that lyse the surrounding cells [176, 181, 200–203]. The effective range of the lysis can span dozens of cells [176], but single-cell resolution is possible if cells are confined by microfluidic structures, such as microchannels [202, 203] or picoliter or femtoliter droplets [204].

Another noteworthy method of optically controlled cell surgery is the photothermal nanoblade, which utilizes a titanium-coated microcapillary to harvest nanosecond laser-pulse energy and convert it into a highly localized cavitation bubble. The rapid fluidic flows and induced shear stress around the bubble punctures transient micrometer-sized pores in cell membrane for delivering cargos with a large size range of 1 nm to 2 μm [205]. This was discussed in more detail in Chapter 9.

Femtosecond laser surgery can also form well-defined holes in the cell membrane [206]. Afterwards, cargo can be transferred by OT through the hole in the membrane. The cargo can be sperm [207] or vectors for cell transfection [187]. Intracellular organelles can also be moved out of the cell [208]. A dual-laser system such as this is also useful in cell fusion [209], cell attachment [210], and the fusion of chemical-laden vesicles, such as liposomes [211].

Optical cell surgery can also be used for many specific applications. Lasers were used for manipulating sperm and eggs as part of a procedure called assisted hatching [212, 213], in which a small area of the protective zona pellucida of the eggs was thinned or removed to facilitate implantation when an early-stage embryo was placed in the womb. Cell fusion was realized by using an infrared Nd:YAG OT to move together two human myeloma cells and a pulsed ultraviolet laser to cut the adjoining cell membranes [209]. Optical cell surgery can also be performed on organelles, including cutting and moving the chromosomal fragments in the cell [214] and severing cytoskeletal filaments [215]. Microscale or nanoscale particles can be optically tweezed and injected into plant cells [216] or mammalian cells [187, 217] using two lasers: one to manipulate the particle and the other to inject the particle into the cell. Another important application is the laser separation and microdissection of individual cells or other small amounts of biomaterial from heterogeneous tissue samples to prepare them for subsequent genomic or proteomic analysis [218, 219]. The dissected specimens can then be collected into a vial for further analysis by laser pressure catapulting [220–222]. Lasers can also be used to inactivate specific proteins or organelles [223, 224]. Proteins or organelles are targeted by antibodies attached to metallic nanoparticles or chromophores, then exposed to laser light. The proteins or organelles are then inactivated or damaged by the absorption of the laser radiation by the nanoparticles or chromophores [223, 224]. This microsurgery technique can reveal the function of the respective proteins and organelles, and can be utilized to study cell proliferation, embryonal development, or cancer mechanisms and therapeutics [190, 223, 224].

Laser surgery can also be applied to cells in vivo for disease treatment. Nanosecond or femtosecond laser pulses focused by an objective lens with moderate numerical aperture can achieve precisely localized laser effects, and can pass through biological media, which is transparent to the laser light at low irradiance [190]. This feature has been utilized for intraocular surgery since the 1980s [225, 226]. For instance, femtosecond laser has been employed for corneal intrastromal refractive surgery [227, 228] and for the creation of corneal flaps in excimer laser refractive surgery (laser-assisted in situ keratomileusis) for decades [229, 230]. More recently, femtosecond lasers have also found usage in cataract surgery [231, 232].

10.6 Conclusion

The potency of optical manipulation for biomedical applications has been described in this chapter. Optical manipulation is especially suited for realizing a stand-alone single-cell analysis platform. All the steps of a single-cell analysis process can be realized using optical manipulation: cell sorting, cell trapping and transportation, cell poration/lysis/surgery, and cell sensing and measurement. Flexible, automated analysis systems can be realized using optical manipulation, and the cost of consumables can be minimized, as the functionality of the platform is realized through optical means. It is now even possible to create reconfigurable microfluidic channels using optical control [233].

One area that can be developed further is the 3D assembly of micro-objects. OT is capable of particle manipulation in 3D, but the workspace is limited. ODEP can assemble objects over a larger workspace, but it is currently limited to 2D assembly, and the resolution of ODEP is more coarse than OT. However, recent work in this area is promising. A deeper understanding of particle dynamics in the ODEP platform could lead the way toward 3D assembly of micro-objects [234], and recent work has been done to enable the ODEP device to manipulate nanoscale objects [235].

Optical manipulation for biomedical applications is still evolving, and more capabilities continue to be added. The possibilities are exciting, and society will continue to benefit from optical technologies in medicine.

References

[1] Pang L., Chen H.M., Freeman L.M., Fainman Y. "Optofluidic devices and applications in photonics, sensing and imaging". *Lab on a Chip.* 2012;**12**(19):3543–51

[2] Huang N.T., Zhang H.L., Chung M.T., Seo J.H., Kurabayashi K. "Recent advancements in optofluidics-based single-cell analysis: optical on-chip cellular manipulation, treatment, and property detection". *Lab on a Chip.* 2014;**14**(7):1230–45

[3] Erickson D., Sinton D., Psaltis D. "Optofluidics for energy applications". *Nature Photonics.* 2011;**5**(10):583–90

[4] Fan X.D., White I.M. "Optofluidic microsystems for chemical and biological analysis". *Nature Photonics.* 2011;**5**(10):591–7

[5] Schmidt H., Hawkins A.R. "The photonic integration of non-solid media using optofluidics". *Nature Photonics.* 2011;**5**(10):598–604

[6] Liang H., Vu K.T., Krishnan P., *et al.* "Wavelength dependence of cell cloning efficiency after optical trapping". *Biophysical Journal.* 1996;**70**(3):1529–33

[7] Liu Y., Sonek G.J., Berns, M.W., Tromberg B.J. "Physiological monitoring of optically trapped cells: assessing the effects of confinement by 1064–nm

laser tweezers using microfluorometry". *Biophysical Journal.* 1996;**71**(4):2158–67

[8] Mohanty S.K., Rapp A., Monajembashi S., Gupta P.K., Greulich K.O. "Comet assay measurements of DNA damage in cells by laser microbeams and trapping beams with wavelengths spanning a range of 308 nm to 1064 nm". *Radiation Research.* 2002;**157**(4): 378–85

[9] Mohanty S.K., Sharma M., Gupta P.K. "Generation of ROS in cells on exposure to CW and pulsed near-infrared laser tweezers". *Photochemical & Photobiological Sciences.* 2006;**5**(1):134–9

[10] Tanase M., Biais N., Sheetz M. "Magnetic tweezers in cell biology". *Methods in Cell Biology.* 2007;**83**:473–93

[11] Ashkin A. "Acceleration and trapping of particles by radiation pressure". *Physical Review Letters.* 1970;**24**(4):156–9

[12] Ashkin A., Dziedzic J.M., Bjorkholm J.E., Chu S. "Observation of a single-beam gradient force optical trap for dielectric particles". *Optics Letters.* 1986;**11**(5):288–90

[13] Ashkin A. "Forces of a single-beam gradient laser trap on a dielectric sphere in the ray optics regime". *Biophysical Journal.* 1992;**61**(2): 569–82

[14] Bowman R.W., Padgett M.J. "Optical trapping and binding". *Reports on Progress in Physics.* 2013;**76**(2):026401

[15] Grier D.G. "A revolution in optical manipulation". *Nature.* 2003;**424**(6950):810–6

[16] Constable A., Kim J., Mervis J., Zarinetchi F., Prentiss M. 'Demonstration of a fiber-optical light-force trap'. 1993;**18**(21):1867–9

[17] Ashkin A., Dziedzic J.M., Yamane T. "Optical trapping and manipulation of single cells using infrared laser beams". *Nature.* 1989;**330**(6150):769–71

[18] Zhang H., Liu K.-K. "Optical tweezers for single cells". *Journal of the Royal Society Interface.* 2008;**5**(24):671–90

[19] Neuman K.C., Chadd E.H., Liou G.F., Bergman K., Block S.M. "Characterization of photodamage to *Escherichia coli* in optical traps". *Biophysical Journal.* 1999;**77**(5):2856–63

[20] McAlinden N., Glass D.G., Millington O.R., Wright A.J. "Accurate position tracking of optically trapped live cells". *Biomedical Optics Express.* 2014;**5**(4):1026–37

[21] Visscher K., Gross S.P., Block S.M. "Construction of multiple-beam optical traps with nanometer-resolution position sensing". *IEEE Journal of Selected Topics in Quantum Electronics.* 1996;**2**(4):1066–76

[22] Fâllman E., Axner O. "Design for fully steerable dual-trap optical tweezers". *Applied Optics.* 1997;**36**(10):2107–13

[23] Neuman K.C., Block S.M. "Optical trapping". *Review of Scientific Instruments.* 2004;**75**(9): 2787–809

[24] Curtis J.E., Koss B.A., Grier D.G. "Dynamic holographic optical tweezers". *Optics Communications.* 2002;**207**(1–6):169–75

[25] Woerdemann M., Alpmann C., Esseling M., Denz C. "Advanced optical trapping by complex beam shaping". *Laser & Photonics Reviews.* 2013;**7**(6):839–54

[26] Bianchi S., Di Leonardo R. "Real-time optical micro-manipulation using optimized holograms generated on the GPU". *Computer Physics Communications.* 2010;**181**(8):1444–8

[27] Di Leonardo R., Ianni F., Ruocco G. "Computer generation of optimal holograms for optical trap arrays". *Optics Express.* 2007;**15**(4):1913–22

[28] Bowman R.W., Gibson G.M., Linnenberger A., *et al.* "Red tweezers: fast, customisable hologram generation for optical tweezers". *Computer Physics Communication.* 2014;**185**(1):268–73

[29] Kirkham G.R., Britchford E., Upton T., *et al.*, "Precision assembly of complex cellular microenvironments using holographic optical tweezers". *Scientific Reports.* 2015;**5**:8577

[30] Grier D.G., Roichman Y. "Holographic optical trapping". *Applied Optics.* 2006;**45**(5):880–7

[31] Lee S.-H., Grier D.G. "Holographic microscopy of holographically trapped three-dimensional structures". *Optics Express.* 2007;**15**(4):1505–12

[32] Martin-Badosa E., Montes-Usategui M., Carnicer A., Andilla J., Pleguezuelos E., Juvells I. "Design strategies for optimizing holographic optical tweezers set-ups". *Journal of Optics A: Pure and Applied Optics.* 2007;**9**(8):S267

[33] Čižmár T., Mazilu M., Dholakia K. "In situ wavefront correction and its application to micromanipulation". *Nature Photonics.* 2010;**4**(6):388–94

[34] Gahagan K.T., Swartzlander G.A. "Optical vortex trapping of particles". *Optics Letters.* 1996;**21**(11):827–9

[35] Ladavac K., Grier D. "Microoptomechanical pumps assembled and driven by holographic optical vortex arrays". *Optics Express.* 2004;**12**(6):1144–9

[36] Paterson L., Papagiakoumou E., Milne G., *et al.* "Light-induced cell separation in a tailored optical landscape". *Applied Physics Letters.* 2005;**87**(12):123901

[37] Garcés-Chávez V., McGloin D., Melville H., *et al.*, "Simultaneous micromanipulation in multiple planes using a self-reconstructing light beam". *Nature.* 2002;**419**(6903):145–7

[38] Fazal F.M., Block S.M. "Optical tweezers study life under tension". *Nature Photonics.* 2011;**5**(6):318–21

[39] Abbondanzieri E.A., Greenleaf W.J., Shaevitz J.W., Landick R., Block S.M. "Direct observation of base-pair stepping by RNA polymerase". *Nature.* 2005;**438**(7067):460–5

[40] Saunter C.D. "Quantifying subpixel accuracy: an experimental method for measuring accuracy in image-correlation-based, single-particle tracking". *Biophysical Journal.* 2010;**98**(8):1566–70

[41] Pavani S.R.P., Greengard A., Piestun R., "Three-dimensional localization with nanometer accuracy using a detector-limited double-helix point spread function system". *Applied Physics Letters.* 2009;**95**(2):021103

[42] Block S.M., Blair D.F., Berg H.C. "Compliance of bacterial flagella measured with optical tweezers". *Nature.* 1989;**338**(6215):514–18
[43] Block S.M., Blair D.F., Berg H.C. "Compliance of bacterial polyhooks measured with optical tweezers". *Cytometry.* 1991;**12**(6):492–6
[44] Ward K.A., Li W.I., Zimmer S., Davis T. "Viscoelastic properties of transformed cells: role in tumor cell progression and metastasis formation". *Biorheology.* 1991;**28**(3–4):301–13
[45] Glenister F.K., Coppel R.L., Cowman A.F., Mohandas N., Cooke B.M. "Contribution of parasite proteins to altered mechanical properties of malaria-infected red blood cells". *Blood.* 2002;**99**(3):1060–3
[46] Mohanty S.K., Uppal A., Gupta P.K. "Self-rotation of red blood cells in optical tweezers: prospects for high throughput malaria diagnosis". *Biotechnology Letters.* 2004:**26**(12):971–4
[47] Guck J., Schinkinger S., Lincoln B., *et al.* "Optical deformability as an inherent cell marker for testing malignant transformation and metastatic competence". *Biophysical Journal.* 2005;**88**(5):3689–98
[48] Lim C.T., Dao M., Suresh S., Sow C.H., Chew K.T. "Large deformation of living cells using laser traps". *Acta Materialia.* 2004;**52**(7):1837–45
[49] Phillips D.B., Simpson S.H., Grieve J.A., *et al.* "Position clamping of optically trapped microscopic non-spherical probes". *Optics Express.* 2011;**19**(21):20622–7
[50] Hayakawa T., Fukada S., Arai F. "Fabrication of an on-chip nanorobot integrating functional nanomaterials for single-cell punctures". *IEEE Transactions on Robotics.* 2014;**30**(1):59–67
[51] Townes-Anderson E., St. Jules R.S., Sherry D.M., Lichtenberger J., Hassanain M. "Micromanipulation of retinal neurons by optical tweezers". *Molecular Vision.* 1998;**4**:12
[52] Lúcio A.D., Santos R.A.S., Mesquita O.N. "Measurements and modeling of water transport and osmoregulation in a single kidney cell using optical tweezers and videomicroscopy". *Physical Review E.* 2003;**68**(4):041906
[53] Zhong M.-C., Wei X.-B., Zhou J.-H., Wang Z.-Q., Li Y.-M. "Trapping red blood cells in living animals using optical tweezers". *Nature Communications.* 2013;**4**:1768
[54] Welsher K., Sherlock S.P., Dai H., "Deep-tissue anatomical imaging of mice using carbon nanotube fluorophores in the second near-infrared window". *Procedings of the National Academy of Sciences of the United States of America.* 2011;**108**(22):8943–8
[55] Akselrod G.M., Timp W., Mirsaidov U., *et al.*, "Laser-guided assembly of heterotypic three-dimensional living cell microarrays," *Biophysical Journal.* 2006;**91**(9):3465–73
[56] Enger J., Goksör M., Ramser K., Hagberg P., Hanstorp D. "Optical tweezers applied to a microfluidic system". *Lab on a Chip.* 2004;**4**(3):196–200
[57] Xiao K., Grier D.G. "Multidimensional optical fractionation of colloidal particles with holographic verification". *Physical Review Letters.* 2010; **104**:028302

[58] Xiao K., Grier D.G. "Sorting colloidal particles into multiple channels with optical forces: prismatic optical fractionation". *Physical Review E.* 2010;**82**(5):051407

[59] Ladavac K., Kasza K., Grier D.G. "Sorting mesoscopic objects with periodic potential landscapes: optical fractionation". *Physical Review E.* 2004;**70**(1):010901(R)

[60] MacDonald M.P., Spalding G.C., Dholakia K. "Microfluidic sorting in an optical lattice". *Nature.* 2003;**426**(6965):421–4

[61] MacDonald M.P., Neale S., Paterson L., Riches A., Spalding G.C., Dholakia K. "Microfluidic optical sorting: particle selection in an optical lattice". *Proceedings of the SPIE.* 2004;**5514**:1–14

[62] Lindström S., Andersson-Svahn H. "Overview of single-cell analyses: microdevices and applications". *Lab on a Chip.* 2010;**10**(24):3363–72

[63] Bowman R.W., Gibson G., Carberry D., Picco L., Miles M., Padgett M.J. "iTweezers: optical micromanipulation controlled by an Apple iPad". *Journal of Optics.* 2011;**13**(4):044002

[64] Grieve J.A., Ulcinas A., Subramanian S., *et al.* "Hands-on with optical tweezers: a multitouch interface for holographic optical trapping". *Optics Express.* 2009;**17**(5):3595–602

[65] Pacoret C., Bergander A., Régnier S. "Haptic feedback of piconewton interactions with optical tweezers". *Haptics: Generating and Perceiving Tangible Sensations Lecture Notes in Computer Science.* 2010;**6192**:333–8

[66] Basdogan C., Kiraz A., Bukusoglu I., Varol A., Doğanay S. "Haptic guidance for improved task performance in steering microparticles with optical tweezers". *Optics Express.* 2007:**15**(18):11616–21

[67] Chan J.W. "Recent advances in laser tweezers Raman spectroscopy (LTRS) for label-free analysis of single cells". *Journal of Biophotonics.* 2013;**6**(1):36–48

[68] Yang A.H.J., Moore S.D., Schmidt B.S., Klug M., Lipson M., Erickson D. "Optical manipulation of nanoparticles and biomolecules in sub-wavelength slot waveguides". *Nature.* 2009;**457**(7225):71–75

[69] Yang A.H.J., Erickson D. "Optofluidic ring resonator switch for optical particle transport". *Lab on a Chip.* 2009;**10**(6):769–74

[70] Mandal S., Serey X., Erickson, D. "Nanomanipulation using silicon photonic crystal resonators". *Nano Letters*, 2010;**10**(1):99–104

[71] Hu J., Lin S., Kimerling L.C., Crozier K. "Optical trapping of dielectric nanoparticles in resonant cavities". *Physical Review A.* 2010:**82**(5):053819

[72] Kühn S., Measor P., Lunt E.J., *et al.* "Loss-based optical trap for on-chip particle analysis". *Lab on a Chip.* 2009;**9**(15):2212–16

[73] Lin S., Crozier K.B. "Planar silicon microrings as wavelength-multiplexed optical traps for storing and sensing particles". *Lab on a Chip.* 2011;**11**(23):4047–51

[74] Jain A., Yang A.H.J., Erickson D. "Gel-based optical waveguides with live cell encapsulation and integrated microfluidics". *Optics Letters.* 2012;**37**(9):1472–4

[75] Soltani M., Lin J., Forties R.A., Inman J.T., Saraf S.N., Fulbright R.M., *et al.* "Nanophotonic trapping for precise manipulation of biomolecular arrays". *Nature Nanotechology*. 2014;**9**(6):448–52
[76] Kovac J.R., Voldman J. "Intuitive, image-based cell sorting using optofluidic cell sorting". *Analytical Chemistry*. 2007;**79**(24):9321–30.
[77] Kim S.B., Yoon S.Y., Sung H.J., Kim, S.S. "Cross-type optical particle separation in a microchannel". *Analytical Chemistry*. 2008;**80**(7):2628–30
[78] Lee K.H., Lee K.S., Jung J.H., Chang C.B., Sung H.J. "Optical mobility of blood cells for label-free cell separation applications". *Applied Physics Letters*. 2013;**102**(14):141911
[79] Jakl P., Arzola A.V., Siler M., Chvatal L., Volke-Sepulveda K., Zemanek P. "Optical sorting of nonspherical and living microobjects in moving interference structures". *Optics Express*. 2014;**22**(24):29746–60
[80] Ashkin A., Dziedzic J. "Radiation pressure on a free liquid surface". *Physical Review Letters*. 1973;**30**(4):139–42
[81] Delville J.-P., de Saint Vincent M.R., Schroll R.D., *et al.* "Laser microfluidics: fluid actuation by light". *Journal of Optics A: Pure and Applied Optics*. 2009;**11**(3):034015
[82] Higurashi E., Sawada R., Ito T. "Optically induced rotation of a trapped micro-object about an axis perpendicular to the laser beam axis". *Applied Physics Letters*. 1998;**72**(23):2951–3
[83] Higurashi E., Ohguchi O., Tamamura T., Ukita H., Sawada R. "Optically induced rotation of dissymmetrically shaped fluorinated polyimide micro-objects in optical traps". *Journal of Applied Physics*. 1997;**82**(6):2773–9
[84] Higurashi E., Ukita H., Tanaka H., Ohguchi O. "Optically induced rotation of anisotropic micro-objects fabricated by surface micromachining". *Applied Physics Letters*. 1994;**64**(17):2209–10.
[85] Galajda P. Ormos P. "Complex micromachines produced and driven by light". *Applied Physics Letters*. 2001;**78**(2):249–51
[86] Galajda P., Ormos P. "Rotors produced and driven in laser tweezers with reversed direction of rotation". *Applied Physics Letters*. 2002;**80**(24): 4653–5
[87] Neale S.L., MacDonald M.P., Dholakia K., Krauss T.F. "All-optical control of microfluidic components using form birefringence". *Nature Materials*. 2005;**4**(7):530–3
[88] Ladavac K., Grier, D. "Microoptomechanical pumps assembled and driven by holographic optical vortex arrays". *Optics Express*. 2004;**12**(6):1144–9
[89] Friese M.E.J., Nieminen T.A., Heckenberg N.R., Rubinsztein-Dunlop H. "Optical alignment and spinning of laser-trapped microscopic particles". *Nature*. 1998; **394**(6691):348–50
[90] Lorenz R.M., Edgar J.S., Jeffries G.D., Zhao Y., McGloin D., Chiu D. T. "Vortex-trap-induced fusion of femtoliter-volume aqueous droplets". *Analytical Chemistry*. 2007;**79**(1):224–8
[91] Garnier N., Grigoriev R.O., Schatz M.F. "Optical manipulation of microscale fluid flow". *Physical Review Letters*. 2003;**91**(5):054501

[92] Farahi R.H., Passian A., Ferrell T.L., Thundat T. "Marangoni forces created by surface plasmon decay". *Optics Letters*. 2005;**30**(6):616–18

[93] Okawa D., Pastine S.J., Zettl A., Fréchet J.M. "Surface tension mediated conversion of light to work". *Journal of the American Chemical Society*. 2009:**131**(15):5396–8.

[94] Rybalko S., Magome N., Yoshikawa K. "Forward and backward laser-guided motion of an oil droplet". *Physical Review E*. 2004;**70**(4):046301

[95] Kotz K.T., Noble K.A., Faris G.W. "Optical microfluidics". *Applied Physics Letters*. 2004;**85**(13): 2658–60

[96] Kotz K.T., Gu Y., Faris G.W. "Optically addressed droplet-based protein assay". *Journal of the American Chemical Society*. 2005;**127**(16):5736–7

[97] Dixit S.S., Kim H., Vasilyev A., Eid A., Faris G.W. "Light-driven formation and rupture of droplet bilayers". *Langmuir*. 2010;**26**(9):6193–200

[98] Baigl D. "Photo-actuation of liquids for light-driven microfluidics: state of the art and perspectives". *Lab on a Chip*. 2012;**12**(19):3637–53

[99] Varanakkottu S.N., George S.D., Baier T., Hardt S., Ewald M., Biesalski M. "Particle manipulation based on optically controlled free surface hydrodynamics". *Angewandte Chemie International Edition*. 2013;**52**(28):7291–5

[100] Kumar A., Williams S.J., Wereley S.T. "Experiments on opto-electrically generated microfluidic vortices". *Microfluidics and Nanofluidics*. 2009;**6**(5): 637–46

[101] Mishra A., Kwon J.S., Thakur R., Wereley S. "Optoelectrical microfluidics as a promising tool in biology". *Trends in Biotechnology*. 2014;**32**(8): 415–22

[102] Williams S.J., Kumar A., Green N.G., Wereley S.T. "Optically induced electrokinetic concentration and sorting of colloids". *Journal of Micromechanics and Microengineering*. 2010;**20**(1):015022

[103] Chiou P.-Y., Chang Z., Wu M.C. "Droplet manipulation with light on optoelectrowetting device". *IEEE Journal of Microelectromechanical Systems*. 2008;**17**(1):133–8

[104] Krogmann F., Qu H., Monch W., Zappe H. "Push/pull actuation using opto-electrowetting". *Sensors and Actuators A*. 2008;**141**(2):499–505

[105] Pohl H. A. "Some effects of nonuniform fields on dielectrics". *Journal of Applied Physics*. 1958;**29**(8):1182–8

[106] Jubery T.Z., Srivastava S.K., Dutta P. "Dielectrophoretic separation of bioparticles in microdevices: a review". *Electrophoresis*. 2014;**35**(5): 691–713

[107] Cetin B., Li D.Q. "Dielectrophoresis in microfluidics technology". *Electrophoresis*. 2011;**32**(18):2410–27

[108] Chiou P.Y., Ohta A.T., Wu M.C. "Massively parallel manipulation of single cells and microparticles using optical images". *Nature*. 2005;**436**(7049):370–2

[109] Wu M.C. "Optoelectronic tweezers". *Nature Photonics*. 2011:**5**(6):322–4

[110] Ohta A.T., Chiou P.-Y., Han T.H., *et al.* "Dynamic cell and microparticle control via optoelectronic tweezers". *IEEE Journal of Microelectromechanical Systems*. 2007;**16**(3):491–9

[111] Liang W., Wang S., Dong Z., Lee G.-B., Li W.J. "Optical spectrum and electric field waveform dependent optically-induced dielectrophoretic (ODEP) micro-manipulation". *Micromachines* 2012;**3**:492–508.

[112] Neale S.L., Mazilu M. Wilson J.I.B., Dholakia K., Krauss T.F. "The resolution of optical traps created by light induced dielectrophoresis (LIDEP)". *Optics Express.* 2007;**15**(20):12619–26

[113] Zarowna-Dabrowska A., Neale S.L., Massoubre D., *et al.* "Miniaturized optoelectronic tweezers controlled by GaN micro-pixel light emitting diode arrays". *Optics Express.* 2011;**19**(3):2720–8

[114] Choi W., Kim S.-H., Jang J., Park J.-K. "Lab-on-a-display: a new microparticle manipulation platform using a liquid crystal display (LCD)". *Microfluidics and Nanofluidics.* 2007;**3**(2):217–25

[115] Hwang H., Choi Y.J., Choi W., Kim S.H., Jang J., Park J.K. "Interactive manipulation of blood cells using a lens-integrated liquid crystal display based optoelectronic tweezers system". *Electrophoresis.* 2008;**29**(6): 1203–12

[116] Valley J.K., Jamshidi A., Ohta A.T., Hsu H.-Y., Wu, M.C. "Operational regimes and physics present in optoelectronic tweezers". *IEEE Journal of Microelectromechanical Systems.* 2008;**17**(2):342–50

[117] Chiou P.-Y., Ohta A.T., Jamshidi A., Hsu H.-Y., Wu M.C. "Light-actuated ac electroosmosis for nanoparticle manipulation". *IEEE Journal of Microelectromechanical Systems.* 2008;**17**(3):525–31

[118] Hwang H., Park, J.K. "Rapid and selective concentration of microparticles in an optoelectrofluidic platform". *Lab on a Chip.* 2009;**9**(2):199–206

[119] Higuchi Y., Kusakab T., Tanemura T., Sugano K., Tsuchiya T., Tabata O. "Manipulation system for nano/micro components integration via transportation and self-assembly". *Proceedings of the IEEE 21st International Conference on Micro Electro Mechanical Systems (MEMS)*; Tucson, AZ, USA, 2008, pp. 836–9

[120] Glaesener S., Esseling M., Denz C. "Multiplexing and switching of virtual electrodes in optoelectronic tweezers based on lithium niobate". *Optics Letters.* 2012;**37**(18):3744–6

[121] Yang S.-M., Yu T.-M., Huang H.-P., *et al.* "Light-driven manipulation of picobubbles on a titanium oxide phthalocyanine-based optoelectronic chip". *Applied Physics Letters.* 2011;**98**(15):153512

[122] Yang S.-M., Tseng S.-Y., Chen H.-P., Hsu L., Liu C.-H. "Cell patterning via diffraction-induced optoelectronic dielectrophoresis force on an organic photoconductive chip". *Lab on a Chip.* 2013;**13**(19):3893–902

[123] Wang W., Lin Y.-H., Wen, T.-C., Guo T.-F., Lee G.-B. "Selective manipulation of microparticles using polymer-based optically induced dielectrophoretic devices". *Applied Physics Letters.* 2010;**96**(11):113302

[124] Lin S.-J., Hung S.-H., Jeng J.-Y., Guo T.-F., Lee, G.-B. "Manipulation of micro-particles by flexible polymer-based optically-induced dielectrophoretic devices". *Optics Express.* 2012;**20**(1):583–92

[125] Huang K.-W., Wu Y.-C., Lee J.-A., Chiou P.-Y. "Microfluidic integrated optoelectronic tweezers for single-cell preparation and analysis". *Lab on a Chip.* 2013;**13**(18):3721–7

[126] Ohta A.T., Chiou P.-Y., Phan H.L., *et al.* "Optically controlled cell discrimination and trapping using optoelectronic tweezers". *IEEE Journal of Selected Topics in Quantum Electronics.* 2007;**13**(2):235–43

[127] Tien M.-C., Ohta A.T., Yu K., Neale S.L., Wu M.C. "Heterogeneous integration of InGaAsP microdisk laser on a silicon platform using optofluidic assembly". *Applied Physics A: Materials Science & Processing.* 2009;**95**(4):967–72

[128] Ohta A.T., Jamshidi A., Pauzauskie P.J., Hsu H.-Y., Yang P., Wu M.C. "Trapping and transport of silicon nanowires using lateral-field optoelectronic tweezers". *Proceedings of the Conference on Lasers and Electro-Optics (CLEO)*; Baltimore, MD, USA, 2007, pp. 828–9

[129] Ohta A.T., Neale S.L., Hsu H.Y., Valley J.K., Wu M.C. "Parallel assembly of nanowires using lateral-field optoelectronic tweezers". *Proceedings of the IEEE/LEOS International Conference on Optical MEMs and Nanophotonics;* Freiburg, Germany, 2008, pp. 7–8

[130] Neale S.L., Fan Z., Ohta A. T., *et al.* "Optofluidic assembly of red/blue/green semiconductor nanowires". *Proceedings of the Conference on Lasers and Electro-Optics (CLEO)*; Baltimore, MD, USA, pp. 2009

[131] Jamshidi A., Pauzauskie P.J., Schuck P.J., *et al.* "Dynamic manipulation and separation of individual semiconducting and metallic nanowires". *Nature Photonics.* 2008;**2**(2):85–9

[132] Shah G.J., Ohta A.T., Chiou E.P.Y., Wu M.C., Kim C.-J. "EWOD-driven droplet microfluidic device integrated with optoelectronic tweezers as an automated platform for cellular isolation and analysis". *Lab on a Chip.* 2009;**9**(12):1732–9

[133] Park S., Pan C., Wu T.-H., *et al.* "Floating electrode optoelectronic tweezers: light-driven dielectrophoretic droplet manipulation in electrically insulating oil medium". *Applied Physics Letters.* 2008;**92**(15):151101

[134] Park S.-Y., Kalim S., Callahan C., Teitell M.A., Chiou E.P.Y. "A light-induced dielectrophoretic droplet manipulation platform". *Lab on a Chip.* 2009;**9**(22):3228–35

[135] Valley J.K., Pei S.N., Jamshidi A., Hsu H.-Y., Wu M.C. "A unified platform for optoelectrowetting and optoelectronic tweezers". *Lab on a Chip.* 2011;**11**(7):1292–7

[136] Hsu H.-Y., Ohta A.T., Chiou P.-Y., Jamshidi A., Neale S.L., Wu M.C. "Phototransistor-based optoelectronic tweezers for dynamic cell manipulation in cell culture media". *Lab on a Chip.* 2010;**10**(2):165–72

[137] Pauzauskie P.J., Hsu H.-Y., Jamshidi A., Valley J.K., Pei S.N., Wu M. C. "Quantifying heat transfer in DMD-based optoelectronic tweezers with infrared thermography". *Proceedings of the SPIE.* 2010;**7596**: 759609

[138] Huang K.-W., Su T.-W., Ozcan A., Chiou P.-Y. "Optoelectronic tweezers integrated with lens-free holographic microscopy for wide-field interactive cell and particle manipulation on a chip". *Lab on a Chip.* 2013;**13**(12): 2278–4

[139] Chiou P.Y., Wong W., Liao J.C., Wu M.C. "Cell addressing and trapping using novel optoelectronic tweezers". *Proceedings of the 17th IEEE International Conference on Micro Electro Mechanical Systems (MEMS);* Maastricht, Netherlands, 2004, pp. 21–4

[140] Neale S.L., Ohta A.T., Hsu H.-Y., Valley J.K., Jamshidi A., Wu M.C. "Trap profiles of projector based optoelectronic tweezers (OET) with HeLa cells". *Optics Express.* 2009;**17**(7):5232–9

[141] Choi W., Nam S.W., Hwang H., Park S., Park J.K. "Programmable manipulation of motile cells in optoelectronic tweezers using a gray scale image". *Applied Physics Letters.* 2008;**93**(14):143901

[142] Lin Y.-H., Chang C.-M., Lee G.-B. "Manipulation of single DNA molecules by using optically projected images". *Optics Express.* 2009;**17**(17):15318–29

[143] Lin Y.-H., Yang Y.-W., Chen Y.-D., Wang S.-S., Chang Y.-H., Wu M.-H. "The application of an optically switched dielectrophoretic (ODEP) force for the manipulation and assembly of cell-encapsulating alginate microbeads in a microfluidic perfusion cell culture system for bottom-up tissue engineering". *Lab on a Chip.* 2012;**12**(6):1164–73.

[144] Lau A.N.K., Ohta A.T., Phan H.L., *et al.* "Antifouling coatings for optoelectronic tweezers". *Lab on a Chip.* 2009;**9**(20):2952–7

[145] Hwang H., Oh Y., Kim J.J., *et al.* "Reduction of nonspecific surface-particle interactions in optoelectronic tweezers". *Applied Physics Letters.* 2008;**92**(2):024108

[146] Jeorrett A.H., Neale S.L., Massoubre D., *et al.* "Optoelectronic tweezers system for single cell manipulation and fluorescence imaging of live immune cells". *Optics Express.* 2014;**22**(2):1372–80

[147] Gascoyne P.R.C., Wang X.-B., Huang Y., Becker F.F. "Dielectrophoretic separation of cancer cells from blood". *IEEE Transactions on Industry Applications.* 1997;**33**(3):670–8

[148] Cheng J., Sheldon E.L., Wu L., Heller M.J., O'Connell J.P. "Isolation of cultured cervical carcinoma cells mixed with peripheral blood cells on a bioelectronic chip". *Analytical Chemistry.* 1998;**70**(11):2321–6

[149] Huang Y., Yang J.M., Hopkins P.J., *et al.* "Separation of simulants of biological warfare agents from blood by a miniaturized dielectrophoresis device". *Biomedical Microdevices.* 2003;**5**(3):217–25

[150] Gascoyne P.R.C., Vykoukal J. V. "Dielectrophoresis-based sample handling in general-purpose programmable diagnostic instruments". *Proceedings of the IEEE.* 2004;**92**(1):22–42.

[151] Pethig R., Talary M.S., Lee R.S. "Enhancing traveling-wave dielectrophoresis with signal superposition". *IEEE Engineering in Medicine and Biology Magazine.* 2003;**22**(6):43–50

[152] Hwang H., Lee D.H., Choi W., Park J.K. "Enhanced discrimination of normal oocytes using optically induced pulling-up dielectrophoretic force". *Biomicrofluidics*. 2009;**3**(1):014103–10

[153] Valley J.K., Swinton P., Boscardin W.J., *et al.* "Preimplantation mouse embryo selection guided by light-induced dielectrophoresis". *PLoS One*. 2010;**5**(4): e10160

[154] Ohta A.T., Garcia M., Valley J.K., *et al.* "Motile and non-motile sperm diagnostic manipulation using optoelectronic tweezers". *Lab on a Chip*. 2009;**10**(23):3213–17

[155] Zhao Y., Liang W., Zhang G., *et al.* "Distinguishing cells by their first-order transient motion response under an optically induced dielectrophoretic force field". *Applied Physics Letters*. 2013;**103**(18):183702

[156] Huang S.-B., Chen J., Wang J., Yang C.-L., Wu M.-H. "A new optically-induced dielectrophoretic (ODEP) force-based scheme for effective cell sorting". *International Journal of Electrochemical Science*. 2012;**7**:12656–67

[157] Huang S.-B., Wu M.-H., Lin Y.-H., *et al.* "High-purity and label-free isolation of circulating tumor cells (CTCs) in a microfluidic platform by using optically-induced-dielectrophoretic (ODEP) force". *Lab on a Chip*. 2013;**13** (7):1371–83

[158] Lin Y.-H., Lee G.-B. "Optically induced flow cytometry for continuous microparticle counting and sorting". *Biosensors and Bioelectronics*. 2008;**24**(4):572–8

[159] Lee G.-B., Wu H.-C., Yang P.-F., Mai J.D. "Optically induced dielectropheresis sorting with automated medium exchange in an integrated optofluidic device resulting in higher cell viability". *Lab on a Chip*. 2014;**14**(15): 2837–43

[160] Neale S.L. Mody N., Selman C., Cooper J.M. "Optoelectronic tweezers for the measurement of the relative stiffness of erythrocytes". *Proceedings of the SPIE*. 2012;**8458**:845827

[161] Zhao Y., Lai H.S.S., Zhang G., Lee G.-B., Li W.J. "Rapid determination of cell mass and density using digitally controlled electric field in a microfluidic chip". *Lab on a Chip*. 2014;**14**(22):4426–34

[162] Bodles-Brakhop A. M., Heller R., Draghia-Akli R. "Electroporation for the delivery of DNA-based vaccines and immunotherapeutics: current clinical developments". *Molecular Therapy*. 2009;**17**(4):585–92

[163] Valley J.K., Neale S., Hsu H.Y., Ohta A.T., Jamshidi A., Wu M.C. "Parallel single-cell light-induced electroporation and dielectrophoretic manipulation". *Lab on a Chip*. 2009;**9**(12):1714–20

[164] Wang C.-H., Lee Y.-H., Kuo H.-T., Liang W.-F., Li W.-J., Lee G.-B. "Dielectrophoretically-assisted electroporation using light-activated virtual microelectrodes for multiple DNA transfection". *Lab on a Chip*. 2014; **14**(3):592–601

[165] Lin Y.-H., Lee G.-B. "An optically induced cell lysis device using dielectrophoresis". *Applied Physics Letters*. 2009;**94**(3):033901

[166] Witte C., Kremer C., Chanasakulniyom M., Reboud J., Wilson R., Cooper J.M., Neale S.L. "Spatially selecting a single cell for lysis using light-induced electric fields". *Small.* 2014;**10**(15):3026–31

[167] Kremer C., Witte C., Neale S.L., Reboud J., Barrett M.P., Cooper J.M. "Shape-dependent optoelectronic cell lysis". *Angewandte Chemie.* 2014;**126**(3):861–65

[168] Palumbo G., Caruso M., Crescenzi E., Tecce M.F., Roberti G., Colasanti A. "Targeted gene transfer in eucaryotic cells by dye-assisted laser optoporation". *Journal of Photochemistry and Photobiology B: Biology.* 1996; **36**(1):41–6

[169] Schneckenburger H., Hendinger A., Sailer R., Strauss W.S., Schmitt M. "Laser-assisted optoporation of single cells". *Journal of Biomedical Optics.* 2002;**7**(3):410–16

[170] Nikolskaya A.V., Nikolski V.P., Efimov I.R. "Gene printer: laser-scanning targeted transfection of cultured cardiac neonatal rat cells". *Cell Communication and Adhesion.* 2006;**13**(4):217–22

[171] Paterson L., Agate B., Comrie M., *et al.* "Photoporation and cell transfection using a violet diode laser". *Optics Express.* 2005;**13**(2):595–600

[172] Stevenson D.J., Gunn-Moore F.J., Campbell P., Dholakia K. "Single cell optical transfection". *Journal of the Royal Society Interface.* 2010: **7**(47):863–71

[173] Fan Q., Hu W., Ohta A.T. "Laser-induced microbubble poration of localized single cells". *Lab on a Chip.* 2014;**14**(9):1572–8

[174] Fan Q., Hu W., Ohta A.T. "Efficient single-cell poration by microsecond laser pulses". *Lab on a Chip.* 2015;**15**(2):581–8

[175] Marmottant P., Raven J.P., Gardeniers H., Bomer J.G., Hilgenfeldt S. "Microfluidics with ultrasound-driven bubbles". *Journal of Fluid Mechanics.* 2006;**568**:109–18

[176] Hellman A.N., Rau K.R., Yoon H.H., Venugopalan V. "Biophysical response to pulsed laser microbeam-induced cell lysis and molecular delivery". *Journal of Biophotonics.* 2008;**1**(1):24–35

[177] Venugopalan V., Guerra III A., Nahen K., Vogel A. "Role of laser-induced plasma formation in pulsed cellular microsurgery and micromanipulation". *Physical Review Letters.* 2002;**88**(7):078103

[178] Wang M., Orwar O., Olofsson J., Weber S.G. "Single-cell electroporation". *Analytical and Bioanalytical Chemistry.* 2010;**397**(8):3235–48.

[179] Tsukakoshi M., Kurata S., Nomiya Y., Ikawa Y., Kasuya T. "A novel method of DNA transfection by laser microbeam cell surgery". *Applied Physics B.* 1984;**35**(3):135–40

[180] Soughayer J.S., Krasieva T., Jacobson S.C., Ramsey J.M., Tromberg B.J., Allbritton N.L. "Characterization of cellular optoporation with distance". *Analytical Chemistry.* 2000;**72**(6):1342–7

[181] Rau K.R., Quinto-Su P.A., Hellman A.N., Venugopalan V. "Pulsed laser microbeam-induced cell lysis: time-resolved imaging and analysis of hydrodynamic effects". *Biophysical Journal.* 2006;**91**(1):317–29

[182] Clark I.B., Hanania E.G., Stevens J., *et al.* "Optoinjection for efficient targeted delivery of a broad range of compounds and macromolecules into diverse cell types". *Journal of Biomedical Optics.* 2006;**11**(1):014034

[183] Le Gac S., Zwaan E., van den Berg A., Ohl C.D. "Sonoporation of suspension cells with a single cavitation bubble in a microfluidic confinement". *Lab on a Chip.* 2007;**7**(12):1666–72

[184] Sankin G.N., Yuan F., Zhong P. "Pulsating tandem microbubble for localized and directional single-cell membrane poration". *Physical Review Letters.* 2010;**105**(7):078101

[185] Li Z.G., Liu A.Q., Klaseboer E., Zhang J.B., Ohl C.D. "Single cell membrane poration by bubble-induced microjets in a microfluidic chip". *Lab on a Chip.* 2013;**13**(6):1144–50

[186] Tirlapur U.K., Konig K. "Targeted transfection by femtosecond laser". *Nature.* 2002;**418**(6895):290–1

[187] Waleed M., Hwang S.U., Kim J.D., Shabbir I., Shin S.M., Lee Y.G. "Single-cell optoporation and transfection using femtosecond laser and optical tweezers". *Biomedical Optics Express.* 2013;**4**(9):1533–47

[188] Baumgart J., Bintig W., Ngezahayo A., *et al.* "Quantified femtosecond laser based opto-perforation of living GFSHR-17 and MTH53 a cells". *Optics Express.* 2008;**16**(5):3021–31

[189] Uchugonova A., König K., Bueckle R., Isemann A., Tempea G. "Targeted transfection of stem cells with sub-20 femtosecond laser pulses". *Optics Express.* 2008;**16**(13):9357–64

[190] Vogel A., Noack J., Hüttman G., Paltauf G. "Mechanisms of femtosecond laser nanosurgery of cells and tissues". *Applied Physics B.* 2005;**81**(8): 1015–47

[191] Kohli V., Acker J.P., Elezzabi A.Y. "Reversible permeabilization using high-intensity femtosecond laser pulses: applications to biopreservation". *Biotechnology and Bioengineering.* 2005;**92**(7):889–99

[192] Torres-Mapa M. L., Gardner J., Bradburn H., King J., Dholakia K., Gunn-Moore F. "Femtosecond optical transfection as a tool for genetic manipulation of human embryonic stem cells". *Proceedings of the SPIE.* 2013;**8611**:861104

[193] Stevenson D., Agate B., Tsampoula X., *et al.* "Femtosecond optical transfection of cells: viability and efficiency". *Optics Express.* 2006;**14** (16):7125–33

[194] Tsampoula X., Garcés-Chávez V., Comrie M., *et al.* "Femtosecond cellular transfection using a nondiffracting light beam". *Applied Physics Letters.* 2007;**91**(5):053902

[195] Cižmár T., Kollárová V., Tsampoula X., *et al.* "Generation of multiple Bessel beams for a biophotonics workstation". *Optics Express.* 2008;**16** (18):14024–35

[196] Zeira E., Manevitch A., Khatchatouriants A., *et al.* "Femtosecond infrared laser—an efficient and safe *in vivo* gene delivery system for prolonged expression". *Molecular Therapy* 2003;**8**(2):342–50

[197] Zeira E., Manevitch A., Manevitch Z., *et al.* "Femtosecond laser: a new intradermal DNA delivery method for efficient, long-term gene expression and genetic immunization". *The FASEB Journal.* 2007;**21**(13):3522–33

[198] Tsen S.W., Wu C.Y., Meneshian A., Pai S.I., Hung C.F., Wu T.C. "Femtosecond laser treatment enhances DNA transfection efficiency *in vivo*". *Journal of Biomedical Science.* 2009;**16**(2):36

[199] Berns M.W., Rounds D.E. "Cell surgery by laser". *Scientific American.* 1970;**222**(2):98–103

[200] Nan L., Jiang Z., Wei X. "Emerging microfluidic devices for cell lysis: a review". *Lab on a Chip.* 2014;**14**(6):1060–73

[201] Rau K.R., Guerra III A., Vogel A., Venugopalan V. "Investigation of laser-induced cell lysis using time-resolved imaging". *Applied Physics Letters.* 2004;**84**(15):2940–2

[202] Lai H.H., Quinto-Su P.A., Sims C.E., *et al.* "Characterization and use of laser-based lysis for cell analysis on-chip". *Journal of The Royal Society Interface.* 2008;**5**(Suppl 2):S113–S121

[203] Quinto-Su P.A., Lai H.H., Yoon H.H., Sims C.E., Allbritton N.L., Venugopalan V. "Examination of laser microbeam cell lysis in a PDMS microfluidic channel using time-resolved imaging". *Lab on a Chip.* 2008;**8**(3):408–14

[204] He M., Edgar J.S., Jeffries G.D., Lorenz R.M., Shelby J.P., Chiu D.T. "Selective encapsulation of single cells and subcellular organelles into picoliter- and femtoliter-volume droplets". *Analytical Chemistry.* 2005;**77**(6):1539–44

[205] Wu T.H., Teslaa T., Kalim S., *et al.* "Photothermal nanoblade for large cargo delivery into mammalian cells". *Analytical Chemistry.* 2011;**83**(4):1321–7

[206] Misawa H., Koshioka M., Sasaki K., Kitamura N., Masuhara H. "Three-dimensional optical trapping and laser ablation of a single polymer latex particle in water". *Journal of Applied Physics.* 1991;**70**(7):3829–36

[207] Schütze K., Clement-Sengewald A., Ashkin A. "Zona drilling and sperm insertion with combined laser microbeam and optical tweezers". *Fertility and Sterility.* 1994;**61**(4):783–6

[208] Shelby J.P., Edgar J.S., Chiu D.T. "Monitoring cell survival after extraction of a single subcellular organelle using optical trapping and pulsed-nitrogen laser ablation". *Photochemistry and Photobiology.* 2005;**81**(4):994–1001

[209] Steubing R.W., Cheng S., Wright W.H., Numajiri Y., Berns M.W. "Laser induced cell fusion in combination with optical tweezers: the laser cell fusion trap". *Cytometry.* 1991;**12**(6):505–10

[210] Katchinskiy N., Godbout R., Goez H.R., Elezzabi A.Y. 'Femtosecond laser-induced cell-cell surgical attachment". *Lasers in Surgery and Medicine.* 2014;**46**(4):335–41

[211] Kulin S., Kishore R., Helmerson K., Locascio L. "Optical manipulation of liposomes as microreactors". *Biophysical Journal.* 2003;**84**(2):196A

[212] Neev J., Tadir Y., Ho P., Berns M.W., Asch R.H., Ord T. "Microscope-delivered ultraviolet laser zona dissection: principles and practices". *Journal of Assisted Reproduction and Genetics.* 1992;**9**(6):513–23

[213] Schütze K., Clement-Sengewald A. "Catch and move-cut or fuse". *Nature.* 1994;**368**(6472):667–9

[214] Liang H., Wright W.H., Rieder C.L., *et al.* "Directed movement of chromosome arms and fragments in mitotic newt lung cells using optical scissors and optical tweezers". *Experimental Cell Research.* 1994;**213**(1):308–12

[215] Kumar S., Maxwell I.Z., Heisterkamp A., *et al.* "Viscoelastic retraction of single living stress fibers and its impact on cell shape, cytoskeletal organization, and extracellular matrix mechanics". *Biophysical Journal.* 2006;**90**(10):3762–73

[216] Buer C.S., Gahagan K.T., Swartzlander G.A., Weathers P.J. "Insertion of microscopic objects through plant cell walls using laser microsurgery". *Biotechnology and Bioengineering.* 1998;**60**(3);348–55

[217] McDougall C., Stevenson D.J., Brown C.T., Gunn-Moore F., Dholakia K. "Targeted optical injection of gold nanoparticles into single mammalian cells". *Journal of Biophotonics.* 2009;**2**(12):736–43

[218] Isenberg G., Bielser W., Meier-Ruge W., Remy E. "Cell surgery by laser micro-dissection: a preparative method". *Journal of Microscopy.* 1976;**107**(1):19–24

[219] Meier-Ruge W., Bielser W., Remy E., Hillenkamp F., Nitsche R., Unsöld R. "The laser in the Lowry technique for microdissection of freeze-dried tissue slices". *The Histochemical Journal.* 1976;**8**(4):387–401

[220] Schütze K., Pösl H., Lahr G. "Laser micromanipulation systems as universal tools in cellular and molecular biology and in medicine". *Cellular and Molecular Biology.* 1998;(**44**):735–46

[221] Kummari E., Guo-Ross S.X., Eells J.B. "Laser capture microdissection a demonstration of the isolation of individual dopamine neurons and the entire ventral tegmental area". *Journal of Visualized Experiments: JoVE.* 2015;(**96**):e52336

[222] Olofsson L., Lundgren A., Brodelius P.E. "Trichome isolation with and without fixation using laser microdissection and pressure catapulting followed by RNA amplification: expression of genes of terpene metabolism in apical and sub-apical trichome cells of Artemisia annua L". *Plant Science.* 2012;**183**:9–13

[223] Jay D.G., Sakurai T. "Chromophore-assisted laser inactivation (CALI) to elucidate cellular mechanisms of cancer". *Biochimica et Biophysica Acta (BBA—Reviews on Cancer).* 1999;**1424**(2–3):M39–M48

[224] Yao C., Rahmanzadeh R., Endl E., Zhang Z., Gerdes J., Hüttmann G. "Elevation of plasma membrane permeability by laser irradiation of selectively bound nanoparticles". *Journal of Biomedical Optics.* 2005;**10**(6):064012

[225] Steinert R.F., Puliafito C.A. "The Nd-YAG laser in ophthalmology: principles and clinical applications of photodisruption". *The British Journal of Ophthalmology*. 1986;**70**(1):79

[226] Vogel A., Hentschel W., Holzfuss J., Lauterborn W. "Cavitation bubble dynamics and acoustic transient generation in ocular surgery with pulsed neodymium: YAG lasers". *Ophthalmology*. 1986;**93**(10):1259–69

[227] Juhasz T., Loesel F.H., Kurtz R.M., Horvath C., Bille J.F., Mourou G. "Corneal refractive surgery with femtosecond lasers". *IEEE Journal of Selected Topics in Quantum Electronics*. 1999;**5**(4):902–10

[228] Ratkay-Traub I., Ferincz I.E., Juhasz T., Kurtz R.M., Krueger R.R. "First clinical results with the femtosecond neodynium-glass laser in refractive surgery". *Journal of Refractive Surgery*. 2002;**19**(2):94–103

[229] Sutton G., Hodge C. "Accuracy and precision of LASIK flap thickness using the IntraLase femtosecond laser in 1000 consecutive cases". *Journal of Refractive Surgery*. 2008;**24**(8):802–6

[230] Knox C.N., Tyrer J.R., Jaycock P.D., Marshall J. "Effects of variation in depth and side cut angulations in LASIK and thin-flap LASIK using a femtosecond laser: a biomechanical study". *Journal of Refractive Surgery*. 2012;**28**(6):419–25

[231] He L., Sheehy K., Culbertson W. "Femtosecond laser-assisted cataract surgery". *Current Opinion in Ophthalmology*. 2011;**22**(1):43–52

[232] Bali S.J., Hodge C., Lawless M., Roberts T.V., Sutton G. "Early experience with the femtosecond laser for cataract surgery". *Ophthalmology*. 2012; **119**(5):891–9

[233] Haulot G., Benahmed A.J., Ho C.M. "Optoelectronic reconfigurable microchannels". *Lab on a Chip*. 2012;**12**(23):5086–92

[234] Zhu X., Yin Z., Ni Z. "Dynamics simulation of positioning and assembling multi-microparticles utilizing optoelectronic tweezers". *Microfluidics and Nanofluidics*. 2012;**12**(1–4):529–44

[235] Ota S., Wang S., Wang Y., Yin X., Zhang X. "Lipid bilayer-integrated optoelectronic tweezers for nanoparticle manipulations". *Nano Letters*. 2013;**13**(6):2766–70

Chapter 11

Polymer-based optofluidic lenses

Liqiang Ren[a], Yusheng Bian[a], Chenglong Zhao[b], Yanhui Zhao[a] and Tony Jun Huang[a]

Optofluidic lenses that can image or manipulate light at the microscale have attracted increased attention in recent years due to their tunability and reconfigurability. The discovery on transparent and elastic polymer materials has contributed to the development of the optofluidic lenses. These polymer-based optofluidic lenses feature better mechanical stability while remaining high tunability. The chapter begins with a brief introduction of optofluidic technologies. This is then followed by a summary of past and recent work on out-of-plane and in-plane optofluidic lenses, which is categorized according to their operation modes. The chapter is concluded with a commentary by the authors, which discusses possible future directions relating to optofluidic lenses.

11.1 Introduction

Optical technologies have been widely used in biological and medical research for sample detection and analysis.[1] Over the past few decades, tremendous academic and industrial attention has been directed at miniaturizing optical components, aiming to develop portable, easy-to-operate, and low-cost systems that can be used in electronics (e.g., cameras, cell phones, tablets), and bioimaging devices such as endoscope and miniaturized microscopes.[2–7] Meanwhile, microfluidics has emerged as an ideal platform that enables the miniaturization and integration of biochemical assays on a small chip, known as lab-on-a-chip assays.[8–19] Although the microfluidic systems can accurately handle small-volume samples on the microscale, many lab-on-a-chip assays require an integrated miniaturized optical system to process the analysis. The precise fabrication, alignment, and actuation of various traditional optical components in the extremely limited space is a challenging task. Thus, in recent years, researchers have been seeking new approaches

[a]Department of Engineering Science and Mechanics, Pennsylvania State University, University Park, PA 16802, USA
[b]Department of Physics, and Electro-Optics Graduate Program, University of Dayton, Dayton, OH 45469, USA

to further miniaturize the optical systems. Years of research efforts have yielded a large catalog of smart microfluidic devices and elements that can manipulate light at the microscale—a discipline now termed optofluidics.[20–23] Optofluidics can offer some unique properties that are not attainable using its solid counterpart, such as tunable refractive index, reconfigurable geometry, and convenient integration. Among the wide variety of optofluidic elements being investigated, optofluidic lenses are particularly important due to their versatile functionality on light manipulation and imaging.[24–38] Thus, these optofluidic lenses have captured the interest of the research community that continues to advance this promising research field.

During the development of optofluidic lenses, polymer materials have played crucial roles in both lens chamber construction and lens functions. In the early years, microfluidic-based devices were usually constructed on either silicon or glass substrates, which resulted in difficulties when one decided to integrate on-chip optical components with microfluidic systems. Silicon cannot transmit visible light, with most incident light reflected, scattered, or absorbed. Glass, on the other hand, allows high transmittance for a wide optical frequency range. However, engineering glass microstructures with optical smoothness is extremely challenging. In the past several years, the fabrication or replication of micro- and nanostructures using elastomeric stamps, molds, and conformable photomasks have been enabled with the emergence of soft-lithography techniques.[39–43] Soft-lithography has provided a new technical route for the development of polymer-based on-chip optical systems/components. One of the most commonly used polymeric materials for soft-lithography is polydimethylsilxane (PDMS),[13,44–48] a gel material that is viscoelastic. It can be cured upon heating and is optically clear, inert, non-toxic, and non-flammable. With such properties, PDMS is now the most commonly used material for microfluidic channel fabrication in academic labs and some industrial applications. Its simple fabrication process and unique properties provide enormous potential for developing PDMS-based on-chip optical devices, especially PDMS-based tunable optofluidic lenses for on-chip imaging and light manipulations. Other than PDMS, polymer materials with reversible optical properties, such as liquid crystal (LC), can also be used to construct optofluidic lenses with tunabilities, leading to versatile optofluidic lens functionalities.

Many excellent reviews on optofluidics have been published in the last decade, and the optofluidic lenses have also been discussed from different aspects.[49–51] Optofluidic lenses can be categorized based on their driving mechanisms, operation modes, materials, and constructions.[52] In this work, we will focus on polymer-based optofluidic lenses and structure our discussion based on their operation modes, namely, out-of-plane optofluidic lenses and in-plane optofluidic lenses. Following each route, we will explore the contributions of the polymer materials to the lens constructions. It is worth mentioning that there are other interesting lens examples that are beyond the scope of our discussion. We therefore encourage our readers to find more detailed information on other interesting optofluidic lens examples by referring to References 53–66.

11.2 Out-of-plane optofluidic lenses

Out-of-plane optofluidic lenses are not simply scaled-down versions of macroscale optical lenses, but a new category of lenses operating at the microscale with additional functionality. Back in 1999, a variable-focus out-of-plane optofluidic lens was implemented by trapping and flowing liquids with different refractive indices into a channel that resembled the crystalline lens structure of the human eye.[67] Such a lens changes its performance when different liquids are flowing inside the lens chamber. The elastic membrane constructed by transparent polymers, on the other hand, offers an alternative design concept by constructing tunable optofluidic lenses through deforming lens curvatures rather than flowing liquids with different refractive indices.[68–83] By deforming lens curvatures, a liquid/membrane lens is formed in such a way that liquids are trapped within a sealed elastic membrane cavity. The sealed membrane is thin and possesses the freedom to bend outward and inward. The curvature of the membrane can then be altered through various means by producing and maintaining temporary convex (focusing) or concave (divergent) structures, thereby adjusting the focal length to achieve the best imaging effect at the target point.

Following this design concept, tunable optofluidic lenses with pneumatic pressures deforming the transparent polymer membrane have been explored and reported extensively in the past decades. One of the design instances was demonstrated by Chronis *et al.* in their reported work on a tunable optofluidic lens array.[68] Elastic PDMS membranes are used to seal pre-defined lens array fluidic chambers with an underneath microfluidic network (Figure 11.1). The PDMS membranes are optically transparent so that they can be used as a lens after deformation. Upon pneumatically regulating the pressure inside the microfluidic network through a pneumatic pump with a pressure regulator, the focal lengths of all the lenses can be simultaneously controlled. However, to achieve maximum tunability of the focal length, a significant refractive index difference between PDMS and contained liquids is required in addition to the PDMS membrane deformations. The focal length can be tuned anywhere from several hundred microns to several millimeters. Figure 11.1(b) shows the variable focusing results characterized by a photograph imaged through the tunable lens, where increased resolution can be achieved by changing the pneumatic pressure.

While most of the pneumatically driven membrane lenses can only achieve a basic function (i.e., focusing of light), compound optofluidic lenses consisting of multiple transparent polymer layers can be constructed with real-time reconfigurable functions of both convex (focusing) and concave (diverging) lenses. One optofluidic compound microlens following this design concept was reported by Fei *et al.* (Figure 11.1(c)).[82] They used a multi-layered PDMS structure to simulate a self-aligned lens group consisting of one bioconvex and two plano-concave lens chambers. Each lens component is controlled independently by programmable pneumatic valves, which can be switched on and off digitally. In such operation modes, liquids with different refractive indices can be pushed into different lens chambers with precise control, thus adjusting the imaging power on demand from centimeters to sub-millimeters. Depending on the liquid selection and chamber shapes, more sophisticated lens functions, including focusing and diverging, can be

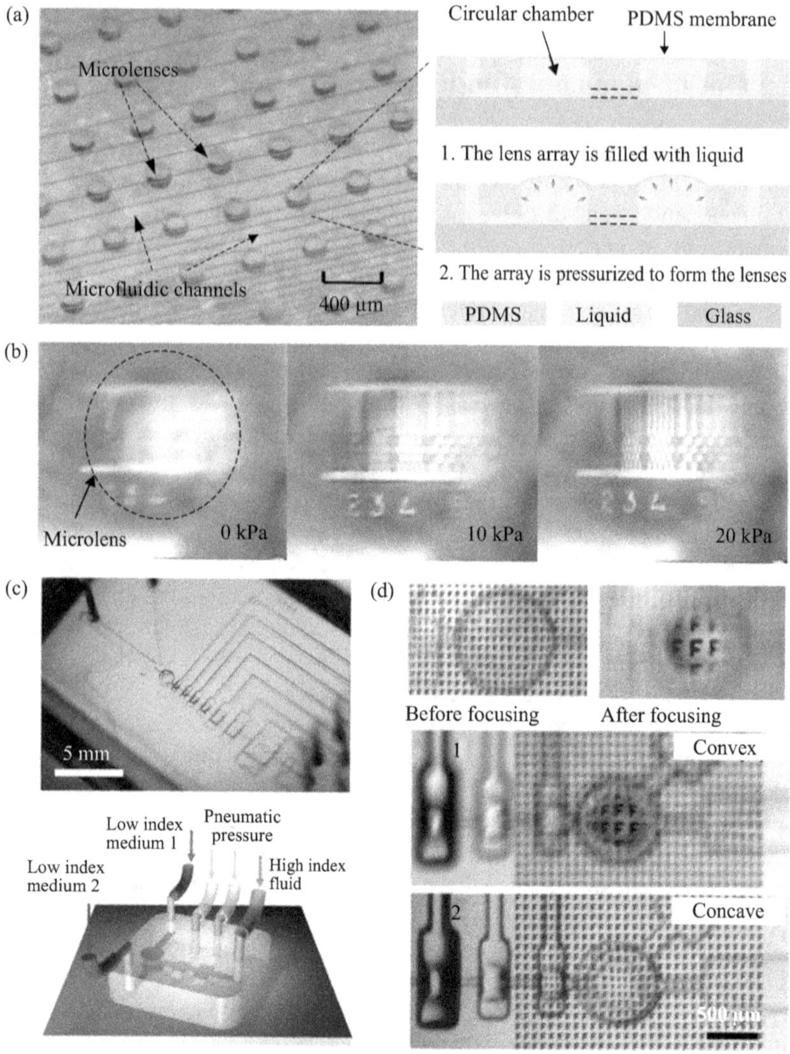

Figure 11.1 *(a) Microscopic image of the microlens array with underlining microfluidic network for pressure regulation (left) and the working mechanism of the microlens array (right). (b) Microscope imaging results corresponding to variable focusing effect under different pressures. (c) Schematic of compound three-layer microfluidic lens. In each layer, different liquids can be pushed into the channel to provide different refractive index contrast. Pneumatic valves are used to increase the liquid pressure in the center layer discretely to adjust the deformation at the liquid/membrane interface. (d) Tunable focusing effect can be controlled precisely using the pneumatic valves. Using different combinations of liquids in center bi-convex chamber and two plano-concave chambers at the top and bottom layers, the device can perform as convex or concave lens (Modified with permission from Chronis (2003) [68] and Fei (2011) [82], copyright the Optical Society and the Royal Society of Chemistry)*

configured in real time, corresponding to the constructions of convex (focusing) and concave (diverging) lenses. The imaging results of variable focusing and lens functioning are shown in Figure 11.1(d).

Furthermore, there are several mechanisms that can be adapted for in-situ control of membrane curvature other than fluidic pressure control. Lee *et al.* introduced a tunable liquid lens with an integrated electromagnetic actuator to control membrane deformation (Figure 11.2(a)).[84] The device consists of a fluidic chamber with two openings. Both openings are sealed with a thin PDMS slab to form two deformable PDMS membranes: one for the lens surface and the other for the driving unit that carries a coiled electrode to generate the Lorentz force for lens actuation. The membrane is placed within a transverse external magnetic field. When current is applied to the electric coil, a Lorentz force acts in the direction normal to the actuating membrane, causing it to deform. If the volume of the fluidic chamber is fixed and the liquid is incompressible, the PDMS membrane for the lens will also deform to offset the volumetric change induced by the deformation of actuating PDMS membrane (Figure 11.2(b)). The deformed membrane and high-refractive-index fluid inside the chamber create a tunable lens whose curvature can be actuated by the current passing through the coil. The results are shown in Figure 11.2(c). Besides lens chambers sealed using elastic polymers similar to PDMS, solid polymer such as Teflon can also be used in constructing polymer-based optofluidic lenses.[85] Figure 11.2(d) shows the schematic of such a lens activated by the electrochemical method. The lens is formed by overfilling a small through-hole (diameter less than 1 mm) in a roughly 3-mm-thick Teflon plate. The non-wetting property of Teflon ensures that the contact lines of the liquid are pinned at the opening of the holes and that the optical axis of the droplet is aligned with the axis of the hole. The hanging meniscus has a spherical shape and can be used to focus light. The curvature of the hanging meniscus is determined by the balance of surface tension of the liquid at the top and bottom of the contacting surface, as well as the hydrostatic pressure. In order to tune the curvature of the lens, a water-soluble ferrocenyl surfactant, whose surface activity can be electrochemically controlled, is added to the water. Surfactant can increase (when oxidized) or decrease (when reduced) surface tension over a total range of 16 dyn/cm (more than 1/5 of the surface tension of pure water). Surface tension variation results in a change of the droplets' balance position and hanging meniscus curvature. Figure 11.2(e) shows a reversible lensing effect, which can be controlled electrochemically by varying the potential of the cathode and anode (ring electrode embedded at the top and bottom of the through-holes). Obvious change in the resolution after electrochemical activation of the liquid lens can be seen in Figure 11.2(f).

Figure 11.3(a) describes an interesting design of an adaptive optofluidic lens developed by Dong *et al.*[86,87] The lens consists of a water-filled stimuli-responsive hydrogel ring that is sandwiched between a glass substrate and an aperture slip, whose opening is aligned with the hydrogen ring. The hydrogen ring is first filled with water through the nearby microfluidic network. Afterward, oil is added to the top of the water meniscus and capped with an additional glass cover slip. Pertaining to aperture design, the surfaces of the sidewall and bottom ('ca' in Figure 11.3(a))

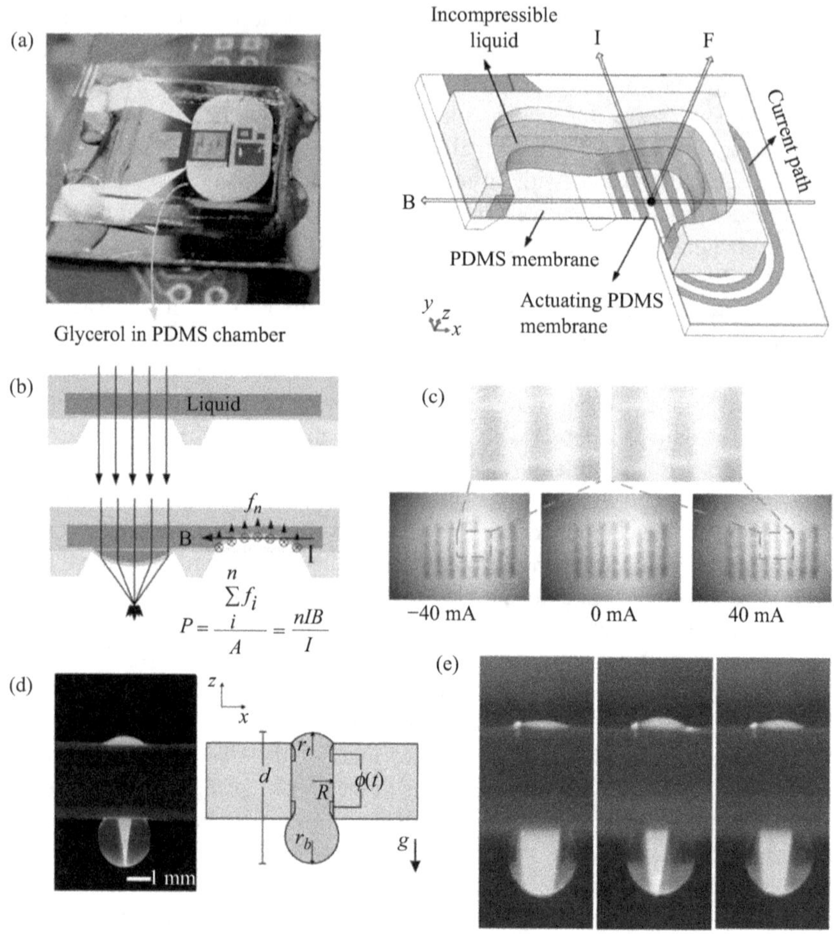

Figure 11.2 (a) Optical image and schematic of an optofluidic lens driven by an electromagnetic actuator. (b) The tuning mechanism of electromagnetic actuator controlled optofluidic lens. (c) Images obtained when different currents are applied on the actuator, confirming the change of the focal length. (d) The image of a green laser source via a redox-surfactant tunable liquid lens (left). The schematic of the liquid lens shows pinned hanging meniscus and the electrode embedded at each contact circle (right). (e) Reversible focusing effect with electrochemical activation. (f) 1 and 2, images of the standard target imaged through the pure water (2 is a magnified view of the square in 1). 3 is the image of the same area after the actuation of the lens (Modified with permission from Lee (2007) [84] and Lopez (2005) [85], copyright AIP Publishing LLC)

(f)

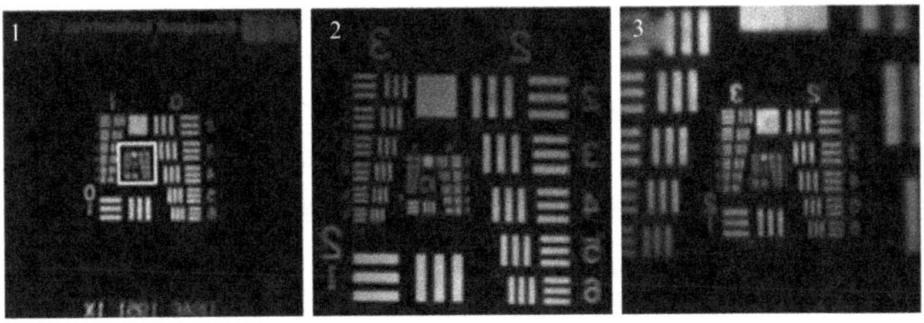

Figure 11.2 (Continued)

were modified to be hydrophilic, while the top surface ('ts' in Figure 11.3(a)) was treated to be hydrophobic. Such treatment is crucial in establishing a stable water–oil interface needed for lens tuning. It ensures that the boundary between water and oil phases is always pinned along the hydrophobic–hydrophilic contact lines (edge of the apertures opening). Figure 11.3(b) shows that when proper stimuli (e.g., pH, temperature, light) are provided, the hydrogel underneath the aperture can expand or shrink, resulting in a volumetric change of the space inside the ring. The volume changes cause the over-flow of water enclosed by the ring and top oil layer, which in turn causes a pressure that acts across the water–oil interface. The curvature of the interface, and hence the focal length of the liquid lens, changes as a response to the pressure difference, as shown in Figure 11.3(c). A similar design concept was also adopted in designing a thermo-pneumatically tunable optofluidic lens.[83] The lens is based on separate pump and optical cavities in a microfluidic system. A liquid-filled lens cavity sealed by a deformable transparent membrane contains four thermal pumps inside air-filled cavities. Once heated, the air in the thermal pump expands, transferring the expansion to the lens cavities to compress the optical liquid and deform the lens membrane (Figure 11.3(d)). Figure 11.3(e) shows images of a test object at two different applied voltages, indicating a tuning effect of the reported optofluidic lens.

In addition to being used as materials for lens container constructions, some polymers have proven to be good candidate materials in forming optofluidic lens cores due to their unique optical responses. Among various polymers and polymer composites that can be potentially used in constructing optofluidic lens materials, polymer networks of LCs are able to exhibit different optical responses upon external stimuli, such as: electric field,[88] thermo-field,[89] optical field,[90] and acoustic field.[91] Hence, it can be used to construct tunable optofluidic lenses that work in out-of-plane configurations. A planar hermaphroditic LC microlens is reported by Ren *et al.*[92] The plano-concave lens is constructed from an ultra-violet (UV) cured polymer-LC composite on a polyimide surface. The polyimide surface is rubbed and UV-cured to generate lens chambers that hold LC. The curved lens chamber is anisotropic and has different refractive indices along x and y directions as depicted in Figure 11.4(a). The polymer network LC is carefully constructed with its ordinary refractive index n_o and extraordinary refractive index n_e meeting

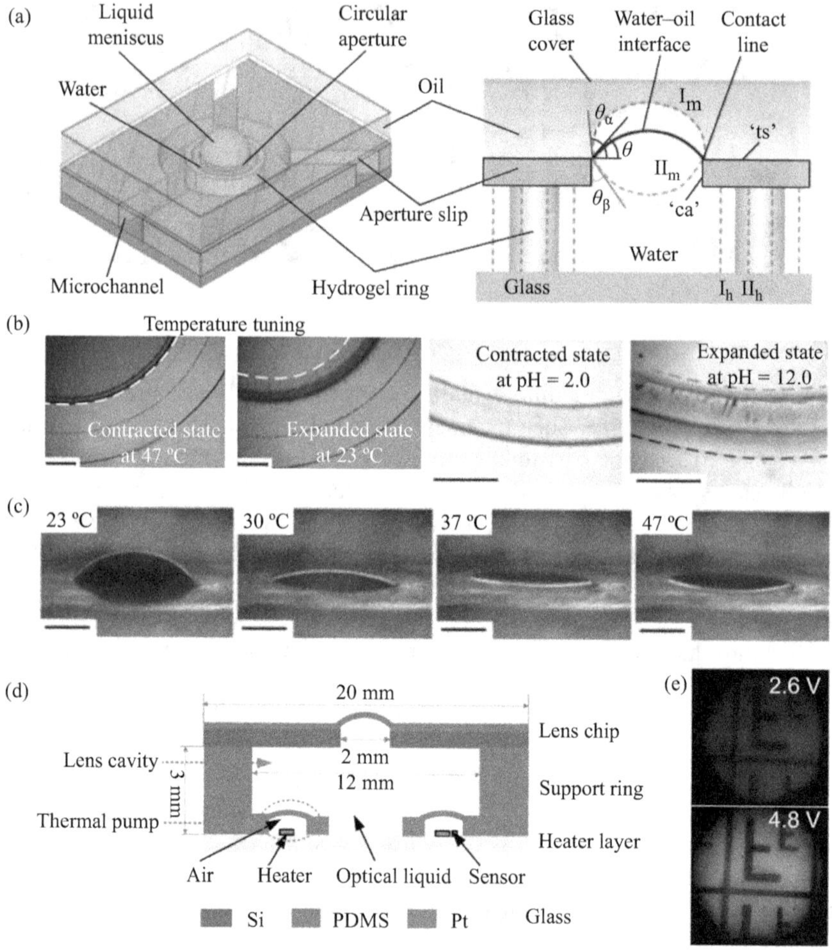

Figure 11.3 *(a) Schematic of the assembly of the hydrogel ring with the microfluidic components for pinning the water–oil interface. (b) Various focusing mechanism that can be achieved through stimuli-responsive hydrogel under stimuli such temperature and pH. (c) Changes of the fluidic lens curvature due to the external stimuli of varying temperatures. (d) Schematic design of the thermally tunable optofluidic lens. Thermal expansion of the air causes compression of the liquid inside lens cavity and deforms the lens membrane. (e) Tunable focusing effect can be observed from thermally tunable optofluidic lens under different thermal expansions from different voltages (Modified with permission from Dong (2006) [86] and Zhang (2011) [83], copyright Nature Publishing Group and the Optical Society)*

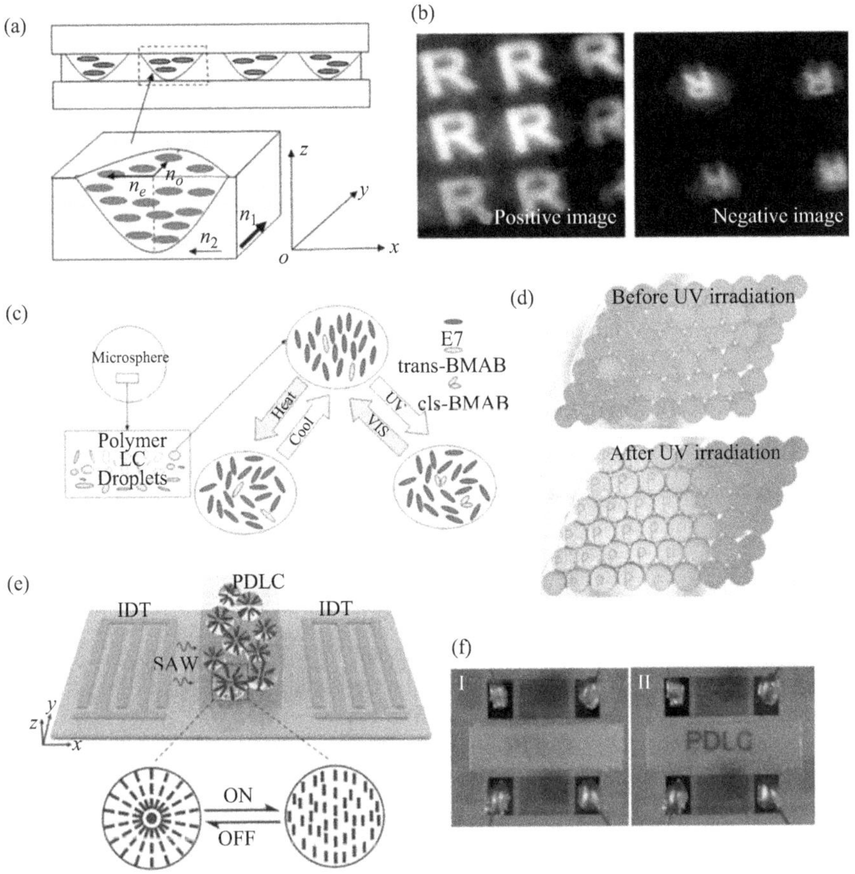

Figure 11.4 *(a) Schematic of hermaphroditic LC optofluidic lens arrays:* n_1 *and* n_2 *are the refractive indices of the molded microlenses along the* y *and* x *axes, respectively, and* n_o *and* n_e *are the ordinary and extraordinary refractive indices of the LC material, respectively. (b) Positive and negative imaging results obtained from hermaphroditic LC microlens array by rotating polarization direction 90° manually or electrically. (c) The tuning mechanism of phototunable optofluidic lens array based on a PDLC network. (d) Optical images of an optofluidic lens array based on PDLC before and after UV irradiations. (e) The mechanism of a tunable PDLC-based optofluidic lens controlled by acoustic field. (f) PDLC-based optofluidic lens can be driven by SAWs and shows an optical shutter effect (Modified with permission from Liu (2011) [91], Ren (2005) [92] and Xiong (2009) [93], copyright the Optical Society and Wiley)*

the relation $n_e \approx n_1 > n_2 \approx n_o$, with the LC alignments between n_e and n_o switchable through an applied electric field. Such a polymer-based LC lens will respond differently to the incident light at different polarizations. This will generate both positive and negative imaging effects toward the same object upon rotating the lens/polarizer, or use an electrically controlled nematic cell (Figure 11.4(b)). Xiong *et al.* reported a polymer-dispersed LC (PDLC) microlens array that can be reversibly reconfigured via light irradiations.[93] The LC (E7) mixtures used in this case are thermo-sensitive and will change their optical responses upon thermal excitation and relaxation; however, to make the PDLC mixture respond to different light irradiation, photoresponsive polymer is needed. 4-Butyl-4-methoxyazobenzene (BMAB) is one of such photoresponsive molecules used in the demonstrated case. It can undergo *trans*- to *cis*-isomerization under UV irradiation and can be recovered through thermal relaxation or visible light illumination, as shown in Figure 11.4(c). A microsphere lens consisting of both PDLC and BMAB, controlled via light irradiation, can achieve a reversible change between opacity (aligned) state and transparent (isotropic) state. Microlens arrays can be fabricated in a close-pack format through a self-assemble method on the target surface with capabilities of imaging multiple objects controlled by light. Optical images of a photoswitchable polymer LC microlens before and after UV irradiations are shown in Figure 11.4(d). Besides electric and optical fields, LC-based optofluidic lenses can also be driven by acoustic field, as reported by Liu and coworkers.[72] Surface acoustic waves (SAWs) generated from interdigital transducers can be precisely controlled on their wavelengths and amplitude and can provide enough power to drive a PDLC. Pre-aligned LC networks can be disturbed through acoustic wave induced shear-flow, switching its state from anisotropic-like opacity to isotropic-like transparency (Figure 11.4(e)). This effect has been demonstrated in Figure 11.4(f), with target characters underneath PDLC networks both hidden and visible, controlled by applied SAW fields.

In general, the out-of-plane optofluidic lenses are compatible with traditional optical systems and have good performance in three-dimensional (3D) manipulation of light. However, the requirement of precise alignment makes their fabrication processes quite complex. Furthermore, these vertical configurations in some situations also present significant challenges for on-chip integration and miniaturization. In this regard, the out-of-plane optofluidic lenses are usually used for imaging purpose.

11.3 In-plane optofluidic lenses

Another category of optofluidic lens, called in-plane optofluidic lens, is constructed and operated within a two-dimensional (2D) plane that confines the microfluidic chips. Limited by dimensions, most constructed optofluidic lenses resemble cylindrical optical lenses that lack the maneuver power to manipulate light along the vertical direction.[94–97] This property of in-plane optofluidic lenses makes them less desirable for imaging purposes; however, the in-plane structures are more amenable to system integration. The capabilities of manipulating light in-plane are much preferred by other lab-on-a-chip applications such as on-chip particle/cell manipulation, sensing and detection. Similar to out-of-plane lenses, the contributions of polymer materials in this case are to provide an optically transparent container or confinement that can form a lens shape with real-time reconfiguration. They are achieved through flow manipulation,[98–107] electric[108,109] and pneumatic methods.[110–114]

Figure 11.5(a) shows a dynamic optofluidic lens with a liquid core–liquid cladding configuration reported by Tang *et al.*[100] The flow pattern of liquid core and liquid cladding will change under different flow rates in a rectangular chamber fluidic channel, and this interface change between the two fluids with different refractive indices provides a lens-like modulation on the incident light from the optical fiber. An obvious focusing effect is shown in Figure 11.5(b) with the focal length dynamically controlled by changing flow rate. A "microfluidic drifting" lens is another smart design, which utilizes a curve-shaped channel as shown in Figure 11.5(c).[98,99] The curved channel offers a centrifugal force applied on the interface of two immiscible liquids. With the help from these curved channels, the force will change upon different applied flow rates, causing an interface control for light focusing. As a final example of an in-plane optofluidic lens, 2D Liquid Gradient-Refractive-Index (L-GRIN) lens is illustrated in Figure 11.5(d).[107] By sandwiching a core flow of high refractive index between two cladding flows of low refractive index, refractive index diffusion gradients are established at their flow interfaces. Then when a light beam is directed down the microfluidic channel, the L-GRIN causes the light to focus at a point downstream of the flow channel. By changing the flow rates of the cladding and core flows, the focal point of the L-GRIN lens can be tuned not only along the lens axis but also orthogonal to the lens axis. Thus, the unique L-GRIN lens achieves tuning of focal point in both dimensions of the device plane, as shown in Figure 11.5(e).

Flow-caused diffusion is the most simple and straightforward way to construct in-plane optofluidic lenses; however, the flow methods suffer drawbacks when the designed lens requires extreme flow conditions. Considering the elastic nature of PDMS channels that are usually used in constructing tunable out-of-plane optofluidic lenses, ultra-high flow rates with built-in pressures tend to deform in-plane lens chambers and are less preferred when designing in-plane optofluidic lenses. Therefore, electric-based tuning methods have been explored and developed to achieve non-contact tuning of in-plane optofluidic lenses. Li *et al.* reported a bi-concave optofluidic lens that can be tuned electro-kinetically under the combined effect of both driving flow and electro-osmosis.[109] A schematic of said optofluidic lens is shown in Figure 11.6(a). The lens is designed as a conventional liquid core–liquid cladding optofluidic lens, as discussed in previous cases, but with constant driving flows. Conduct flows and non-conduct flows are used alternatively as core fluids and cladding fluids, having their interface respond differently toward external electric fields to form a focusing lens or diverging lens. The diverging and focusing lens effects are shown in Figure 11.6(b) and 6(c), respectively, under different applied voltages. An alternative design to relief pressures applied on the

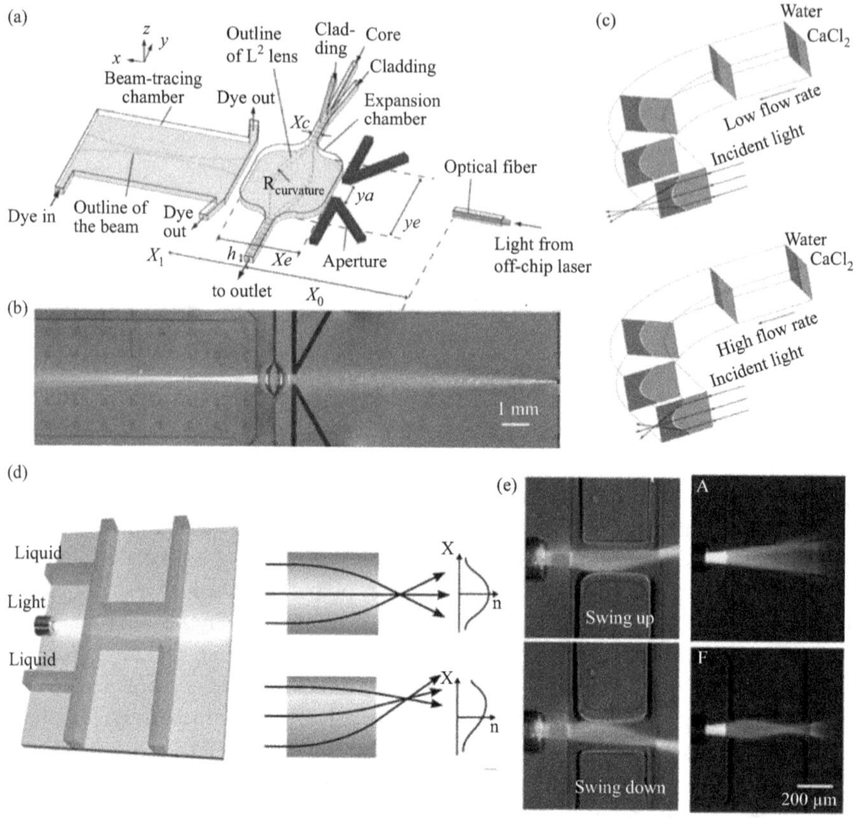

Figure 11.5 *(a) Schematic of the liquid-core/liquid cladding tunable optofluidic lens. The lens is built upon a specially design PDMS channel to shape the liquid/liquid interface through diffusion and flow rate control. (b) Optical image of the tunable optofluidic lens in action. (c) Mechanism of in-plane tunable fluid–fluid lens for focusing light in the z-direction with respect to device plane (where the plane of the device is the xy plane). The tuning of the lens curvature is via taking advantage of the "Dean flow" effect. Change of the lens curvature and focal distance can be realized through changing the flow rate. (d) The schematic of the L-GIRN device that utilizes the diffusion to establish the light-focusing and light-deflecting refractive index profile. (e) The focusing and deflecting power of the L-GRIN lens can be conveniently adjusted by changing the flow conditions (Modified with permission from Mao (2007) [98], Tang (2008) [100] and Mao (2009) [107], copyright the Royal Society of Chemistry)*

polymer chamber is to guide flow pressure to open through special channel designs. This concept can also be used to construct in-plane optofluidic lenses. Such a lens configuration using an air–liquid interface inside a PDMS channel is shown in Figure 11.7(a).[113] The PDMS microchannel is specially designed such that the

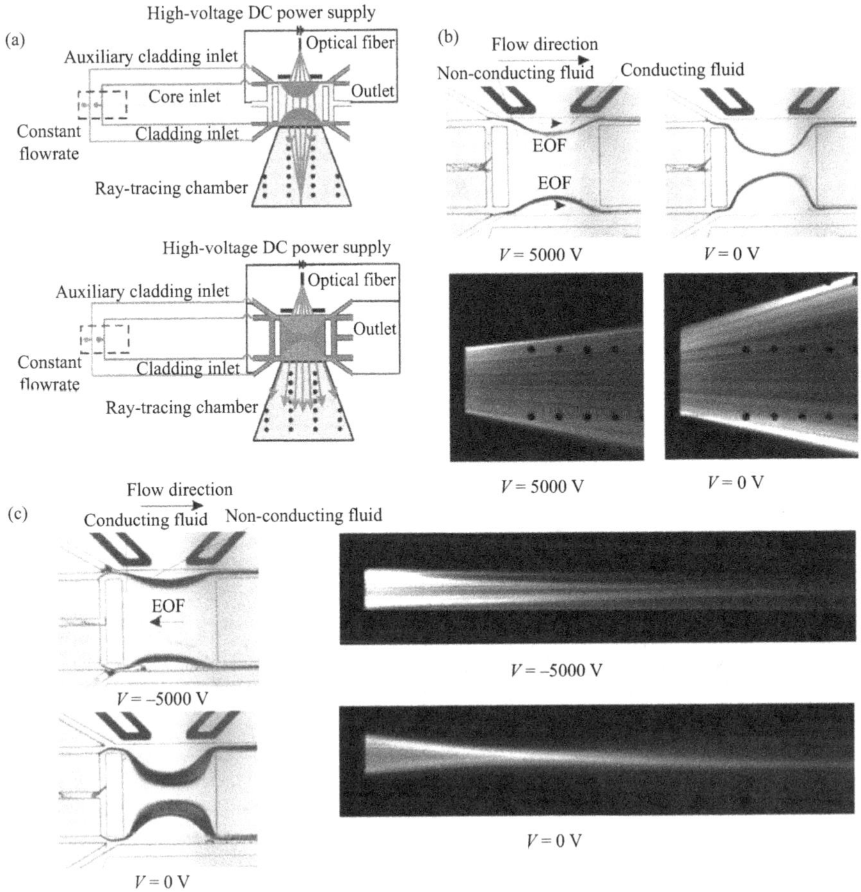

Figure 11.6 *(a) Schematic of bi-concave lens in focusing mode and diverging mode. Different lens functions are achieved through interchanging the liquids for the core, cladding, and auxiliary cladding streams. The flow rate is kept constant, while positive and negative voltage biases are applied to achieve desired lens functions. (b) The interfacial shape and ray-tracing path of the optofluidic bi-concave lens in diverging mode based on the different applied voltages and fixed flow rates. (c) The interfacial shape and ray-tracing path of the optofluidic bi-concave lens in focusing mode based on the different applied voltages and fixed flow rates (Modified with permission from Song (2012) [109], copyright the Royal Society of Chemistry)*

interface is in-line with the optical fiber and modulates the laser beam coming out from the fiber tips. The air–liquid interface can be finely adjusted by simply changing the flow rate of the liquids inside the chamber. Furthermore, since the interface is not as stiff as the channel wall and cannot withstand the pressure from

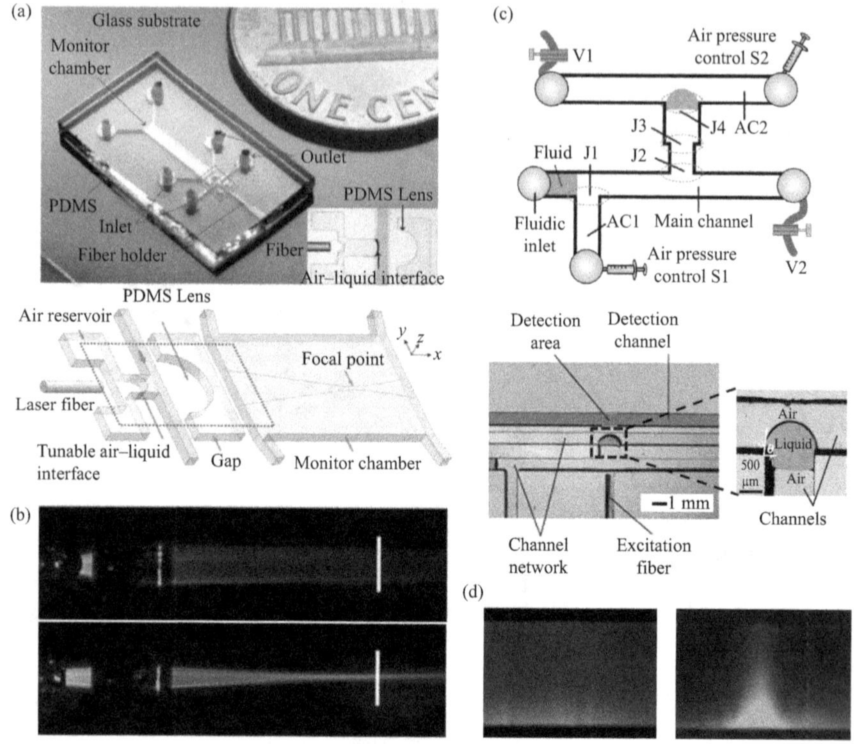

Figure 11.7 *(a) Schematic and optical image of pneumatically driven in-plane optofluidic lens. The inset is a microscopic image of the lens construction. (b) Ray-tracing experiments to characterize the variable focal length and light intensity at different flow rates of DI waters. (c) Schematic and optical image of microfluidic setup for the in situ construction of an optofluidic tunable lens driven by pressure difference. The curved liquid/air interface of the optofluidic lens is pinned at a T-shape junction. The contact angle as well as the lens position can be actively tuned by changing the pressure difference across the optofluidic lens. (d) Fluorescence emission detected inside microfluidic channel without (left) and with (right) the optofluidic lens in action (Modified with permission from Shi (2009) [113] and Dong (2007) [115], copyright Springer and AIP Publishing LLC)*

the high flow rate, it will be pushed toward the airside, causing itself to deform and thereby modulate the light beams impinged on the interface. Together with another PDMS curvature design as light beam collector, the air–liquid interface and PDMS channel design form a microfluidic lens system with tunable lens performance through adjustable flow rate (Figure 11.7(b)). Figure 11.7(c) shows another design of in-plane optofluidic lens based on a liquid droplet.[115] The droplet is confined in microfluidic channels with a built-in pressure difference on its two sides (Figure 11.7(c)). By taking advantage of the surface tensions in microscale, the pressure difference applied on the liquid droplet will deform the droplet shape, thus changing the lens performance. The light focusing with and without the liquid droplet lens is shown in Figure 11.7(d), indicating obvious focusing effect with the presence of the lens.

Capabilities of light manipulations of in-plane optofluidic lenses are not limited within 3D light waves, but can be extended to 2D surface waves that are termed surface plasmon polaritons (SPPs). SPPs are surface-confined electron oscillations that can only exist when strict excitation conditions are met. The studies on SPP generate a new research direction called plasmonics,[116–120] which have led to many groundbreaking research outcomes thus far. The studies on plasmonics required manipulation of SPP at challenging conditions below the diffraction limit of light.[121,122] This is an especially challenging task considering the surface confinement and decay nature of the SPP. The manipulation of SPP is conceptually similar to light manipulation of in-plane optofluidic lenses in the 2D plane, and efforts have been made to use poly(methyl methacrylate) (PMMA) to manipulate SPP in recent studies. A plasmonic Luneburg and Eaton lens made of PMMA on a metal surface can focus SPP effectively, as shown in Figure 11.8(a).[123] By adiabatically modifying the topology of the PMMA layer, the effective mode index of SPPs changed gradually rather than abruptly in these lenses. This makes the PMMA layer work as a lens toward SPPs, and their propagation losses caused by scattering can be significantly reduced. Inset of Figure 11.8(a) shows a SEM image of the plasmonic Luneburg lens. Figure 11.8(b) shows both experimental and simulation results that SPP was directed by the plasmonic Luneburg lens and maintained its well-focusing state after passing through this lens.

Research interests on integrating optofluidics with plasmonics are leading to the formation of new research branches of plasmofluidics.[124,125] Although still in its early stage, this field has shown enormous potential as it has the advantages of both fluids and plasmonics. For example, Zhao *et al.* invented a new method of using thermal bubbles induced and controlled through a laser source to manipulate SPPs.[124] When a low-power diode laser is focused onto a gold–water interface, the gold film is quickly heated up and a vapor surface bubble quickly forms on the gold surface. Figure 11.8(c) shows a schematic of the proposed plasmofluidic lens design and two surface bubbles with different sizes that were generated on a metal film. The surface bubble introduces the local index change for surface plasmons. By controlling the size and location of the surface bubbles, as well as the phase of the incident surface plasmons, they have demonstrated using the same lens to focus, diverge, and collimate SPPs. Figure 11.8(d) shows how the focal point of SPP was dynamically changed as the size of the surface bubble is altered by the laser power. The advantage of using a bubble to manipulate SPPs lies in the fact that the bubble size, position and even the shape can be easily tuned and controlled by a laser.

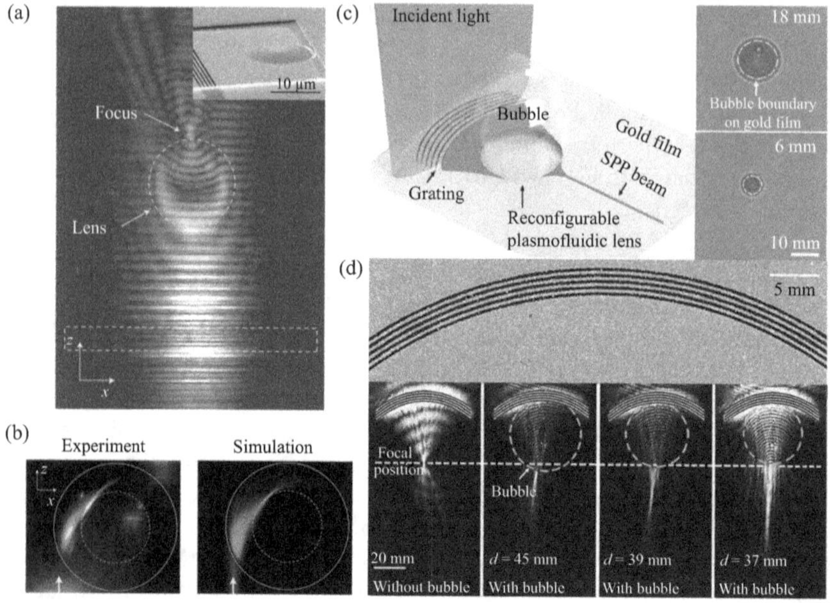

Figure 11.8 (a) Intensity image obtained from leakage radiation microscope of SPP passing through a PMMA Luneburg lens. The focusing size of SPP about 70 nm can be achieved as seen from the figure. The inset is a scanning electron microscope (SEM) image of the PMMA plasmonic Luneburg lens. (b) Fluorescence microscopy image and corresponding simulation of the fluorescence intensity of the Eaton lens for SPPs propagating in the positive z-direction and bending to the right side when passing through the lens. Arrows indicate the launching position and direction of the SPPs. Solid lines mark the outer diameter of the lens, and dashed lines the high index region, which was set to 1.54. (c) Schematic of the reconfigurable plasmofluidic lens, wherein a laser-induced surface bubble is used to control the propagation of SPPs at the metal surface. Surface bubbles with diameters of 18 and 6 mm are generated on the gold film. The dashed circle represents the surface bubble boundary on gold film. Scale bar is 10 mm. (d) Experimental demonstration of SPP divergence by the reconfigurable plasmofluidic lens. The leakage radiation images of SPPs are recorded when surface bubbles are absent and with different diameters (45, 39, and 37 mm). The dashed circle represents the surface bubble boundary on gold film. Scale bar for the SEM image and optical images are 5 and 20 µm, respectively (Modified with permission from Zentgraf (2011) [123] and Zhao (2013) [124], copyright Nature Publishing Group)

11.4 Prospective and conclusion

Polymer-based optofluidic lenses have gone through a long route to reach their current state, and a promising future is expected. However, much work remains to be done before the field can fulfill its potential. For example, how to utilize various polymer materials to maximize the potentials of optofluidic lenses is a question that has yet been answered, and we would like to share our thoughts in this section to inspire our readers to pursue their own research in this field.

As a category of stimuli-responsive polymer materials, smart polymer materials have been a focus of polymer research for several decades.[126–130] What make such materials attract great interests are their capability to respond to external stimulus and small changes in environment conditions. The unique property of smart materials makes them promising candidates for applications in microfluidics and lab-on-a-chip systems. We have already seen some examples on tunable optofluidic lenses that utilized the "smart" properties of polymer materials from our discussion above. However, the tremendous potential for smart polymer materials on constructing tunable optofluidic lenses has only been scratched on its surface. While the tunabilities of optofluidic lenses are usually achieved from the mechanical deformation and some of the chemical/electrical stimulus. The smart polymer materials have demonstrated their responses to a much wider range of stimuli, including but not limiting to: pH,[131,132] ion concentration,[133] additional solvents/gases,[134,135] redox reactions,[136] voltage,[137] temperature,[138,139] light-irradiation,[140–142] mechanical stress[143,144] and some biochemical stimulus such as antigen, enzyme and ligands. With such versatile responses from the smart polymer materials, the new types of tunable optofluidic lenses will be beneficial in two aspects: new lens chamber materials and new lensing materials that alter their optical properties upon stimulus.

The booming of microfluidic research starts from the inventions of revolutionary soft-lithography techniques and PDMS. They have significantly simplified the channel fabrication techniques based on glass or silica, providing several advantages such as biocompatibility, chemical stability, optical transparency, adaptability and low-cost that make microfluidic research an enabling platform for clinical diagnosis and analysis. In terms of optofluidic lens constructions, PDMS is considered as an excellent material due to its transparency and excellent elasticity. On the other hand, PDMS has properties (such as robustness) that make it difficult to develop PDMS-based commercially viable products. Future development on polymer synthesis may lead to new types of smart polymer material that have excellent optical and mechanical properties and can replace PDMS as lens materials. These new material properties could not only reshape the research field of optofluidic lenses but may also lead to breakthroughs in the field of microfluidics and lab on a chip.

The stimulus-responsive smart polymers themselves can also be used as lens materials due to their reversibility on optical properties and shapes. LCs are very good candidates for tunable optofluidic lenses seen from the examples we provided in this review. We believe that other smart polymer materials with similar reversible optical properties will be developed and can hopefully provide some novel optical properties. Although not demonstrated yet, optofluidic lenses may also be constructed from microvesicles that are self-assembled from azopyridine-containing diblock copolymers.[145,146] The microvesicle will change its shape reversibly through a photo-controlled swelling–shrinking process, affecting the size of the optical windows which can lead to continuous tuning on optical behavior, such as beam focusing and energy concentration.

In summary, we have reviewed recent progresses made on constructing tunable optofluidic lenses by using polymers that exhibit excellent optical, mechanical, electrical, and chemical properties. Polymer-based optofluidic lenses have shown great potential in microfluidic-based diagnostics, sensing, and detections. Their potentials in preparing the next-generation optofluidic lenses with superior functions, such as high-resolution imaging, nano-object manipulation and detection are yet to be explored. Seeking alternative polymers to improve the performance and functionality of the current optofluidic lens is another important task to further push the integration of polymer science, microfluidics and optics.

Acknowledgments

This research was supported by National Science Foundation (IDBR-1455658 and CBET-1438126), National Institutes of Health (1 R01 GM112048-01A1 and 1R33EB019785-01), and the Penn State Center for Nanoscale Science (MRSEC) under grant DMR-1420620.

References

1 X. Fan and I. M. White, *Nature Photonics*, 2011, **5**, 591–597.
2 F. B. Myers and L. P. Lee, *Lab on a Chip*, 2008, **8**, 2015–2031.
3 A. W. Martinez, S. T. Phillips, G. M. Whitesides and E. Carrilho, *Analytical Chemistry*, 2010, **82**, 3–10.
4 X. Mao and T. J. Huang, *Lab on a Chip*, 2012, **12**, 1412–1416.
5 S. O. Isikman, A. Greenbaum, M. Lee, *et al.*, *Analytical Cellular Pathology (Amsterdam)*, 2012, **35**, 229–247.
6 H. Zhu, S. O. Isikman, O. Mudanyali, A. Greenbaum and A. Ozcan, *Lab on a Chip*, 2013, **13**, 51–67.
7 O. Mudanyali, S. Dimitrov, U. Sikora, S. Padmanabhan, I. Navruz and A. Ozcan, *Lab on a Chip*, 2012, **12**, 2678–2686.
8 Y. Zhao, D. Chen, H. Yue, *et al.*, *Lab on a Chip*, 2013, **13**, 2183–2198.
9 G. M. Whitesides, *Nature*, 2006, **442**, 368–373.
10 J. Wu, X. Wu and F. Lin, *Lab on a Chip*, 2013, **13**, 2484–2499.
11 D. Di Carlo, *Lab on a Chip*, 2009, **9**, 3038–3046.
12 J. Wang, L. Ren, L. Li, *et al.*, *Lab on a Chip*, 2009, **9**, 644–652.

13 W. Zhang, S. Lin, C. Wang, *et al.*, *Lab on a Chip*, 2009, **9**, 3088–3094.
14 I. Barbulovic-Nad, H. Yang, P. S. Park and A. R. Wheeler, *Lab on a Chip*, 2008, **8**, 519–526.
15 B. Weigl, G. Domingo, P. Labarre and J. Gerlach, *Lab on a Chip*, 2008, **8**, 1999–2014.
16 S.-Y. Teh, R. Lin, L.-H. Hung and A. P. Lee, *Lab on a Chip*, 2008, **8**, 198–220.
17 S. K. Sia and L. J. Kricka, *Lab on a Chip*, 2008, **8**, 1982–1983.
18 Y. Gao, P. Li and D. Pappas, *Biomedical Microdevices*, 2013, **15**, 907–915.
19 L. Ren, X. Wu, M. Li, X. Zhang, L. Liu and L. Xu, *Optics Letters*, 2012, **37**, 3873–3875.
20 V. R. Horowitz, D. D. Awschalom and S. Pennathur, *Lab on a Chip*, 2008, **8**, 1856–1863.
21 C. Monat, P. Domachuk and B. J. Eggleton, *Nature Photonics*, 2007, **1**, 106–114.
22 D. Erickson, D. Sinton and D. Psaltis, *Nature Photonics*, 2011, **5**, 583–590.
23 H. Schmidt and A. R. Hawkins, *Nature Photonics*, 2011, **5**, 598–604.
24 R. Won, *Nature Photonics*, 2011, **5**, 578–578.
25 S. O. Isikman, W. Bishara, S. Mavandadi, *et al.*, *Proceedings of the National Academy of Sciences of the United States of America*, 2011, **108**, 7296–7301.
26 S. O. Isikman, W. Bishara, U. Sikora, O. Yaglidere, J. Yeah and A. Ozcan, *Lab on a Chip*, 2011, **11**, 2222–2230.
27 O. Mudanyali, E. McLeod, W. Luo, *et al.*, *Nature Photonics*, 2013, **7**, 240–247.
28 K.-W. Huang, T.-W. Su, A. Ozcan and P.-Y. Chiou, *Lab on a Chip*, 2013, **13**, 2278–2284.
29 X. Cui, L. M. Lee, X. Heng, *et al.*, *Proceedings of the National Academy of Sciences of the United States of America*, 2008, **105**, 10670–10675.
30 G. Zheng, S. A. Lee, S. Yang and C. Yang, *Lab on a Chip*, 2010, **10**, 3125–3129.
31 W. Song and D. Psaltis, *Optics Express*, 2010, **18**, 16561–16566.
32 S. Pang, C. Han, L. M. Lee and C. Yang, *Lab on a Chip*, 2011, **11**, 3698–3702.
33 H. Zhu, S. Mavandadi, A. F. Coskun, O. Yaglidere and A. Ozcan, *Analytical Chemistry*, 2011, **83**, 6641–6647.
34 S. A. Lee, R. Leitao, G. Zheng, S. Yang, A. Rodriguez and C. Yang, *PLoS ONE*, 2011, **6**, e26127.
35 W. Song and D. Psaltis, *Biomicrofluidics*, 2011, **5**, 441101–4411011.
36 L. Pang, H. M. Chen, L. M. Freeman and Y. Fainman, *Lab on a Chip*, 2012, **12**, 3543–3551.
37 Y. Zhao, Z. S. Stratton, F. Guo, *et al.*, *Lab on a Chip*, 2013, **13**, 17–24.
38 Y. C. Seow, S. P. Lim and H. P. Lee, *Lab on a Chip*, 2012, **12**, 3810–3815.
39 B.-H. Jo, L. M. Van Lerberghe, K. M. Motsegood and D. J. Beebe, *Journal of Microelectromechanical Systems*, 2000, **9**, 76–81.
40 J. C. McDonald and G. M. Whitesides, *Accounts of Chemical Research*, 2002, **35**, 491–499.

41 K. Stephan, P. Pittet, L. Renaud, *et al.*, *Journal of Micromechanics and Microengineering*, 2007, **17**, N69–N74.
42 N. Sundararajan, D. Kim and A. A. Berlin, *Lab on a Chip*, 2005, **5**, 350–354.
43 J. C. Love, J. R. Anderson and G. M. Whitesides, *MRS Bulletin*, 2011, **26**, 523–528.
44 R. Pantoja, J. M. Nagarah, D. M. Starace, *et al.*, *Biosensors & Bioelectronics*, 2004, **20**, 509–517.
45 X. Zhang, L. Ren, X. Wu, H. Li, L. Liu and L. Xu, *Optics Express*, 2011, **19**, 22242–22247.
46 A. Dodge, E. Brunet, S. Chen, *et al.*, *The Analyst*, 2006, **131**, 1122–1128.
47 K. S. Lee and R. J. Ram, *Lab on a Chip*, 2009, **9**, 1618–1624.
48 I. K. Dimov, A. Riaz, J. Ducrée and L. P. Lee, *Lab on a Chip*, 2010, **10**, 1468–1471.
49 U. Levy and R. Shamai, *Microfluidics and Nanofluidics*, 2007, **4**, 97–105.
50 D. Erickson, X. Heng, Z. Li, *et al.*, *Proc. SPIE 5908*, Optical Information Systems III, 59080S, September 10, 2005.
51 D. Psaltis, S. R. Quake and C. Yang, *Nature*, 2006, **442**, 381–386.
52 N.-T. Nguyen, *Biomicrofluidics*, 2010, **4**, 031501.
53 A. Tripathi, T. V. Chokshi and N. Chronis, *Optics Express*, 2009, **17**, 19908–19918.
54 S. Calixto, M. E. Sánchez-Morales, F. J. Sánchez-Marin, M. Rosete-Aguilar, A. M. Richa and K. A. Barrera-Rivera, *Applied Optics*, 2009, **48**, 2308–2314.
55 H. Huang, X. Mao, S.-C. S. Lin, B. Kiraly, Y. Huang and T. J. Huang, *Lab on a Chip*, 2010, **10**, 2387–2393.
56 V. Presnyakov, K. Asatryan, T. Galstian and A. Tork, *Optics Express*, 2002, **10**, 865–870.
57 S. Kuiper and B. H. W. Hendriks, *Applied Physics Letters*, 2004, **85**, 1128–1130.
58 S. Kwon and L. P. Lee, *Optics Letters*, 2004, **29**, 706–708.
59 B. H. W. Hendriks, S. Kuiper, M. A. J. VAN As, C. A. Renders and T. W. Tukker, *Optical Review*, 2005, **12**, 255–259.
60 C.-C. Cheng, C. A. Chang, C.-H. Liu and J. A. Yeh, *Journal of Optics A: Pure and Applied Optics*, 2006, **8**, S365–S369.
61 C.-C. Cheng, C. A. Chang and J. A. Yeh, *Optics Express*, 2006, **14**, 4101–4106.
62 M. Ye, B. Wang, T. Takahashi and S. Sato, *Optical Review*, 2007, **14**, 173–175.
63 H. T. Dai, Y. J. Liu, X. W. Sun and D. Luo, *Optics Express*, 2009, **17**, 4317–4323.
64 D. Koyama, R. Isago and K. Nakamura, *Optics Express*, 2010, **18**, 25158–25169.
65 H. Ren, S. Xu and S.-T. Wu, *Optics Express*, 2010, **18**, 11904–11910.
66 J. Y. Kim, C. Martin-Olmos, N. S. Baek and J. Brugger, *Journal of Materials Chemistry C*, 2013, **1**, 2152–2157.
67 S.-H. Ahn and Y.-K. Kim, *Sensors and Actuators A: Physical*, 1999, **78**, 48–53.

68 N. Chronis, G. Liu, K.-H. Jeong and L. Lee, *Optics Express*, 2003, **11**, 2370–2378.

69 D.-Y. Zhang, V. Lien, Y. Berdichevsky, J. Choi and Y.-H. Lo, *Applied Physics Letters*, 2003, **82**, 3171–3172.

70 M. Agarwal, R. A. Gunasekaran, P. Coane and K. Varahramyan, *Journal of Micromechanics and Microengineering*, 2004, **14**, 1665–1673.

71 J. Chen, W. Wang, J. Fang and K. Varahramyan, *Journal of Micromechanics and Microengineering*, 2004, **14**, 675–680.

72 K.-H. Jeong, G. L. Liu, N. Chronis and L. P. Lee, *Optics Express*, 2004, **12**, 2494–2500.

73 D.-Y. Zhang, N. Justis and Y.-H. Lo, *Applied Physics Letters*, 2004, **84**, 4194.

74 L. Pang, U. Levy, K. Campbell, A. Groisman and Y. Fainman, *Optics Express*, 2005, **13**, 9003–9013.

75 H. Ren and S.-T. Wu, *Applied Physics Letters*, 2005, **86**, 211107.

76 A. Werber and H. Zappe, *Applied Optics*, 2005, **44**, 3238–3245.

77 H. Ren, D. Fox, P. A. Anderson, B. Wu and S.-T. Wu, *Optics Express*, 2006, **14**, 8031–8036.

78 H. Ren and S.-T. Wu, *Optics Express*, 2007, **15**, 5931–5936.

79 G.-R. Xiong, Y.-H. Han, C. Sun, L.-G. Sun, G.-Z. Han and Z.-Z. Gu, *Applied Physics Letters*, 2008, **92**, 241119.

80 G.-H. Feng and Y.-C. Chou, *Sensors and Actuators A: Physical*, 2009, **156**, 342–349.

81 H. B. Yu, G. Y. Zhou, F. S. Chau, F. W. Lee, S. H. Wang and H. M. Leung, *Optics Express*, 2009, **17**, 4782–4790.

82 P. Fei, Z. He, C. Zheng, T. Chen, Y. Men and Y. Huang, *Lab on a Chip*, 2011, **11**, 2835–2841.

83 W. Zhang, K. Aljasem, H. Zappe and A. Seifert, *Optics Express*, 2011, **19**, 2347–2362.

84 S. W. Lee and S. S. Lee, *Applied Physics Letters*, 2007, **90**, 121129.

85 C. A. López, C.-C. Lee and A. H. Hirsa, *Applied Physics Letters*, 2005, **87**, 134102.

86 L. Dong, A. K. Agarwal, D. J. Beebe and H. Jiang, *Nature*, 2006, **442**, 551–554.

87 L. Dong, A. K. Agarwal, D. J. Beebe and H. Jiang, *Advanced Materials*, 2007, **19**, 401–405.

88 Q. Hao, Y. Zhao, B. K. Juluri, *et al.*, *Journal of Applied Physics*, 2011, **109**, 084340.

89 B. Zhang, Y. Zhao, Q. Hao, *et al.*, *Optics Express*, 2011, **19**, 15221–15228.

90 Y. Zhao, Q. Hao, Y. Ma, *et al.*, *Applied Physics Letters*, 2012, **100**, 053119.

91 Y. J. Liu, X. Ding, S.-C. S. Lin, *et al.*, *Advanced Materials*, 2011, **23**, 1656–1659.

92 H. Ren, J. R. Wu, Y.-H. Fan, Y.-H. Lin and S.-T. Wu, *Optics Letters*, 2005, **30**, 376–378.

93 G.-R. Xiong, G.-Z. Han, C. Sun, H. Xu, H. Wei and Z.-Z. Gu, *Advanced Functional Materials*, 2009, **19**, 1082–1086.

94 Z. Wang, J. El-Ali, M. Engelund, *et al.*, *Lab on a Chip*, 2004, **4**, 372–377.
95 J. Godin, V. Lien and Y.-H. Lo, *Applied Physics Letters*, 2006, **89**, 061106.
96 M. Rosenauer and M. J. Vellekoop, *Biomicrofluidics*, 2010, **4**, 43005.
97 M. Rosenauer, W. Buchegger, I. Finoulst, P. Verhaert and M. Vellekoop, *Microfluidics and Nanofluidics*, 2010, **10**, 761–771.
98 X. Mao, J. R. Waldeisen and T. J. Huang, *Lab on a Chip*, 2007, **7**, 1260–1262.
99 X. Mao, J. R. Waldeisen, B. K. Juluri and T. J. Huang, *Lab on a Chip*, 2007, **7**, 1303–1308.
100 S. K. Y. Tang, C. A. Stan and G. M. Whitesides, *Lab on a Chip*, 2008, **8**, 395–401.
101 M. Rosenauer and M. J. Vellekoop, *Lab on a Chip*, 2009, **9**, 1040–1042.
102 C. Song, N.-T. Nguyen, S.-H. Tan and A. K. Asundi, *Lab on a Chip*, 2009, **9**, 1178–1184.
103 C. Song, N.-T. Nguyen, S.-H. Tan and A. K. Asundi, *Journal of Micromechanics and Microengineering*, 2009, **19**, 085012.
104 C. Song, N.-T. Nguyen, S.-H. Tan and A. K. Asundi, *Microfluidics and Nanofluidics*, 2010, **9**, 889–896.
105 C. Song, T.-D. Luong, T. F. Kong, N.-T. Nguyen and A. K. Asundi, *Optics Letters*, 2011, **36**, 657–659.
106 C. Song, N.-T. Nguyen, Y. F. Yap, T.-D. Luong and A. K. Asundi, *Microfluidics and Nanofluidics*, 2010, **10**, 671–678.
107 X. Mao, S.-C. S. Lin, M. I. Lapsley, J. Shi, B. K. Juluri and T. J. Huang, *Lab on a Chip*, 2009, **9**, 2050–2058.
108 H. Li, T. N. Wong and N.-T. Nguyen, *Microfluidics and Nanofluidics*, 2010, **10**, 1033–1043.
109 H. Li, C. Song, T. D. Luong, N.-T. Nguyen and T. N. Wong, *Lab on a Chip*, 2012, **12**, 3680–3687.
110 V. Lien, Y. Berdichevsky and Y.-H. Lo, *Applied Physics Letters*, 2003, **83**, 5563.
111 L. Dong and H. Jiang, *Journal of Microelectromechanical Systems*, 2008, **17**, 381–392.
112 S.-K. Hsiung, C.-H. Lee and G.-B. Lee, *Electrophoresis*, 2008, **29**, 1866–1873.
113 J. Shi, Z. Stratton, S.-C. S. Lin, H. Huang and T. J. Huang, *Microfluidics and Nanofluidics*, 2009, **9**, 313–318.
114 Y. Liu, B. Aldalali and H. Jiang, *Journal of Micromechanics and Microengineering*, 2012, **22**, 105010.
115 L. Dong and H. Jiang, *Applied Physics Letters*, 2007, **91**, 041109.
116 W. L. Barnes, A. Dereux and T. W. Ebbesen, *Nature*, 2003, **424**, 824–830.
117 A. V. Zayats, I. I. Smolyaninov and A. A. Maradudin, *Physics Reports*, 2005, **408**, 131–314.
118 E. Ozbay, *Science*, 2006, **311**, 189–193.

119 J. A. Schuller, E. S. Barnard, W. Cai, Y. C. Jun, J. S. White and M. L. Brongersma, *Nature Materials*, 2010, **9**, 193–204.
120 M. L. Brongersma and V. M. Shalaev, *Science*, 2010, **328**, 440–441.
121 Y. Zhao, S.-C. S. Lin, A. A. Nawaz, *et al.*, *Optics Express*, 2010, **18**, 23458–23465.
122 J. S. T. Smalley, Y. Zhao, A. A. Nawaz, *et al.*, *Optics Express*, 2011, **19**, 15265–15274.
123 T. Zentgraf, Y. Liu, M. H. Mikkelsen, J. Valentine and X. Zhang, *Nature Nanotechnology*, 2011, **6**, 151–155.
124 C. Zhao, Y. Liu, Y. Zhao, N. Fang and T. J. Huang, *Nature Communications*, 2013, **4**, 2305.
125 J. S. Donner, G. Baffou, D. McCloskey and R. Quidant, *ACS Nano*, 2011, **5**, 5457–5462.
126 F. D. Jochum and P. Theato, *Chemical Society Reviews*, 2013, **42**, 7468–7483.
127 Y. Liu, H. Lv, X. Lan, J. Leng and S. Du, *Composites Science and Technology*, 2009, **69**, 2064–2068.
128 J. Hu and S. Liu, *Macromolecules*, 2010, **43**, 8315–8330.
129 Y. Zhao, *Macromolecules*, 2012, **45**, 3647–3657.
130 S. Das, S. Varghese and N. S. S. Kumar, *Langmuir: The ACS Journal of Surfaces and Colloids*, 2010, **26**, 1598–1609.
131 S. Dai, P. Ravi and K. C. Tam, *Soft Matter*, 2008, **4**, 435–449.
132 N. Murthy, J. Campbell, N. Fausto, A. S. Hoffman and P. S. Stayton, *Bioconjugate Chemistry*, 2003, **14**, 412–419.
133 J. P. Magnusson, A. Khan, G. Pasparakis, A. O. Saeed, W. Wang and C. Alexander, *Journal of the American Chemical Society*, 2008, **130**, 10852–10853.
134 Y. Lvov, A. A. Antipov, A. Mamedov, H. Möhwald and G. B. Sukhorukov, *Nano Letters*, 2001, **1**, 125–128.
135 D. Han, X. Tong, O. Boissière and Y. Zhao, *ACS Macro Letters*, 2012, **1**, 57–61.
136 A. Akhoury, L. Bromberg and T. A. Hatton, *ACS Applied Materials & Interfaces*, 2011, **3**, 1167–1174.
137 Q. Yan, J. Yuan, Z. Cai, Y. Xin, Y. Kang and Y. Yin, *Journal of the American Chemical Society*, 2010, **132**, 9268–9270.
138 J. E. Betancourt and J. M. Rivera, *Journal of the American Chemical Society*, 2009, **131**, 16666–16668.
139 I. Dimitrov, B. Trzebicka, A. H. E. Müller, A. Dworak and C. B. Tsvetanov, *Progress in Polymer Science*, 2007, **32**, 1275–1343.
140 Y. Zhao, *Journal of Materials Chemistry*, 2009, **19**, 4887.
141 F. Ercole, T. P. Davis and R. A. Evans, *Polymer Chemistry*, 2010, **1**, 37–54.
142 J.-M. Schumers, C.-A. Fustin and J.-F. Gohy, *Macromolecular Rapid Communications*, 2010, **31**, 1588–1607.

143 C. Weder, *Nature*, 2009, **459**, 45–46.

144 S. S. M. Konda, J. N. Brantley, C. W. Bielawski and D. E. Makarov, *The Journal of Chemical Physics*, 2011, **135**, 164103.

145 W. Su, K. Han, Y. Luo, Z. Wang, Y. Li and Q. Zhang, *Macromolecular Chemistry and Physics*, 2007, **208**, 955–963.

146 K. Han, W. Su, M. Zhong, Q. Yan, Y. Luo, Q. Zhang and Y. Li, *Macromolecular Rapid Communications*, 2008, **29**, 1866–1870.

Chapter 12

Nanostructured aluminum oxide-based optical biosensing and imaging

Long Que

This chapter covers the basics of anodic aluminum oxide or nanostructured aluminum oxide (NAO), how to fabricate the wafer-scale NAO thin film and the NAO micropatterms on glass substrate, the optical properties of the NAO thin film, and its applications for label-free biosensing, fluorescence-based detection, and imaging.

12.1 Introduction

Single nanopore and ordered nanopores have been the widely used nanostructures for a variety of applications, spanning from life science, biology, biomedical engineering, and template-based nanofabrication to energy harvesting and storage [1–5]. Ordered nanopores are particularly interesting nanostructures due to their unique optical properties, and thus offer a new technical platform for optical sensing and detection.

Nanopores can be fabricated using a top-down nanofabrication process [6, 7]. Examples include the nanopores fabricated from silicon wafer using electron-beam lithography (EBL), followed by a dry etching process or using a focused ion beam (FIB) nanolithography. The resulting nanopores have been used for DNA sequencing successfully [7, 8] among other applications. However, the top-down EBL or FIB nanofabrication is quite time consuming and expensive due to their serial fabrication nature, preventing the high-volume production of nanopores in a cost-effective manner. The discovery of the formation of ordered nanopores in silicon and in aluminum by an electrochemical process (i.e., anodization process) [9, 10] opens up alternative approaches to fabricate ordered nanopores efficiently and inexpensively. There are many advantages of the electrochemical/anodization fabrication method in comparison with the nanolithography-based top-down fabrication methods, such as cost effectiveness, wide and simple accessibility,

Department of Electrical and Computer Engineering, Iowa State University, Ames, IA, USA

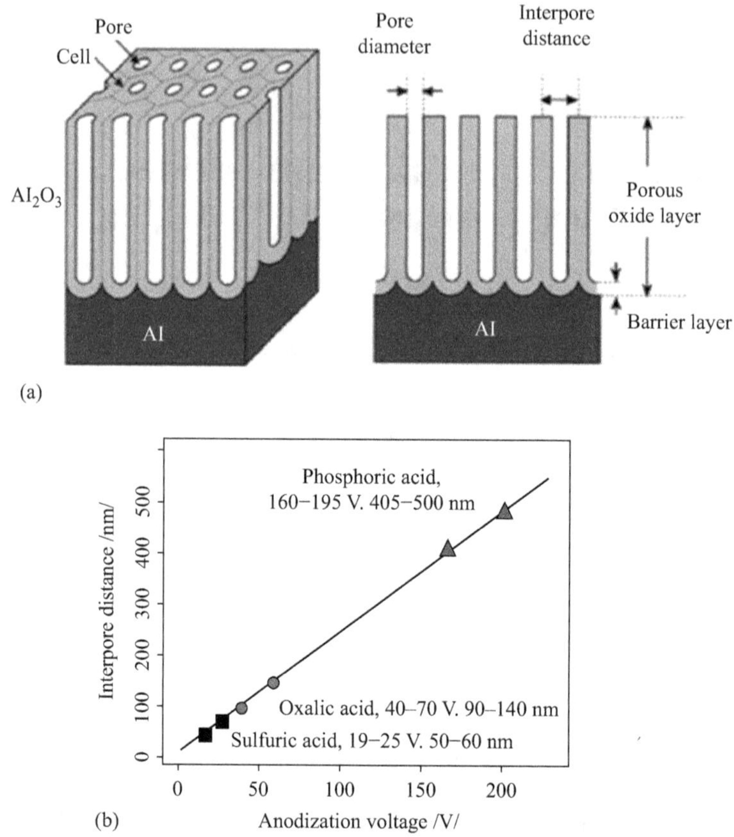

Figure 12.1 (a) Schematic of NAO structure prepared by electrochemical anodization of Aluminum; (b) summary: anodization voltage and corresponding interpore distance of NAO fabricated by three electrolytes including sulfuric, oxalic, and phosphoric. Reprinted from Reference 19 with permission

capability of achieving nanoscale precision, and feasibility for fabricating high aspect ratio nanostructures [11].

In this chapter, we only focus on the nanopores formed by anodizing aluminum (Al). The details of the nanopores formed in silicon by electrochemical process can be found in the literatures [9, 12]. The formation of nanopores in anodized aluminum was first discovered in 1990s [13]. The nanostructures of anodic aluminum oxide (AAO) or NAO are basically a close-packed hexagonal array of parallel cylindrical nanopores perpendicular to the surface on top of the underlying Al substrates (Figure 12.1(a)). The main parameters of the nanopores are pore organization (hexagonal), pore diameter, interpore distance, wall thickness, pore length, and barrier layer thickness. By controlling the anodization conditions, these parameters can be precisely tuned in the range of about 10–400 nm for pore diameter,

50–600 nm for interpore distance, pore aspect ratio of 10–5000, a thickness of the porous layer of 10 nm to 150 μm, a pore density of 10^9 to 10^{11} cm^2 and porosity of 5% to 50% [14]. Due to its remarkable properties such as chemical, thermal stability, hardness, and high surface area [14, 15], NAO has been extensively used for many emerging applications over the past decades. The applications include molecular separation, chemical/biological sensing devices, cell adhesion, catalysis, energy storage, and drug delivery vehicles, just to name a few [16–18].

On the other hand, over the past decades, extensive studies have also been devoted to exploring the anodization parameters for aluminum, including the applied voltage, current, temperature, pH and electrolyte type, electrotype composition, and concentration, for improving the controllability for fabricating the NAO with desirable and tunable dimensions of the highly ordered nanopores. In terms of the electrolytes, sulfuric, oxalic, and phosphoric acids are three commonly used ones for anodization processes. Typical anodization voltages and the resulting interpore distance by the three widely used electrolytes are given in Figure 12.1(b). Basically, depending on the requirements of the resulting nanopores, different electrolytes can be selected for different pore diameters and the interpore distance. Research efforts have also found that other acids such as chromic, boric, citric, and tartaric acid can be used for fabricating NAO as well, but the pore ordering is usually not as good as that with the three common electrolytes. Some very good reviews are also available for the fabrication of the NAO with controlled nanopore diameters and the characterization of the fabricated NAO [19]. Besides the electrolytes, there are two methods to obtain the NAO: one-step anodization and two-step anodization. In one-step anodization method, the aluminum is anodized just one time to obtain the NAO. More details on the resulted NAO will be covered in the following sections. In contrast, in two-step anodization method, the aluminum is anodized twice to obtain the NAO, offering a much higher degree of nanopore order in comparison with one-step anodization method. In this chapter, both methods are covered and discussed. Which method is used to fabricate the NAO is highly dependent on the nanopore requirements and applications of the NAO thin film.

12.2 General fabrication process of NAO thin film

The NAO is usually fabricated from an Al foil or plate directly [13]. However, the fabrication process of the NAO from an Al foil or plate is not compatible with a standard lithography-based micro- or nanofabrication process. As a result, it is very challenging and almost impossible to fabricate arrayed NAO-based nanosensors in a cost-effective manner. Basically the fabrication approach from an Al foil or plate is unsuitable to build large-scale NAO-based nanosensors for potential high throughput biological or biomedical applications. Furthermore, the NAO film fabricated from an Al foil or plate is usually so thick that it is not optically transparent, also limiting its applications. Hence, a new method to fabricate NAO thin film on a solid substrate such as silicon or glass substrate is becoming necessary.

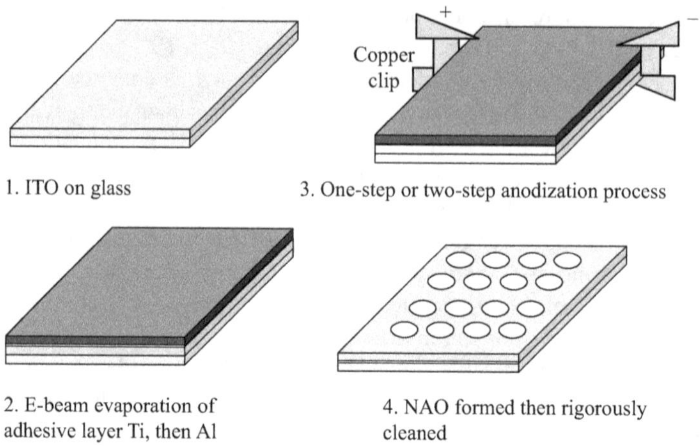

Figure 12.2 Fabrication process flow: Wafer-scale AAO thin film fabrication

Recently some techniques have been developed to fabricate NAO thin film from Al thin film deposited on a variety of solid substrates [20–27], which indicates the possibility to render the fabrication process of NAO thin film compatible with a standard top-down lithography-based micro- or nanofabrication process. In the reported literatures [20–27], typically the thickness of Al thin film spans from tens of nanometers to several micrometers. In this chapter, only the NAO thin film fabrication on a glass substrate is described in detail, while on other substrates such as silicon can be found in related citations [20]. Basically, the fundamental principle of the fabrication process of NAO thin film on different substrates is similar. One specific example is NAO thin film fabricated on an indium–tin oxide (ITO) glass. In this case, typically the ITO glass substrate (Nanocs, Inc.) has a sheet resistance of ITO of 100 Ω/square, and its surface roughness is less than 5 nm. It should be noted that the roughness of the substrate must be several nanometers for successfully fabricating the NAO from Al thin film deposited on the substrate.

The fabrication process flow is illustrated in Figure 12.2. Starting from an ITO glass substrate, it is cleaned by deionized (DI) water, acetone, ethanol, and then DI water for 20 minutes with each solution. Using electron-beam evaporation, 0.5 to 2.2 μm Al with high quality can be routinely deposited on the substrate with 10 nm Ti as an adhesion layer. The measured surface roughness of Al thin film is in the range of 6 to 12 nm, which is smooth enough for carrying out the anodization process and forming NAO thin film. In contrast, in order to fabricate NAO from an Al foil or an Al plate, the mechanical and chemical surface polishing of the Al foil or plate is always required to reduce its surface roughness down to several nanometers [10, 14, 23]. Since the Al thin film deposited by electron-beam evaporation on a smooth ITO glass substrate does not require the surface polishing steps, it makes the NAO thin film fabrication actually much simplified.

As aforementioned, three commonly used acids are used for anodization including sulfuric, oxalic, and phosphoric acids. Anodization can be carried out on

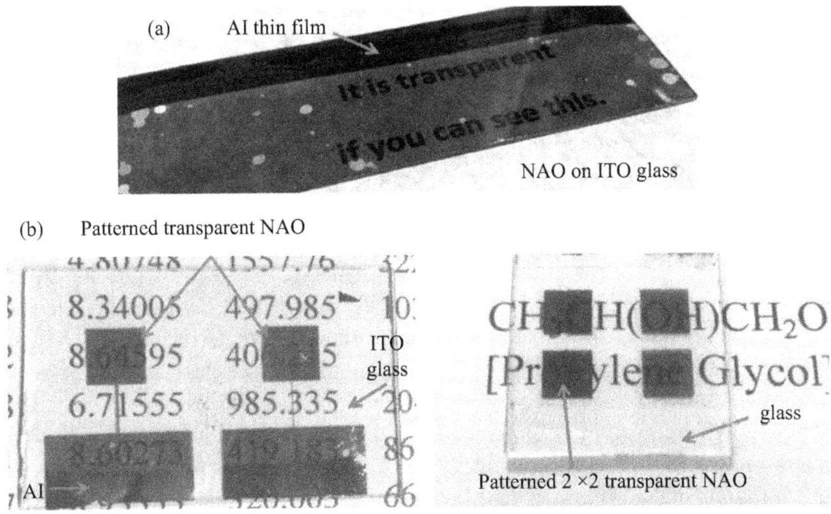

Figure 12.3 *(a,b) Images of the fabricated NAO thin film on glass substrate. Reprinted from Reference 25 with permission*

the Al thin film either once or twice, depending on the quality requirements of the NAO. Specifically, the quality of the formed nanopores is not very good in terms of size and distribution uniformity after just one-time or one-step anodization. Using two-time anodization or two-step anodization, as aforementioned, the size and distribution of nanopores become almost perfectly uniform. For two-step anodization, the Al thin film samples are anodized in an acid solution (either 0.3 M oxalic acid at 2 °C at 50 V, or 0.3 M sulfuric acid at 2 °C at 17 V) for 10–15 minutes (the first step anodization) through the copper clips by applying a voltage source. Then the anodized oxide layer alumina is etched away by immersing the samples into a mixture of phosphoric acid (0.4 M) and chromic oxide (0.2 M) at 65 °C for 10 minutes or 60 minutes. After this step, the second anodization step is carried out for 30–45 minutes under the same conditions as the first anodization step. Finally, the NAO samples are rinsed rigorously using DI water before their usage.

Some photographs of the NAO thin film are shown in Figure 12.3. In Figure 12.3(a), the NAO thin film is fabricated using a one-step anodization process on a wafer-scale ITO glass substrate, and the acid for anodization is oxalic acid. The NAO thin film is evidently optically transparent since a sentence underneath can be seen clearly, indicating that the 10 nm Ti layer has also been anodized and oxidized as transparent TiO_2 [22]. Figure 12.3(b) gives the image of NAO thin film fabricated using a two-step anodization process and the acid for anodization is sulfuric acid. Similarly the numerical numbers and letters can be clearly seen through the NAO thin film, indicating its optical transparency, even though the degree of transparency is lower than that of ITO glass on the same substrate. The degree of transparency depends on the final thickness, the nanopore size, and density of the NAO thin film. For comparison, the Al thin film not anodized on the same substrate is totally opaque as shown in Figure 12.3(a,b).

In order to compare the surface morphologies and profiles of the fabricated NAO thin films using one-step and two-step anodization process, all the NAO thin films have been examined by using a scanning electron microscope (SEM; HITACHI, S-4800). The SEM images of the fabricated NAO thin film on ITO glass substrates under different process conditions are given in Figure 12.4. Figure 12.4 (a) gives the image of NAO thin film formed by one-step anodization process using sulfuric acid. In this case, the alumina nanoscale grains in the film can still be clearly observed. In other words, the Al thin film is actually not an extended and continuous thin film on the substrate, rather it contains many Al nanoscale grains of sizes in the range of 100–500 nm before the anodization. Figure 12.4(b) gives the image of NAO thin film formed by one-step anodization process using oxalic acid. The nanopores are even more sparsely formed in the alumina nanoscale grains than those in Figure 12.4(a). As shown in Figure 12.4(a,b), the nanopores after one-step anodization are rather sparsely distributed in the alumina nanoscale grains and do not have high degree of periodicity. The average nanopore size and interspacing are relatively larger from NAO thin film fabricated by oxalic acid (30 ± 5 nm for nanopore size and 125 ± 5 nm for the interspacing) than those fabricated by sulfuric acid (10 ± 2 nm for nanopore size and 35 ± 2 nm for the interspacing).

The SEM images of NAO thin film fabricated using two-step anodization process and oxalic acid for anodization are given in Figure 12.4(c,d). Under this condition, the alumina nanoscale-grain domains totally disappear. In Figure 12.4 (c), the time to remove the alumina layer after the first-step anodization is 10 minutes by immersing the samples into a mixture of phosphoric acid (0.4 M) and chromic oxide (0.2 M) at 65°C, while in Figure 12.4(d), the time is 60 minutes. As can be seen, the nanopore size, the interspacing among nanopores, and overall periodicity are different for these two cases, confirming that the nanopore size and periodicity of the AAO can be tuned by using different fabrication parameters and acids. Specifically, for the case in Figure 12.4(c), the alumina layer formed in the first-step anodization has only been partly removed after 10-minute wet etching, the nanopores formed in the second-step anodization are simply the continuation and widen of the nanopores formed in the first-step anodization. Hence the positions of nanopores are essentially the same as those in the first-step anodization process. But the size of the nanopores increases, and thus the interspacing among them decreases [23]. In contrast, for the case in Figure 12.4(d), the alumina layer formed in the first-step anodization has been totally removed by 60-minute wet etching. Then hemispherical nanopatterns exist on top of the underneath aluminum layer, due to the self-assembly characteristics of the anodizing process. Hence the nanopores formed in second-step anodization are essentially new nanopores initiated from the existing hemispherical nanopatterns and thus are more uniform [10, 14, 23]. After completing the second-step anodization, the nanopore size can be further widened if the samples are immersed in phosphoric acid (H_3PO_4) for several to tens of minutes [19, 23]. It should be noted that the diameter of the hemispherical nanopatterns can be easily controlled from tens to hundreds of nanometers by

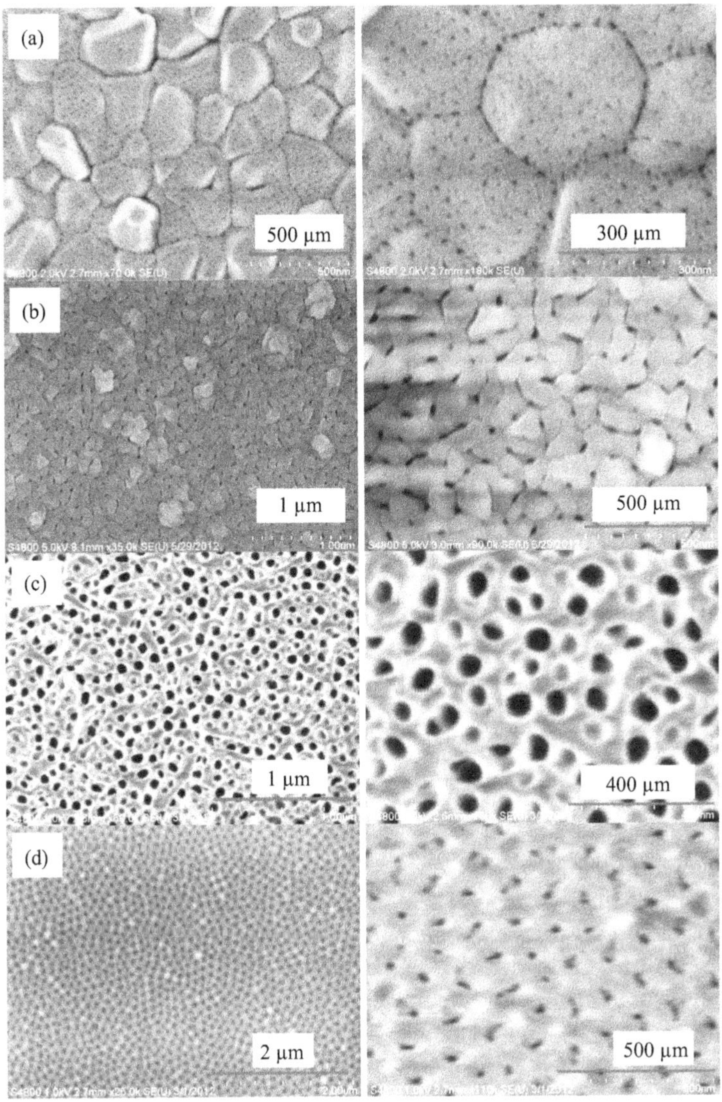

Figure 12.4 SEMs of AAO nanostructure thin film on ITO glass substrate (a) fabricated using one-step anodization process with sulfuric acid; (b) fabricated using one-step anodization process with oxalic acid; (c) fabricated using two-step anodization process with oxalic acid. The time to remove the alumina between the two anodization steps is 10 minutes; (d) fabricated using two-step anodization process with oxalic acid. The time to remove the alumina between the two anodization steps is 60 minutes. Reprinted from Reference 25 with permission

applying different anodization voltage and by changing the type and temperature of the anodization acids, which have been studied extensively for the past decades. Therefore, the size and interspacing among nanopores can be tuned, and thus the quality and the uniformity of the nanopores can be optimized using the two-step anodization process.

In order to carry out anodization on the Al thin film on the ITO glass substrate successfully to fabricate NAO thin film, the purity of the Al thin film is critical. For instance, the Al should not contain any oxidized Al [25, 26]. Hence, care should be taken to deposit Al thin film on ITO glass substrate during the electron-beam evaporation to avoid any potential oxidation. Systematic experiments have found that even slightly or partially oxidized Al cannot be used to obtain NAO successfully [25, 26].

12.3 Fabrication and integration of NAO thin film micropatterns

A thin film with periodically distributed nanopores such as NAO thin film has been proved to be a powerful nanostructure for many applications. However, in order to be integrated with microdevices or microfluidic devices for both label-free biodetection and fluorescence-based biosensing, the NAO thin film needs to be micropatterned. For the past decades, several techniques have been developed to fabricate micropatterned NAO film [24–27]. One technique is to fabricate NAO micropatterns by utilizing a dry etching or an ion-milling process after the NAO film has been fabricated from an Al plate [24] . However, this kind of approach is really complicated and relatively expensive, and is not easy to be scaled up for high-volume production. Furthermore, due to the NAO fabricated from an Al plate, the NAO micropatterns are difficult to be integrated with microdevices or microfluidic devices. Other work has been reported to fabricate the NAO from Al thin film deposited on a variety of solid substrates [20–22], which may be adapted to render the fabrication process of NAO compatible with a standard top-down lithography-based micro- or nanofabrication process. In the following two sections, two complementary techniques developed to fabricate NAO micropatterns on glass substrate are described.

12.3.1 Technique 1: Lift-off-based process

In this technique, the Al thin film micropatterns are fabricated first using a lift-off process, thereafter the Al micropatterns are converted to NAO micropatterns by either one-step anodization process or two-step anodization process [25]. The fabrication process flow is illustrated in Figure 12.5. This process starts from an ITO glass substrate. It is cleaned by DI water, acetone, ethanol, and then DI water for 20 minutes with each solution. Then a lift-off process is used to form the Al micropatterns. Specifically, a layer of LOR7b (Microchem, Inc.) followed by a layer of Shipley PR1813 is spun on the substrate and then is patterned using optical lithography. Using electron-beam evaporation, 0.5–2.2 μm Al is deposited on the

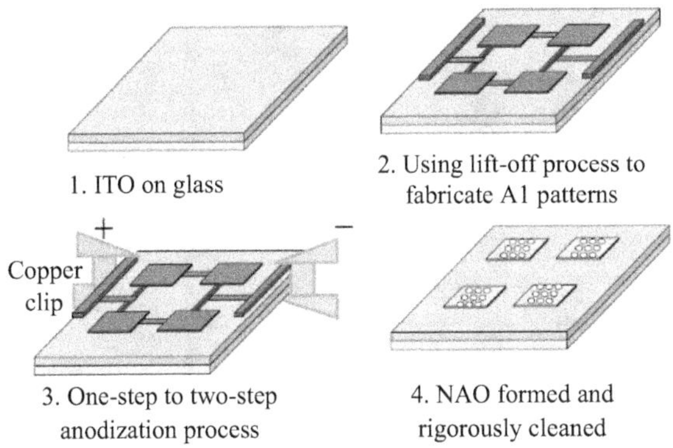

Figure 12.5 Fabrication process flow of arrayed optically transparent NAO micropatterns on an ITO glass substrate (1) start with a clean ITO glass substrate; (2) Al micropatterns fabricated using a lift-off process; (3) NAO thin film micropatterns formed using one-step or two-step anodization process; (4) the Al lines removed and the NAO thin film micropatterns cleaned by acetone and DI water. Reprinted from Reference 25 with permission

substrate with 10 nm Ti as an adhesion layer. Thereafter, the photoresist is removed by immersing the whole wafer into an EBR PG (Positive Radiation Resist Edge Bead Remover) solution. As a result, Al patterns are formed and connected with each other with some Al lines as shown in the step 2 as shown in Figure 12.5. The Al lines connecting Al patterns are crucial for fabricating lithographically patterned NAO thin film since the Al lines can warrant sufficient current flow inside the Al patterns during the anodization steps. Experiments found that the width of the Al lines of 100 μm is sufficient to completely anodize the Al patterns with size of ~6 mm × 6 mm. Otherwise if the width of Al line is too small, the Al lines would have been totally anodized themselves and the current could not flow through them before the anodization of the Al patterns has been completed. If the size of Al patterns to be anodized is scaled down, the width of the Al lines connecting them might be scaled down as well. However, the scaling ratio should be carefully selected.

Once the Al patterns have been fabricated and cleaned with acetone, ethanol, and then DI water, one-step or two-step anodization can be carried out to form NAO patterns as shown in the step 3 in Figure 12.5.

The optical and SEM images of several different shapes of the fabricated NAO micropatterns are given in Figure 12.6(a–c). A variety of shapes of the NAO micropatterns can be fabricated from several ten micrometers to several hundred micrometers. As can be seen clearly from Figure 12.6(d), the nanostructures (i.e., nanopores) are formed and distributed among the nanoscale grains inside the NAO micropatterns. Typical size of the nanopores inside the NAO micropatterns for our experiments is in the range of 10–30 nm. Again, the size of the nanopores

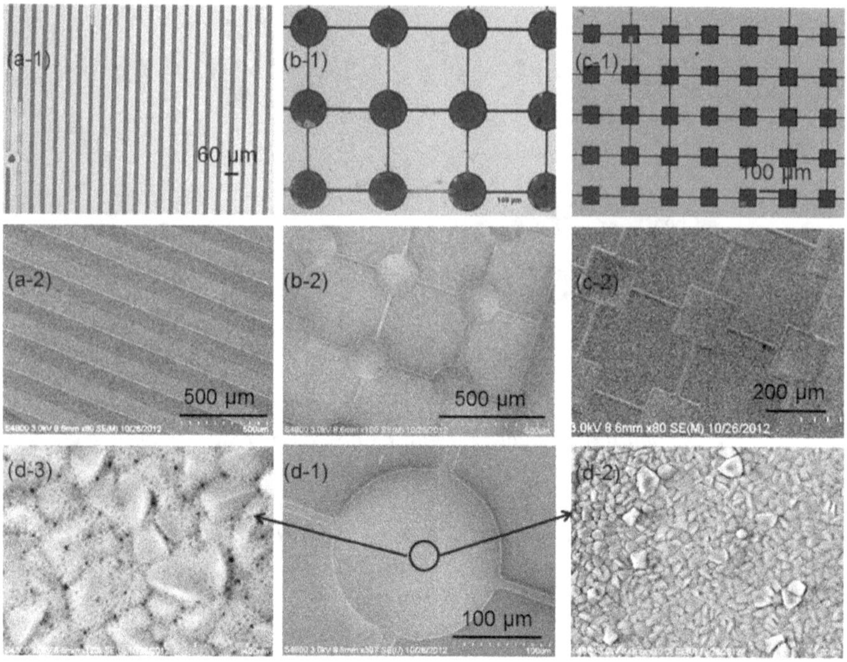

Figure 12.6 (a–c) Representative optical micrographs and SEM images of the fabricated NAO micropatterns with different shapes and sizes using one-step anodization process; (d-1) SEM images of a circular NAO micropattern; (d-2) and (d-3) SEM images of the close-up of the NAO, showing nanoscale aluminum oxide grains with nanopores distributed among them. Reprinted from Reference 26 with permission

and the distance among them can be readily tuned by changing the anodization parameters and the electrolytes.

However, this fabrication process has the following limitations: (1) Al microlines are required to connect the Al micropatterns to ensure that the current can always flow to each micropattern during the anodization process. If the Al microlines have been anodized before the Al micropatterns are anodized, no current can flow through the Al micropatterns anymore. As a result, the anodization process in Al micropatterns stops and only part of them can be anodized. It has been reported that the anodization rate is related to the current, the selection of the anodization current is thus critical. (2) Since each micropattern has to be connected by Al microlines for anodization, the choices of shapes of the micropatterned NAO are limited. Otherwise, the micropatterns cannot be properly anodized. This limitation reduces the flexibility tremendously for designing different shapes of NAO micropatterns for some specific applications, and it could be a problem to integrate them into microdevices and microfluidic devices. (3) Due to the reason discussed in

(1), the feature size of the micropatterns is significantly limited. Experiments found that the minimum size of the NAO micropatterns, which can be fabricated with good yield using this fabrication technique, is ~15 μm [25, 26]. If smaller NAO micropatterns are needed, an alternative fabrication process becomes necessary. In order to address these aforementioned issues, another process for fabricating micropatterned NAO with a greater flexibility has been developed and described in the following section. And this process is more compatible with the standard fabrication process for MEMS and microfluidics.

12.3.2 Technique 2: NAO thin film microlithography and etching based process

In this technique, the NAO thin film will be firstly fabricated at a wafer scale, followed by fabricating NAO micropatterns [27]. The fabrication process flow is illustrated in Figure 12.7. The ITO glass substrate is washed thoroughly with four steps: it is washed by DI water, followed by Acetone, IPA, and DI water in sequence. After a 5-minute baking of the cleaned ITO glass, about 2-μm thick aluminum layer is deposited by E-beam evaporation with 10 nm Ti as an adhesion layer. As aforementioned, the quality of the deposited Al is critical for successful anodization. The deposited Al should be totally oxide-free, which means during E-beam evaporation, any oxidation of Al has to be avoided, otherwise the as-deposited Al cannot be anodized.

Then anodization process (either one-step or two-step) in acid solution (0.3 M oxalic acid) with 45 V DC voltage at 2°C to form NAO on the ITO glass substrate is carried out. In this step, a layer of NAO is formed on the whole substrate. For the one-step anodization process to form NAO, we only carry out one-step anodization on Al samples. In other words, the Al should be totally anodized. In contrast, for the two-step anodization process to form NAO, only partial Al layer is anodized after the step-1 anodization, the anodized Al layer is then etched by a mixture of phosphoric acid (0.4 M) and chromic acid (0.2 M) at 65°C for 30 mins, followed by a 40-minute step-two anodization in 0.3 M oxalic acid with the same experimental condition as the step-one anodization. The wafer is then rinsed by DI water rigorously. Thereafter, a 150-nm thick aluminum layer is deposited on the NAO surface by thermal evaporation. Photolithography is performed on the Al-coated NAO

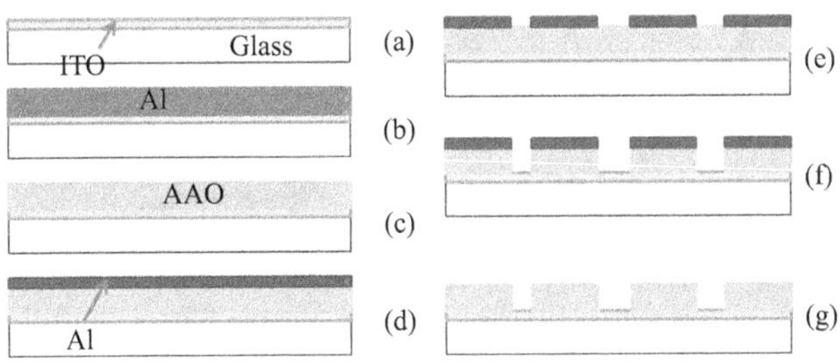

Figure 12.7 Sketch of the fabrication process flow for fabricating NAO micropatterns. Reprinted from Reference 27 with permission

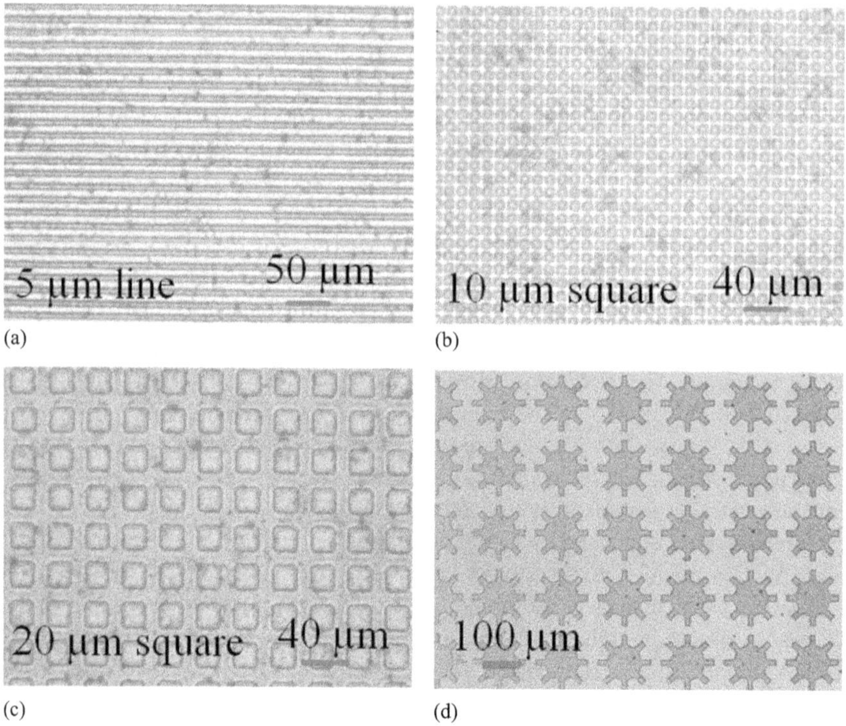

Figure 12.8 Different shapes of the NAO micropatterns: (a) 5 µm microlines; (b) 10 µm square micropatterns; (c) 20 µm square micropatterns; and (d) gear micropatterns. Reprinted from Reference 27 with permission

substrate. During this step, care should be taken to avoid any over-etching of the Al since the patterned Al layer serves as the mask for etching the NAO. Then the substrate is immersed in a mixture of phosphoric acid (0.4 M) and chromic acid (0.2 M) at 20°C for 100 minutes to etch away the unprotected NAO and transfer the patterns into the NAO layer. Al is chosen as the mask for the NAO layer instead of the photoresist since it has been found that in most cases the photoresist cannot serve as a robust mask due to the long etching time (100 minutes) of the NAO layer. As a result, the micropatterns cannot be properly transferred directly from the photoresist to the NAO layer.

Some representative optical micrographs of the patterned NAO are shown in Figure 12.8. The microlines of NAO can be as narrow as ~5 µm in width (Figure 12.8(a)). The NAO square micropatterns with dimensions of 10 µm × 10 µm and 20 µm × 20 µm, respectively, are shown in Figure 12.8(b) and Figure 12.8(c), respectively. The NAO gear micropatterns are given in Figure 12.8(d). All these NAO micropatterns are optically transparent. Using this fabrication procedure, a variety of NAO micropatterns with different shapes can be readily fabricated.

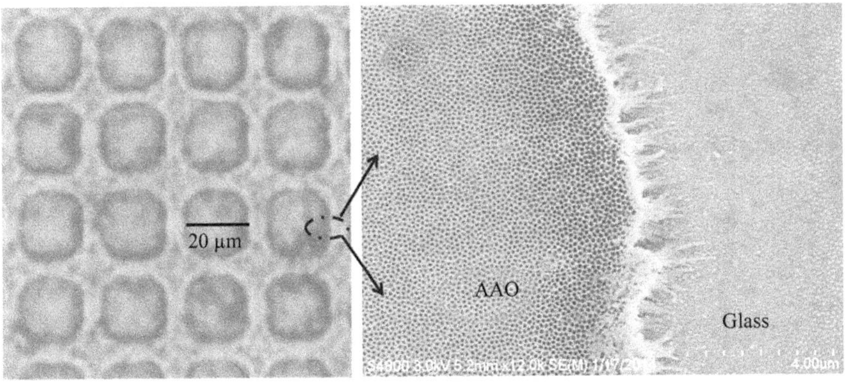

(a) (b)

Figure 12.9 *(a) Optical micrograph of the over-etched NAO 20 μm × 20 μm square micropatterns, (b) SEM image showing the interface between glass and NAO micropattern. Reprinted from Reference 27 with permission*

The critical dimensions of the micropatterns can be readily fabricated as small as 5 μm. However, if the etching time is not properly controlled during the NAO etching step, over-etching of the NAO underneath Al will occur. One example of the over-etched 20 μm × 20 μm NAO square patterns is shown in Figure 12.9(a). It is clear that the sharp corners of the square micropatterns become round. The SEM image in Figure 12.9(b) gives the close-up of the interface between the NAO patterns and the glass substrate. Some remaining traces of NAO are not totally etched away at the interface, indicating that the improvement and optimization of the NAO etching step are required.

Using this technique, it has also been found that the NAO micropatterns can be successfully fabricated directly on glass substrate rather than the ITO coated glass substrate.

12.4 Optical properties of NAO thin film

12.4.1 Optical interference signals from NAO thin film

The unique optical properties of periodic nanopores have been utilized for a great deal of applications. For instance, the arrayed silicon nanopores have been utilized for detecting the hybridization of DNAs and offers very high sensitivity [28]. The NAO thin film offers another alternative to the nanoporous Si thin film.

One important optical property of the NAO thin film is the optical interference signals. Given the half transparency of the NAO thin film, both transmitted and reflected interference signals can be obtained, and thus both can be used for transducing signals for the NAO thin film–based sensors. For these measurements, the optical source is a tungsten halogen source (Ocean Optics, Inc.), the

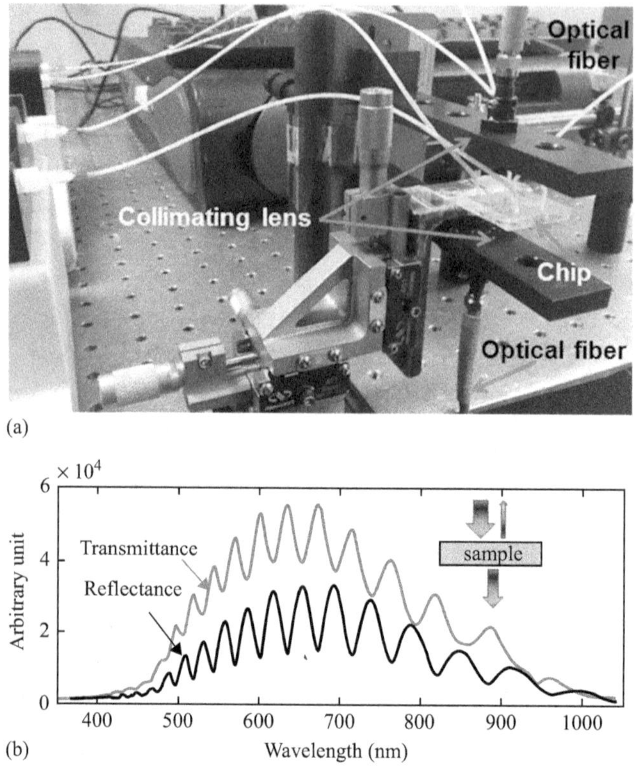

Figure 12.10 (a) Optical setup for the measurement: A white light source is delivered to the chip through an optical fiber, the transmitted signal is collected by another optical fiber, leading to a spectrometer; (b) measurement of transmittance and reflectance signals of a 2.2-µm thick NAO thin film. Reprinted from Reference 25 with permission

spectrometer is USB4000-VIS-NIR (Ocean Optics, Inc.) with a detection spectral range covering 350–1050 nm. A photograph showing the transmittance measurement is given in Figure 12.10a. The incident light beam is collimated and normally incident to the surface of the NAO thin film, the other lens will be used to couple the transmitted signals into an optical fiber probe, leading to the spectrometer. For the reflectance measurement, the optical fiber probe is replaced with the fiber optic reflection/backscattering probes (Ocean Optics, Inc.) [29]. Figure 12.10(b) gives a measurement of the transmittance and reflectance of a 2.2-µm thick NAO thin film fabricated by one-step anodization process. As expected, the profiles of the transmittance and reflectance optical signals are essentially complementary to each other in terms of the interference fringes. In other words, the interference peaks of the transmission signal always correspond to the interference valleys of the reflectance signal. Similarly, the interference valleys of the transmission signal always correspond to the interference peaks of reflectance signal. The optical performance (e.g., finesse) of the NAO-based thin film interferometer can be evaluated by analyzing either the transmission signals or the reflection signals [30].

It has been found that the number of the interference fringes of the optical signals is related to the thickness of the NAO thin film. In Figure 12.11a, the measured reflectance signals from the NAO nanostructure thin films fabricated using one-step anodization process with different thicknesses are plotted. Based on a series of experiments, it has been found that when the thickness of NAO thin film is 0.5 μm or lower, no interference fringes can be observed no matter the NAO thin film is fabricated by one-step or two-step anodization process. In contrast, the interference fringes are clearly observed by increasing the thickness of NAO thin film to 1.5 μm (8 interference peaks) and 2.2 μm (12 interference peaks). Basically the number of the interference peaks increases with the thickness (e.g., at least in the range of 0.5–2.2 μm) of the NAO thin film. The exact thickness (e.g., the threshold thickness) of the NAO thin film has not been determined for the emergence of the interference fringes. It is quite clear that the threshold thickness should be between 0.5 and 1.5 μm.

Figure 12.11b gives the measured reflectance signals of the NAO thin films fabricated with one-step and two-step anodization process. For these three samples, the thicknesses of the lithographically patterned Al thin film are the same of 2.2 μm before the anodization. After one-step anodization, the thickness has little change and is still ~2.2 μm, but the thickness is reduced to ~1.9 μm after two-step anodization. In this case, measurements have found that the reflectance signals have similar basic profiles (e.g., reflectance intensity) even though the pore sizes, the interspacing among them, and the thickness of the nanostructure thin film have some difference. But for samples with one-step anodization, the number of the interference peaks is 12. After two-step anodization, the number of the interference peaks is reduced to 9 since the thickness of the NAO thin film is reduced from ~2.2 μm to ~1.9 μm. In Figure 12.11c, the reflectance signals from the NAO thin films fabricated from Al with the same thickness of 2.2 μm, but using different acids (sulfuric and oxalic acid) by one-step anodization process are plotted. During the oxalic acid anodization, the temperature is 2°C and voltage is 50 V. While during the sulfuric acid anodization, the temperature is 2°C, and the voltage is 17 and 30 V, respectively.

It is well known that the finesse is a critical parameter to determine the resolving power of a Fabry–Perot interferometer (FPI), which is defined as [30, 31]: $F = FSR/\Delta\lambda$, where FSR is the free spectral range, $\Delta\lambda$ is the full-width half-maximum. For the NAO thin film–based FPI, by examining the optical signals (e.g., interference fringes) in Figure 12.11c, it is clear that the average finesse of the NAO thin film fabricated by sulfuric acid is bigger than that fabricated by oxalic acid. The finesse measurements on several samples are averaged and summarized in Figure 12.11d. The average finesse is 2.09 ± 0.05 for NAO thin film samples fabricated by sulfuric acid, and the finesse is 1.79 ± 0.05 for NAO thin film samples fabricated by oxalic acid. As shown in Table 12.1, the nanopore size and

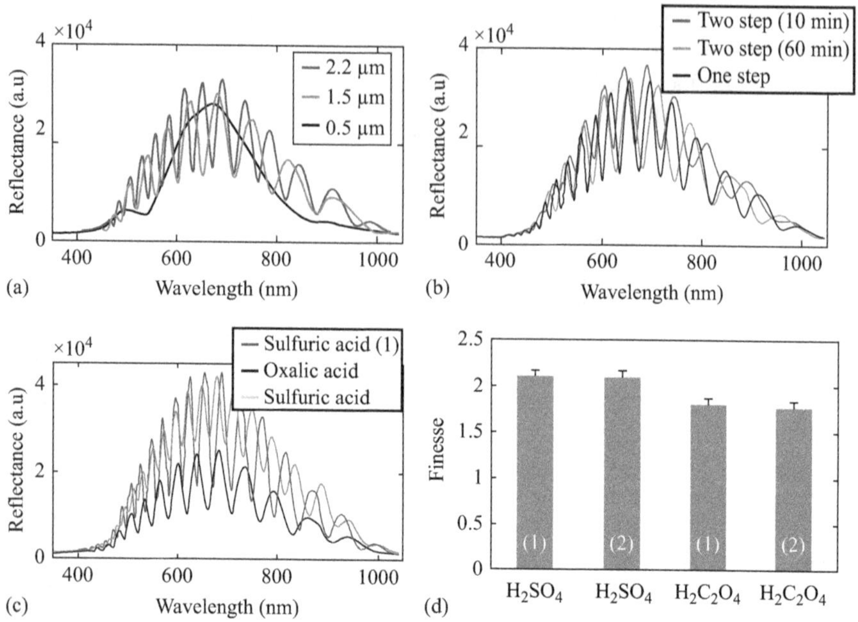

Figure 12.11 *(a) Measured reflectance signals from NAO thin film with different thicknesses of 0.5, 1.5, and 2.2 μm using one-step anodization process (0.3 M oxalic at 2°C at 50 V for 15 minutes); (b) Measured reflectance signals from NAO thin film fabricated with one-step anodization process (0.3 M oxalic at 2°C at 50 V for 15 minutes), then the alumina layer etched by immersing the NAO substrate in a mixture solution of phosphoric acid (0.4 M) and chromic acid (0.2 M) at 65°C for 10 or 60 minutes followed by 2-step anodization process (0.3 M oxalic at 2°C at 50 V for 45 minutes). In the plot, two step (10 minutes): the alumina layer etching time is 10 minutes; two step (60 minutes): the alumina layer etching time is 60 minutes; (c) measured reflectance signals from NAO thin film fabricated from Al with the same thickness of 2.2 μm but with different acids with one-step anodization process. In the plot, sulfuric acid (1): 0.3 M sulfuric acid at 2°C at 17 V for 15 minutes; sulfuric acid (2): 0.3 M sulfuric acid at 2°C at 30 V for 15 minutes; oxalic acid: 0.3 M oxalic at 2°C at 50 V for 15 minutes; (d) summary of finesse of the NAO thin film in (c) fabricated and anodized by sulfuric acid and oxalic acid. Reprinted from Reference 25 with permission*

interspacing among nanopores are different for the NAO thin film by one-step anodization using sulfuric and oxalic acid, thereby resulting in different finesse. The bigger the finesse offered by the NAO thin film, the higher the sensitivity for this type of NAO thin film-based interferometer.

Table 12.1 Typical size of nanopores and the interspacing among nanopores fabricated from 2.2 μm thick Al

One or two-step anodization	Electrolyte	Etching alumina time prior to the second step anodization	Anodization voltage/ temperature	Average pore size (nm)	Average interspacing among pores (nm)
One-step	Sulfuric acid		17 V/2°C	10 ± 2	35 ± 2
One-step	Oxalic acid		50 V/2°C	30 ± 5	125 ± 5
Two-step	Oxalic acid	10 minutes	50 V/2°C	50 ± 5	50 ± 5
Two-step	Oxalic acid	60 minutes	50 V/2°C	25 ± 2	125 ± 2

12.4.2 Optical emission from NAO thin film under UV irradiance

Besides the unique optical interference signals from the NAO thin film, the NAO also exhibits strong blue emission upon exposure to a UV light irradiance [32]. It has been found that the intensity of the blue emission increases with the increased anodization time, due to the increased amounts of NAO formation and the increased density of the formed oxygen vacancies in NAO. This phenomenon is very useful for determining the anodization level of Al, particularly convenient for monitoring the anodization level of Al micropatterns. The widely used determination technique of the NAO formation is based on the current change during the anodization process. While this method is quite dependable for NAO formation in bulk Al sheet, it is not always viable for NAO formation in Al micropatterns due to the nonuniformity of anodization current during an anodization process. Some Al micropatterns might not be anodized at all. The optical characterization offers an alternative approach for rapid and nondestructive technique to determine if the Al has been anodized and converted into NAO in the Al micropatterns.

The optical characterization of the NAO micropatterns fabricated by using a one-step anodization under a UV irradiance has been carried out [33]. In Figure 12.12a (the first row in Figure 12.12 from the left to the right), the SEM image of the as-deposited Al using e-beam evaporation, the SEM images after the Al anodized with 25, 35, and 45 minutes are given, respectively. Before anodization, the surface of the Al is quite smooth and the nanoscale domains are clearly visible. The measured surface roughness is 6–12 nm. With the increased anodization time, clearly the surface shows distinct changes and nanopores become clearer and clearer, indicating the gradual and increasing formation of the NAO. As expected, the nanopores are not as well-formed and uniform as those nanopores in the NAO fabricated by using a two-step anodization process. The corresponding bright field optical images of the Al and the NAO microlines from Figure 12.12a are shown in Figure 12.12b (the second row in Figure 12.12 from the left to the right). The corresponding optical images under UV light (330–385 nm) irradiance of the Al and the NAO microlines from Figure 12.12b are shown in Figure 12.12c

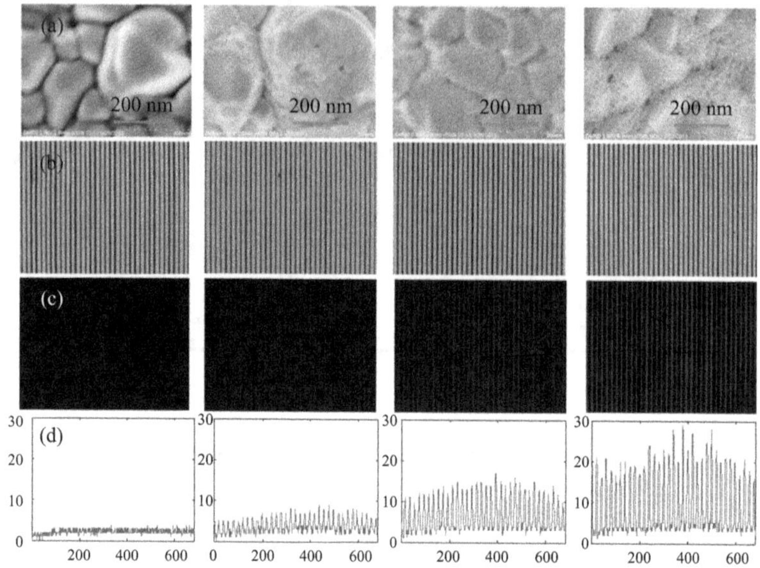

Figure 12.12 *Row (a) to row (d): (a) SEM images; (b) bright field optical images; (c) optical micrographs under UV irradiance; (d) cutline plots of the blue emission intensity under UV irradiance from (c) for the Al microlines, and anodized Al microlines for 25, 35, and 45 minutes from left to right, respectively. Width of the microlines is ~15 μm. Reprinted from References 27, 33 with permission*

(the third row in Figure 12.12 from the left to the right). The corresponding cutline-plots of the blue emission intensity from the Al and the NAO microlines in Figure 12.12c are given in Figure 12.12d (the fourth row in Figure 12.12 from the left to the right). As shown, it is clear that no blue emission from Al microlines has been observed. In contrast, the intensity of blue emission from anodized Al increases with the anodization time from 25 to 45 minutes, indicating the increased density of the oxygen vacancies in the sample, since more NAO has been formed. Furthermore, it has been found that the intensity of the blue emission becomes saturated once all Al has been anodized. This technique clearly offers a simple and nondestructive method to determine if the Al has been totally converted into NAO at a micrometer scale. This method is also applicable for NAO micropatterns formed by two-step anodization.

12.5 NAO-enabled optical biosensing

12.5.1 NAO thin film–based label-free biosensing

Since no fluorophores are required to tag the biomolecules, label-free biosensing has become increasingly a very efficient and important technique for biomedical engineering, environmental monitoring, biology, and life science research [34–36].

For the micro- or nanoscale label-free biosensors, there are three main categories in terms of transducing mechanisms, which include electrical, mechanical, and optical responses [37–39]. Electrical transduction is enabled by semiconducting carbon nanotubes (CNTs) or semiconducting nanowires (NWs). The electrical conductance of the CNTs or NWs changes upon the binding between the receptors immobilized on them and the targets [37, 40]. Mechanical transduction is achieved by micro-cantilevers. The binding between the targets and the receptors immobilized on the cantilever surface changes the cantilever's surface stress. As a result, the cantilever bends and its resonant frequency shifts. Both of these responses of the cantilever have been utilized for label-free biosensing [38, 41].

The major optical label-free techniques are propagating surface plasmon resonance (P-SPR) [42], Raman spectroscopy [43], and surface enhanced Raman spectroscopy (SERS) [44], localized SPR (L-SPR) [45], liquid core optical ring resonator (LCORR) technology [46], and photonic crystal (PC) nanostructures [39]. In addition, porous silicon and alumina oxide thin film surface interferometry for label-free DNA and protein biodetection has been demonstrated [47–49]. However, all these label-free optical techniques are plagued by some of the following issues: expensive nanofabrication processes, complicated and expensive optical setup for operation, and difficult miniaturization for microfabricating arrayed sensors for multiplexed detection. These issues contribute to the difficulty of making the platform cost-effective and compact. Recently disposable label-free optical biosensors based on micromachined polymer Fabry–Perot interferometers (μFPIs) have been developed [50]. This type of μFPIs is fabricated using polydimethylsiloxane (PDMS) and glass. The PDMS and glass serves as the two FPI planar plates. The sensing surface is the glass plate inside the FPI cavity, which is about $100 \times 100\ \mu m^2$ in area. However, for biochemical sensing, a conventional μFPI has the following intrinsic limitations. First, it has a limited sensing surface area ($100 \times 100\ \mu m^2$), which means that binding sites available for biomolecules are limited, resulting in low transducing signals. Second, the optical sensitive range/area is essentially limited only on the surface of the planar glass plate. And finally, it has a limited intensity of the transducing signals due to the optical power losses of the reflected light from the FPI at the interface of air and the FPI plates.

In order to address these issues, a nanostructured FPI has been developed [51, 52]. A schematic of a white-light source-operated nanostructured-μFPI is given in Figure 12.13(a). Different from nanopore-thin film biosensors [46–49], each nanostructured-μFPI consists of a PDMS plate and an Au-coated NAO layer that is anchored on a glass plate inside the FPI cavity. Typical gap size of 50 μm is formed between the PDMS plate and the nanopore layer. Compared to a conventional μFPI, the nanopore layer greatly increases the sensing area since the binding sites for biomolecules include not only the top surface of the nanopore layer but also the sidewall of the nanopores as shown in Figure 12.13(a,b). In addition, this type of nanostructure extends the optical sensitive depth/range up to the thickness of the nanopore layer, typically up to 1–5 μm, which is up to 100× larger than that of a P-SPR-based biosensor. The layer of nanopores is coated with tens of angstroms of Au thin film for optical signal amplification. Furthermore, since it is easy to

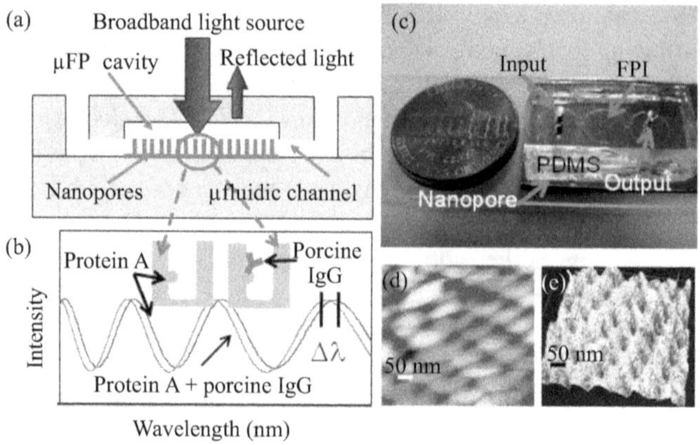

Figure 12.13 (a) Schematic of a nanostructured FPI, showing a layer of nanopores inside its cavity; (b) operational principle of the nanostructured FPI: immobilization of capture proteins (e.g., Protein A) on the surface of nanopores, and the binding between the proteins and IgG (e.g., porcine IgG) will change the transducing signals-the shift of the reflected interference fringes; (c) optical image of a fabricated PDMS-based nanostructured FPI; (d) AFM image of topside view of the nanopore layer; (e) 3-D AFM image of the nanopore layer, showing the nanopores uniformly distributed. Reprinted from Reference 52 with permission

fabricate arrayed FPI cavities and a microfluidic network in a PDMS layer, arrayed nanostructured FPIs can be readily fabricated by bonding the PDMS replica with the nanopore layer, facilitating the multiplexed biosensing. Using this type of sensor, it has been demonstrated that detection of the binding interaction between Protein A and porcine IgG in real time can be realized, and the detectable concentration of porcine IgG can be as low as 10 pg/mL.

12.5.2 Cancer protein biomarker detection

The detection of cancers at their early stage is critical for the survival of the patients. In the field of genomics and proteomics, a variety of technologies have been developed for biomarker discovery and early detection, such as various DNA microarrays [53], DNA sequence methods [54], enzyme-linked immunosorbent assay (ELISA) [55], two-dimensional polyacrylamide gel electrophoresis [56], mass spectrometry [57], proteomic pattern diagnostics [58], and protein/antibody microarrays [59]. However, most of them suffer from a complicated fluorescent dye labeling process, bulk instrumentation, and low sensitivity, and are not applicable for trace detection of biomarkers at the early stage of tumor diseases. For instance, sandwich ELISA is the gold standard in the detection and quantification of protein/cancer biomarkers. But sandwich assays are not convenient for achieving large-scale multiplexed measurements. On the other hand, most fluorescence

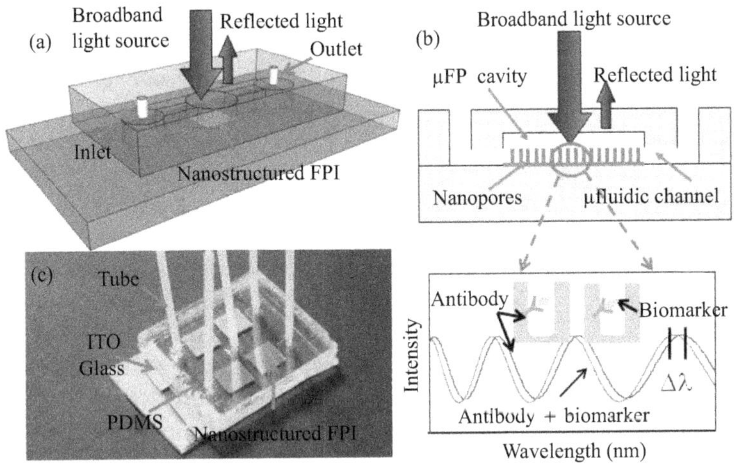

Figure 12.14 (a,b) Schematic of the nanostructured FPI and the sketch showing the immobilization of antibody on the surface of the nanostructures and the binding between antibody and biomarker will change the effective index of refraction and thus effective optical thickness (EOT), causing the shift of the reflected interference fringes. (c) Photograph of a fabricated 2 × 2 nanostructured FPI devices with assembled microfluidic interface. Reprinted from Reference 67 with permission

imaging–based techniques are incapable of providing sufficient sensitivity to monitor biomarker affinity at concentrations of around 10 pM or lower without some forms of signal amplification [60]. The concentration of many clinically relevant biomarkers residing in biofluids (i.e., blood) is usually at picomolar concentrations or lower [61], about five to seven orders of magnitude lower than the most abundant plasma protein. Since there is no universal ultrasensitive enzymatic amplification method for proteins like polymerase chain reaction for the detection of nucleic acid, a diagnostic tool with ultrasensitivity (picomolar, femtomolar concentration or lower) and high specificity is required.

A variety of different detection technologies based on micro and nanotechnologies have been developed in the past decades for the measurement of the tumor markers at a low concentration level [62–66], showing great promise for potential applications in point-of-care diagnostics and clinic settings. Among these technologies, label-free technique is particularly attractive for biosensing since no fluorescent dyes are needed to be attached to the biomolecules. As a result, the experimental cost can be reduced dramatically, and the possible perturbation of the properties of the biomolecules can be totally avoided.

Using the nanostructured FPI devices fabricated from the micropatterned NAO thin film on ITO glass substrates, detection of prostate cancer biomarker free prostate-specific antigen (f-PSA) have been demonstrated [67]. Figure 12.14(a,b) illustrates the schematic and operational principle of the nanostructured FPI microchip. When the biomolecules (e.g., antibody) are immobilized on the nanostructures and the binding between biomolecules (e.g., between antibody and biomarker) occur, the fringes of the reflected signals will shift due to the change of the effective refractive index inside the cavity. A photograph of the fabricated 2 × 2 nanostructured FPI chips is shown in Figure 12.14(c).

In order to immobilize the antibodies on the Au-coated NAO surface, the surface functionalization is performed step by step following the protocol illustrated in Figure 12.15. It is a well-established method to form a mixed self-assembled monolayer (SAM) of alkanethiols by the adhesion reaction of the thiol group on gold surface. The monolayer is well packed and the tethered carboxylic acid is easy to be functionalized for biological molecule immobilization. Upon the presence of EDC/NHS, the carboxylic groups form active O-acylisourea intermediates and readily react with primary amine groups which exist at the N-terminus of each polypeptide chain and in the side chain of lysine (Lys, K) residues. Because of their positive charge at physiologic conditions, primary amines are usually outward-facing (i.e., on the outer surface) of proteins; hence, they are usually accessible for conjugation without denaturing protein structure. In such a way, the detector mAbs for f-PSA are covalently attached to the top of the mixed SAMs. The remaining active O-acylisourea intermediate groups are deactivated by the amino acid glycine to avoid nonspecific biological attachment caused

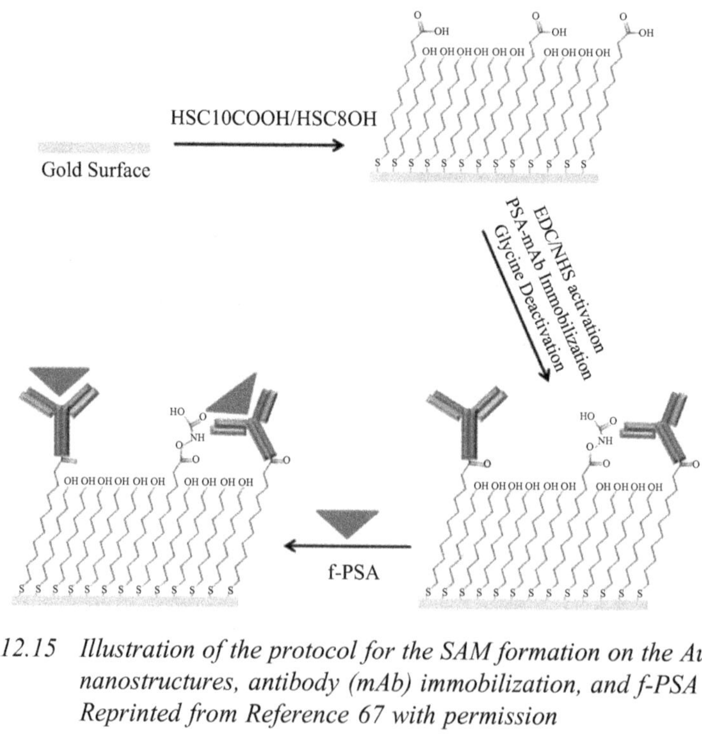

Figure 12.15 Illustration of the protocol for the SAM formation on the Au-coated nanostructures, antibody (mAb) immobilization, and f-PSA testing. Reprinted from Reference 67 with permission

by the intermediates. At this stage, the mAbs are conjugated to the nanostructured surface and ready for f-PSA detection.

PSA is a protein produced by cells of the prostate gland and has been used to detect prostate cancer at its early stage. Two forms of PSA, free or attached to a protein molecule, circulate in the blood. Usually a PSA level below 4.0 ng/mL is regarded as normal. An elevated PSA level (> 4 ng/ml) might be a sign of prostatitis, enlarged prostate, or prostate cancer. Under benign prostate conditions, there is more f-PSA. In men with prostate cancer, the ratio of f-PSA to total PSA is decreased. It has been found that the risk of cancer increases if the f-PSA to total PSA ratio is less than 25%. The lower the ratio, the greater the probability of prostate cancer.

Using nanostructured FPI microchips, it has been demonstrated that the chemical and biochemical functionalization of the nanostructured sensing surface can be monitored in real time. The upper dynamic detection range can be changed by varying the amount of capture antibodies immobilized on the sensing surface. The limit-of-detection for f-PSA is about 10 pg/mL or lower, which could be further lowered down by optimizing the optical properties (e.g., finesse) of the microchips. Experiments have also demonstrated the high specificity and selectivity of the immunoassay used in the biosensing, indicating the great promise for the detection of cancer biomarkers at trace levels in biofluids. Finally, due to the feasibility of fabricating hundreds of nanostructured FPI microdevices on a single chip, this technical platform offers great potential for highly multiplexed label free biodetection for the diagnosis of various cancers or diseases in a clinic setting and for anticancer drug screen and discovery applications.

12.5.3 Circulating tumor cell (CTC) detection

It has been reported that up to 15% (~8 million) of total deaths worldwide is because of different type of cancer, mostly due to the metastatic disease [68]. Metastatic disease, the spread of a cancer or disease from one organ to another, occurs as a result of cancer cells from the primary tumor entering the bloodstream, and hence initiating and forming new tumors in various peripheral locations. These spreading tumor cells in the bloodstream are known as CTCs, which are most often observed in patients with disease of epithelial origin, such as pancreatic, melanoma, and bladder cancer. In order to improve the survival rate of the patients, during the initial stages of metastatic disease, the detection and quantification of CTCs in blood are critical in diagnosis, monitoring, and treatment of cancer diseases.

However, the major challenge for detecting CTCs is the small number of CTCs in blood. It has been found that even for patients at an advanced stage of metastasis, the number of CTCs may only be few tens to thousands per milliliter of blood. To make the detection even more challenging, there are millions of leukocytes and billions of erythrocytes in blood [69]. To address this issue, different methods [70–74], including density gradients, size filtration, immunomagnetic separation, dielectrophoresis, micropillars embedded in a microfluidic device, and the nanostructured surface, have been utilized or developed for sorting and enumerating CTCs from blood samples. As a result, the enrichment of the CTCs can be

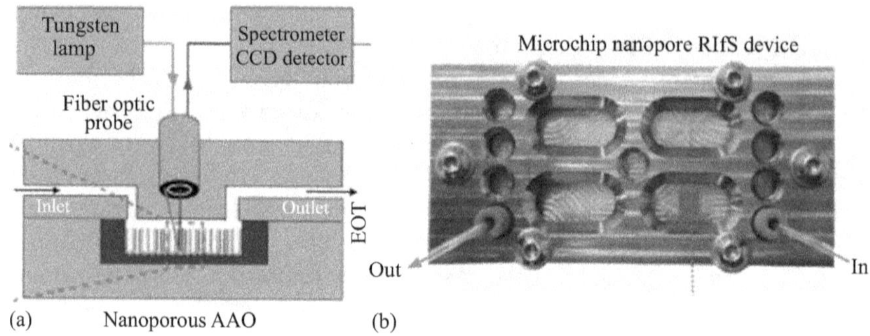

Figure 12.16 (a) Schematic of the microchip and the detection system: an NAO nanopore layer as sensing platform integrated into a microfluidic chip, fiber optic probe connected to a light source, and CCD detector. The interaction of light with the pore structures generates a Fabry–Perot interference pattern and the shift of fringe pattern as a result of binding molecules or cells on the NAO surface is resolved into EOT/nm by applying Fast Fourier Transform analysis and used as detection signal; (b) a photograph of a fabricated microchip. Reprinted from Reference 79 with permission

achieved. The identification of CTCs then can be realized by using techniques such as flow cytometry and fluorescence microscopy. But these techniques are labor-intensive and time-consuming. In order to address these limitations, label-free detection techniques of CTCs without enrichment and separation steps have been developed. Examples include electrochemical impedance, quartz crystal micro-balance, surface Plasmon resonance spectroscopy, and optical waveguide light mode spectroscopy [75–78]. Recently, the NAO-based devices have also been explored to detect CTCs [79]. The device and the setup is schematically given in Figure 12.16(a), and the photograph of the fabricated device is given in Figure 12.16(b). The basic operational principle is similar to those used for detecting protein biomarkers [67].

In this case, the biotinylated anti-EpCAM antibody that specifically binds to human cancer cells of epithelial origin such as pancreatic cancer cells (PANC-1) is firstly covalently attached to an SAM layer on the NAO surface through multiple surface functionalization steps [79]. Whole-blood or phosphate-buffered saline spiked with low numbers of pancreatic cancer cells is then flowed into the NAO-based reflectance interference sensor through a specifically designed microfluidic interface (Figure 12.16(b)). Upon the binding between the anti-EpCAM antibody and the CTCs, the optical interference fringes shift. As a result, the CTCs in the samples are successfully detected, eliminating any labor-intensive fluorescence labeling and/or pre-enrichment process of the CTCs. The reported results show that the NAO-based sensor with a specially functionalized surface can selectively detect cancer cells within a concentration range of 1000–100,000 cells/mL, with a detection limit of <1000 cells/mL. The response time for the detection is

<5 minutes, and the required sample volume is about 50 μL. The NAO-based sensing method shows great potential for achieving rapid and cost-effective point-of-care diagnostic detection of CTCs in patients with metastatic cancer.

12.6 Fluorescence detection and imaging

12.6.1 Background

Even though the labeling of fluorescent dyes to biomolecules is a relatively complicated work, due to its high sensitivity and multiplexing capability, fluorescence sensing and detection has been one of the most widely used technologies in many different fields such as the medical imaging (e.g., cancer cell imaging), biological detection and sensing (e.g., DNA arrays and gene sequencing), just to name a few [80–81]. While the optimization of the performance of the fluorophores or utilization of quantum dots is one approach to further enhance the sensitivity and stability of the fluorescence technology, the exploration of the advanced nanomaterial and nanostructure-based substrates as enhancers of the fluorescent signals is another viable approach to achieve the same goal. Both approaches are useful for the detection of biomolecules at an ultra-low concentration, the latter approach would be particularly useful to facilitate the on-chip fluorescence detection and imaging if the advanced fluorescence substrate can be readily integrated with microdevices. Among the advanced fluorescence substrates, the metallic (Au, Ag, Al) nanostructure substrates [82] and some semiconducting or metallic oxide nanostructure substrates [83–88] show great promise. For instance, Au, Ag, or Al nanoparticles or nanostructures have been widely used as fluorescent enhancers. The physical mechanism for the metal-enhanced fluorescence (MEF) is due to the interaction of the excited fluorophores with surface Plasmon resonances in metals. However, the fluorophores have to be located in close proximity to but cannot be directly on the surface of the metallic nanostructures in order to avoid the quenching effect. Hence, a thin layer of dielectric material (e.g., SiO_2 of 5 nm) usually has to be used as a spacer between the metallic nanostructures and the fluorophores. Examples of MEF substrates include silver fractal nanostructures [82], gold surface nanoscale grating [86], aluminum bow tie nanoapertures [87], and gold plasmonic nano-antenna-dots [88]. However, some of the advanced fluorescence substrates have the following shortcomings. One is the expensive material such as Au or Ag used for constructing the substrates. The other is that the expensive nanofabrication process such as EBL or FIB milling is required to fabricate the nanostructure substrates, thereby preventing the production of inexpensive nanostructure substrates for fluorescence enhancement in a cost-effective manner. On the other hand, some nonmetallic nanomaterials and nanostructures, which have been also discovered and exploited to increase the fluorescent signals and sensitivity, may offer alternative cost-effective technical platforms in that no expensive nanofabrication process is necessarily required. Some representative examples such as the nanoscaled zinc oxide (ZnO) substrate [83], ZnO/SiO2 core/shell nanorod substrate [84], and nanoscaled SnO2 substrate [85] have been demonstrated for fluorescence enhancement.

12.6.2 Fluorescence enhancement by NAO surface

Recently it has been found that the fluorescent signals can be greatly enhanced by directly depositing the fluorescent dyes on the NAO surface [89]. It has also been demonstrated that the NAO surface can achieve consistent fluorescence enhancement for different types of fluorophores, even though the enhancement factors have some difference. In Figure 12.17a, the same amount of R6G is uniformly coated on NAO surface and ITO glass substrate, and 10 nm Au-coated NAO surface and ITO glass substrate. As shown clearly, the fluorescence of R6G on NAO surface has been enhanced significantly compared to that on the ITO glass. But the fluorescence of R6G on Au-coated NAO surface has somewhat less enhancement, which is due to the quenching effect since the R6G is directly coated on the Au thin film. In Figure 12.17(b), the fluorescence enhancement of Calcein AM on NAO surface is also clearly observed. In addition, fluorescence enhancement of three different fluorophores on four patterned NAO surfaces is shown in Figure 12.17(c).

In order to quantify the fluorescence enhancement of the NAO substrate compared to a glass substrate, the fluorescence spectra of R6G and FSS on an NAO

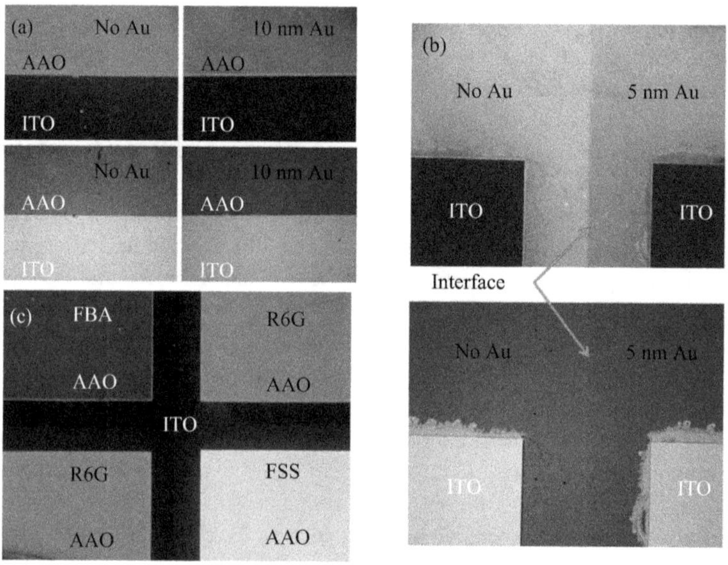

Figure 12.17 (a) Top row: fluorescent images of R6G on NAO surface and ITO substrate and Au-coated NAO surface and ITO substrate; bottom row: corresponding bright field optical images of the substrates; (b) top: fluorescent images of Calcein AM on NAO surface and ITO substrate and Au-coated NAO surface and ITO substrate; bottom: corresponding bright field optical image of the same substrate; (c) fluorescent images of three different dyes: FBA, R6G, and FSS on NAO surface and ITO substrates. Reprinted from Reference 89 with permission

substrate and a glass substrate of the same size have been measured. The optical scattering spectra of the same NAO surface and glass substrate prior to attaching any fluorophores have been measured as a reference. After subtracting the optical scattering spectra from the fluorescence spectra, the corrected fluorescence spectra of R6G and FSS are obtained in Figure 12.18. The fluorescence enhancement factor of the NAO substrate compared to the glass substrate for both R6G and FSS is in the range of two orders of magnitude, which is comparable to that of the ZnO nanostructure substrate [83].

Control experiments have been carried out to understand the fluorescence enhancement mechanism. As shown in Figure 12.19(a), the fluorescence enhancement on totally anodized Al (pure NAO) is somewhat lower than that on the partially anodized Al (partial NAO). The partial NAO is not optically transparent as the

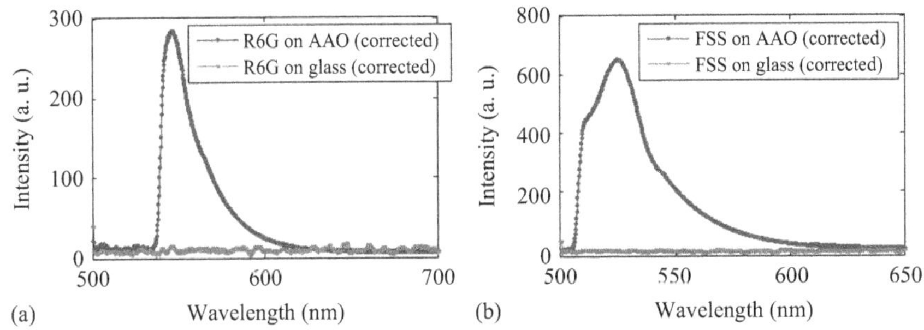

Figure 12.18 Fluorescence spectra of R6G and FSS on NAO surface and glass substrate corrected by subtracting the optical scattering spectra of bare NAO surface and glass substrate. Reprinted from Reference 89 with permission

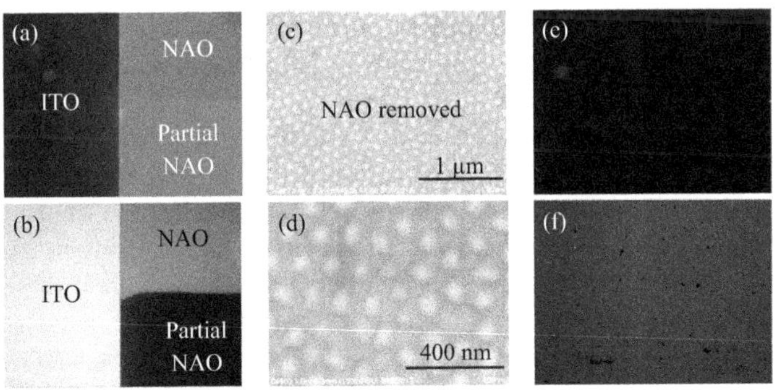

Figure 12.19 (a) Fluorescent and (b) corresponding bright field image of R6G on NAO, partial NAO and ITO glass substrate; (c,d) SEM images of the substrate after etching NAO layer; (e) fluorescent and (f) corresponding bright field image of R6G on a substrate after NAO layer being etched. Reprinted from Reference 89 with permission

pure NAO as shown in the bright field optical image (Figure 12.19(b)). A layer of *opaque* Al below a layer of NAO in the partial NAO causes an enhanced reflectance of the excitation light, thereby resulting in an enhanced fluorescence. In order to confirm the critical role of NAO in the fluorescence enhancement, the NAO layer has been etched away by immersing the NAO substrate in a mixture solution of phosphoric acid (0.4 M) and chromic acid (0.2 M) at 65°C overnight, followed by rigorous DI water rinse. The SEM image of the substrate after the NAO layer etching is given in Figure 12.19(c,d). The remaining on the substrate might be anodized Ti. However, it is too thin to be determined using an X-ray photoelectron spectroscopy. The fluorescent dye R6G is then uniformly coated on the substrate. The fluorescent images are shown in Figure 12.19(e). No fluorescence enhancement can be observed.

The surface area effect of the NAO surface on the fluorescence enhancement has been evaluated. First, the same amount of R6G and FSS is spin-coated on two NAO surfaces and two glass substrates of the same size, respectively. The fluorescent images and spectra of the four substrates are taken after they dried. Then the substrates are immersed in DI water of the same amount in four beakers and are ultrasonically rinsed for several hours. Thereafter, the fluorescent images and spectra of the four washed substrates are measured. After washing, essentially no R6G remains on the NAO surface. This is confirmed by the fluorescent image as shown in Figure 12.20(a). The fluorescent spectra of R6G on NAO after washing in Figure 12.20(b) further confirm that no R6G was left on NAO surface. For FSS, very little amount of fluorescent dyes remains on the NAO as confirmed in Figure 12.20(a). The measured fluorescent spectra of the R6G and FSS solution washed from NAO nanostructure substrates and glass substrates are given in Figure 12.20(c) and Figure 12.20(d), indicating that the amount of R6G and FSS attached on the NAO nanostructure substrate is actually lower than that on glass substrate before washing. These results suggest that the surface area of NAO for attaching fluorophores does not increase compared to that of a glass substrate with the same size, indicating that the intrinsic property of NAO nanostructures plays a critical role in this enhancement, probably similar to that of the ZnO and SnO_2 nanostructures [83–85].

The possible physical mechanisms of the NAO thin film for fluorescence enhancement could be probably mainly related to the following reasons. First, optical scattering of the NAO may play a very important role in the fluorescence enhancement. According to the Mie scattering theory, the scattering property of the NAO is related to the nanopore size, the distance among pores, the excitation optical wavelength, and the oxygen-deficient defects in NAO, which has been confirmed by the photoluminescence measurements of the blank NAO substrates [90, 91]. The surface-scattering effects of the NAO cause the redistribution of the electromagnetic fields with high-surface intensities, resulting in the enhanced fluorescence. Second, the fluorescence enhancement may also result from the evanescent electrical field from the surface of the nanoscale NAO grains, similar to other reported metal oxide nanoscale materials [83–85]. This assumption has been validated by evaluating the effect on the fluorescent signals from the gap between

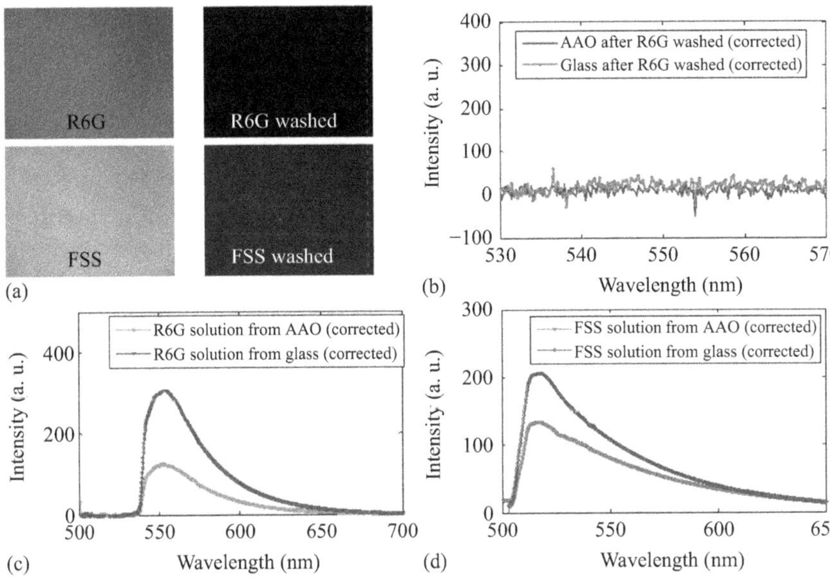

Figure 12.20 *(a) Fluorescent images of R6G and FSS on NAO substrate before and after ultrasonic solution wash; (b) fluorescent spectra of R6G on an NAO substrate and a glass substrate after ultrasonic washing; (c) fluorescent spectra of R6G solution ultrasonically washed from an NAO substrate and a glass substrate of the same size; (d) fluorescent spectra of FSS solution ultrasonically washed from an NAO substrate and a glass substrate of the same size. Reprinted from Reference 89 with permission*

the surface of the NAO and the fluorescent dyes [26, 92]. As observed experimentally, the significant enhancement of the fluorescence occurs only in the range of 100 nm from the surface of the NAO substrates. At and beyond 100 nm from the surface of the NAO, the fluorescence enhancement becomes significantly weak. It should be noted that different from noble metal (i.e., Au, Ag), evanescent field-based fluorescence enhancement technology such as total internal reflection fluorescence microscopy [93], no dielectric layer is needed to separate the fluorescent dyes from the NAO substrates to avoid the quenching effect. Instead, the fluorescent dyes can be directly deposited on the NAO substrates, thereby significantly simplifying the experimental procedure. Another important experimental observation is that the enhancement in signal to noise ratio (SNR), a key performance metric and an important parameter for the detection limit of the NAO substrates, has some difference for different fluorescent dyes but in the similar order. This difference might result from the following reasons: (1) different fluorescent dyes have different quantum efficiency/yield; and (2) one type of NAO substrate is more favorable for one specific fluorescent dye than the other due to its specific optical properties, which are determined by the nanopore size and interspacing and the anodization process. This observation indicates the possibility to optimize the enhancement in SNR for one specific fluorescent dye by using a proper NAO substrate with the optimized optical properties, which can be achieved by tuning the nanopore size and interspacing in NAO.

12.6.3 Fluorescence enhancement on NAO micropatterns

The fluorescence enhancement capability of the fabricated NAO micropatterns has also been demonstrated. The NAO has been fabricated using the two-step anodization process [26, 27]. The SEM images of the NAO are given in Figure 12.21(a,b). The fluorescent images of the NAO micropatterns are shown in Figure 12.21(c,f). For these demonstrations, the fluorescent dyes R6G, FSS, FITC, and FBA are coated on the NAO gear micropatterns and the glass substrate uniformly.

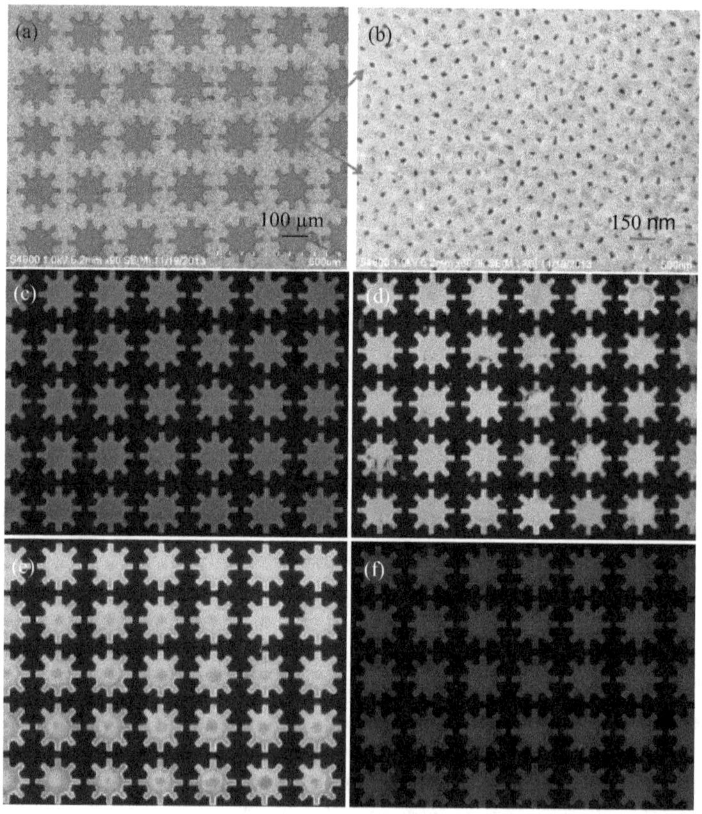

Figure 12.21 (a) SEM image of NAO micropatterns; (b) SEM image of the nanopores in the NAO micropatterns; (c–f) fluorescence images of the arrayed NAO micropatterns after being coated with different fluorescent dyes (c) R6G, (d) FITC, (e) FSS, (f) FBA; Reprinted from Reference 27 with permission

As shown, the fluorescent signals of dyes on the NAO gear micropatterns have been significantly enhanced compared to the glass. Typical enhancement is in the range of two orders of magnitude over the glass substrate. This suggests that the etching chemicals for Al and NAO in the fabrication process do not affect the property of the NAO. The capability to batch-fabricate many NAO micropatterns on a single chip indicates the feasibility to fabricate arrayed fluorescence micro-sensors cost-effectively, paving a way for developing a disposable platform for fluorescence-based multiplexing biosensing platform.

12.6.4 Fluorescence protein sensor based on NAO micropatterns

Based on the unique property of the NAO surfaces, a microfluidic NAO fluorescence sensing platform with a programmable sample delivery microfluidic interface is developed [94]. This platform integrates arrayed fluorescence sensors using NAO as the fluorescence enhancement substrates, and a microfluidic concentration generator of the biological samples on a single chip. These functions make it possible to develop disposable arrayed fluorescence biosensors for high-throughput applications. Figure 12.22(a) gives a schematic of arrayed fluorescence sensors, and a photograph of the chip is shown in Figure 12.22(b). Each sensor consists of a PDMS microfluidic chamber and a micropatterned NAO on ITO glass substrate. The NAO micropattern is a microline pattern, located inside a PDMS chamber. The arrayed sensors are connected to a microfluidic network, which can generate

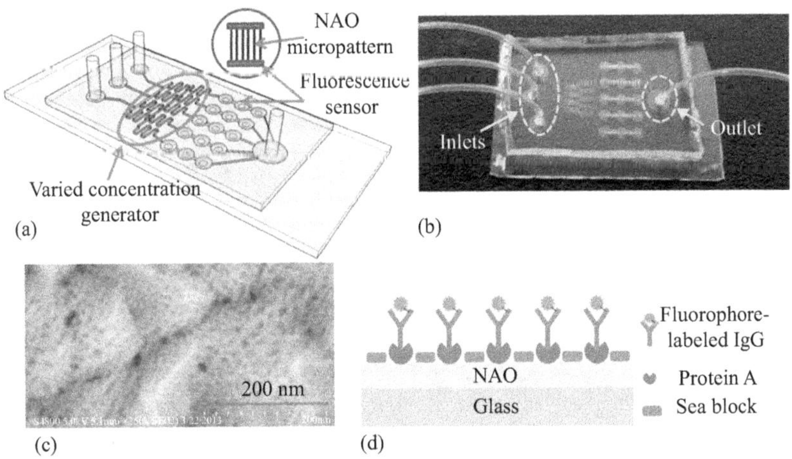

Figure 12.22 (a) 3-D sketch of the fluorescence sensors with a programmable sample concentration generating and delivery microfluidic interface; (b) photograph of a fabricated device; (c) SEM image of NAO inside the fluorescence sensor; (d) sketch of the immunoassay on the fluorescence sensor. Reprinted from Reference 94 with permission

and thus provide programmable concentrations of fluorophore-labeled proteins to the sensors. An SEM image of the micropatterned NAO is given in Figure 12.22(c), fabricated using one-step anodization process. For technical demonstration, Protein A is used as a capture probe, while fluorophore-labeled IgG as a target as schematically shown in Figure 12.22(d).

In Figure 12.23, two representative different concentration profiles of the IgG have been generated and then delivered to the fluorescence sensors, thereby resulting in different fluorescence imaging profiles. Fluorescent images of the arrayed sensors under symmetric flowing scheme are shown in Figure 12.23(a-2). Rhodamine-labeled rabbit IgG is used in this demonstration. The fluorescence intensity varies with different concentrations of IgG. As expected, the sensor in the

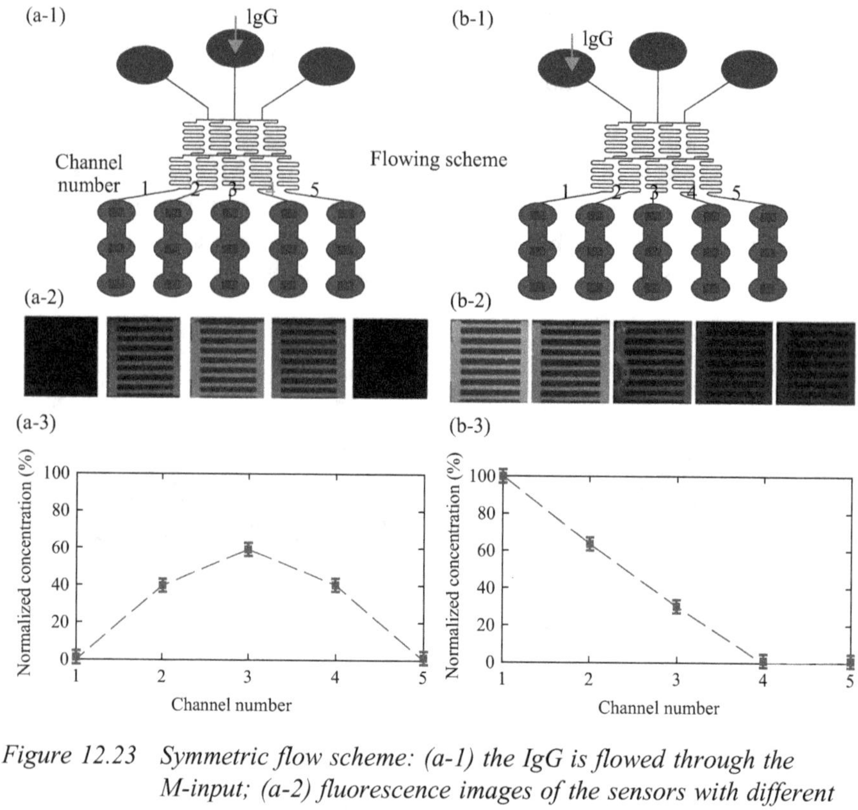

Figure 12.23 Symmetric flow scheme: (a-1) the IgG is flowed through the M-input; (a-2) fluorescence images of the sensors with different concentrations of IgG; (a-3) the measured concentration profile of IgG delivered to each sensor based on the fluorescence intensity. Asymmetric flow scheme: (b-1) the IgG is flowed through the L-input; (b-2) fluorescence images of the sensors with different concentrations of IgG; (b-3) the measured concentration profile of IgG delivered to each sensor based on the fluorescence intensity. Reprinted from Reference 94 with permission

middle connected to channel 3 has the highest fluorescence intensity, while the fluorescence intensities of other four sensors reduce symmetrically around the middle sensor to essentially negligible. The fluorescent images of the arrayed sensors under asymmetric flowing scheme are shown in Figure 12.23(b-2). FITC-labeled human IgG is used for this test. As shown, the fluorescence intensities decrease from the left sensor, connected to channel 1, to the right one, connected to channel 5. The sensor connected to channel 1 has the highest fluorescence intensity. The fluorescence intensities of the two sensors, connected to channels 4 and 5, are found identical to that of the blank NAO sensor. The very low fluorescence emission from these two sensors is just the scattering background from the NAO. Since negligible IgG has been flowed to channels 4 and 5, there is essentially no IgG applied to these two sensors. The fluorescence intensity on each sensor has been first obtained by subtracting the NAO background scattering intensity, then normalized to the fluorescence intensity of the IgG at its original concentration (i.e., the concentration of IgG before flowing into the device) on an identical sensor. Based on the measured fluorescence intensity, the normalized IgG concentration for each channel can be obtained as shown in Figure 12.23(a-3,b-3), which is very consistent ($\pm$5% offset) with the prediction by the equivalent electronic circuit model [95].

These experiments indicate the possibility to quantify the biosamples applied on the sensors. The resulting concentration profiles of IgG can also be easily tuned by simply changing the flowing rates of the samples and/or changing the input for the samples. As such, a variety of preknown concentration profiles of IgG can be generated and delivered to the fluorescence sensors, offering a great deal of flexibility for fluorescence-based bioassay, particularly useful for high-throughput applications. Furthermore, it is also possible to generate and deliver a specific/distinct concentration of the IgG to each single sensor by using a modified microfluidic interface, such as a combination of the microfluidic concentration generator with 2-D or 3-D microfluidic network of controlled valves and pumps [96].

12.6.5 Fluorescence DNA sensor based on NAO substrate

DNA sensor is a critical platform for understanding fundamental biological process and unrevealing the causes of diseases [97]. One of the promising DNA sensors is to use molecular beacons (MBs) for sensing due to their high sensitivity and reliability [98]. Usually the MB is a hairpin DNA functionalized at one end with a fluorophore and at the other end with a quenching agent [99]. In recent years, some efforts have used a metal-coated substrate to quench the fluorescent signals of the fluorophore attached to hairpin DNA [100]. For both approaches, without the target DNA, the quencher is in close proximity to the fluorophore. As a result, no fluorescent signal is generated. In contrast, by adding a complementary target DNA and upon the hybridization with the hairpin DNA, the loop of hairpin DNA is unfolded, the fluorophore thus moves away from the quencher, resulting in signal generation or enhancement.

Recently, a different type of MB sensor with different operation mechanism has been introduced [92]. Specifically, the principle of the new MB biosensor is

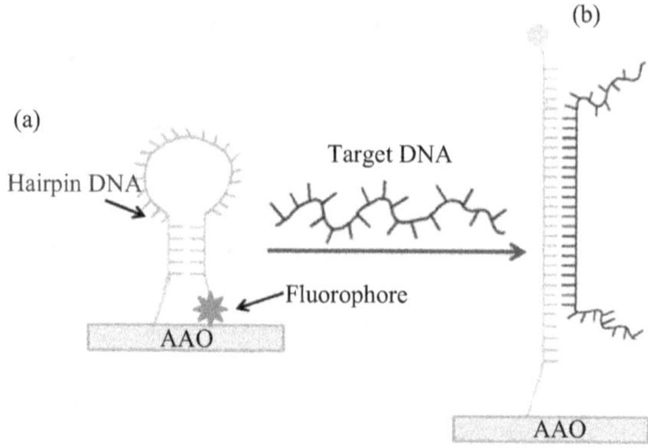

Figure 12.24 Operational principle of the NAO-substrate based MB biosensor: (a) the fluorophore attached to one end of a hairpin DNA is in proximity to the AAO surface, resulting in the enhanced fluorescent signal; (b) upon hybridization with a complementary *target DNA, the loop of the hairpin DNA is open, causing the fluorophore to move away from the AAO surface, resulting in significantly reduced signal. Reprinted from Reference 92 with permission*

opposite to the aforementioned techniques [99, 100] and illustrated in Figure 12.24. In this type of sensor, the fluorescent signal enhancement by NAO substrate and the enhancement factor related to the distance between NAO surface and fluorophore are utilized. The fluorescent signals from the fluorophores attached to hairpin DNA is initially enhanced by staying proximity to the NAO surface. After being bound with a complementary target DNA, the opening of the loop of the hairpin DNA cause the fluorophores to move away from the NAO surface, resulting in the decrease of the fluorescent signals.

The NAO surface is used to immobilize the hairpin DNA and also serves as the enhancer for the fluorescent signals. In order to improve the immobilization and binding of the hairpin DNA on NAO surface, the following chemical treatment of the NAO surface is performed: the NAO surface is immersed with H_2O_2 solution (70%) for 20 minutes and then 3-(triethoxysilyl)propylisocyanate solution (assay 95%) for 20–30 minutes [92]. Thereafter, a drop of DNA hairpin solution is applied on the chemically treated NAO surface. After 20-minute incubation, allowing the DNA hairpin's 5′-end bound with NAO's surface, the surface is rinsed by hot water (90°C or higher), followed by adding buffer PE to keep the surface wet. Then the target DNA will be applied and the corresponding changes of the fluorescent signals can be monitored in real time. One representative experimental result is given in Figure 12.25. In Figure 12.25a, a hairpin DNA (H1 DNA: 5′-/5AmMC6/ACA CGC TCA TCA TAA CCT TCA GCA AGC TTT AAC TCA TAG TGA GCG TGT/3TYE563/-3′) with one end attached with a fluorophore emits strong fluorescent signal when it is immobilized on the NAO substrate and its loop is closed. In contrast, in Figure 12.25b, the fluorescent signal reduces dramatically when the

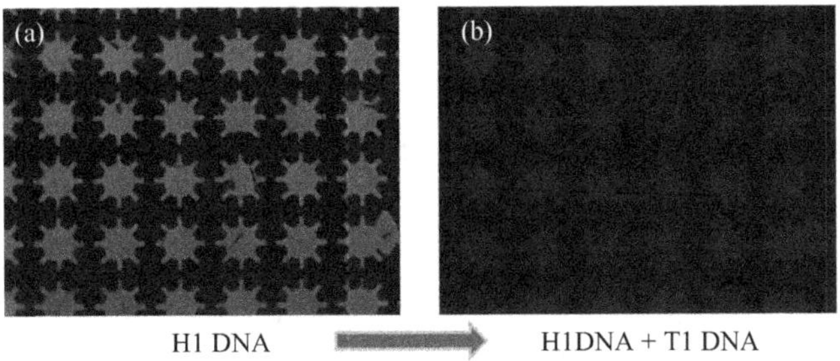

Figure 12.25 Measured result: fluorescent image (a) after the immobilization of the hairpin DNA (H1 DNA) on the sensor surface; (b) after the complementary target DNA (T1 DNA) has been applied to the sensor. Reprinted from Reference 92 with permission

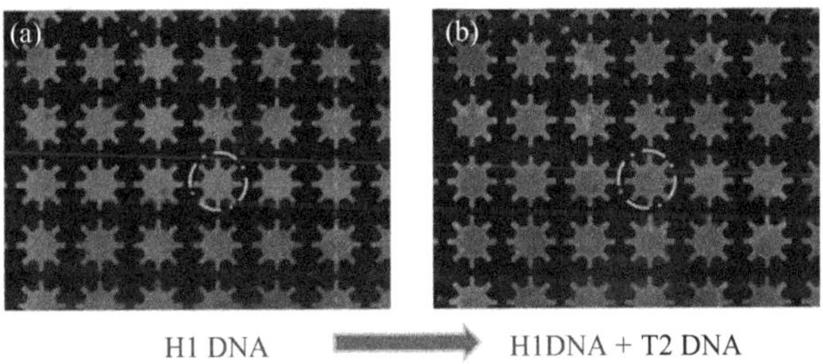

Figure 12.26 Measured result: fluorescent image (a) after the immobilization of the hairpin DNA (H1 DNA) on the sensor surface; (b) after the non-complementary target DNA (T2 DNA) has been applied to the sensor. Reprinted from Reference 92 with permission

loop is open if it is hybridized with a complementary target DNA (T1 DNA: 5′-ACG CTC ACT AGT TAA AGC TTG CTG AAG GTT ATG A -3′) after it is applied 30 minutes later.

In a control experiment in Figure 12.26, a target DNA (T2 DNA: 5′-TAT GGA GGT GTA GAA GGT GTT ATC ATC ATC ATT-3′) is not complementary to H1 DNA. After T2 DNA is applied and incubated even over one and half an hour, the fluorescent signals show very small changes as shown in Figure 12.26(b) in comparison with the fluorescent signals (Figure 12.26(a)) before T2 DNA is applied. This experiment demonstrates the excellent specificity and selectivity of the sensor.

These results indicate this type of sensor is rapid, stable, and highly specific. Experiments have found that the concentration detection limit of the DNA is ~10 nM without any sensor optimization [92]. This type of sensor has great potential for a great deal of biological applications.

12.7 Summary

In this chapter, the fabrication, characterization, and applications of the NAO, one of the most widely used nanostructures, the NAO thin film and the NAO micropatterns on glass substrates have been covered. Specifically, the fabrication processes of the NAO thin film at a wafer scale and the NAO micropatterns on the glass substrate have been described in detail. The optical properties such as the optical interference and optical emission of the NAO thin film and micropatterns have also been evaluated, which offers essentially the same quality of those from the NAO fabricated directly from an Al plate or foil. In addition, the fluorescence enhancement capability of the NAO and the possible physical mechanisms behind the enhancement capability have been discussed in detail. The optical blue emission of the NAO micropatterns can facilitate the determination of the level of the anodization of the Al micropatterns, while the optical interference signal from and fluorescence enhancement of the NAO micropatterns can be exploited for a variety of biosensing applications.

By utilizing the unique optical interference signal from the NAO thin film, label-free biodetection can be achieved. In this chapter, the NAO-based label-free sensors for the detection of the binding between proteins, for the detection of cancer biomarkers, and for the monitoring of the CTCs have been described in detail. On the other hand, by taking advantage of the fluorescence enhancement capability of the NAO, highly sensitive fluorescence biosensing can be achieved. In this chapter, the NAO-based fluorescence sensors for detecting the interaction between proteins and for sensing the DNA hybridization such as MB-based DNA sensing have been described. The significant fluorescence enhancement provided by NAO thin film may open up a lot of opportunities for fluorescence-based biological and biomedical applications with ultra sensitivity. At current stage, the physical mechanisms of the NAO substrates for fluorescence enhancement are not well understood, and thus still require further studies. These studies are the key for optimizing the NAO substrate for further improving the amplification of the fluorescence signals.

The lithography-based fabrication process for NAO micropatterns at a cost-effective manner paves the way for building arrayed sensors for both label-free biodetection and fluorescence-based bioassay with high throughput, which indicate the great potential and impact in the emerging field of optofluidics, DNA sequencing, disease screening and detection, as well as drug discovery.

References

1. Clarke J., Wu H., Jayasinghe L., Patel A., Reid A., Bayley H. 'Continuous base identification for single-molecule nanopore DNA sequencing'. *Nature Nanotechnology*. 2009; **4**(4): 265–270
2. Guo Z., Wang J., Wang E. 'Selective discrimination of small hydrophobic biomolecules based on ion-current rectification in conically shaped nanochannel'. *Talanta*. 2013; **89**: 253–257

3. Desai T., Hansford D., Kulinsky L., *et al.* 'Nanopore technology for biomedical applications'. *Biomedical Microdevices.* 1999; **2**(1):11–40
4. Taşaltın N., Öztürk S., Kılınç N., Yüzer H., Öztürk Z. 'Fabrication of vertically aligned Pd nanowire array in AAO template by electrodeposition using neutral electrolyte'. *Nanoscale Research Letters.* 2010; **5**(7): 1137–1143
5. Liu C., Gillette E., Chen X., *et al.* 'An all-in-one nanopore battery array'. *Nature Nanotechnology.* 2014; **9**: 1031–1039
6. Storm A., Chen J., Ling X., Zandbergen H., Dekker C. 'Fabrication of solid-state nanopores with single-nanometre precision'. *Nature Materials.* 2003; **2** (8): 537–540
7. Li J., Stein D., McMullan C., Branton D., Aziz M., Golovchenko J. 'Ion-beam sculpting at nanometre length scales'. *Nature.* 2001; **412**(6843): 166–169
8. Shendure J., Ji H. 'Next-generation DNA sequencing'. *Nature Biotechnology.* 2008; **26**: 1135–1145
9. Doan V., Sailor M. 'Luminescent color image generation on porous silicon'. *Science.* 1992; **256**(5065): 1791–1792
10. Li A., Müller F., Birner A., Nielsch K., Gösele U. 'Fabrication and microstructuring of hexagonally ordered two-dimensional nanopore arrays in anodic alumina'. *Advanced Materials.* 1999; **11**: 483–487
11. Rabin O., Herz P., Lin Y., Akinwande A., Cronin S., Dresselhaus M. 'Formation of thick porous anodic alumina films and nanowire arrays on silicon wafers and glass'. *Advanced Functional Materials.* 2003; **13**: 631–637
12. *Fabrication and characterization of porous silicon layers.* Chapter 5. Available from: www.tdx.cat/bitstream/10803/8458/5/Cap5Fabrication5.pdf. Accessed March 30, 2015
13. Masuda H., Fukuda K. 'Ordered metal nanohole arrays made by a two-step replication of honeycomb structures of anodic'. *Science.* 1995; **268**(5216): 1466–1468
14. Poinern G., Ali N., Fawcett D. 'Progress in nano-engineered anodic aluminum oxide membrane development'. *Materials.* 2011; **4**: 487–526
15. Thompson G.E. 'Porous anodic alumina: fabrication, characterisation and application'. *Thin Solid Films.* 1997; **297**: 192–201
16. La Flamme K., Popat K., Leoni L., *et al.* 'Biocompatibility of nanoporous alumina membranes for immunoisolation'. *Biomaterials.* 2007; **28**: 2638–2645
17. Jirage K., Hulteen J., Martin C. 'Nanotubule-based molecular filtration membrane'. *Science.* 1997; **278**: 655–658
18. Gultepe E., Nagesha D., Sridhar S., Amiji M. 'Nanoporous inorganic membranes or coatings for sustained drug delivery in implantable devices'. *Advanced Drug Delivery Reviews.* 2010; **62**: 305–315
19. Mutalib Md Jani A., Dusan Losic D., Voelcker N. 'Nanoporous anodic aluminium oxide: Advances in surface engineering and emerging applications'. *Progress in Materials Science.* 2013; **58**: 636–704

20. Hourdakis E., Nassiopoulou A. 'High-density MIM capacitors with porous anodic alumina dielectric'. *IEEE Transactions on Electron Devices.* 2010; **57** (10): 2679–2683
21. Robinson A., Burnell G., Hu M., MacManus-Driscoll J. 'Controlled, perfect ordering in ultrathin anodic aluminium oxide templates on silicon'. *Applied Physics Letters.* 2007, **91**: 1431231–1431233
22. Xiao L., Duan L., Chai J., *et al.* 'Fabrication of large area of anodic aluminum oxide ultrathin film directly onto an ITO electrode with a Ti buffer layer'. *Acta Physico-Chimica Sinica.* 2011; **27**(3):749–753
23. Sulka G., Stroobants S., Moshchalkov V., Borghs G., Celis J. 'Synthesis of well-ordered nanopores by anodizing aluminum foils in sulfuric acid'. *Journal of the Electrochemical Society.* 2002; **149**(7): D97–D103
24. Choi S., Daugherty M. 'Growth and engineering of high aspect-ratio AAO templates integrated on silicon substrates'. *Proceedings of the 2007 Nanotechnology Conference.* 2007, pp. 514–515
25. He Y., Li X., Que L. 'Fabrication and characterization of lithographically patterned and optically transparent anodic aluminum oxide (AAO) nanostructure thin film'. *Journal of Nanoscience and Nanotechnology.* 2012; **12** (10): 7915–7921
26. Li X., He Y., Que L. 'Fluorescence detection and imaging of biomolecules using the micropatterned nanostructured aluminum oxide'. *Langmuir.* 2013; **29**(7): 2439–2445
27. Yin H., Li X., Que L. 'Fabrication and characterization of the arrayed aluminum oxide thin film micropatterns on glass substrate'. *Microelectronic Engineering.* 2014; **128**: 66–70
28. Lin V., Motesharei K., Dancil K., Sailor M., Ghadiri M. 'A porous silicon-based optical interferometric biosensor'. *Science.* 1997; **278**: 840–843
29. Zhang T., Gong Z., Giorno R., Que L. 'Signal sensitivity and intensity enhancement for a polymer-based Fabry–Perot interferometer with embedded nanostructures in its cavity'. *Proceedings of 15th International Conference on Solid-State Sensors, Actuators & Microsystems* (*Transducers'09*). 2009, pp. 2310–2313
30. Born M., and Wolf E. *Principal of Optics.* 7th edn. Cambridge CB2 2RU, UK: Cambridge University Press; 1999
31. Goodman J. *Introduction of Fourier Optics.* 3rd edn. Englewood, Colorado: Roberts & Co Publishers; 2005
32. Huang G., Wu X., Mei Y., Shao X., Siu G. 'Strong blue emission from anodic alumina membranes with ordered nanopore array'. *Journal of Applied Physics.* 2003; **93**(1): 582–585
33. Li X., Yin H., Wang T., Que L. 'Optical characterization of micropatterned anodic aluminum oxide (AAO) using UV light for its fluorescence applications'. *Proceedings of the IEEE Sensors Conference.* 2013, pp. 1208–1211
34. Burbaum J., Sigal N. 'New technologies for high-throughput screening'. *Current Opinion in Chemical Biology.* 1997; **1**(1): 72–78

35. Maraldo D., Rijal K., Campbell G., Mutharasan R. 'Method for label-free detection of femtogram quantities of biologics in flowing liquid samples'. *Analytical Chemistry*. 2007; **79**: 2762–2770
36. Ramachandran N., Larson D., Stark P., Hainsworth E., LaBaer J. 'Emerging tools for real-time label-free detection of interactions on functional protein microarrays'. *FEBS Journal*. 2005; **272**: 5412–5425
37. Zheng G., Patolsky F., Cui Y., Wang W., Lieber C. 'Multiplexed electrical detection of cancer markers with nanowire sensor arrays'. *Nature Biotechnology*. 2005; **23**(10): 1294–1301
38. Wu G., Datar R., Hansen K., Thundat T., Cote R., Majumdar A. 'Bioassay of prostate-specific antigen (PSA) using microcantilever'. *Nature Biotechnology*. 2000; **19**: 856–860
39. Chan L., Gosangari S., Watkin K., Cunningham B. 'Label-free imaging of cancer cells using photonic crystal biosensors and application to cytotoxicity screening of a natural compound library'. *Sensors and Actuators B*. 2008; **132**: 418–425
40. Star A., Tu E., Niemann J., Gabriel J., Joiner C., Valcke C. 'Label-free detection of DNA hybridization using carbon nanotube network field-effect transistors'. *Proceedings of the National Academy of Sciences of the United States of America*. 2006; **103**(4): 921–926
41. Gupta A., Nair P., Akin D., *et al*. 'Anamolous resonance in a nanomechanical biosensor'. *Proceedings of National Academy of Sciences of the United States of America*. 2006; **103**: 13362–13367
42. Karlsson R. 'SPR for molecular interaction analysis: a review of emerging application areas'. *Journal of Molecular Recognition*. 2004; **17**: 151–161
43. Zribi A., Kennerly S., Claydon G., *et al*. *Compact, hand-held Raman spectrometer microsystem on a chip*. US Patent 7,505,128
44. Haynes C., McFarland A., Van Duyne R. 'Surface-enhanced Raman spectroscopy'. *Analytical Chemistry*. 2005; 338–346
45. Zhao J., Zhang X., Yonzon C., Haes A., Van Duyne R. 'Localized surface plasmon resonance biosensor'. *Nanomedicine*. 2006; **1**(2): 219–228
46. White I., Oveys H., Fan X., Smith T., Zhang J. 'Integrated multiplexed biosensors based on liquid core optical ring resonators and anti-resonant reflecting optical waveguides'. *Applied Physics Letters*. 2006; **89**: 1911061–1911063
47. Janshoff, A., Dancil K., Steinem C., *et al*. 'Macroporous p-type silicon Fabry–Perot layers. Fabrication, characterization, and applications in biosensing'. *Journal of the American Chemical Society*. 1998; **120**(46): 12108–12116
48. Kim D., Kerman K., Saito M., *et al*. 'Label-free DNA biosensor based on localized surface plasmon resonance coupled with interferometry'. *Analytical Chemistry*. 2007; **79**: 1855–1864
49. Pan S., Rothberg L. 'Interferometric sensing of biomolecular binding using nanoporous aluminum oxide templates'. *Nano Letters*. 2003; **3**: 811–814

50. Zhang T., Talla S., Gong Z., Karandikar S., Giorno R., Que L. 'Biochemical sensing with a polymer-based micromachined Fabry-Perot sensor'. *Optics Express.* 2010; **18**(17): 18394–18400
51. Zhang T., Gong Z., Giorno R., Que L. 'A nanostructured Fabry-Perot interferometer'. *Optics Express.* 2010; **18**(19): 20282–20288
52. Zhang T., Pathak P., Karandikar S., Giorno R., Que L. 'A polymer nanostructured Fabry-Perot interferometer based biosensor'. *Biosensors and Bioelectronics.* 2011; **30**(1):128–132
53. Maskos U., Southern E. 'Oligonucleotide hybridizations on glass supports: a novel linker for oligonucleotide synthesis and hybridization properties of oligonucleotides synthesized in situ'. *Nucleic Acids Research.* 1992; **20**(7): 1679–84
54. Farwell L., Joshi V. 'DNA sequencing of cancer-related genes for biomarker discovery'. *Methods in Molecular Biology.* 2009; **520**: 205–220
55. Lequin R. 'Enzyme immunoassay (EIA)/enzyme-linked immunosorbent assay (ELISA)'. *Clinical Chemistry.* 2005; **51**(12): 2415–2418
56. O'Farrell P. 'High resolution two-dimensional electrophoresis of proteins'. *The Journal of Biological Chemistry.* 1975; **250**(10): 4007–4021
57. Ackermann B., Hale J., Duffin K. 'The role of mass spectrometry in biomarker discovery and measurement'. *Current Drug Metabolism.* 2006; **7**: 525–539
58. Petricoin E., Rajapaske V., Herman E., *et al.* 'Toxicoproteomics: serum proteomic pattern diagnostics for early detection of drug induced cardiac toxicities and cardioprotection'. *Toxicologic Pathology.* 2004; **32**(Suppl 1): 122–130
59. Stoevesandt O., Taussig M., He M. 'Protein microarrays: high-throughput tools for proteomics'. *Expert Review of Proteomics.* 2009; **6**(2):145–157
60. Bartlett J., Stirling D. 'A short history of the polymerase chain reaction'. PCR Protocols. *Methods in Molecular Biology.* **226** (2nd ed.). 2003; 3–6
61. Lee H., Wark A., Corn R. 'Microarray methods for protein biomarker detection'. *Analyst.* 2008; **133**(8): 975–983
62. Mukundan H., Xie H., Anderson A., Grace W., Shively J., Swanson B. 'Optimizing a waveguide-based sandwich immunoassay for tumor biomarkers: evaluating fluorescent labels and functional surfaces'. *Bioconjugate Chemistry.* 2009; **20**(2): 222–230
63. Gaster R., Hall D., Nielsen C., *et al.* 'Matrix-insensitive protein assays push the limits of biosensors in medicine'. *Nature Medicine.* 2009; **15**: 1327–1332
64. Wilson B., Nie, W. 'Multiplex measurement of seven tumor markers using an electrochemical protein chip'. *Analytical Chemistry.* 2006; **78**: 6476–6483
65. Yanik A., Cetin A., Huang M., *et al.* 'Seeing protein monolayers with naked eye through plasmonic Fano resonances'. *Proceedings of the National Academy of Sciences of the United States of America.* 2005; **108**(29): 11784–11789
66. Gao X., Cui Y., Levenson R., Chung L., Nie S. 'In vivo cancer targeting and imaging with semiconductor quantum dots'. *Nature Biotechnology.* 2004; **22**: 969–976

67. Zhang T., He Y., Wei J., Que L. 'Nanostructured optical microchips for cancer biomarker detection'. *Biosensors and Bioelectronics*. 2012; **38**(1): 382–388
68. Jemal A., Bray F., Center M., Ferlay J., Ward E., Forman D. 'Global cancer statistics'. *CA: A Cancer Journal for Clinicians*. 2011; **61**: 69–90
69. Fan T., Zhao Q., Chen J., Chen W., Pearl M. 'Clinical significance of circulating tumor cells detected by an invasion assay in peripheral blood of patients with ovarian cancer'. *Gynecologic Oncology*. 2009; **112**: 185–191
70. Baker, M., Mikhitarian K., Osta W., *et al.* 'Molecular detection of breast cancer cells in the peripheral blood of advanced-stage breast cancer patients using multimarker real-time reverse transcription-polymerase chain reaction and a novel porous barrier density gradient centrifugation technology'. *Clinical Cancer Research*. 2003; **9**: 4865–4871
71. Zheng S., Lin H., Liu J., *et al.* 'Membrane microfilter device for selective capture, electrolysis and genomic analysis of human circulating tumor cells'. *Journal of Chromatography A*. 2007; **1162**: 154–161
72. Talasaza A., Powellc A., Huber D., *et al.* 'Isolating highly enriched populations of circulating epithelial cells and other rare cells from blood using a magnetic sweeper device'. *Proceedings of the National Academy of Sciences of the United States of America*. 2008; **196**: 3970–3975
73. Gascoyne P., Noshari J., Anderson T., Becker F. 'Isolation of rare cells from cell mixtures by dielectrophoresis'. *Electrophoresis*. 2009; **30**: 1388–1398
74. Kurkuri M., Al-Ejeh F., Shi J., *et al.* 2011. 'Plasma functionalized PDMS microfluidic chips: towards point-of-care capture of circulating tumor cells'. *Journal of Materials Chemistry*. **2011**; 21: 8841–8848
75. Weng J., Zhang Z., Sun L., Wang J. 'High sensitive detection of cancer cell with a folic acid-based boron-doped diamond electrode using an AC impedimetric approach'. *Biosensors and Bioelectronics*. 2011; **26**: 1847–1852
76. Pan Y., Guo M., Nie Z., *et al.* 'Selective collection and detection of leukemia cells on a magnet-quartz crystal microbalance system using aptamer-conjugated magnetic beads'. *Biosensors and Bioelectronics*. 2010; **25**: 1609–1614
77. Myung J., Gajjar K., Pearson R., Launiere C., Eddington, D., Hong S. 'Direct measurements on CD24-mediated rolling of human breast cancer MCF-7 cells on E-selectin'. *Analytical Chemistry*. 2011; **83**: 1078–1083
78. Fang Y. 'Non-invasive optical biosensor for probing cell signaling'. *Sensors*. 2007; **7**(10):2317–2329
79. Kumeria T., Kurkuri M., Diener K., Parkinson L., Losic D. 'Label-free reflectometric interference microchip biosensor based on nanoporous alumina for detection of circulating tumour cells'. *Biosensors and Bioelectronics*. 2012; **35**(1):167–173
80. Michalet X., Pinaud F., Bentolila L., *et al.* 'Quantum dots for live cells, in vivo imaging, and diagnostics'. *Science*. 2005; **307**: 538–544
81. Li X., Que L. 'Fluorescence enhancement enabled by nanomaterials and nanostructured substrates: A brief review'. *Reviews in Nanoscience and Nanotechnology*. 2014; **3**(3): 161–176

82. Parfenov A., Gryczynski I., Malicka J., Geddes C., Lakowicz J. 'Enhanced fluorescence from fluorophores on fractal silver surface'. *Journal of Physical Chemistry B*. 2003; **107**: 8829–8833
83. Dorfman A., Kumar N., Hahm J. 'Highly sensitive biomolecular fluorescence detection using nanoscale ZnO platforms'. *Langmuir*. 2006; **22**(11): 4890–4895
84. Zhao J., Wu L., Zhi J. 'Fabrication of micropatterned ZnO/SiO2 core/shell nanorod arrays on a nanocrystalline diamond film and their application to DNA hybridization detection'. *Journal of Materials Chemistry*. 2008; **18**(21): 2459–2465
85. Gu C., Huang J., Ni N., Li M., Liu J. 'Detection of DNA hybridization based on SnO_2 nanomaterial enhanced fluorescence'. *Journal of Physics D: Applied Physics*. 2008; **41**(17): 1751031–1751036
86. Hung Y., Smolyaninov I., Davis C., Wu H. 'Fluorescence enhancement by surface gratings'. *Optics Express*. 2006; **14**(22): 10825–10830
87. Lu G., Li W., Zhang T., *et al.* 'Plasmonic-enhanced molecular fluorescence within isolated bowtie nanoapertures'. *ACS Nano*. 2012; **6**(2): 1438–1448
88. Zhou L., Ding F., Chen H., Ding W., Zhang W., Chou S. 'Enhancement of immunoassay's fluorescence and detection sensitivity using three-dimensional plasmonic nano-antenna-dots array'. *Analytical Chemistry*. 2012; **84**(10): 4489–4495
89. Li X., He Y., Zhang T., Que L. 'Aluminum oxide nanostructure-based substrates for fluorescence enhancement'. *Optics Express*. 2012; **20**(19): 21272–21277
90. Gaponenko S. *Introduction to Nanophotonics*. Cambridge University Press, West Nyack, NY, 10994–12133: 2010
91. Zhang Z., Zheng H., Dong J., Yan X., Sun Y., Xu H. 'Surface enhanced fluorescence by porous alumina with nanohole arrays'. *Science China Physics, Mechanics and Astronomy*. 2012; **55**(5): 767–771
92. Che X., He Y., Yin H., Que L. 'A molecular beacon biosensor based on the nanostructured aluminium oxide surface'. *Biosensors and Bioelectronics*. 2015; **72**: 255–260
93. Axelrod D. 'Total internal reflection fluorescence microscopy in cell biology'. *Traffic*. 2001; **2**(11): 764–774
94. Li X., Yin H., Que L. 'A nanostructured aluminum oxide-based microfluidic device for enhancing immunoassay's fluorescence and detection sensitivity'. *Biomedical Microdevices*. 2014; **16**(5): 771–777
95. Jeon N., Dertinger S., Chiu D., Choi I., Stroock A., Whitesides G. 'Generation of solution and surface gradients using microfluidic systems'. *Langmuir*. 2000; **16**: 8311–8316
96. Unger M., Chou H., Thorsen T., Scherer A., Quake S. 'Monolithic microfabricated valves and pumps by multilayer soft lithography'. *Science*. 2000; **288**: 113–116
97. Drummond T., Hill M., Barton J. 'Electrochemical DNA sensors'. *Nature Biotechnology*. 2003; **21**(10): 1193–1199

98. Tyagi S., Kramer F. 'Molecular beacons: probes that fluoresce upon hybridization'. *Nature Biotechnology.* 1996; **14**(3): 303–308
99. Fang X., Liu X., Schuster S., Tan W. 'Designing a novel molecular beacon for surface-immobilized DNA hybridization studies'. *Journal of the American Chemical Society.* 1999; **121**: 2921–2922
100. Du H., Disney M., Miller B., Krauss T. 'Hybridization-based unquenching of DNA hairpin on Au surface: prototypical "Molecular Beacon" biosensors'. *Journal of the American Chemical Society.* 2003; **125**: 4012–4013

Index

www.ingramcontent.com/pod-product-compliance
Lightning Source LLC
LaVergne TN
LVHW011902200926
848484LV00002B/503
9781849198974